Food Science: Research and Technology

Food Science: Research and Technology

Editor: Lisa Jordan

RCALLISTO
REFERENCE

www.callistoreference.com

Callisto Reference,
118-35 Queens Blvd., Suite 400,
Forest Hills, NY 11375, USA

Visit us on the World Wide Web at:
www.callistoreference.com

ISBN: 978-1-63239-946-5 (Hardback)

Cataloging-in-Publication Data

Food science : research and technology / edited by Lisa Jordan.
    p. cm.
Includes bibliographical references and index.
ISBN 978-1-63239-946-5
1. Food. 2. Food--Research. 3. Food industry and trade--Technological innovations.
I. Jordan, Lisa.
TX357 .F66 2018
641.3--dc23

Table of Contents

Preface

Food science is concerned with the study of food. It studies the nature of food and the principles fundamental to food processing. Some of the sub-disciplines of food science are food chemistry, food engineering, food microbiology, food preservation, food technology and molecular gastronomy. This book is compiled in such a manner, that it will provide in-depth knowledge about the theory and practice of food science. A number of latest researches have been included to keep the readers up-to-date with the global concepts in this area of study.

This book has been the outcome of endless efforts put in by authors and researchers on various issues and topics within the field. The book is a comprehensive collection of significant researches that are addressed in a variety of chapters. It will surely enhance the knowledge of the field among readers across the globe.

It gives us an immense pleasure to thank our researchers and authors for their efforts to submit their piece of writing before the deadlines. Finally in the end, I would like to thank my family and colleagues who have been a great source of inspiration and support.

Editor

An Invasive Vector of Zoonotic Disease Sustained by Anthropogenic Resources: The Raccoon Dog in Northern Europe

Karmen Süld, Harri Valdmann, Leidi Laurimaa, Egle Soe, John Davison, Urmas Saarma*

Department of Zoology, Institute of Ecology and Earth Sciences, University of Tartu, Tartu, Estonia

Abstract

The raccoon dog (*Nyctereutes procyonoides*) is an introduced species in Europe with a continually expanding range. Since the species is capable of affecting local ecosystems and is a vector for a number of severe zoonotic diseases, it is important to understand its food habits. Raccoon dog diet was studied in Estonia by examining the contents of 223 stomach samples collected during the coldest period of the year, August to March, in 2010–2012. The most frequently consumed food categories were anthropogenic plants (e.g. cereals, fruits; FO = 56.1%) and carrion (e.g. carcasses of artiodactyls and carnivores; FO = 48.4%). Carrion was also the only food category that was consumed significantly more frequently by raccoon dogs exhibiting symptoms of sarcoptic mange than by uninfected animals. Small mammals, which represent intermediate hosts for the zoonotic tapeworm *Echinococcus multilocularis*, were more commonly recorded in samples also containing anthropogenic plants than expected by chance. Comparison of raccoon dog and red fox (*Vulpes vulpes*) diet in Estonia revealed higher overlap than found elsewhere in Europe, with 'carrion' and 'anthropogenic plants' making up the bulk of both species' diet; however, raccoon dogs were more omnivorous than red foxes. Our results suggest that while the use of most food categories reflects the phenology of natural food sources, 'anthropogenic plants' and 'carrion' provide an essential resource for raccoon dogs during the coldest period of the year, with the latter resource especially important for individuals infected with sarcoptic mange. Since both of these food categories and small mammals are often found at supplementary feeding sites for wild boar (*Sus scrofa*), this game management practice may facilitate high densities of mesocarnivores and promote the spread of some severe zoonotic diseases, including alveolar echinococcosis, trichinellosis, rabies and sarcoptic mange.

Editor: Tara C. Smith, Kent State University, United States of America

Funding: This work was supported by grants from the Estonian Research Council (grants IUT 20-32 and ESF 8793), from the European Union through the European Regional Development Fund (Centre of Excellence FIBIR), and from the Estonian Doctoral School of Ecology and Environmental Sciences. The funders had no role in study design, data collection and analysis, decision to publish, or preparation of the manuscript.

Competing Interests: The authors have declared that no competing interests exist.

* E-mail: Urmas.Saarma@ut.ee

Introduction

Introduced terrestrial vertebrates are often species with generalist feeding habits and short generation times, allowing them to adapt to different environments and cause a range of ecological and economic problems [1]. Genovesi *et al.* [2] reported 117 alien mammal species in Europe, of which at least 58 are known to cause environmental or health problems, and 27 have been found to directly affect native species. One of the most successful invasive carnivores in Europe is the raccoon dog (*Nyctereutes procyonoides*) [3], which is suspected of causing damage to native fauna through its feeding habits. Although insects, plants and small mammals are the main food objects for raccoon dogs in Europe [4], the species is an opportunistic feeder and is considered a potential threat to ground nesting birds and amphibians [5,6,7,8].

The raccoon dog is also an important vector of zoonotic disease [3]. In Estonia it has been identified as a vector for rabies [9], trichinellosis [10], sarcoptic mange and alveolar echinococcosis. Rabies has been eradicated from Estonia following a successful anti-rabies vaccination campaign, initiated in 2005 [11]. However, in recent years not only has the raccoon dog population started to increase, but so has the number of animals infected with sarcoptic mange, a highly contagious zoonosis caused by the burrowing mite *Sarcoptes scabiei*. The parasite infects a wide variety of wild and domesticated mammals, including humans [12]. Sarcoptic mange is known to impose considerable nutritional stress on individual animals [13] and ultimately severely reduce population densities of wild canids, especially red foxes and raccoon dogs [3].

Recently, the raccoon dog has been identified as an important host species for the tapeworm *Echinococcus multilocularis*, causative agent for the severe infectious disease, alveolar echinococcosis, which can be fatal for humans [14]. In Estonia, this parasite has been found in red foxes [15] and recently also in raccoon dogs (Laurimaa *et al.*, unpublished). The principal mode of transmission for *E. multilocularis* involves an intermediate mammal host, usually in the form of a small mammal. Since earlier studies have indicated that small mammals and carrion represent potentially important food items for raccoon dogs in Estonia [5,16], detailed studies of the species' feeding habits are essential to evaluate its potential to transmit alveolar echinococcosis and other zoonotic diseases.

Since the introduction of raccoon dogs to Estonia in 1950 [17] three studies of the species' diet have been carried out. According to the two earliest studies, by Naaber [5] and Laanetu [6], small mammals and plants formed the bulk of raccoon dog diet. However, both studies targeted relatively small study areas. The most recent study, by Rätsepp [16], suggested that ungulate carrion and anthropogenic plants (e.g. cereal crops and fruit) are important food sources, but unfortunately the sample size was too small to draw firm conclusions. Wildlife management has changed considerably in Estonia during recent decades, and it is reasonable to assume that food resources and therefore also the feeding habits of raccoon dogs have also changed. One potentially significant factor influencing raccoon dog diet in recent years is the increased number of artificial feeding sites created for wild boar (*Sus scrofa*), which provide animals with greater access to anthropogenic plants and carrion.

We examined raccoon dog stomach contents collected throughout Estonia to establish an overview of the species' current feeding habits and dietary differences between the sexes and in relation to infection with sarcoptic mange. We also compared the diet of the raccoon dog with that of the red fox (*Vulpes vulpes*), another ecologically important mesopredator and vector for zoonotic disease. We hypothesize that anthropogenic food items have become an important source of supplementary food for raccoon dogs, and may be especially critical during winter, when natural food is in short supply and conditions are energy demanding (in Estonia, raccoon dogs spend part of the winter in hibernation, but emerge relatively frequently to search for food). Moreover, as ungulate numbers and hunting bags have increased considerably since the 1960s, we predict that carrion may now play an important role in raccoon dog diet. We also hypothesize that there is a difference in the feeding habits of individuals infected with sarcoptic mange, compared to non-infected individuals: due to significant energy loss, infected animals might prefer energy-rich and easily accessible food items.

Materials and Methods

Carcasses of 252 raccoon dogs were collected in Estonia between August and March, 2010–2012 (Figure S1; Table S1). Of these, 249 samples were collected from animals legally harvested by hunters for purposes other than this project, while three were from animals killed in traffic accidents. After weighing and sexing, stomachs were removed and stored at -80°C for at least five days before proceeding with further analyses, in order to avoid the risk of *Echinococcus* tapeworm infection [18], since both *E. multilocularis* [15] and *E. granulosus* [19] have recently been found in Estonia.

For analysis, stomachs were opened and the contents washed through a sieve (1 mm mesh size). To quantify diet composition and calculate food niche breadth, we measured the volume (ml) of each component by water displacement [20]. In total, 223 stomach contents contained some food items, while 26 were empty.

Identifiable material was sorted into nine food categories: 1) 'small mammals' (Rodentia, Insectivora), 2) 'carrion' (Cervidae, Suidae, Carnivora), 3) 'birds', 4) 'amphibians', 5) 'fish', 6) 'invertebrates', 7) 'anthropogenic plants' (cereals, fruits), 8) 'natural plants' (wild berries, grass, etc.), 9) 'garbage' (remains of vegetables, sausage skins, etc.). Mammal and bird remains were identified to the lowest taxon possible. For identification of mammals we analysed hair cuticular scale pattern, medulla and cross-section according to Teerink [21] and a hair reference collection. Teeth were identified according to Kaal [22]. Reference materials were also used for identification of bird

Table 1. Frequency of occurrence (FO) of different food items in raccoon dog stomachs in different seasons: autumn (August–November), winter (December–March).

Food category	Total FO%	Autumn FO%	Winter	Autumn/Winter χ²	Autumn/Winter p-value	(χ²)	p-value
Small mammals	30(29)	21(21)	34(22)	5.80	0.016	(4.58)	0.03
Carrion	48(45)	33(33)	61(61)	23.7	<0.001	(13.9)	<0.001
Birds	13(13)	8(7)	17(20)	3.84	0.050	(7.33)	0.007
Amphibians	7(7)	14(14)	0(0)	17.4	<0.001	(12.5)	<0.001
Fish	4(4)	4(3)	4(6)	n.s		n.s	
Invertebrates	29(32)	54(52)	5(7)	65.1	<0.001	(44.2)	<0.001
Anthropogenic plants	56(56)	51(52)	59(63)	2.81	0.093	(2.33)	0.127
Natural plants	27(28)	39(38)	14(14)	18.66	<0.001	(13.3)	<0.001
Garbage	14(13)	7(8)	20(19)	7.1	0.008	(5.15)	0.023
n	223(190)	108(102)	111(84)				

In parentheses: uninfected individuals (raccoon dogs with symptoms of sarcoptic mange excluded). Total: uninfected and infected individuals.

Table 2. Frequency of occurrence (FO) and the number of different mammal taxa (n) found in the stomachs of raccoon dogs in Estonia in autumn and winter 2010–2012 (n = 223).

Taxon	FO(%)	n
A. Small mammals		
Cricetidae	**14.8**	**33**
Microtus sp.	8.1	18
Arvicola amphibius	0.9	2
Myodes glareolus	5.7	13
Muridae	**4.4**	**10**
Apodemus agrarius	2.6	6
Apodemus flavicollis	1.3	3
Unidentified	0.4	1
Soricomorpha	**10.5**	**24**
Sorex araneus	0.9	2
Neomys fodiens	0.4	1
Unidentified Soricidae	6.6	15
Talpa europaea	1.3	3
Unidentified	1.3	3
B. Carrion		
Carnivora	**17.5**	**39**
Nyctereutes procyonoides	9.0	20
Canis familiaris	0.9	2
Vulpes vulpes	0.4	1
Unidentified Canidae	1.7	4
Felis catus	0.9	2
Lynx lynx	0.4	1
Unidentified Felidae	0.9	2
Mustela nivalis	0.4	1
Neovison vison	0.4	1
Unidentified Mustelidae	1.3	3
Unidentified	2.6	6
Artiodactyla	**34.5**	**77**
Sus scrofa	18.4	41
Capreolus capreolus	9.0	22
Unidentified	5.7	13
C. Other remains (unidentified)	**12.7**	**29**

feathers. It is important to consider the possibility that conspecific hair in diet studies could be due to grooming [23]. With this in mind, we included raccoon dog hairs as a dietary component only when the volume of hairs in a stomach was >20 ml, which we consider unlikely to result from grooming. Moreover, the time of intensive moulting does not coincide with our study period [24].

We calculated the frequency of occurrence of each food type (FO = number of stomachs containing a specific food item/total number of stomachs). To calculate food niche breadth we used Levins index which is based on volumes of food items [25]. To examine seasonal differences in diet composition the data were grouped into two temporal categories: 1) autumn (August–November), a period of intensive foraging and fat accumulation; and 2) winter (December–March), the period when raccoon dog activity is reduced due to cold weather. As the hunting time of four raccoon dogs was unknown, we excluded these individuals from

the seasonal analysis. We could not obtain study material from the summer period since raccoon dogs are rarely hunted then in Estonia. Of all analysed stomach contents, 33 belonged to raccoon dogs exhibiting symptoms of sarcoptic mange. Differences in diet composition between sexes and seasons and in relation to sarcoptic mange infection were tested using PERMANOVA (with Bray-Curtis distance; function adonis in R package vegan) [26] and visualised using non-metric multidimensional scaling (function metaMDS in R package vegan) [26]. Individual food items that varied in relation to infection, sex or season were identified using χ^2-tests. To determine the potential impact of infected individuals on our results, detailed analysis of season and sex differences was carried out on two datasets: 1) 'total' (uninfected plus infected individuals, n = 223); and 2) only 'uninfected' individuals (n = 190).

To assess the co-occurrence of food categories in raccoon dog diet, we calculated the C-score [27] for all pairs of food types. To

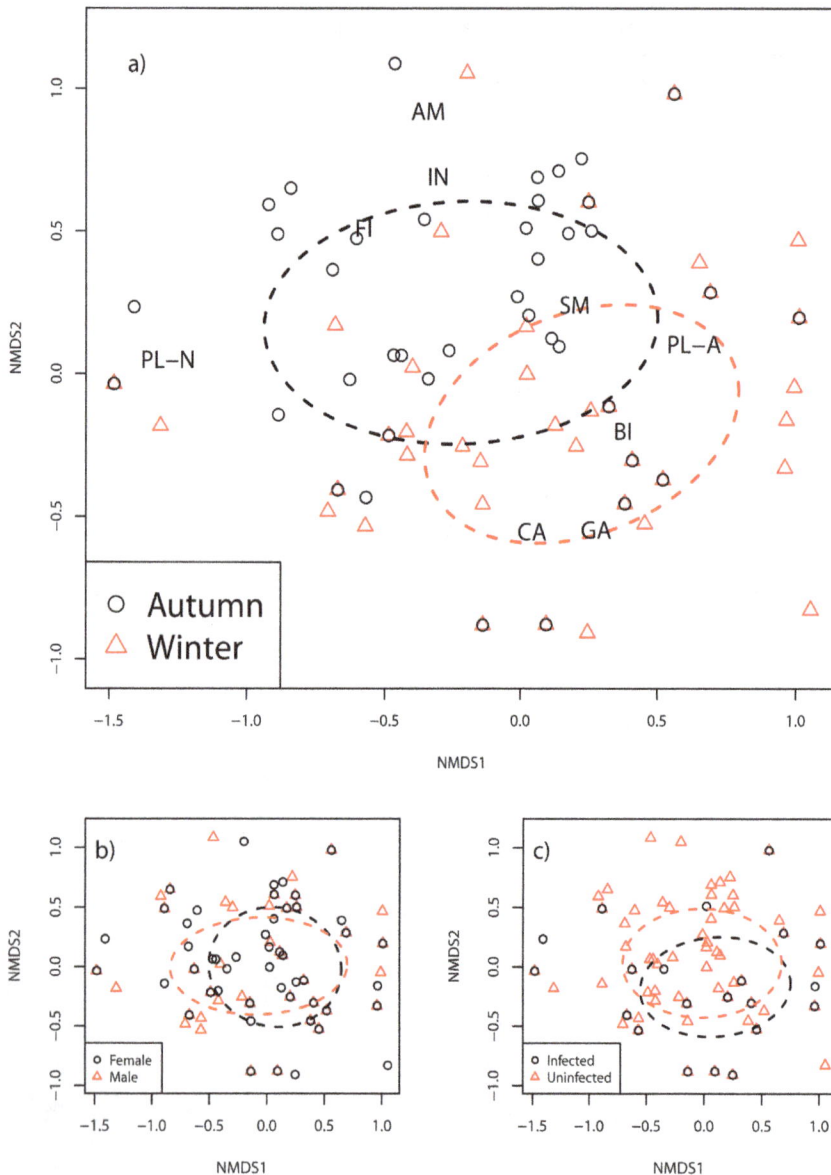

Figure 1. Non-metric multi-dimension scaling (NMDS; stress = 0.12) plots of raccoon dog diet between August and March in Estonia: a) samples are distinguished according to season (autumn or winter); b) the sex of the animal; or c) the mange infection status of the animal (infected or uninfected). Dashed ellipses indicate one standard deviation around the multivariate centroid of sample groups. PL-A – 'anthropogenic plants', PL-N – 'natural plants', BI – 'birds', SM – 'small mammals', CA – 'carrion', AM – 'amphibians', FI – 'fish', IN – 'invertebrates', GA – 'garbage'.

generate a distribution of C-scores that could be expected if food types were distributed randomly with respect to one another, we generated 999 random matrices with fixed row and column occurrences, and recalculated all pairwise C-scores for each matrix. Observed C-scores were standardised by subtracting the mean of the randomised C-scores and dividing by the standard deviation of the randomised values. The significance of effects was estimated from the number of randomised C-scores more extreme than the observed value and corrected using the false discovery rate approach [28]. Analyses were carried out on the full data set and separately for autumn and winter samples.

We also compared our data with the results of earlier diet studies performed in Estonia by Rätsepp [16] and Naaber [5] (see Figure S2; Information S1).

Raccoon dog diet from autumn and winter in Estonia was also compared with red fox diet (stomach contents, n = 92) from the same periods and areas (Soe *et al.*, unpublished) using PERMA-NOVA. Dietary overlap between male and female raccoon dogs, and between raccoon dogs and red foxes was evaluated using Pianka's index [29]. A Mann-Whitney U-test was used to test for differences in stomach volume between sexes. For these analyses, small mammals and carrion were considered as one food category 'mammals', as it was not possible to estimate the volume of hairs separately for these two categories.

Table 3. Frequency of occurrence (FO) of different food items in the stomachs of raccoon dogs uninfected and infected with sarcoptic mange.

Food category	Uninfected	Infected	Uninfected/Infected	
	FO%		χ^2	p-value
Small mammals	28.9	33.3	0.26	0.610
Carrion	45.3	66.7	**5.16**	**0.023**
Birds	12.6	15.2	n.s	
Amphibians	7.4	3.0	n.s	
Fish	4.2	3.0	n.s	
Invertebrates	31.6	12.1	**5.20**	**0.023**
Anthropogenic plants	56.3	54.5	0.04	0.85
Natural plants	28.4	21.2	0.74	0.391
Garbage	12.6	21.2	n.s	

Results

General Diet Composition

The most frequently consumed food items were 'anthropogenic plants' (FO = 56.1%) and 'carrion' (FO = 48.4%); 'small mammals', 'invertebrates' and 'natural plants' were recorded nearly two times less frequently (Table 1). 'Anthropogenic plants' included cereals (rye *Secale cereale*, wheat *Triticum* spp., oat *Avena sativa*), whereas 'carrion' consisted mainly of artiodactyls (FO = 34.5%), such as wild boar (*Sus scrofa*) and roe deer (*Capreolus capreolus*), but also carnivores (FO = 17.5%). Among carnivores, the most frequently found remains were from raccoon dogs (FO = 9.0%) (Table 2). Among 'small mammals' (FO = 29.6%), voles (*Microtus* spp., and especially the bank vole *Myodes glareolus*) and shrews (*Sorex* spp.) were most frequently present (Tables 1 and 2). Of bird feathers, 37.9% were identified as Passeriformes, and 6.9% belonged to the hazel grouse (*Tetrastes bonasia*); the remaining feathers could not be identified due to severe degradation. Based on the nine food categories presented in Table 1, the food niche breadth was 2.83.

Seasonal Changes

Raccoon dog diet differed significantly between autumn and winter (PERMANOVA, pseudo-F = 28.2, R^2 = 11.7, P = 0.001; Figure 1). The largest differences between seasons were in the consumption of 'amphibians' and 'invertebrates', which were consumed significantly more in autumn, whereas 'small mammals', 'carrion', 'natural plants' and 'garbage' were more frequent in the winter diet. 'Anthropogenic plants' was the only category that was consumed frequently in both autumn and winter (Table 1).

Comparison of Males and Females

The only individual food category consumed significantly differently by female and male raccoon dogs was 'birds' (χ^2 = 6.08, df = 1, p = 0.01; females: FO = 19.4%; males: FO = 8%) (Table S2). No significant difference was found in overall diet composition (PERMANOVA, pseudo-F = 1.86, R^2 = 0.01, P = 0.15; Figure 1). The most frequently consumed food categories for both sexes were 'anthropogenic plants', which occurred in 56% of stomachs of both sexes, followed by carrion: FO = 52.4 for females and FO = 46.9 for males. Moreover, we did not find any significant difference in the volume of stomach food

contents between male (n = 113; 84.2 ml ± 101.9 SD) and female raccoon dogs (n = 103; 108.7 ml ± 119.3 SD) (W = 12668, p = 0.27). Food niche breadth was very similar for the sexes, 2.79 for females and 2.65 for males, and food niche overlap between the sexes was very high (0.99).

Evaluating the Impact of Sarcoptic Mange on Raccoon Dog Food Habits

Overall diet composition did not differ between raccoon dogs infected with sarcoptic mange (n = 33) and uninfected individuals (n = 190) (PERMANOVA, pseudo-F = 0.25, R^2 = 0.001, P = 0.79; Figure 1). However, in comparison with uninfected animals, individuals with symptoms of sarcoptic mange consumed 'carrion' significantly more frequently during the whole study period (FO = 45.3 vs FO = 66.7; χ^2 = 5.61, df = 1, p = 0.023), while healthy animals consumed significantly more invertebrates (FO = 31.6 vs FO = 12.1; χ^2 = 5.20, df = 1, p = 0.023) (Table 3); note that invertebrates were almost exclusively consumed in autumn (data not shown). No significant differences were found in the consumption of other food categories.

When infected individuals were excluded from the seasonal comparison (autumn versus winter) the consumption of 'birds' became significantly different, being higher in winter (χ^2 = 7.33, p = 0.007; Table 1). Exclusion of infected animals had no significant effect on other food categories.

Co-occurrence of Food Items in Raccoon Dog Diet

'Natural plants' and 'anthropogenic plants' never co-occurred in the same stomach sample (Table 4). However, 'anthropogenic plants' were found alongside 'small mammals' significantly more often than expected by chance. 'Anthropogenic plants' also commonly co-occurred with 'carrion', though this relationship did not deviate significantly from random. Overall, individual raccoon dog stomachs tended to contain either broadly anthropogenic ('anthropogenic plants, 'garbage' and 'carrion' commonly co-occurred) or broadly natural ('invertebrates', 'amphibians' and 'natural plants' co-occured) items, but mixtures of these categories were less frequent than expected (Table 4; Table S3; Figure S3). To some extent this reflected the availability of different food items in the different seasons, but the distinction remained largely intact within each season (Table S4).

Table 4. Co-occurrence of food items in raccoon dog diet.

	PL-A	PL-N	BI	SM	CA	AM	FI	IN	GA
PL-A(125)		**12.39**	-1.51	**-2.38**	-0.88	-1.21	0.44	-0.83	-1.17
PL-N(61)	0		0.99	1.91	1.32	-0.84	-1.43	**-2.1**	**2.09**
BI(29)	19	5		-1.36	-0.49	-0.3	1.08	0.15	**-3.36**
SM(66)	43	11	11		-1.32	**-1.78**	-0.51	-0.55	0.14
CA(108)	60	23	14	34		**2.59**	1.42	**3.58**	**-3.06**
AM(15)	10	5	2	7	2		0.75	**-5.79**	**1.55**
FI(9)	4	4	0	3	2	0		-0.52	1.13
IN(64)	36	22	7	19	18	15	3		0.97
GA(31)	19	3	10	8	22	0	0	6	

Numbers of samples in which a particular food type was recorded is shown in parentheses. Cells below the diagonal show the number of samples in which food types co-occurred. Cells above the diagonal show the standardised C-score (values below zero indicate co-occurrence; values above zero indicate separation). C-scores that deviate significantly from a random null model are shown in bold typeface. PL-A – anthropogenic plants, PL-N – natural plants, BI - birds, SM – small mammals, CA - carrion, AM - amphibians, FI - fish, IN - invertebrates, GA – garbage.

Differences in Food Habits between Raccoon Dog and Red Fox

Overall autumn and winter diet composition differed significantly between raccoon dogs and red foxes (PERMANOVA, pseudo-$F = 61.5$, $R^2 = 0.17$, $P = 0.001$), and food niche overlap between the two species was 0.86. When food categories were analysed separately, consumption of 'small mammals', carrion, 'invertebrates' and 'plants' differed significantly between raccoon dogs and red foxes (Table 5). The most frequently consumed food categories for red foxes were mammals ('small mammals' and 'carrion'), whereas for raccoon dogs it was 'plants'.

Discussion

Temporal Variation in Raccoon Dog Diet: The Increasing Importance of Anthropogenic Plants and Carrion

Based on the results of this study, raccoon dogs in Estonia exhibit an opportunistic feeding strategy, and the consumption of different food items depends largely on their phenology. Carrion and anthropogenic plants appear to be the most important food sources for raccoon dogs in both autumn and winter, irrespective of sex. In comparison with earlier studies from Estonia there are marked differences in the consumption of small rodents and carrion (Figure S2; Information S1). Ungulate numbers and the hunting bag have increased during recent decades, which presumably explain the observed trends. Frequent consumption of ungulates can probably be also attributed to the high mortality of ungulates during the harsh winters in 2010–2012 (The Estonian Environment Information Centre). This interpretation is supported by the strong decline observed in the hunting bag size and winter track indices of both roe deer and wild boar in these winters (http://www.keskkonnainfo.ee). Ungulate carcasses are primarily left by large carnivores - ungulates are preferred prey of grey wolf (*Canis lupus*) and Eurasian lynx (*Lynx lynx*) in Estonia [30,31] - but also by hunters. Previous studies from Lithuania [32,33], Belarus [34], Poland [35], Russia [36] and Finland [37] have also shown that during their short active periods in winter, raccoon dogs mainly rely on ungulate carcasses. Carrion derived from dead carnivores was also found to form a significant food source for raccoon dogs in our study (17.5%). Unlike other scavengers, raccoon dogs do not avoid carnivore carcasses as a food source [38]. Moreover, hairs of raccoon dogs were found in 9% of stomachs, indicating that they consume the carcasses of conspecifics. Indeed, raccoon dog carcasses have long been used in Estonia as trap-line baits to lure other raccoon dogs.

The number of supplementary feeding sites for wild boar has increased in Estonia from approximately 2000 to 4000 since the early 1990s. Such sites are regularly provisioned with cereals, vegetables and fruit and sometimes also with the remains of hunted animals. The effect of this game management practice is clearly seen in the consumption of anthropogenic plants and perhaps also of carrion by raccoon dogs. While the plants consumed by raccoon dogs in the 1960s mainly consisted of the fruits and vegetative parts of naturally occurring plants [5,39], both our study and that of Rätsepp [16] revealed that anthropogenic plants have come to represent the bulk of raccoon dog diet in Estonia during the past decade (FO>50%). A significant contribution of anthropogenic plants in the autumn and winter diet of raccoon dogs has also been demonstrated in Finland (FO = 55%) [37] and Germany (FO = 22.4%) [4]. Though natural food is available in autumn, the anthropogenic plants at supplementary feeding sites almost certainly represent a more concentrated and energy-rich food source.

Table 5. Frequency of occurrence (FO) of different food items in raccoon dog and red fox stomachs in autumn and winter.

Food category	Raccoon dog	Red fox		
	FO%		χ^2	p-value
Small mammals	28.6	53.4	13.98	<0.001
Carrion	48.5	64.8	6.53	0.01
Birds	13.6	17.0	1.02	0.31
Amphibians	6.3	-	-	-
Fish	4.3	5.7	n.s	
Invertebrates	25.2	2.3	21.70	<0.001
Plants	82.5	44.3	48.90	<0.001
Garbage	14.1	11.4	1.69	0.688

The Raccoon Dog as a Predator of Native Fauna

It has been suggested that the introduced raccoon dog may pose a threat to populations of ground nesting birds and amphibians, especially during their respective breeding seasons [5,6,7,8]. However, the results of Naaber [5], Rätsepp [16] and this study suggest that the quantity of amphibians and birds in raccoon dog diet in Estonia depends on the season and that outside their respective breeding seasons, consumption of birds and amphibians generally remains low (though amphibians were a fairly important food source in autumn: FO = 13.9%). As most of the bird remains in this study belonged to passerines, raccoon dogs probably fed on carcasses rather than hunting the birds themselves. At a biogeographical scale, the consumption of birds by raccoon dogs has been shown to increase with latitude [4], and a general increase in carnivory with latitude has been reported for other omnivorous mammals [40].

A phenological effect is also evident in the consumption of invertebrates, which form an important part of raccoon dog diet in autumn (FO = 53.7%), but are rarely consumed in winter due to their low availability. In comparison with the 1960s [5], small mammals are consumed less frequently (46% vs 29.6%). The reduced importance of rodents and insectivores in raccoon dog diet found in this study probably reflects the increased availability of alternative supplementary food.

Feeding Habits of Male and Female Raccoon Dogs

The food habits of female and male raccoon dogs were very similar. However, despite the fact that male and female pairs are strictly monogamous and usually forage together, the intake of birds was significantly higher in females, especially in early spring (March), when the difference was up to fivefold (data not shown). It is unclear why females might prefer birds more than males do, but there could be a nutritional benefit that is particularly important for female fecundity. Food niche breadth was also slightly broader for females, perhaps reflecting elevated energy demands in females prior to the mating season. The condition of female raccoon dogs is known to influence their litter size [41].

Comparison of Dietary Habits with the Red Fox

Although dietary overlap between raccoon dogs and red foxes was significantly higher than in previous studies conducted in Europe [33,35], raccoon dogs still exhibited more omnivorous and red fox more carnivorous feeding habits. Foxes consumed significantly more small mammals and carrion, whereas raccoon dogs ate more invertebrates and plants. Nevertheless, carrion and plants made up the bulk of both species' diets in autumn and winter.

Influence of Sarcoptic Mange on Raccoon Dog Food Habits

According to our study, overall raccoon dog diet did not vary significantly in relation to infection with sarcoptic mange. However, there were differences in the consumption of specific items. Individuals infected with sarcoptic mange consumed significantly fewer invertebrates (12.1% vs. 31.6%), but more carrion (66.7% vs 45.3%) in comparison with uninfected animals. This suggests that sarcoptic mange may influence foraging decisions or the ability to locate or catch live prey. It is probable that some of the carrion was consumed at or near to supplementary feeding sites, as carrion and anthropogenic plants tended to co-occur in stomachs more than expected by chance. As shown recently by Oja [42], infected raccoon dogs visit supplementary feeding sites more often than healthy individuals. The concentrated food sources available at feeding sites are probably more attractive to nutritionally impoverished infected animals than to healthy animals, which may prefer a diversity of food sources or choose not to visit the anthropogenically disturbed area as regularly.

The effect of sarcoptic mange was also apparent in the autumn-winter comparison, when uninfected individuals consumed significantly more birds in winter than in autumn. Carcasses of birds are likely to be more available during the winter (mortality due to the harsh conditions is higher in winter), and uninfected individuals are able to find them more easily; infected individuals are either less efficient foragers or concentrate more on supplementary feeding sites.

Raccoon Dog Diet and Zoonotic Diseases

Since supplementary feeding sites attract rodents [42] they may facilitate the spread of another important zoonotic disease, alveolar echinococcosis, which is caused by the tapeworm *Echinococcus multilocularis*. In Estonia, about 30% of red foxes are infected with the parasite [15] and infected raccoon dogs have also been found recently (Laurimaa *et al.*, unpublished data). We found that in over half of the cases where rodents appeared in stomach contents, cereals were also present, and the two items co-occurred in stomach samples significantly more often than expected by chance. Since rodents and anthropogenic food are also heavily consumed by foxes, feeding sites may promote inter- and intraspecific contacts and as a consequence increase the prevalence of the parasite. Similarly, frequent visits to supplementary feeding sites by animals with sarcoptic mange can increase contacts between infected and healthy individuals and therefore promote the spread of sarcoptic mange within and between species.

Considering that consumption of mammalian carcasses is the principle method for nematode transmission, and that *Trichinella* spp. are prevalent in raccoon dog populations in Estonia [43], the hunting practice of leaving the entrails and skins of quarry at supplementary feeding sites and in the forest could contribute to the spread of trichinellosis. Although rabies has been eradicated from Estonia following a successful anti-rabies vaccination campaign initiated in 2005, the disease remains at the border with Russia, and presents a genuine risk of spill-over. Thus, by mitigating the energetic demands imposed by harsh winter conditions and by promoting intra- and interspecific interactions, supplementary feeding sites and hunted remains may facilitate the spread of rabies and other severe infectious zoonoses.

Supporting Information

Figure S1 Sampling locations and numbers of raccoon dogs from different hunting districts in Estonia (the two-letter combinations represent the two first letters of the corresponding hunting district, for their full names see Table S1).

Figure S2 Comparison of raccoon dog autumn and winter diet between three different study periods in Estonia ('plants' = 'anthropogenic plants' and 'natural plants; 'other animals' = 'invertebrates', 'amphibians', and 'fish').

Figure S3 Food categories that significantly co-occurred, or were significantly separate (based on Table S3).

Table S1 Data for the sampled raccoon dogs in Estonia (see also Figure S1).

Table S2 Frequency of occurrence (FO) of different food items in female and male raccoon dog stomachs.

Table S3 Food categories that significantly co-occurred, or were significantly separate (summary based on Table 4).

Table S4 Co-occurrence of food items in raccoon dog diet in autumn and winter.

Information S1 Comparison with two earlier studies performed in Estonia.

Acknowledgments

We thank Raul Melsas and Mati Martinson for their generous help.

Author Contributions

Conceived and designed the experiments: KS HV US. Performed the experiments: KS LL ES. Analyzed the data: KS HV JD US. Contributed reagents/materials/analysis tools: HV. Wrote the paper: KS HV LL ES JD US.

References

1. Erlich PR (1989) Atributes of Invaders and the Invading Processes: Vertebrates. In: Drake, J.A., Mooney, H.A., Castri, F. et al. (eds) Biological Invasions: a Global Perspective. Wiley, New York, 315–328.
2. Genovesi P, Carnevali L, Alonzi A, Scalera R (2012) Alien mammals in Europe: updated numbers and trends, and assessment of the effects on biodiversity. Integr Zool 7: 247–253.
3. Kauhala K, Kowalczyk R (2011) Invasion of the raccoon dog Nyctereutes procyonoides in Europe: History of colonization, features behind its success, and threats to native fauna. Curr Zool 57: 584–598.
4. Sutor A, Kauhala K, Ansorge H (2010) Diet of the raccoon dog Nyctereutes procyortoides - a canid with an opportunistic foraging strategy. Acta Theriol 55: 165–176.
5. Naaber J (1971) Kährikkoer. Eesti Loodus 14: 449–455 (in estonian).
6. Laanetu N (1986) Ondatra - kiskja saakloom. Eesti Ulukid 4: 15–30 (in estonian).
7. Kauhala K (1996) Introduced carnivores in Europe with special reference to central and northern Europe. Wildlife Biol 2: 197–204.
8. Neronov VM, Khlyap LA, Bobrov VV, Warshavsky AA (2008) Alien species of mammals and their impact on natural ecosystems in the biosphere reserves of Russia. Integr Zool 3: 83–94.
9. Niin E, Laine M, Guiot AL, Demerson JM, Cliquet F (2008) Rabies in Estonia: Situation before and after the first campaigns of oral vaccination of wildlife with SAG2 vaccine bait. Vaccine 26: 3556–3565.
10. Pozio E, Bandi C, La Rosa G, Järvis T, Miller I, et al. (1995) Concurrent infection with sibling Trichinella species in a natural host. Int J Parasitol 25: 1247–1250.
11. Pärtel A (2013) Self-declaration by Estonia on the recovery of its rabies-free status. Bull OIE 3: 58–61.
12. Bornstein S, Mörner T, Samuel WM (2001) Sarcoptes scabiei and Sarcoptic Mange. In Samuel, W.M., Pybus, M.J. & Kocan, A.A. Parasitic Diseases of Wild Mammals. 107–119.
13. Newman TJ, Baker PJ, Harris S (2002) Nutritional condition and survival of red foxes with sarcoptic mange. Can J Zoolog 80: 154–161.
14. Davidson RK, Romig T, Jenkins E, Tryland M, Robertson LJ (2012) The impact of globalisation on the distribution of Echinococcus multilocularis. Trends Parasitol 28: 239–247.
15. Moks E, Saarma U, Valdmann H (2005) Echinococcus multilocularis in Estonia. Emerg Infect Dis 11: 1973–1974.
16. Rätsepp M (2005) Kährikkoera (Nyctereutes procyonoides) ja punarebase (Vulpes vulpes) talvine toitumine Eestis. Bachelor Thesis, University of Tartu (in estonian).
17. Aul J, Ling H, Paaver K (1957) Eesti NSV imetajad. Eesti Riiklik Kirjastus, Tallinn, 256–257 (in estonian).
18. Eckert J, Schantz PM, Gasser RB, Torgerson PR, Bessonov AS, et al. (2001) WHO/OIE Manual on Echinococcosis in Humans and Animals: a Public Health Problem of Global Concern World Organisation for Animal Health, Paris.
19. Moks E, Jõgisalu I, Saarma U, Talvik H, Järvis T, et al. (2006) Helminthological survey of the wolf (Canis lupus) in Estonia, with an emphasis on Echinococcus granulosus. J Wildlife Dis 42: 359–365.
20. Sato Y, Mano T, Sieki T (2000) Applicability of the point-frame method for quantitative evaluation of bear diet. Wildlife Soc B 28: 311–316.
21. Teerink BJ (1991) Hair of West-Europaean Mammals, Atlas and identification key. Cambridge, Univ. Press, Cambridge, 223.
22. Kaal M (ed.) (1981) Eesti imetajate määramistabelid. Eesti NSV Põllumajandusministeeriumi Informatsiooni Juurutamise Valitsus, Tallinn.
23. Remonti L, Balestrieri A, Domenis L, Banchi C, Valvo T, et al. (2005) Red fox (Vulpes vulpes) cannibalistic behaviour and the prevalence of Trichinella britovi in NW Italian Alps. Parasitol Res 97: 431–435.
24. Xiao Y (1995) Seasonal Moulting in Adult Male Raccoon Dog (Nyctereutes procyonoides). Acta Agr Scand a-An 45: 186–190.
25. Krebs CJ (1999) Ecological Methodology. 2nd edn. Benjamin Cummings, Menlo Park, California.
26. Oksanen J, Blanchet FG, Kindt R, Legendre P, Minchin PR, et al. (2012) vegan: Community Ecology Package. R package version 2.0–5. http://CRAN.R-project.org/package=vegan.
27. Stone L, Roberts A (1990) The checkerboard score and species distributions. Oecologia 85: 74–79.
28. Benjamini Y, Hochberg Y (1995) Controlling the false discovery rate: a practical and powerful approach to multiple testing. J R Stat Soc B 57: 289–300.
29. Pianka ER (1973) The Structure of Lizard Communities. Ann Rev Ecol Syst 4: 53–54.
30. Kübarsepp M, Valdmann H. (2003) Winter diet and movements of wolf (Canis lupus) in Alampedja Nature Reserve, Estonia. Acta Zool Lit 13: 28–33.
31. Valdmann H, Andersone-Lilley Z, Koppa O, Ozolins J, Bagrade G (2005) Winter diets of wolf Canis lupus and lynx Lynx lynx in Estonia and Latvia. Acta Theriol 50: 521–527.
32. Baltrunaite L (2002) Diet composition of the red fox (Vulpes vulpes L.), pine marten (Martes martes L.) and raccoon dog (Nyctereutes procyonoides Gray). Acta Zool Lit 12: 362–368.
33. Baltrunaite L (2006) Diet and winter habitat use of the red fox, pine marten and raccoon dog in Dzukija National Park, Lithuania. Acta Zool Lit 16: 46–60.
34. Sidorovich VE (2011) Seasonal and annual variation in the diet of raccoon dog in Paazerje forest: the role of habitat type and family group. In: Analysis of vertebrate predator-prey community. Tesey, Minsk, 429–440.
35. Jedrzejewska B, Jedrzejewski W (1998) Predation in vertebrate communities. The Bialowieza Primeval Forest as a case study. Vol 135. Ecological studies. Springer, Berlin.
36. Tumanov IL (2003) Biological characteristics of carnivores mammals of Russia. Nauka, Saint-Petersburg.
37. Kauhala K, Kaunisto M, Helle E (1993) Diet of the Raccoon Dog, Nyctereutes procyonoides, in Finland. Z Saugetierkd 58: 129–136.
38. Selva N, Jedrzejewska B, Jedrzejewski W (2005) Factors affecting carcass use by a guild of scavengers in European temperate woodland. Can J Zoolog 83: 1590–1601.
39. Naaber J (1974) Rebane ja kährikkoer meie looduses. In: Jaht ja ulukid. Eesti NSV Jahimeeste Seltsi aastaraamat 1969–1972. Valgus, Tallinn, 102–115 (in estonian).

40. Vulla E, Hobson KA, Korsten M, Leht M, Martin AJ, et al. (2009) Carnivory is positively correlated with latitude among omnivorous mammals: evidence from brown bears, badgers and pine martens. Ann Zool Fenn 46: 395–415.

41. Helle E, Kauhala K (1995) Reproduction of the raccoon dog in Finland. J Mammal 76: 1036–1046.

42. Oja R (2011) Metssea (*Sus scrofa*) lisasöötmise kõrvalmõjud maaspesitsevatele lindudele, teistele imetajatele ja taimedele. MSc Thesis, University of Tartu (in estonian).

43. Miller I, Järvis T (2004) Wild animal trichinellosis in Estonia. Main achievements and perspectives of parasitology development (174–175). Moscow: IP RAN.

2

Increased Set Shifting Costs in Fasted Healthy Volunteers

Heather M. Bolton[1,2]*, **Paul W. Burgess**[1], **Sam J. Gilbert**[1], **Lucy Serpell**[1,3]

1 Division of Psychology and Language Sciences, University College London, London, United Kingdom, **2** South London and Maudsley NHS Foundation Trust, Bethlem Royal Hospital, Beckenham, United Kingdom, **3** North East London NHS Foundation Trust, Porters Avenue Health Centre, Dagenham, Essex, United Kingdom

Abstract

We investigated the impact of temporary food restriction on a set shifting task requiring participants to judge clusters of pictures against a frequently changing rule. 60 healthy female participants underwent two testing sessions: once after fasting for 16 hours and once in a satiated state. Participants also completed a battery of questionnaires (Hospital Anxiety and Depression Scale [HADS]; Persistence, Perseveration and Perfectionism Questionnaire [PPPQ-22]; and Eating Disorders Examination Questionnaire [EDE-Q6]). Set shifting costs were significantly increased after fasting; this effect was independent of self-reported mood and perseveration. Furthermore, higher levels of weight concern predicted a general performance decrement under conditions of fasting. We conclude that relatively short periods of fasting can lead to set shifting impairments. This finding may have relevance to studies of development, individual differences, and the interpretation of psychometric tests. It also could have implications for understanding the etiology and maintenance of eating disorders, in which impaired set shifting has been implicated.

Editor: Alessio Avenanti, University of Bologna, Italy

Funding: SJG is supported by a Royal Society University Research Fellowship. The funder had no role in study design, data collection and analysis, decision to publish, or preparation of the manuscript.

Competing Interests: The authors have declared that no competing interests exist.

* Email: boltonheather@gmail.com

Introduction

Cognitive flexibility is an executive function that allows us to behave adaptively in accordance with shifting goals and intentions. Neuropsychological studies have operationalized this flexibility using a variety of set shifting (or task switching) paradigms. In such paradigms, participants must, on certain trials, update the rules they use to determine the correct response to each stimulus. Performance can then be compared between shift trials (where task rules differ from the previous trial) and repeat trials (where the task is repeated). Performance is typically poorer on shift than repeat trials [1,2]. fMRI studies have indicated that set shifting abilities are linked to particular neural structures, such as the medial prefrontal cortex and anterior cingulate cortex [3]. Furthermore, the "shift cost" has been found to be enhanced, compared with healthy young adults, in older adults, participants with acquired brain damage, and a variety of clinical populations [4–6].

Individuals with Anorexia Nervosa (AN) constitute one such population. AN is a severe and chronic eating disorder with the highest mortality rate of any psychiatric illness [7,8] and treatment efficacy is limited [9,10]. Cognitive inflexibility has been cited repeatedly as a significant contributing factor in the maintenance of AN [11,12], manifesting as a rigid focusing of attention on food and weight, obsessive traits and a perfectionistic, inflexible personality style [13,14].

Numerous studies have demonstrated that patients with AN are impaired on a range of set shifting tasks relative to healthy controls [13,15–17] with corresponding abnormalities in frontal brain areas [18]. This cognitive rigidity cannot be explained as reflecting a general intellectual deficit or diminished information processing. Individuals with AN tend to have higher than average IQs [19],

suggesting a relatively specific impairment, hindering the ability to change responses in accordance with shifting contingencies [20].

Although the existence of cognitive rigidity has been well-replicated in studies of currently ill AN patients, and has been cited as a maintaining factor and a barrier to psychological treatment [21], there is some debate over the extent to which this impairment is the result of a temporary starvation-induced state (a result of the physiological effects of fasting), or represents a more enduring impairment.

It has been argued that set shifting deficits in AN exist at least partially independently of nutritional status [11] and may have genetic and neurobiological underpinnings [15,22–24]. In some studies, set shifting deficits have been shown to persist in patients following recovery from AN [17,25–29]. However, others have reported that such deficits improve with recovery [30,31] and are related to the degree of fasting [32,33]. In a preliminary study, Merwin et al. [34] report that psychological flexibility improved in parallel with symptom remission from anorexia nervosa. Danner and colleagues [35] caution that there are large individual differences in the degree of impairment shown in AN and that set shifting difficulties may only occur in a subset of patients. Furthermore, most studies in adolescents with AN show no set shifting abnormalities [36;37], suggesting the difficulties may be related to chronic starvation. Whilst some neuroimaging studies have suggested that abnormalities persist following weight restoration [38–40], others have demonstrated that the brain changes observed in patients with AN can normalise with full weight restoration [41;42].

Given this continued debate over whether set shifting is a cause or a symptom of AN, and some criticisms directed at existing research as lacking control of confounding factors [20], the present study attempted to provide some greater clarification of the potential interaction between starvation and set shifting. By investigating the impact of temporary food restriction on the baseline set shifting ability of healthy volunteers, we aimed to clarify whether set shifting deficits could be influenced by the effects of food restriction.

Previous studies have examined the impact of food restriction or fasting on cognition and behaviour. In these, fasting has led to observable perseverative behaviour in humans and laboratory animals [30,43–45]. Furthermore one study [46] reported that after fasting for 24 hours, healthy participants' brains showed differential patterns of activation in the insula, dorsolateral prefrontal cortex and inferior occipito-temporal cortex compared to a satiated state, in response to food-related stimuli.

A systematic review of the literature examining the impact of experimental starvation on cognition in general [47] found 4 studies which included a measure of cognitive flexibility. Three studies used a Stroop task to measure flexibility; two of these [48 & 49] found non-significant trends towards a decrement in performance amongst fasting participants, whilst the third [50] found no differences in cognitive flexibility. However, the use of the Stroop task may be criticised as it appears to measure a number of different aspects of executive function, not just cognitive flexibility [51]. Piech and colleagues [45] examined executive functioning in terms of set-shifting ability. Using a modified Wisconsin Card Sorting Test (WCST), they found that fasted participants were slower and less accurate at set shifting, but only when they were primed by viewing food images.

The current study aims to establish whether performance on a novel measure of set shifting, is impacted by fasting. Given recent evidence of differential neural responses in eating disorders to food and non-food stimuli [52,53] we decided to develop a novel set-shifting task which included both food and non-food images. Furthermore, we aimed to assess the extent to which anxiety and depression may moderate any deficits in set shifting.

If fasting is shown to have a direct effect on cognitive flexibility, this will raise important questions to be answered for our understanding of AN. Firstly, it will be important to determine the extent to which the set shifting deficit characteristic of AN is accounted for by the day-to-day effects of severe calorie restriction or fasting. Furthermore, it will be important to determine whether food deprivation interacts in some way with a pre-existing tendency to rigidity [54], exaggerating inflexibility in an individual with rigid traits. If this is the case then this might inform a theory of AN that accounts for the role of fasting in the maintenance of AN (as in the model proposed by Treasure and Schmidt [14]) and eventually leads to more effective clinical treatment.

By using a within subjects design, whereby participants are randomised to be fasted on one occasion and satiated on the other, it is possible to control for many of the individual differences which might also affect set shifting performance. In any study examining rigidity in AN, it is important to control for the impact of low mood as AN shares a high comorbidity with depression [55] and there is an extensive evidence base demonstrating that depression is linked to weak set shifting [56–58]. Of particular note, Wilsdon and Wade [28] established that set shifting ability in patients with AN was mediated by depression scores (see also [59]).

The present study was designed to explore whether set shifting difficulties emerge under conditions of fasting, while accounting for symptoms of depression. A novel rule change paradigm was devised for the current study in order to measure set shifting ability in a controlled manner. The paradigm was intended to allow a comparison between trials in which participants were required to shift their mental set, versus trials in which no set shift was required. The difference in response time (RT) between switch and repeat trials was used as a measure of set shifting ability, assessed under conditions of fasting and satiety. Furthermore, both food and non-food stimulus items were used in the task, to investigate potential effects of food cues on set shifting ability. The study exclusively recruited women since AN is significantly more common in women [60] and women may be more sensitive to the effects of short-term fasting than men [46].

Hypotheses

1. Mean RTs will be slower on trials where a shift of set is required compared to non-shift trials, and this shift cost will be exacerbated by fasting (i.e. on trials which require a shift of set, fasting will increase the shift cost relative to when switches are made under satiated conditions).

2. In line with previous studies where food-related stimuli compromised ability, we predict that reaction times will be slowed on trials where the stimuli consist of food items compared to non-food items, particularly in the fasting condition.

3. Elevated depression, perseveration and eating disorder symptoms, assessed by relevant questionnaire measures, will each be associated with impaired set shifting ability.

Method

Ethics

Ethical approval was obtained from the University College London Research Ethics Committee (reference 1699/001). All participants provided informed written consent before taking part.

Participants

Power analysis was conducted using G*Power [61], based on a medium effect size, as reported by Wilsdon and Wade [28] in their examination of WCST perseverative errors made by AN patients. It was calculated that 52 participants would be required in order to achieve 80% power. Sixty adult female volunteers were recruited through a poster campaign. Each received £15 compensation for their time. Participants were unaware of the specific aims of the study, although they understood that it sought to contribute to the understanding of eating disorders. Eligibility requirements included not currently receiving treatment for a medical condition, not being pregnant or diabetic, having normal or corrected to normal vision and fluency in English. Handedness was not assessed as part of the recruitment criteria.

Design

The study adopted a within-subjects repeated measures design comprising two sessions spaced approximately one week apart. All participants were required to abstain from eating or drinking (other than water) for 16 hours prior to one session (fasted condition) and to eat regular meals for the other (satiated condition). The order of these sessions was randomised. At each session, participants undertook the rule change task and completed self-report questionnaires.

Measures

Diary ratings. Several days prior to each testing session, participants were sent an email containing a self-report diary measure. They were required to print this out and bring it with them to the testing session, after filling it in at five set time points (6pm on the evening before testing, 11pm, 8am or on waking,

11am, and on arrival at the testing session). This required ratings of hunger, food preoccupation, mood and irritability, using a Likert scale ranging from 1 (not at all) to 7 (very much so).

Questionnaires. At the fasted session participants completed the Hospital Anxiety and Depression Scale (HADS) [62]. At the satiated session, participants completed the HADS, the Persistence, Perseveration and Perfectionism Questionnaire (PPPQ-22 [63] and the Eating Disorder Examination Questionnaire 6.0 (EDE-Q6) [64]. The PPPQ-22 is a 22-item measure, divided into 3 subscales (persistence, perseveration and perfectionism) and has been shown to have adequate test-retest reliability and internal consistency [63]. The EDE-Q6 is a self-report version of the Eating Disorder Examination [65] which is proven to be effective in the identification and assessment of disturbed eating and displays good internal consistency [66].

Rule change task. The rule change (RC) task was administered on a laptop using the Cogent toolbox (http://www.vislab.ucl.ac.uk/cogent.php) running under Matlab version 6.5 (MathWorks). Figure 1 shows an example of the experimental stimuli. On each trial, a set of identical photographs was presented in the centre of the screen in a subitized fashion (as on the face of a die) in conjunction with a written question. The question took one of four forms (*Odd?*, *Even?*, *High?* or *Low?*), and required participants to press one of two keys to answer either yes or no, in accordance with the number of pictures on screen. At each presentation the number of photographs on the screen varied from between one and six. If either one, two or three pictures appeared on screen, the question "*Low?*" would require a "Yes" response whilst the correct response to "*High?*" would be to press the "No" key. One, three or five pictures should elicit a press of the "Yes" key to the question "*Odd?*" and "No" to "*Even?*", and vice versa with two, four or six pictures. The question periodically changed without warning (with a 33% probability on each trial), requiring a shift of mental set. A low probability of shift was chosen in order that participants could not anticipate a shift in advance. On this basis, each trial was classified as either a "shift" trial or a "stay" trial. Shifts took the form of both inter- (e.g. "*High?*" to "*Even?*") and intra-dimension (e.g. "*High?*" to "*Low?*") changes, with equal probability. On each trial the stimuli remained on screen until a response key was pressed, followed by a random delay of 250–500 ms and then the next set of stimuli. No feedback was provided about accuracy of responses.

Photographic stimuli were taken from a database created for neuropsychological studies of AN [46,53]. Each photograph showed either a foodstuff or an inedible item, forming a collection of 118 individual stimuli. Foods consisted of sweet and savoury high-calorie items (e.g. a pizza or a donut) and were photographed on white plates over a blue background. Inedible items were household objects (e.g. a ball of string) which were presented on white circular shapes resembling the plates used in the food pictures, with the same blue background. On each trial there was a 50% probability of a food stimulus or an inedible stimulus being shown.

The computer task was presented on a laptop with a standard keyboard, positioned approximately 60 cm from the seated participant.

Experimental procedure

Following randomisation for the order of conditions (fasted vs. satiated first), each participant was informed of the particular order of their two sessions by e-mail and was given written instructions for fasting. They were instructed to stop eating at least 16 hours prior to the fasted session and in each case this involved fasting overnight. Participants were encouraged to consume water

during the period of fasting, but were required to abstain from alcohol and sugary drinks. In order to ensure compliance with the instructions, participants were informed that their urine would be tested for ketones (a by-product of fasting) at each session.

At each session the diary measure was collected and then the experimental task was administered, followed by the questionnaires and urine test. Participants were presented with written instructions for the task and underwent a practice block of 20 trials followed by 4 blocks of 100 trials. They were instructed to respond as quickly and as accurately as possible, using their dominant hand. At the satiated session, demographic measures were collected and measurements of weight and height were taken using digital scales and a portable stadiometer, for the purposes of calculating body mass index (BMI). At the fasted session, participants were asked to report the length of time for which they had fasted and to rate their subjective ease of fasting on a 5-point scale (1 = easy, 5 = hard).

Results

Characteristics of the sample

The mean age of the sample was 27.4 years (S.D. = 7.41; range: 19–55). The mean BMI was 22.3 kg/m^2 (S.D. = 2.9; range: 17.8–29.3) and all participants' BMIs fell above the 17.5 kg/m^2 clinical cut-off for AN (WHO, 1992). Age and BMI did not significantly correlate (r = .254, p = .059). Thirty-six participants (60%) were working professionals, with the remainder either undergraduate or postgraduate students. Seventy-seven percent of the sample was White and the remainder Asian (8%), Chinese (7%), Black (5%) and mixed race (3%).

Participants reported fasting for an average of 16.67 hours (S.D. = 1.12; range 15.25–21). The mean subjective ease of fasting was rated at 2.42 (S.D. = .85; range 1–5; where 1 = easy, 5 = difficult). There were no significant correlations between BMI, subjective ease of fasting and reported length of fast (p>.576). Two participants admitted to having eaten during the period in which they were expected to be fasting and their data were excluded from further analyses, leaving a total sample of n = 58.

The urinalysis was not sensitive enough to detect ketones as a longer period of fasting is required before ketones are produced. Nevertheless, as participants believed that urine testing would reveal the extent of their fasting, it served as a tool to ensure compliance with fasting requirements.

Forty-five participants (77.6%) returned both sets of diary measures (see Table 1). Paired samples t-tests were carried out on these diary ratings, comparing ratings at time 1 (T1; the evening before testing) to time 5 (T5; arrival at the testing session). The subjective feeling of hunger significantly increased during the starvation period (mean hunger ratings at T1 = 2.87, and T5 = 4.9; t = 6.27, df = 44, p<.001, d = 1.9). The reverse effect occurred in the satiated condition, as mean hunger levels at T5 were significantly less than T1 (T1 = 2.80, T5 = 2.20; t = 2.20, df = 44, p = .033, d = .66).

Fasting also led to worsening in subjective mood across time (T1 = 2.38, T5 = 2.91; t = 2.63, df = 44, p = .012, d = .79, where 1 = better mood), an increase in irritability (T1 = 2.02, T5 = 2.96; t = 3.30, df = 44, p = .002, d = .99) and an increase in food preoccupation (T1 = 2.91, T5 = 4.49; t = 4.79, df = 44, p<.001, d = 1.44). Under satiated conditions, food preoccupation significantly decreased over time (T1 = 2.75, T5 = 1.98; t = 3.30, df = 44, p = .002, d = .99), while mood did not significantly change (T1 = 2.53, T5 = 2.27; t = 1.60, df = 44, p = .12, d = .48). There were also no significant changes in irritability (T1 = 1.91,

Odd?

Figure 1. Example trial of the rule change task.

T2 = 1.64; t = 1.37, df = 44, p = .18, d = .41) across time in the satiated condition.

Questionnaire data

Table 2 shows the mean HADS scores for the sample. In relation to healthy population norms [67], the mean level of anxiety within the sample was comparable, but the prevalence of depression symptoms was lower than conventional levels found in community samples. Neither age nor BMI significantly correlated with HADS-anxiety (p>.244) or HADS-depression (p>.211).

Mean scores on each of the PPPQ-22 subscales were comparable to a non-clinical population assessed by Serpell et al. [63]. Within the subscales, PPPQ-perseveration did not correlate with either persistence (r = .23, p = .084) or perfectionism (r = .13, p = .350). Persistence and perfectionism were strongly positively correlated (r = .42, p<.001). Neither age nor BMI were significantly correlated with any PPPQ-22 subscales (p>.425).

All subscales of the EDE-Q6 were highly intercorrelated (r> .646, p<.001) and mean group scores were comparable to a non-clinical sample [68]. Neither age nor BMI correlated significantly with any subscales. Although restraint bore no significant relationship to the other measures, weight, shape and eating concern of the EDE-Q6 shared moderate positive correlations

Table 1. Mean diary ratings during fasted and satiated conditions.

	Fasted		Satiated	
	T1 (day prior to testing)	T5 (immediately prior to testing)	T1	T5
Hunger	2.87[***]	4.93[***]	2.80[*]	2.20[*]
	(SD = 1.93)	(SD = 1.59)	(SD = 1.65)	(SD = 1.38)
Food preoccupation	2.91[***]	4.49[***]	2.75[*]	1.98[*]
	(SD = 1.59)	(SD = 1.60)	(SD = 1.43)	(SD = 1.07)
Mood	2.38[*]	2.91[*]	2.53	2.27
	(SD = 1.13)	(SD = 1.18)	(SD = 1.08)	(SD = 1.07)
Irritability	2.02[*]	2.96[*]	1.91	1.64
	(SD = 1.29)	(SD = 1.59)	(SD = 1.26)	(SD = 1.09)

SD = standard deviation
[*] = significant difference between T1 and T5 (p<.05)
[***] = significant difference between T1 and T5 (p<.001)

Table 2. Mean anxiety and depression scores on the HADS in fasted and satiated conditions.

	Fasted	Satiated	Normed sample (Crawford et al., 2001)
Anxiety	5.86	6.29	6.14
	(SD = 3.59)	(SD = 3.58)	
Depression	2.17	2.9	3.68
	(SD = 2.46)	(SD = 3.22)	

with HADS depression (r>.348, p<.008) and PPPQ-perseveration (r>.366, p<.005).

Experimental task performance

Incorrect responses and trials where RT<150 ms or >3000 ms were excluded from all analyses to eliminate outliers (where participants might have been inattentive or responded prematurely to stimuli). Error rates were generally low (mean 6%), except for 2 participants whose accuracy on the task was comparatively low (>3 S.D. from mean accuracy score). Their data were excluded from the analysis, along with the two participants who had admitted to having eaten during the fasting period, because they did not adequately comply with the experimental instructions, making it difficult to interpret results. Seeing as there were only four participants that did not adequately comply with task instructions – too few for a separate analysis – data from these participants were not considered further. Therefore, n = 56 for the main analysis. There were no significant differences between inter- and intra-question shifts (for instance from *high* to *low*, versus from *high* to *odd*) and so these data were collapsed in the results below.

To test the first two hypotheses and explore the effects of fasting, shifting and food stimuli, a general linear model (GLM) was constructed consisting of 3 within-subject factors, each with 2 levels: fasted vs. satiated; shift vs. stay; and food vs. inedible. The GLM investigating RTs indicated a main effect of fast (F (1, 55) = 5.00, p = .029, η^2 = .08), a main effect of shift (F (1, 55) = 403, p = <.001, η^2 = .88) and a significant fast x shift interaction (F (1, 55) = 4.22, p = .045, η^2 = .07). This shows that the shift cost was more pronounced under conditions of fasting. There was no main effect of food pictures (F (1, 55) = .015, p = .90, η^2 < .01) and no significant interactions with this factor (F (1, 55) < 3.3, p > .08, η^2 < .06). These results are summarised in Figure 2 and Table 3.

Analysis of accuracy data revealed significantly lower accuracy in the fasting (93.6%) than the satiated condition (94.9%; F(1, 55) = 7.5, p = .008, η^2 = .12). Accuracy was also significantly lower on shift trials (92.7%) than repeat trials (95.9%; F(1, 55) = 91, p < .001, η^2 = .62). No other effects were significant (F(1, 55) < 2.2, p > .14, η^2 < .04).

In order to test the third hypothesis, which predicted that questionnaire scores would relate to task performance, correlational analysis was undertaken. From the rule change task, three variables were entered into the correlation analyses: 1) fast vs. satiated (i.e. overall difference in RT between fasted and satiated sessions); 2) shift cost (i.e. overall difference in RT between switch and repeat trials); 3) fast x shift interaction (i.e. difference in shift cost between fasted and satiated sessions).

Pearson correlations revealed that none of the 3 variables shared significant relationships with subscales of the PPPQ-22 or HADS. Only the EDE-Q6 scale of weight concern significantly correlated with the *fast vs. satiated* variable (r = .348, p = .009). There were no correlations between the 3 variables and any of the

demographic measures of BMI, age, length of fast or subjective ease of fasting (p>.19).

Discussion

This study used a novel set shifting paradigm to examine the influence of short-term fasting on cognitive flexibility in a non-clinical population. In line with our first hypothesis, short-term food deprivation impaired set shifting ability in our healthy sample. Our second hypothesis – that food-related stimuli would impair performance and exacerbate shift costs – was not supported: we found that the content of the stimuli did not have a significant impact on RT. Counter to our third hypothesis, the interaction between fasting and set shifting occurred independently of self-reported symptoms of depression, perseveration and disordered eating, aside from the association with weight concern.

The finding of an interaction between fasting and set shifting is intriguing, especially given the theory that the cognitive inflexibility observed in patients with AN is at least partially due to the biological effects of short-term food restriction [31,33]. Whilst this study has been conducted within a non-clinical population and thus cannot tell us anything directly about AN, it is noteworthy that even short-term starvation had a significant impact on set shifting.

Although this study aimed to examine the impact of food restriction on cognitive performance in a nonclinical sample, it is important to note that short-term fasting is not equivalent to the chronic undernourishment seen in AN. Chronic starvation leads to neurochemical, metabolic, and structural brain changes in patients with AN [69,70], some of which may persist after weight restoration [30]. Fasting for sixteen hours is unlikely to lead to such changes. Future research comparing AN patients with healthy controls who had been fasting for at least sixteen hours, as well as with individuals who limit their food intake for other reasons, may help us to understand further whether the changes in performance seen in this study are comparable with the cognitive profile of AN patients. It will also be important to include a range of cognitive tasks to tease out whether the effect is related specifically to cognitive flexibility, or more broadly linked to changes in inhibition or attention.

Even without a predisposition towards cognitive inflexibility, set shifting difficulties might emerge under conditions of fasting triggered by other factors, such as dieting, physical illness, loss of appetite and/or low mood. Following this initial compromise in flexibility during food restriction, cognitive and behavioural rigidity might then be further exaggerated and manifest as symptoms of AN, such as more deliberate food restriction and preoccupation with thinness. Moreover, a perseverative style may also make it difficult for the individual to terminate efforts at food restriction once started.

The weight concern scale of the EDE-Q6 was correlated with a general performance decrement under fasted conditions. Although

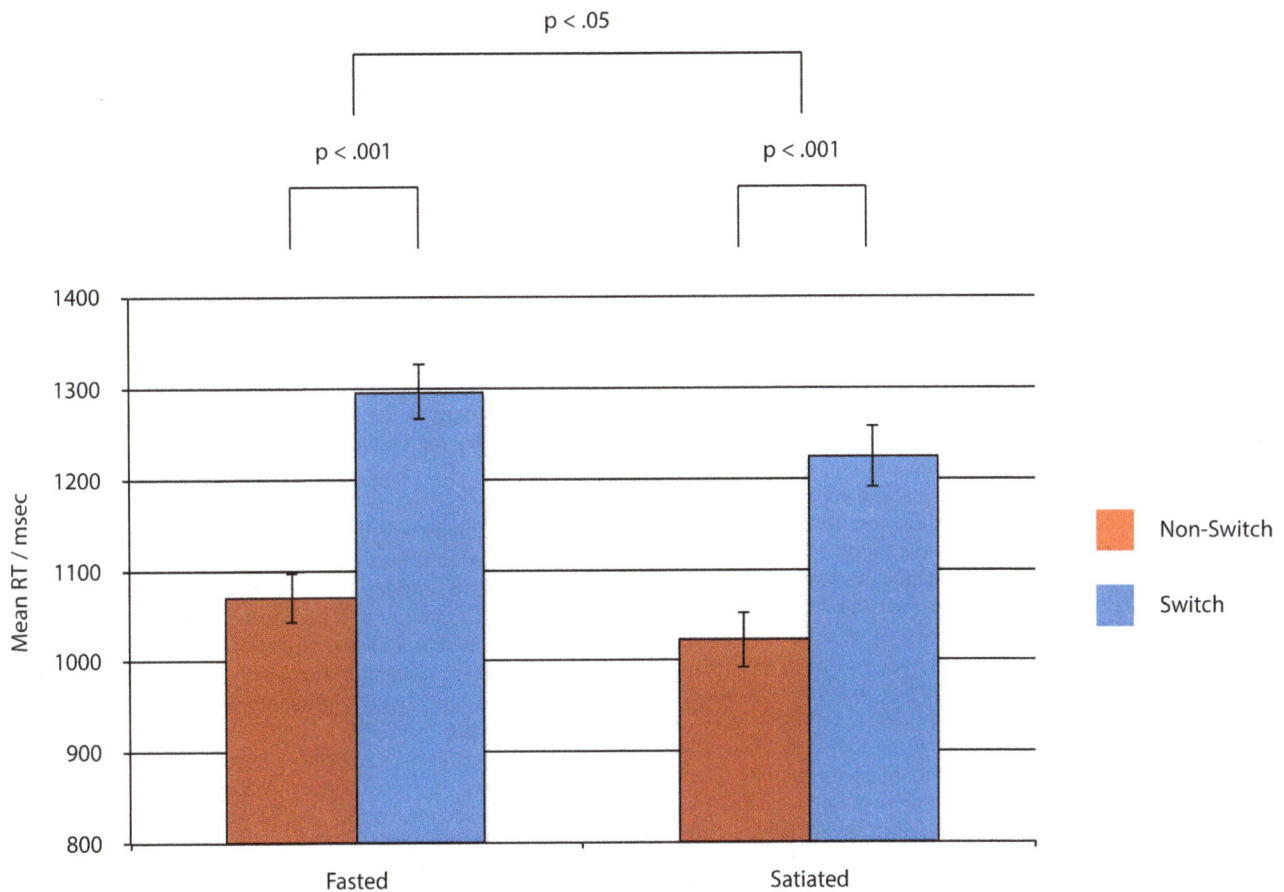

Figure 2. Mean reaction times in the rule change task. Error bars indicate standard errors.

this was not specific to set shifting, it illustrates that the very individuals who are more concerned over their weight are those whose cognitive processes are most detrimentally affected by food deprivation. Clinically this might indicate that some vulnerability exists in those with weight concern, in that these are people who are likely to be restricting their food intake frequently.

Counter to expectations, self-reported depression scores did not have any significant bearing on task performance. This suggests that depressive symptomatology did not impact upon cognitive flexibility in the sample. However, the levels of reported depression in the sample were below population norms [67] and therefore a sample with a broader range of levels of depression might allow for a more accurate test of the hypothesis that low mood would relate to poorer task performance.

Clinical implications

The results provide tentative support for the view that short-term fasting may contribute to the expression of an individual's propensity towards rigidity, either exaggerating existing set shifting difficulties or evoking a novel, temporary set shifting impairment. Such an explanation would be in line with the argument that the cognitive deficits in AN are associated with the biological effects of food restriction [31]. It might be that individuals with AN show an abnormally high rate of perseveration in response to fasting and those whose flexible cognitive processes are more resilient to the effects of short-term fasting would be at a lesser risk of developing AN. This could contribute to explanations of why food restriction does not lead to the development of AN in the majority of individuals who diet or miss meals.

Table 3. Mean reaction times (RT) and percent accuracy in the rule change task.

	Fasted		Satiated	
	Non-switch	**Switch**	**Non-switch**	**Switch**
RT (msec)	1070	1296	1023	1224
	(SD = 201)	(SD = 228)	(SD = 224)	(SD = 249)
Accuracy (%)	95	92	97	93
	(SD = 3)	(SD = 5)	(SD = 3)	(SD = 5)

Another implication is that fasting may potentially account for some of the rigid thoughts and behaviours observed in patients with AN. If this finding is confirmed in future research, it might support an approach to treatment that focuses on initial nutritional rehabilitation before attempting to address cognitive or emotional factors. This is especially important considering that the cognitive inflexibility inherent in AN reduces the ability to fully engage in therapy [28].

Limitations of the present study and future research needs

The current study was designed to be exploratory and has generated further hypotheses that merit testing. However, there are several limitations to its design and we acknowledge that these may limit the degree to which we can extrapolate our results. Firstly, we cannot draw conclusions about the more general issue of cognitive processing, (i.e. attention and cognitive and motor flexibility) following fasting, as our measurement was restricted to set shifting specifically. Furthermore, as we did not take physiological or neural measures, we cannot draw any conclusions about the biological processes that might mediate or explain the effects of dietary restriction on cognitive flexibility. The exact mechanisms by which fasting led to increased shift costs is also a matter requiring further investigation. This is true both at psychological and neurophysiological levels. For example, at a psychological level it is unclear how far a fasting-induced set shifting deficit should be understood in terms of distraction/preoccupation with feeling of hunger, difficulty inhibiting inappropriate response tendencies, interpreting and updating task rules, or a combination of these and other factors. At a neurophysiological level, the impact of fasting on regional brain activity and distinct neuromodulatory systems also merits further study.

In addition, we acknowledge that sixteen hours is a relatively short period for fasting. However, other recent studies in the literature have investigated the impact of fasting periods as short as two hours on neuropsychological functioning [71]. Future studies aiming to build on our results might benefit from a lengthier period of starvation in order to maximise the impact of food restriction. Linked to this, it was not possible to fully establish whether participants had adhered to the fasting requirements, other than relying on their belief that urinalysis would reveal noncompliance and the encouragement of honest reporting. However, if any participants ate surreptitiously without disclosing it to the experimenter, and the extent of fasting was underestimated in the analysis, this would make the findings more striking.

It might be argued that using an established measure of set shifting (e.g. the WCST) would have provided more compelling results and allowed for comparison with other clinical studies.

However, the ecological validity of such tests has been questioned [72], and it is not clear that they have suitable psychometric properties to detect variation in nonclinical samples, in comparison with the RT measures investigated in the present study. Furthermore, the present task was designed to allow comparison between food and non-food stimuli. Nevertheless, it would be helpful for future fasting studies to include multiple measures of set shifting, including more traditional tasks such as the WCST. Practice effects within the rule change task might have lessened the within-subject differences and therefore it may also be useful for future studies to employ a more complex shifting task to reduce the incidence of practice effects. Clinically, the existence of practice effects may add support to interventions such as cognitive remediation therapy [73] which focus on improving cognitive flexibility, as it suggests that cognitive inflexibility can be overcome with effort under conditions of fasting.

Conclusions

Given the impact of short-term fasting, it is highly recommended that future studies examining cognition in AN investigate the effects of food restriction in their participant samples. Further studies of healthy controls who have fasted for hours or days would also help to establish a clearer picture of the effects of (at least short-term) food restriction on cognition. Furthermore, an explicit examination of the factors that interact with rigid thinking might also be important.

Given that relatively brief dietary restriction had selective effects on performance of the rule change task (i.e. greater effect on switch than repeat trials), it remains to be explored how individual differences in diet over a longer time period (including a developmental timespan) may impact on performance of specific tasks. This has clear implications for the study of development and individual differences. Furthermore, our results suggest that caution should be used in interpreting results from tests conducted in conditions where dietary restriction is common, for example in a hospital setting.

Acknowledgments

The authors would like to thank Miss Angelina Samuel for assistance with data collection and Dr. Rudolf Uher for access to his team's database of photographic stimuli.

Author Contributions

Conceived and designed the experiments: HMB PWB SJG LS. Performed the experiments: HMB. Analyzed the data: HMB SJG. Contributed reagents/materials/analysis tools: SJG. Wrote the paper: HMB PWB SJG LS.

References

1. Monsell S (2003) Task switching. Trends Cogn Sci 7: 134–140.
2. Sandson J, Albert ML (1987) Perseveration in behavioural neurology. Neurology 37: 1736–1741.
3. Bissonette GB, Powell EM, Roesch MR (2013) Neural structures underlying set-shifting: Roles of medial prefrontal cortex and anterior cingulate cortex. Behav Brain Res 250: 91–101.
4. Aron AR, Monsell S, Sahakian BJ, Robbins TW (2004) A componential analysis of task-switching deficits associated with lesions of left and right frontal cortex. Brain 127: 1561–1573.
5. Reimers S, Maylor EA (2005) Task switching across the life span: Effects of age on general and specific switch costs. Dev Psychol 41: 661–671.
6. Gu BM, Park JY, Kang HD, Lee SJ, Yoo SY, et al. (2008) Neural correlates of cognitive inflexibility during task-switching in obsessive-compulsive disorder. Brain 131: 155–164.
7. Sullivan PF (1995) Mortality in anorexia. Am J Psychiatry 152: 1073–107.
8. Papadopoulos FC, Ekbom A, Brandt L, Ekselius L (2009) Excess mortality, causes of death and prognostic factors in anorexia nervosa. Br J Psychiatry 194: 10–17.
9. Berkman ND, Lohr KN, Bulik CM (2007) Outcomes of eating disorders: a systematic review of the literature. Int J Eat Disord 40: 293–309.
10. Agras WS, Robinson AH (2008) Forty years of progress in the treatment of the eating disorders. Nord J Psychiatry 62: 19–24.
11. Roberts ME, Tchanturia K, Treasure JL (2010) Exploring the neurocognitive signature of poor set-shifting in anorexia and bulimia nervosa. J Psychiatr Res 44: 964–970.
12. Tchanturia K, Harrison A, Davies H, Roberts ME, Oldershaw A, et al. (2011) Cognitive flexibility and clinical severity in eating disorders. PloS ONE 6(6): e20462.
13. Fassino S, Piero A, Daga GA, Leobruni P, Mortara P, et al. (2002) Attentional biases and frontal functioning in anorexia nervosa. Int J E Disord 31: 274–283.

14. Treasure J, Schmidt U (2013) The Cognitive-Interpersonal Maintenance Model of Anorexia Nervosa Revisited: A summary of the evidence for cognitive, socio-emotional and interpersonal predisposing and perpetuating factors. J Eat Disord 1: 13.

15. Roberts M, Tchanturia K, Stahl D, Southgate L, Treasure J (2007) A systematic review and meta-analysis of set-shifting ability in eating disorders. Psychol Med 37: 1075–1084.

16. Tchanturia K, Davies H, Roberts M, Harrison A, Nakazato M, et al. (2012) Poor cognitive flexibility in eating disorders: examining the evidence using the Wisconsin Card Sorting Task. PLoS ONE 7(1): e28331.

17. Tenconi E, Santonastaso P, Degortes D, Bosello R, Titton F, et al. (2010) Set-shifting abilities, central coherence, and handedness in anorexia nervosa patients, their unaffected siblings and healthy controls: Exploring putative endophenotypes. World J Biol Psychiatry 11: 813–823.

18. Zastrow A, Kaiser S, Stippich C, Walther S, Herzog W, et al. (2009) Neural correlates of impaired cognitive-behavioural flexibility in anorexia nervosa. Am J Psychiatry 166: 608–616.

19. Lopez C, Stahl D, Tchanturia K (2010) Estimated intelligence quotient in anorexia nervosa: a systematic review and meta-analysis of the literature. Ann Gen Psychiatry 9: 1–10.

20. Steinglass J, Walsh BT (2006) Habit learning and anorexia nervosa: a cognitive neuroscience hypothesis. Int J Eat Disord 39: 267–275.

21. Tchanturia K, Lock J (2011) Cognitive remediation therapy for eating disorders: development, refinement and future direction. In: R Adan, W Kaye, editors. Behavioral Neurobiology of Eating Disorders. Berlin: Springer. pp. 269–287.

22. Holliday J, Tchanturia K, Landau S, Collier D, Treasure J (2005) Is impaired set shifting an endophenotype of anorexia nervosa? Am J Psychiatry 162: 2269–2275.

23. Kanakam N, Raoult C, Collier D, Treasure J (2013). Set shifting and central coherence as neurocognitive endophenotypes in eating disorders: A preliminary investigation in twins. World J Biol Psychiatry 14: 1–12.

24. Sato Y, Saito N, Utsumi A, Aizawa E, Shoji T, et al. (2013) Neural Basis of Impaired Cognitive Flexibility in Patients with Anorexia Nervosa. PloS ONE 8(5): e61108.

25. Gillberg CI, Billstedt E, Wentz E, Anckarsäter H, Råstam M, et al. (2010) Attention, executive functions, and mentalising in anorexia nervosa eighteen years after onset of eating disorder. J Clin Exp Neuropsychol 32: 358–365.

26. Lindner SE, Fichter MM, Quadflieg N (2014) Set-shifting and its relation to clinical and personality variables in full recovery of anorexia nervosa. Eur Eat Disord Rev 22: 252–259.

27. Nikendei C, Funiok C, Pfuller U, Zastrow A, Aschenbrenner S, et al. (2011) Memory performance in acute and weight-restored anorexia nervosa patients. Psychol Med 41: 829–838.

28. Wilsdon A, Wade TD (2006) Executive functioning in anorexia nervosa: exploration of the role of obsessionality, depression and fasting. J Psychiatr Res 40: 746–754.

29. Tchanturia K, Morris R, Surguladze S, Treasure J (2002) An examination of perceptual and cognitive set shifting tasks in acute Anorexia Nervosa and following recovery. Eat Weight Disord 7: 312–315.

30. Kingston K, Szmukler G, Andrewes D, Tress P, Desmond P (1996) Neuropsychological and structural brain changes in anorexia nervosa before and after refeeding. Psychol Med 26: 15–28.

31. Duchesne M, Mattos P, Fontenelle LF, Veiga H, Rizo L, et al. (2004) Neuropsychology of eating disorders: a systematic review of the literature. Rev Bras Psiquiatr 26: 107–117.

32. Pollice C, Kaye WH, Greeno CG, Weltzin TE (1997) Relationship of depression, anxiety, and obsessionality to state of illness in anorexia nervosa. Int J Eat Disord 21: 367–376.

33. Zakzanis KK, Campbell Z, Polsinelli A (2010) Quantitative evidence for distinct cognitive impairment in anorexia nervosa and bulimia nervosa. J Neuropsychol 4: 89–106.

34. Merwin RM, Timko CA, Moskovich AA, Ingle KK, Bulik C (2011) Psychological inflexibility and symptom expression in anorexia nervosa. Eat Disord 19: 62–82.

35. Danner UN, Sanders N, Smeets PA, van Meer F, Adan RA, et al. (2012) Neuropsychological weaknesses in anorexia nervosa: Set-shifting, central coherence, and decision making in currently ill and recovered women. Int J Eat Disord 45: 685–694.

36. Lang K, Stahl D, Espie J, Treasure J, Tchanturia K (2013) Set shifting in children and adolescents with anorexia nervosa: An exploratory systematic review and meta-analysis. Int J Eat Disord 47: 394–399.

37. Shott ME, Filoteo JV, Bhatnagar KA, Peak NJ, Hagman JO, et al. (2012) Cognitive Set-Shifting in Anorexia Nervosa. Eur Eat Disord Rev 20: 343–349.

38. Kojima S, Nagai N, Nakabeppu Y, Muranaga T, Deguchi D, et al. (2005) Comparison of regional cerebral blood flow in patients with anorexia nervosa before and after weight gain. Psychiat Res 140: 251–258.

39. Oberndorfer TA, Kaye WH, Simmons AN, Strigo IA, Matthews SC (2011) Demand-specific alteration of medial prefrontal cortex response during an inhibition task in recovered anorexic women. Int J Eat Disord 44: 1–8.

40. Råstam M, Bjure J, Vestergren E, Uvebrant P, Gillberg IC, et al. (2001) Regional cerebral blood flow in weight-restored anorexia nervosa: a preliminary study. Dev Med Child Neurol 43: 239–242.

41. McCormick LM, Keel PK, Brumm MC, Bowers W, Swayze V, et al. (2008). Implications of fasting-induced change in right dorsal anterior cingulate volume in anorexia nervosa. Int J Eat Disord 41: 602–610.

42. Wagner A, Greer P, Bailer UF, Frank GK, Henry SE, et al. (2006) Normal brain tissue volumes after long-term recovery in anorexia and bulimia nervosa. Biol Psychiatry 59: 291–293.

43. Keys A, Brozek J, Henschel A, Mickelsen O, Taylor HL (1950) The Biology of Human Fasting. Minneapolis: The University of Minnesota Press.

44. Epling WF, Pierce WD, Stefan L (2006) A theory of activity-based anorexia. Int J Eat Disord 3: 27–46.

45. Piech RM, Hampshire A, Owen AM, Parkinson JA (2009) Modulation of cognitive flexibility by hunger and desire. Cogn Emot 23: 528–540.

46. Uher R, Treasure J, Heining M, Brammer MJ, Campbell IC (2006) Cerebral processing of food-related stimuli: effects of fasting and gender. Behav Brain Res 169: 111–119.

47. Benau E, Orloff N, Janke E, Serpell L, Timko CA (2014) A Systematic Review of the Effects of Experimental Fasting on Cognition. Appetite 77: 52–61.

48. Doniger G, Simon ES, Zivotofsky AZ (2006) Comprehensive computerized assessment of cognitive sequelae of a complete 12–16 hour fast. Behav Neurosci 120: 804–816.

49. Stewart SH & Samoluk SB (1997) Effects of short-term food deprivation and chronic dietary restraint on the selective processing of appetitive-related cues. Int J Eat Disord 21: 129–35.

50. Owen L, Scholey A, Finnegan Y, Sünram-Lea SI (2013) Response variability to glucose facilitation of cognitive enhancement. Br J Nutr 110: 1873–1884.

51. Stuss DT, Floden D, Alexander MP, Levine B, Katz D (2001) Stroop performance in focal lesion patients: dissociation of processes and frontal lobe lesion location. Neuropsychologia 39: 771–786.

52. Brooks SJ, Owen GO, Uher R, Friederich HC, Giampietro V, et al. (2011) Differential neural responses to food images in women with bulimia versus anorexia nervosa. PLoS ONE 6(7): e22259.

53. Uher R, Murphy T, Brammer MJ, Dalgleish T, Phillips ML, et al. (2004) Medial prefrontal cortex activity is associated with symptom provocation in eating disorders. Am J Psychiatry 161: 1238–1246.

54. Brecelj-Anderluh M, Tchanturia K, Rabe-Hesketh S, Treasure J (2003) Childhood obsessive-compulsive personality traits in adult women with eating disorders: defining a broader eating disorder phenotype. Am J Psychiatry 160: 242–247.

55. Salbach-Andrae H, Lenz K, Simmendinger N, Klinkowski N, Lehmkuhl, et al. (2008) Psychiatric comorbidities among female adolescents with anorexia nervosa. Child Psychiatry Hum Dev 39: 261–272.

56. Ilonen T, Taiminen T, Karlsson H, Lauerma H, Tuimala P, et al (2000) Impaired Wisconsin card sorting test performance in first-episode severe depression. Nord J Psychiatry 54: 275–280.

57. Grant MM, Thase ME, Sweeney JA (2001) Cognitive disturbance in outpatient depressed younger adults: evidence of modest impairment. Biol Psychiatry 50: 35–43.

58. Snyder HR (2013) Major depressive disorder is associated with broad impairments on neuropsychological measures of executive function: A meta-analysis and review. Psychol Bull 139: 81–132

59. Giel KE, Wittorf A, Wolkenstein L, Klingberg S, Drimmer E, et al. (2012) Is impaired set-shifting a feature of "pure" anorexia nervosa? Investigating the role of depression in set-shifting ability in anorexia nervosa and unipolar depression. Psychiatry Res 200: 538–543.

60. Gowers S, Bryant-Waugh R (2004) Management of child and adolescent eating disorders: the current evidence base and future directions. J Child Psychol Psychiatry 45: 63–83.

61. Erdfelder E, Faul F, Buchner A (1996) GPOWER: a general power analysis program. Behav Res Methods Instrum Comput 28: 1–11.

62. Zigmond AS, Snaith RP (1983) The Hospital Anxiety and Depression Scale. Acta Psychiatr Scand 67: 361–370.

63. Serpell L, Waller G, Fearon P, Meyer C (2009) The roles of persistence and perseverance in psychopathology. Behav Ther 40: 260–271.

64. Fairburn CG, Beglin S (2008) The Eating Disorder Examination Questionnaire (EDE-Q6.0). In Fairburn CG, editor. Cognitive Behaviour Therapy and Eating Disorders (appendix II). Guilford Press: New York

65. Fairburn CG, Cooper Z (1993) The Eating Disorder Examination (12th Ed.). In Fairburn CG, Wilson GT, editors. Binge Eating: Nature, Assessment and Treatment. New York: Guilford Press. pp. 317–360.

66. Luce KH, Crowther JH (1999) The reliability of the Eating Disorder Examination – self-report questionnaire version (EDE-Q). Int J Eat Disord 25: 349–351.

67. Crawford JR, Henry JD, Crombie C, Taylor EP (2001) Brief report: normative data for the HADS from a large non-clinical sample. Br J Clinical Psychol 40: 429–434.

68. Mond JM, Hay PJ, Rodgers B, Owen C (2005) Eating Disorder Examination Questionnaire (EDE-Q): Norms for young adult women. Behav Res Ther 44: 53–62.

69. Mühlau M, Gaser C, Ilg R, Conrad B, Leibl C, et al. (2007) Gray matter decrease of the anterior cingulate cortex in anorexia nervosa. Am J Psychiatry 164: 1850–1857.

70. Sidiropoulos M (2007) Anorexia nervosa: The physiological consequences of starvation and the need for primary prevention efforts. McGill J Med 10: 20–25.

71. Owen L, Scholey AB, Finnegan Y, Hu H, Sunram-Lea SI (2012) The effect of glucose dose and fasting interval on cognitive function: A double-blind, placebo-controlled, six-way crossover study. Psychopharmacol 220: 577–589.

72. Burgess PW, Alderman N, Forbes C, Costello A, Coates LMA, et al. (2006) The case for the development and use of "ecologically valid" measures of executive function in experimental and clinical neuropsychology. J Int Neuropsycholog Soc 12: 194–209.

73. Tchanturia K, Davies H (2010) Cognitive Remediation Therapy for Anorexia Nervosa. Cambridge University Press: Cambridge.

3

Vegetable Exudates as Food for *Callithrix* spp. (Callitrichidae): Exploratory Patterns

Talitha Mayumi Francisco[1], Dayvid Rodrigues Couto[2], José Cola Zanuncio[1]*, José Eduardo Serrão[3], Ita de Oliveira Silva[1], Vanner Boere[4]

1 Departamento de Biologia Animal, Universidade Federal de Viçosa, 36570-900, Viçosa, MG, Brazil, 2 Departamento de Botânica/Museu Nacional, Universidade Federal do Rio de Janeiro, 20940-040, Rio de Janeiro, RJ, Brazil, 3 Departamento de Biologia Geral, Universidade Federal de Viçosa, 36570-900, Viçosa, Brazil, 4 Departamento de Bioquímica e Biologia Molecular, Universidade Federal de Viçosa, 36570-900, Viçosa, Brazil

Abstract

Marmosets of the genus *Callithrix* are specialized in the consumption of tree exudates to obtain essential nutritional resource by boring holes into bark with teeth. However, marmoset preferences for particular tree species, location, type, and other suitable factors that aid in exudate acquisition need further research. In the current study, the intensity of exudate use from *Anadenanthera peregrina* var. *peregrina* trees by hybrid marmosets *Callithrix* spp. groups was studied in five forest fragments in Viçosa, in the state of Minas, Brazil. Thirty-nine *A. peregrina* var. *peregrina* trees were examined and 8,765 active and non-active holes were analyzed. The trunk of *A. peregrina* var. *peregrina* had a lower number of holes than the canopy: 11% were found on the trunk and 89% were found on the canopy. The upper canopy was the preferred area by *Callithrix* spp. for obtaining exudates. The intensity of tree exploitation by marmosets showed a moderate-to-weak correlation with diameter at breast height (DBH) and total tree height. The overall results indicate that *Anadenanthera peregrina* var. *peregrina* provides food resources for hybrid marmosets (*Callithrix* spp.) and these animals prefer to explore this resource on the apical parts of the plant, where the thickness, location, and age of the branches are the main features involved in the acquisition of exudates.

Editor: Roscoe Stanyon, University of Florence, Italy

Funding: Funding provided by "Conselho Nacional de Desenvolvimento Científico e Tecnológico (CNPq)", "Coordenação de Aperfeiçoamento de Pessoal de Nível Superior (CAPES)" and "Fundação de Amparo à Pesquisa do Estado de Minas Gerais (FAPEMIG)". The funders had no role in study design, data collection and analysis, decision to publish, or preparation of the manuscript.

Competing Interests: The authors have declared that no competing interests exist.

* Email: zanuncio@ufv.br

Introduction

Vegetable exudates, such as saps and gums, are constitutive or induced compounds, which are essential components of the diet of many primates [1]. To date, at least 69 species of primates, representing 12 families, are known to consume plant exudates [2], being either 'nonspecialists' or 'specialists' feeders [3]. Specialist species include marmosets *Callithrix* spp. Erxleben, 1777 (Cebidae); *Mico* Lesson, 1840 (Cebidae); *Cebuella pygmaea* Spix, 1823 (Cebidae); *Phaner* Gray, 1870 (Cheirogaleidae); *Euoticus* Gray, 1863 (Galagidae) and slow lorises *Nycticebus* spp. É. Geoffroy, 1812 (Lorisidae) [2,4,5]. These genera are obligatory consumers of plant exudates [1,3]; they have evolved anatomical adaptations of their teeth and of the bones and muscles in their skull to perforate plant branches to stimulate the release of exudates [6,7]. This behavior has been termed 'gouging.'

Within the Primates order, members of the New World subfamily Callitrichinae perhaps consume the largest amount of exudates [3,4,8]. In particular, the genera *Callithrix*, *Cebuella*, and *Mico* are specialized in the acquisition of this nutritional resource. During specific seasons, exudates can comprise up to 70% of the diet of *Callithrix* spp. [9,10], with 30% of the daily activity of these species spent gouging [3,11]. *Callithrix* species have evolved functional characteristics that enable them to extract exudates from plants [12,14], including a specialized lower dentition [15,16], with modification of the architecture of the mandible [7,17] and bones of the masticatory apparatus [6]. Accordingly, tree-gouging is characterized by the use of the maxillary teeth to puncture and hold onto a plant branch and the mandibular incisors to scrape the branch [12,13]. These characteristics enable gouging animals to drill holes of various shapes and sizes to reach ducts in the plant tissue and release the exudates [12,13,18–20]. Additionally, marmosets have evolved a specialized system for digesting exudates through fermentation, as their cecum and colon are disproportionately large compared to the rest of their body [14,21]. This is an important feature because fermentation by microorganisms is essential for the extraction of energy from the complex polysaccharides contained within plant exudates [11,22].

The use of exudates as a food has implications for the ecology and social organization of primates [1,3,18]. The predictability of periodic amounts of gums confers a competitive advantage to these primates, who have relatively high energy demands because of their small size [1]. The availability of exudates favors the social use of a food resource [3], with more exudativorous species inhabiting smaller areas because of the predictable and more or less constant presence of their food source. Apart from metabolism, exudativory also appears to have an impact on the development of psychological differences among species [23].

Exudates are significant sources of complex carbohydrates, proteins and certain minerals, especially calcium [1,4,24–26]. *Callithrix* species often extract exudates from trees of the genus *Anadenanthera* (Fabaceae) especially *Anadenanthera peregrina* var. *peregrina* (L.) Speg. [19,20,27–29]. Species of the genus *Anadenanthera* are widely distributed in the phytogeographical areas of the Caatinga, Cerrado (Brazilian savannah), and Atlantic Forests in the Northeast, Midwest, and Southeast regions of Brazil [30].

The exploitation of *Anadenanthera* spp. by marmosets is surprising because this plant can produce tannin, dyes, timber, medicinal, and psychoactive products [31–35] and has relatively hard bark. *Anadenantgera* spp. is sometimes covered with prickles, and high tannin levels [31,34,36], both which can repel primates [37]. However, these characteristics do not appear to dismay certain *Callithrix* species in obtaining exudate a from this plant. Such species include *C. penicillata* É Geoffroy, 1812 [18,20,38], *Callithrix jacchus* L., 1758 [18,19,28,29] and *Callithrix flaviceps* Thomas, 1903 (Primates: Cebidae) [27,39]. *Callithrix* species are well known for consuming exudates from plants, but their exploitation of gum produced by trees, in particular, is poorly studied. However, this information is important for understanding the relationships between the behavioral and ecological traits of marmosets and their target tree species.

The study of plant–animal interactions has a key role in ecological theories and in the understanding issues related to the conservation of biodiversity. This is the first study to describe the ecological relationships between marmosets and their target trees and this work contributes to a better understanding of the exudativory pressure exerted by these primates. Here we specifically evaluate the intensity of exploitation of exudates from *A. peregrina* var. *peregrina* by *Callithrix* spp. in forest fragments of the Atlantic Forest in the state of Minas Gerais, Brazil. We also determine the preferred foraging location on the tree and the number of holes made to obtain exudates, and link this information to dendrometric data (DBH diameter at 1.3 m height and total height) of *Anadenanthera peregrina* var. *peregrina*.

Materials and Methods

Study areas

The study was conducted in areas used by five marmosets groups, each with six to twelve hybrid individuals with intermediate characteristics of *C. penicillata* × *C. jacchus* and *C. penicillata* × *Callithrix geoffroyi*. The study of tree orifices gouged by hybrid marmosets is complementary to previous studies of the behavioral ecology of these animals [12,40]. Moreover, populations of hybrid marmosets are increasing in number in the Atlantic Forest as a result of anthropogenic environmental disturbance and illegal trafficking of wild animals [41].

Our study areas referred to as Fragments 1 (20°45′34.71″S, 42°51′57.84″W); 2 (20°45′22.28″S, 42°52′23.81″W); 3 (20°45′11.16″S, 42°52′16.80″W); 4 (20°45′13.85″S, 42°52′26.90″W), and 5 (20°46′17.41″S, 42°52′37.03″W) (Fig. 1) were located on the campus of the Universidade Federal de Viçosa, Viçosa, in the state of Minas Gerais State, Brazil, at 675 to 709 m above sea level.

The region studied has montane seasonal semideciduous forest [42] and a highland tropical climate with rainy summers and cold, dry winters, which is classified as 'Cwb' based on the Köppen climate classification [43]. The area's average annual rainfall and temperature are 1221 mm and 19°C, respectively [44].

Data collection and analysis

Anadenanthera peregrina var. *peregrina* was selected as the focal study tree because it is the only tree species exploited by the groups of *Callithrix* spp. present within the study area.

Thirty-nine *A. peregrina* var. *peregrina* trees exploited by marmosets were sampled (Fig. 2). These trees were marked with sequentially numbered aluminum plates and the tree holes were measured and quantified with rock climbing equipment and techniques adapted for canopies [45]. This method enabled data collection from the higher and less accessible parts of the tree. The total tree height was measured with a hypsometer (Suunto PM 5, Finland) and the DBH with a diameter tape.

Plants with holes, which characterize its use by marmoset were collected for identification, except those that were difficult to access.

The fertile parts (branches with flowers and fruits) of four voucher specimens were collected in the field, herborized with floristic methodology, and sent to relevant experts for identification. This material was incorporated into the collection of the Herbarium of the Universidade Federal de Viçosa with the numbers VIC, 38241; 38240; 38239; and 38615.

Individuals of *A. peregrine* var. *peregrina* were divided into two ecological zones: the trunk and the canopy [46] (Fig. 3a), with the canopy subdivided into three segments: lower, middle, and upper canopy [46,47] (Fig. 3b). All holes on the trunk were counted and measured, whereas those in the canopy were counted and measured as high up as it was safe for the researcher to go. Therefore, the percentage of holes was estimated for the canopy. The basal and distal diameter and the total length of the branches of the canopy sampled were also measured and an average diameter was obtained for the sampled branches.

Gouged tree holes were classified as inactive, characterized by presence of scar tissue and as active, without presence of scar tissue. The size of holes was measured with a digital caliper and their height (to the bark, in a vertical stroke), width (up to the bark, in a horizontal stroke), and depth (deepest portion in the hole) recorded. These parameters were obtained for active holes only, because these were assumed to be currently exploited by the marmosets. All the holes were counted, including those with scar tissue and those that were not being used by the animals. The intensity of marmoset exploitation of *A. peregrina* var. *peregrina* plant excudates was assessed by: (1) the number of active and nonactive holes, and (2) height, width, and depth of the active holes.

The relation between the total number of gouged holes and the trunk diameter at 1.3 m from the soil (DBH) and tree height was analyzed by using the Pearson correlation test. The chi-square test was applied to determine whether the number of active and nonactive holes differed between the three canopy segments. Analysis of variance (ANOVA) and *post hoc* Tukey test was used to assess whether the average diameter of the branches of each canopy segment and the dimensions of the active holes differed between these segments. Differences between the volumes of active holes in the trunk versus those in the canopy were evaluated with Student's t-test. All analyses were carried out using the computer program R 3.0.1. [48].

No specific permits were required for to study *A. peregrina* var. *peregrina* and *Callithrix* spp. in Brazil. The field studies did not involve endangered or protected species.

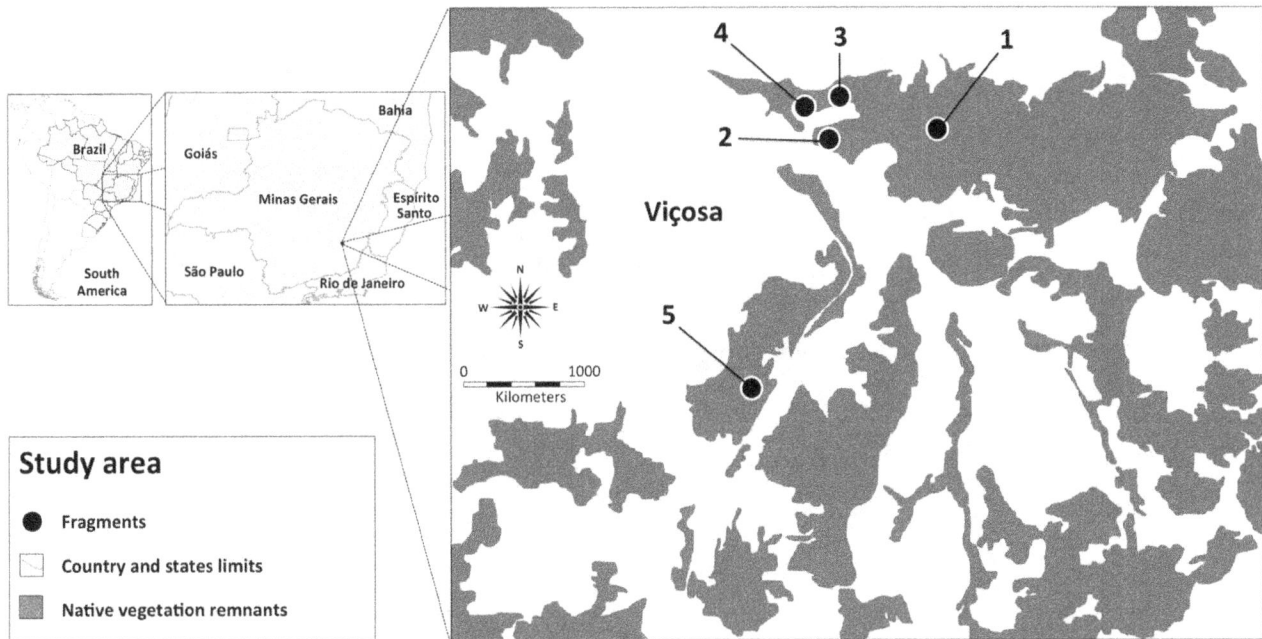

Figure 1. Location of the five forest fragments of the Atlantic Forest in Viçosa, Minas Gerais State, Brazil, where the study was conducted.

Results

Measurements from 39 *A. peregrina* var. *peregrina* trees recorded ranges of DBH and total height of 6.8–64.9 cm and 6.9–35.5 m, respectively. We counted a total of 8,765 gouged holes, with 970 (11%) of holes located in the trunk and 7,795 (89%) holes located in the canopy. The total number of holes per tree ranged from 8 to 2288.

Hybrid marmosets preferentially obtained exudates from branches in the canopy of *A. peregrina* var. *peregrina*. The upper canopy showed the higher number ($x^2 = 143.38$; p<0.001) of both active and inactive holes (48%) followed by middle (30%) and lower canopy (22%).

A separate analysis of active gouge holes showed the same type of variation in the three regions of the tree canopy (Figure 3), showed that there was a significant difference in means of tree holes present in the three different portions of the canopy in the upper. (ANOVA, df = 2, F = 53.45, p<0.001). In the upper canopy, the average branch length was smaller (32.19±0.95; p< 0,001) than those in the middle canopy (39.36±1.33 cm) and lower canopy (53.30±2.03 cm) respectively.

Of the 915 active holes recorded, 810 (89%) were in the canopy and 105 (11%) were in the trunk (Table 1). The lower canopy had a smaller number and dimension of holes than the upper canopy (Table 1). The marmosets exploited the upper canopy more than the lower canopy to obtain exudates (Table 2).

The total number of gouged holes showed weak to moderate correlation with the DBH (Pearson correlation; $r^2 = 0.530$) and total height (Pearson correlation; $r^2 = 0.435$ of trees). These correlations suggest a positive relationship between concentration of gouged holes and their location in the canopy and that marmoset preferably obtain exudates from canopy branches in the canopy of *A. peregrina* var. *peregrina*.

Discussion

The feeding by marmosets on gum from holes in trees depends on various tree features. *Anadenanthera peregrina* var. *peregrina* are among the 80 species identified as sources of exudates exploited by *Callithrix* spp. [2]. The high number of scarifications on this tree species recorded in this study highlights it as a preferred source of food for marmosets; in addition, no scarifications were found on the other plant species such as *Tapirira guianensis* Aubl., *Allophylus edulis* Radlk. ex Warm.; *Astronium fraxinifolium* Schott, which are used by *C. jacchus*. However, the use of only one plant species differs from the results in the Caatinga and Cerrado, where *Callithrix jacchus* and *C. penicillata* exploited a larger number of gum trees species [20,28,29,49]. The greater number and size of holes in the canopy compared with those in the trunk can be explained by the smaller branch diameter in the canopy. The positive correlation between the number of scarifications with DBH and tree height may be explained by a larger area to be exploited to obtain exudates.

The exclusive use of *A. peregrina* var. *peregrina* by hybrid marmosets to obtain exudates in the five forest fragments and the absence of scarification on other tree species merits further study. Our results particularly raise the question if there are particular nutrients present in *A. peregrina* that are lacking other plants to explain the exclusive use of just a single plant species. Further study of this question would help us understand the marmosets selection and exploitation of exudate resources. It is known that the exudates of *Anadenanthera* have high concentrations of polysaccharides and calcium [50,51] and exudate polysaccharides are an important energy source for marmosets [52]. Calcium plays an important role to maintain the calcium/phosphorus metabolism balance of organisms [53]. Female marmosets typically give birth to twins twice a year, and calcium may be especially important during pregnancy and milk production for these animals [18,54]. Thus, one possibility for the exclusive utilization

Figure 2. Exploration of *Anadenanthera peregrina* **var.** *peregrina* **(Fabaceae) exudates by** *Callithrix* **spp. (Callitrichidae) hybrids in fragments of Atlantic Forest in Minas Gerais State, Brazil, where: (a) Hybrid** *Callithrix* **spp. scarifying hole, (b) detail of the canopy branches extensively explored, (c) branches with exudate released (d and e) detail of the holes with exudate.**

of *Anadenanthera* by marmosets in our study may be a unique role of the tree species to fulfill the dietary and energetic needs of these primates.

The use of exclusive trees by *Callithrix* spp. to obtain exudates is uncommon given that, generally, these species use a large number of gum tree species to obtain exudates. *Phaner furcifer* Blainville, 1839 in Madagascar [55], *Nycticebus coicang* Boddaert, 1875 in West Malaysia (Manjung District, Perak State), *Cebuella pygmaea* in Northeastern Ecuador [56] and *C. pygmaea* in Iquitos, Peru [57] fed on 10, 9, 18 and 58 plant species, respectively. In the Brazilian Cerrado, *Callithrix penicillata* used 14 gum tree species [2,20,58,59] and this same number was reported to have been used by *Callithrix jacchus* in the Caatinga [28,29]. The scarce availability of resources in these ecosystems, with more extreme climatic conditions, could explain the use of different plant species [18]. However, *C. jacchus* explores mainly exudates from *Anadenathera peregrina* [28] in the Caatinga, a biome with poor gum tree diversity. Therefore, further phytosociological studies of *A. peregrina* in different habitats are necessary to explain patterns of marmoset preference for gum trees.

The marmosets evaluated in this study fed only exudates of *A. peregrina* var. *peregrina*, despite the fact that the study areas were inhabited by other plant gum trees, e.g., *Tapirina guianensis* Aubl. used for *C. jacchus* [13,18]; *Callithrix kuhlii* Coimbra Filho, 1985 [60]; *C. penicillata* [49]; *Mico melanurus* (É. Geoffroy in Humboldt, 1812) [61], *Piptadenia gonoacantha* (Mart.) J.F. Macbr. used for *C. flaviceps* [62] and *Astronium fraxinifolium* Schott. used for *C. jacchus* [18,63]; which show a preference of marmosets for *A. peregrina* var. *peregrina*.

The absence of scarification on some *A. peregrina* var. *peregrina* plants is similar to the pattern for other plant gums as *Vochysia pyramidalis* Mart., *Callisthene major* Mart. & Zucc. and *Tapirina guianensis* Aubl. used by *C. penicillata* [2] and on *Anacardium occidentale* L. [18], *Anadenhantera peregrina* (L.) Speg., *Astronium fraxinifolium* Schott, *Enterolobium contortisili-quum* (Vell.) Morong, and *Coccoloba* sp. by *C. jacchus* [28].

The preference for certain trees of the same species can be explained by differences between the trees, such as increased production and nutritional quality of exudates, smaller amount of

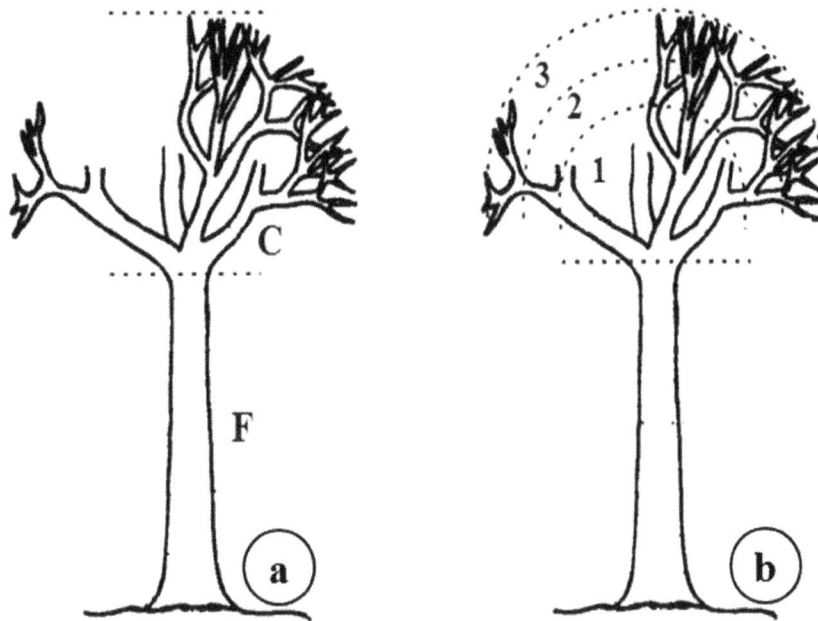

Figure 3. Scheme of the division of *Anadenanthera peregrina* **var.** *peregrina* **(Fabaceae) trees in ecological zones to quantify the exploitation of exudates by** *Callithrix* **spp. (Callitrichidae) in fragments of Atlantic Forest in Minas Gerais State, Brazil.** Division of the tree (a) in two ecological zones (F- trunk, C- canopy) and of the canopy (b) in three segments (1- lower canopy; 2- midlle canopy; 3- upper canopy).

secondary metabolites, such as tannins, and greater protection from predators [1,26,55,58].

The preference of the animals for the canopy, as shown by the number and dimensions of the holes between the ecological zones of the tree (trunk and canopy) in *A. peregrina* found in this study, might be related to the presence of thinner branches in this region, which are preferred by marmosets to exploit this resource. This preference also might be the result of physical and mechanical factors, such as a thicker bark, and properties that facilitate scarification by these animals [6]. The quantity and nutritional quality of exudates in the canopy might also be better because of a higher metabolic rate in this part of the tree and a lower quantity of secondary compounds [31,34]. However, in ecosystems with more extreme environmental conditions, such as, the Caatinga and Cerrado, *Callithrix* spp. (*C. jacchus* and *C. penicillata*) use both the trunk and canopy of gum trees [18,20,28,49,59], with a greater number of holes found in the trunk [28,49]. Thus, the strategy of using such tree species by *C. jacchus* and *C. penicillata* might be affected by environmental conditions [28,59]. Similarly,

Nycticebus spp. *pygmaeus* Bonhote, 1907 (Primates: Lorisidae) made more use of the trunk than of the canopy for exudate feeding in a mixed deciduous forest in the Seima Protection Forest, Eastern Cambodia [5].

The increased preference of marmosets for the apical segments of the canopy, with a higher percentage of exudates and larger scarified holes, agrees with observations made between canopy segments. The increased branch use in the external parts of the canopy, where younger branches are found, reflects a preference for plant parts that are less thick, facilitating scarification. The preference of marmosets for tree segments with flatter branches horizontally and thinner bark reduces the energy used for foraging and effort in obtaining exudates. Marmoset species such as *C. penicillata* and *C. jacchus* have a bimodal pattern of exudate exploitation, with peaks in the morning and late afternoon [38,64]. After consuming the exudates, the marmosets scarify the bark and holes in parts of the tree that have a higher metabolic rate, justifying the exploration of more exposed and thinner branches of the canopy. The benefits of acquiring the exudates must be higher

Table 1. Height, width, depth and number (Num.) of active holes scarified by hybrid marmosets in the stem and canopy of *Anadenanthera peregrina* var. *peregrina* (Fabaceae).

Parameters	Tree Zone	Num.	Average (mm) ± standard error
Height	Trunk	105	10.21±0.18
	Canopy	810	11.70±0.10
Width	Trunk	105	14.13±0.28
	Canopy	810	18.49±0.24
Depth	Trunk	105	4.98±0.26
	Canopy	810	7.05±0.09

Student's t-test (p<0.01) (n=915).

Table 2. Height, width, depth and number (Num.) of active holes scarified by hybrid marmosets in the lower, middle and upper canopy of *Anadenanthera peregrina* var. *peregrina* (Fabaceae).

Parameters	Canopy segments	Num.	Average (mm) ± standard error
Height	Lower	145	10.35±0.19
	Midlle	377	11.63±0.15
	Upper	288	12.45±0.16
Width	Lower	145	16.01±0.49
	Midlle	377	17.92±0.29
	Upper	288	20.48±0.48
Depth	Lower	145	5.73±0.17
	Midlle	377	6.93±0.14
	Upper	288	7.88±0.17

ANOVA, Tukey test, *post hoc* ($p<0.01$) (n = 810).

than the associated costs, which will depend on the energy required to keep the marmosets in the canopy and involved in scarification. These factors both have an impact on the energy spent by marmosets in obtaining exudates, as reported for other primate species) [65,66].

The forest fragments used by marmosets are inside an anthropogenic matrix, surrounded by urbanized areas and/or through-ways for humans, which could have a negative impact on the animals [67–70]. The marmosets might have a lower predation risk in these areas, facilitating the exploration of exudates in the canopy, especially those in the outside part of the trees. Furthermore, exploration of the external canopy to acquire exudates is favored by the small size of the animal.

The positive relation between the number of scarified holes and the dendrometric data of trees (DBH and total height) agrees with results observed previously for *C. penicillata* and *C. jacchus* on *Qualea parviflora* Mart [59], *Astronium fraxinifolium*; *Enterolobium contortisiliquum*, and *Anadenanthera peregrina* [28]. This result suggests that larger trees tend to be more heavily scarified by *Callithrix*, and this choice is related to the ability of these animals to climb trees and the larger area available for scarification. However, the low correlation between the number of holes and DBH and tree height indicates that this preference is not determined by tree size, but by other factors, such as bark thickness and the chemical composition of the exudates [1,26].

The preference of *Callithrix* sp. for *A. peregrina* var. *peregrina* exudates in the Atlantic Forest differs from the pattern for these animals in the Caatinga and Cerrado forests, where primates, in general, exploit many species of gum trees [28,29,49,58]. The external canopy is further explored by these animals to obtain this resource with a positive correlation between the number of scarifications and DBH and tree height.

The high number of scarifications highlights the importance of *A. peregrina* var. *peregrina* as a source of exudates for *Callithrix* spp. These results contribute to understanding of the selective pressures exerted by marmosets on certain tree species and individuals to obtain this valuable resource.

Acknowledgments

To "Conselho Nacional de Desenvolvimento Científico e Tecnológico (CNPq)," "Coordenação de Aperfeiçoamento de Pessoal de Nível Superior (CAPES)" and "Fundação de Amparo à Pesquisa do Estado de Minas Gerais (FAPEMIG)." Asia Science edited and corrected the English of this manuscript.

Author Contributions

Conceived and designed the experiments: TMF DRC IOS VB. Performed the experiments: TMF DRC. Analyzed the data: TMF VB. Contributed reagents/materials/analysis tools: DRC VB IOS. Contributed to the writing of the manuscript: TMF JES JCZ VB.

References

1. Nash LT (1986) Dietary, behavioral, and morphological aspects of gummivory in Primates. Yearb Phys Anthropol 29: 113–137.
2. Smith AC (2010) Influences on gum feeding in primates. In: Burrows A, Nash L, editors. The evolution of exudativory in primates. New York: Springer. 109–122.
3. Harrison ML, Tardif SD (1994) Social implications of gummivory in marmosets. Am J Phys Anthropol 95: 399–408.
4. Power ML (2010) Nutrional and digestive challenges to being a gum feeding Primates. In: Burrows A, Nash L, editors. The evolution of exudativory in Primates. New York: Springer. 25–44.
5. Starr C, Nekaris KAI (2013). Obligate exudativory characterizes the diet of the Pygmy Slow Loris *Nycticebus pygmaeus*. Am J Primatol 75: 1054–1061.
6. Vinyard CJ, Wall CE, Williams SH, Hylander WL (2003) Comparative functional analysis of skull morphology of tree-gouging Primates. Am J Phys Anthropol 120: 153–170.
7. Taylor AB, Vinyard CJ (2004) Comparative analysis of masseter fiber architecture in tree-gouging (*Callithrix jacchus*) and non-gouging (*Saguinus oedipus*) callitrichids. J Morph 261: 276–285.
8. Sussman RW, Kinzey WG (1984) The ecological role of the Callitrichidae. Am J Phys Anthropol 64: 49–419.
9. Martins MM, Setz EZF (2000) Diet of buffy tufted-eared marmosets (*Callithrix aurita*) in a forest fragment in Southeastern Brazil. Int J Primatol 21: 467–476.
10. Passamani M, Rylands AB (2000) Feedind behavior of Geoffroy's Marmostes (*Callithrix geoffroyi*) in an Atlantic Forest fragment of south-eastern Brazil. Primates 41: 27–38.
11. Power ML, Oftedal OT (1996) Differences among captive callitrichids in the digestive responses to dietary gum. Am. J. Primatol 40: 131–144.
12. Coimbra-Filho AF, Mittermeier RA (1976) Exudate eating and tree-gouging in marmosets. Nature 262: 630–632.
13. Coimbra-Filho AF, Mittermeier RA (1977) Exudate-eating, and the "short-tusked" condition in *Callithrix* and *Cebuella*. In: Kleiman DG, editors. The biology and conservation of the Callitrichdae. Washington: Smithsonain Institution Press. 105–115.
14. Coimbra-Filho AF, Rocha NC, Pissinatti A (1980) Morfofisiologia do ceco e sua correlação com o tipo odontológico em Callitrichidae (Platyrrhini, Primates). Rev Bras Biol 40: 177–185.
15. Rosenberger AL (1978) Loss of incisor enamel in marmosets. J Mammal 59: 207–208.
16. Natori M, Shigehara N (1992) Interspecific differences in lower dentition among eastern Brazilian marmosets. J Mammal 73: 668–671.

17. Eng CM, Ward SR, Vinyard CJ, Taylor AB (2009) The mechanics of the masticatory apparatus facilitate muscle force production at wide jaw gapes in tree-gouging common marmosets (*Callithrix jacchus*). J Exp Biol 212: 4040–4055.

18. Stevenson MF, Rylands AB (1988) The marmosets, genus *Callithrix*. In: Mittermeier RA, Rylands AB, Coimbra-Filho AF, Fonseca GAB, editors. Ecology and behavior of Neotropical Primates. Washington: World Wildlife Fund. 131–222.

19. Coimbra-Filho AF, Aldrighi AD, Martins HF (1973) Nova contribuição ao restabelecimento da fauna do Parque Nacional de Tijuca, GB, Brasil. Bras Florest 4: 7–25.

20. Rizzini CT, Coimbra-Filho AF (1981) Lesões produzidas pelos saguis, *Callithrix p. penicillata* (E. Geoffroy, 1982), em árvores do Cerrado (Callitrichadae, Primates). Rev Bras Biol 41: 579–583.

21. Ferrari SF, Martins ES (1992) Gummivory and gut morphology in two sympatric callitrichids (*Callithrix emiliae* and *Saguinus fuscicollis weddelli*) from western Brazilian Amazonia. Am J Phys Anthropol 88: 97–103.

22. Canton JM, Hill DM, Hume JD, Crook GA (1996) The digestive strategy of the common marmoset, *Callithrix jacchus*. Comp Biochem Phys A 144: 1–8.

23. Stevens JR, Hallinan EV, Hauser MD (2005) The ecology and evolution of patience in two New World monkeys. Biol Lett 1: 223–226.

24. Bearder SK, Martin RD (1980) Acacia gum and its use by bushbabies, *Galago senegalensis* (Primates: Lorisidae). Int J Primatol 1: 103–128.

25. Ushida K, Fugita S, Ohashi G (2006) Nutritional significance of the selective ingestion of *Albizia zygiagum* exudate by wild chimpanzees in Bossou, Guinea. Am J Primatol 68: 143–151.

26. Smith AC (2000) Composition and proposed nutritional importance of exudates eaten by saddleback (*Saguinus fuscicollis*) and mustached (*Saguinus mystax*) tamarins. Int J Primatol 21: 69–83.

27. Coimbra-filho AF, Mittermeier RA, Constable ID (1981) *Callithrix flaviceps* (Thomas, 1903) recorded from Minas Gerais, Brazil (Callitrichidae, Primates). Rev Bras Biol 41: 141–147.

28. Thompson CL, Robl NJ, Melo LCO, Valença-Montenegro MM, Valle YBM, et al. (2013) Spatial distribution and exploitation of trees gouged by common marmosets (*Callithrix jacchus*). Int J Primatol 34: 65–85.

29. Amora TD, Beltrão-Mendes R, Ferrari SF (2013) Use of alternative plant resources by common Marmosets (*Callithrix jacchus*) in the Semi-Arid Caatinga Scrub Forests of Northeastern Brazil. Am J Primatol 75: 333–341.

30. Lorenzi H (1992) Árvores brasileiras. Manual de Identificação e cultivo de plantas arbóreas nativas do Brasil. Nova Odessa, SP. Ed. Plantarum. 384p.

31. Carneiro ACO, Vital BR, Castro AFNM, Santos RCS, Castro RVO, et al. (2012) Parâmetros cinéticos de adesivos produzidos a partir de taninos de *Anadenanthera peregrina* e *Eucalyptus grandis*. Rev Arvore 36: 767–775.

32. Carvalho PER (1994) Espécies Florestais Brasileiras: Recomendações silviculturais, potencialidades e uso da madeira. Colombo: EMBRAPA-CNPF/SPI. Brasília. 639p.

33. Pereira ZV, Fernandes SSL, Sangalli A, Mussury RM (2012) Usos múltiplos de espécies nativas do bioma Cerrado no assentamento Lagoa Grande, Dourados, Mato Grosso do Sul. Rev Bras Agroecol 7: 126–136.

34. Paes JB, Santana GM, Barbosa KL, Azevedo TKB, Morais RM, et al. (2010) Substâncias tânicas presentes em várias partes da árvore angico-vermelho (*Anadenanthera colubrina* (Vell.) *Brenan*. var. *cebil* (Gris.) Alts.). Sci For 38: 441–447.

35. Mori CLSO, Mori FA, Mendes LM, Silva JRM (2003) Caracterização da madeira de angico-vermelho (*Anadenanthera peregrina* (Benth) Speng) para confecção de móveis. Bras Florest 23: 29–36.

36. Lorenzi H (2008) Árvores brasileiras: manual de identificação e cultivo de plantas arbóreas nativas do Brasil. Nova Odessa, SP. Instituto Plantarum. 384p.

37. Taiz L, Zeiger E (2009) Fisiologia vegetal. 4. ed. Porto Alegre: Artmed. 819p.

38. Miranda GHB, De Faria DS (2001) Ecological aspects of black-pincelled marmoset (*Callithrix penicillata*) in the Cerradão and dense Cerrado of the Brazilian Central Plateau. Braz J Biol 61: 397–404.

39. Ferrari SF, Corrêa HKM, Coutinho PEG (1996) Ecology of the "Southern" marmosets (*Callithrix aurita* and *Callithrix flaviceps*). In: Norconk MA, Rosenberger AL, Garber PA, editors Adaptive radiations of Neotropical Primates. New York: Plenum Press. 157–171.

40. Lacher TE Jr, Fonseca GAB, Alves C Jr, Magalhães-Castro B (1981) Exudate-feeding, scent marking and territoriality in wild population of marmosets. Animal Behav 291: 306–307.

41. Nogueira DM, Ferreira AMR, Goldschmidt B, Pissinatti A, Carelli JB, et al. (2011) Cytogenetic study in natural hybrids of *Callithrix* (Callitrichidae: Primates) in the Atlantic forest of the state of Rio de Janeiro, Brazil. Iheringia Sér. Zool. 101: 156–160.

42. Veloso HP, Rangel Filho ALR, Lima JCA (1991) Classificação da vegetação brasileira, adaptada a um sistema universal. Rio de Janeiro, Brasil: IBGE. 123p.

43. Golfari L (1975) Zoneamento ecológico do Estado de Minas Gerais. Centro de Pesquisa Florestal da Região do Cerrado. Belo Horizonte, MG: PRODE-PEF.65p.

44. Vianello RL, Alves AR (1991) Meteorologia básica e aplicações. Viçosa, MG: Universidade Federal de Viçosa. 448p.

45. Perry DR (1978) A method of access into the crowns of emergent and canopy trees. Biotropica 10: 155–157.

46. Giongo C, Waechter JL (2004) Composição florística e estrutura comunitária de epífitos vasculares em florestas de galeria na Depressão Central do Rio Grande do Sul. Rev. Bras Bot 27: 563–572.

47. Johansson DR (1974) Ecology of vascular epiphytes in West African rain forest. Acta Phytogeogr suec 59: 1–129.

48. R Core Team (2012) R: A language and environment for statistical computing.R Foundation for Statistical Computing. Vienna, Austria. Avaliable at: http:// www.R-project.org/.

49. Lacher TE, Fonseca GAB, Alves C, Magalhaes-Castro B (1984) Parasitism of trees by marmosets in a central Brazilian gallery forest. Biotropica 16: 202–209.

50. Rylands AB (1984) Exudate-eating and tree-gouging by marmosets (Callitrichidae, Primates). In: Chavdwick AC, Sutton SL editors. Tropical Rain Forest: The Leeds Symposium Leeds Philosophical and Literary Society. 155–158.

51. Paula RCM, Budd PM, Rodrigues JF (1997) Characterization of *Anadenanthera macrocarpa* Exudate Polysaccharide. Polym Int 44: 55–60.

52. Rylands AB, Faria DS (1993) Habitats, feeding ecology and home range size in the genus *Callithrix*. In: Rylands AB editor. Marmosets and tamarins: systematics. behavior and ecology. Oxford Science Publications. 262–272.

53. Garber PA, Teaford MF (1986) Body weights in mixed species troops of *Saguinus mystax mystax* and *Saguinus fusiicollis nigrifrons* in Amazonian Peru. Am J Primatol 71: 331–336.

54. Fleagle JG (1999) Primate Adaptation and Evolution. 2. ed. Academic Press: San Diego. 596p.

55. Schülke O (2003) To breed or not to breed–food competition and other factors involved in female breeding decisions in the pair-living nocturnal fork-marked lemur (*Phaner furcifer*). Behav Ecol Sociobiol 55: 11–21.

56. Yépez P, Torre S, Snowdon CT (2005) Interpopulation differences in exudate feeding of Pygmy Marmosets in Ecuadorian Amazonia. Am J Primatol 66: 145–158.

57. Soini P (1988) The pygmy marmosets, genus *Cebuella*. In: Mittermeier RA, Rylands AB, Coimbra-Filho AF, Fonseca GA. editors. Ecology and behavior of Neotropical primates. Washington: World Wildlife Fund. 79–129.

58. Fonseca GAB, Lacher TE (1984) Exudate-feeding by *Callithrix jacchus penicillata* in Semideciduous Woodland (Cerradão) in Central Brazil. Primates 25: 441–450.

59. Passamani M (1996) Uso de árvores gomíferas por *Callithrix penicillata* no Parque Nacional da Serra do Cipó, MG. Bol Mus Biol Mello Leitao, Nova Ser 4: 25–31.

60. Raboy BE, Canale GR, Dietz JM (2008) Ecology of *Callithrix kuhlii* and a review of eastern Brazilian Marmosets. Int J Primatol 29: 449–467.

61. Rylands AB (1984) Exudate-eating and tree-gouging by marmosets (Callitrichidae, Primates). In: Chadwick AC, Sutton SL, editors. Tropical rain forest: The Leeds Symposium Leeds Philosophical and Literary Society, Leeds. Pp. 155–168.

62. Ferrari SF, Corrêa HKM, Coutinho PEG (1996) Ecology of the "Shouthern" marmosets (*Callithrix aurita* and *Callithrix flaviceps*). In: Norconk MA, Rosenberg AL, Garber PA, editors. Adaptative radiations of neotropical primates. Plenum Press, New York. Pp 157–171.

63. Garber PA, Porter LM (2010) The ecology of exudate production and exudate feeding in *Saguinus* and *Callimico*. In: Burrows AM, Nash LT, editors. The evolution of exudativory in Primates. New York: Springer. 89–108.

64. Lazaro-Perea C, Snowdon CT, Arruda MF (1999) Scent-marking behavior in wild groups of common marmosets (*Callithrix jacchus*), Behav Ecol Sociobiol 46: 313–324.

65. Warren RD, Crompton RH (1997) Locomotor ecology of *Lepilemur edwardsi* and *Avahi occidentalis*. Am J Phys Anthropol 104: 471–486.

66. Hanna JB, Schmitt D (2011) Locomotor energetics in Primates: gait mechanics and their relationship to the energetics of vertical and horizontal Locomotion. Am J Phys Anthropol 145: 43–54.

67. Patterson ME, Montag JM, Williams DR (2003) The urbanization of wildlife management: Social science, conflict, and decision making. Urban For & Urban Greening 1: 171–183.

68. Pontes ARM, Normande IC, Fernandes ACA, Ribeiro PFR, Soares ML (2007) Fragmentation causes rarity in common marmosets in the Atlantic forest of northeastern Brazil. Biodivers Conserv 4: 1175–1182.

69. Millsap A, Bear C (2000) Density and reproduction of burrowing owl along an urban development gradient. J Wildl Manage 64: 33–41.

70. Otoni I, Oliveira FFR, Young RJ (2009) Estimating the diet of urban birds: The problems of anthropogenic food and food digestibility. Appl Anim Behav Sci 117: 42–46.

4

Personnel and Participant Experiences of a Residential Weight-Loss Program

Unni Dahl[1,2], Marit By Rise[2]*, Bård Kulseng[3], Aslak Steinsbekk[2]

1 Central Norway Health Authority, Stjørdal, Norway, **2** Department of Public Health and General Practice, Norwegian University of Science and Technology, Trondheim, Norway, **3** Regional Centre for Obesity Treatment, St. Olav's University Hospital, Trondheim, Norway

Abstract

Background: Residential weight-loss programs aim to help persons with obesity lose weight and maintain a long-term healthy lifestyle. Knowledge is needed on the different actors' perceptions and experiences from such programs. The aim of this study was to describe how personnel argued for and perceived a residential weight-loss program, to investigate how the participants experienced the program, and to contrast these perspectives.

Methods: This qualitative study took place in an 18-week residential weight-loss program. Exercise, diet, and personal development were the main components in the program. Data was collected through participant observation and individual and focus group interviews with participants and personnel.

Results: Program personnel characterized persons with obesity in specific terms, and these formed the basis of the educational aims, teaching principles, and content of the program. According to personnel, persons with obesity typically had problems acknowledging their own resources, lived unstructured lives, had a distorted relationship to food, experienced a range of social problems and featured a lack of personal insight. Program participants reported enthusiasm about their experiences of exercise and appreciated measures of success with the exercise program. They had, however, very different experiences regarding the usefulness and appropriateness of the parts of the program focused on social and personal development. Some felt that weight loss required an engagement with personal development while others viewed it as unnecessary and inappropriate.

Conclusion: The reliance in personnel accounts on particular characteristics of persons with obesity as a rationale for the program might lead to stigmatizing and stereotyping. Program activities focused on social and personal development need to be better understood by participants if they are to be viewed as helpful. To achieve this personnel must carefully consider how these parts of the program are communicated and conducted.

Editor: Christy Elizabeth Newman, The University of New South Wales, Australia

Funding: The authors have no support or funding to report.

Competing Interests: The authors have declared that no competing interests exist.

* E-mail: marit.b.rise@ntnu.no

Introduction

Obesity is a growing problem in the Western world [1], and many diseases and illnesses accompany it [2,3]. Recently surgery has become more common to achieve weight-loss. A new systematic review and meta-analysis on the outcomes and risks of bariatric surgery showed that surgery was more effective than non-surgical interventions regarding weight-loss [4]. Although risks of complications exist, death rates were lower than in previous meta-analyses. Meta-analysis of observational studies also indicates that bariatric surgery might lead to a reduced risk of cardiovascular disease and mortality for persons with obesity [5], but this has yet to be investigated in clinical trials. Martins and colleagues compared bariatric surgery to three conservative treatments - including the weight-loss camp that served as the setting for this qualitative study [6]. They found that although all treatments led to significant weight-loss, patients who had undergone bariatric surgery had lost significantly more weight one year after the interventions.

Although weight-loss surgery is increasingly used, the combination of exercise, diet, and behaviour therapy in comprehensive weight-loss programs seems to be the most commonly available intervention [7,8]. Although many clinical studies of weight-loss programs show that the participants lose weight during and after the intervention [9–13], some authors have stated that non-surgical treatments of obesity are generally ineffective in long-term weight-control [4]. One study investigating the long-term maintenance of weight-loss after an extensive lifestyle intervention - at the weight-loss camp investigated in the qualitative study presented here - showed for example that a 15% weight reduction after the intervention was reduced to 5.3% reduction of the initial body weight 2–4 years afterwards [12]. Thus, one of the main challenges in weight-loss programs is to achieve long-term sustainable changes.

Soderlund and colleagues [8] conducted a systematic review of randomized controlled trials investigating long-term effects of physical exercise with or without behavior-change therapy and/or dieting. The authors conclude that the programs with the best results involve different professional groups, and include a combination of diet, behavioral-modification therapy, and exercise. Other studies have also shown that adding exercise or behavior therapy to diet improved weight-loss and reduced the health risk factors connected to obesity [9,11,14].

Weight-loss programs thus aim to help the participants modify their health behaviors. The main argument for residential programs is to fortify a sustainable long-term lifestyle change. Residential weight-loss programs constitute a rather new approach [7], and research on the effects of such programs on adults is scarce [15]. Weight-loss programs include behaviour therapies that aims to help the participants achieve the necessary skills to reach a more healthy weight [16]. Lifestyle interventions usually include many behavioural techniques such as self-monitoring, modelling, environmental restructuring, and support, both in groups and individually [13]. Other central terms are self-regulation, goal setting, and problem solving [17,18]. This makes it difficult to recognize the particular elements which might be supporting learning and behavioural changes. In addition, most behaviour therapy interventions are poorly defined and described in the literature [8]. Many authors have emphasized that interventions in general often are implemented without a deep understanding of how different initiatives are supposed to lead to change [19]. More detailed knowledge about the content of weight-loss programs is therefore required, as well as a deeper understanding of what really happens in lifestyle interventions for persons with obesity.

A residential program differs from out-patient intervention by length and intensity, and in that the participants are removed from their everyday lives. A residential program also offers more time for communication between participants and personnel [20]. Therefore, residential weight-loss programs are arenas for close interaction between participants and personnel, and with a broad range of opportunities provided for personnel to teach and support participants in their endeavour to change behaviour. Although lifestyle-modification programs often combine exercise, diet and behaviour change activities, a good relationship that allows for open communication is generally recommended [21]. Several have also emphasized the importance of motivation to make lifestyle changes [8,17,18]. Since motivation to alter lifestyle is crucial to achieve long-term changes, support and help from personnel during such programs would be important [8]. Many have argued that personnel skills and attitudes should be characterized by trust, respect and acceptance when caring for persons with chronic diseases [22,23]. Respect has been described as unconditionally recognizing the patient as a valuable person, and as an integral part of health professionals' work [24].

Very little research has explored different actors' perceptions of and experiences from weight-loss programs. It has been emphasized that to achieve good results different stakeholders in an intervention should share an understanding of how change is supposed to be implemented and sustained [25]. Including both personnel and participants in the same study would contribute to the understanding of the stakeholders' rationale for an intervention, as well as their views on potential outcome. Therefore, the aim of this study was to describe how personnel argued for and perceived a residential weight-loss program, to investigate how the participants experienced the program, and to contrast these perspectives.

Methods

Ethics Statement

The Regional Ethics Committee for Research in Medicine in Central Norway approved of the study. All participants received oral and written information about the study and signed a written consent form. None of the authors have any contact with or interest in the Danish weight-loss camp outside this research project.

Design

This was a qualitative study consisting of on-site participant observation, in-depth semi-structured individual interviews, and focus group interviews. A combination of methods was chosen to ensure data triangulation. Observation was chosen to provide data on real life interaction between participants and personnel and to gain insight into the participants' immediate reactions to the program. Focus group interviews were conducted to explore the interaction and discussions between participants regarding their experiences and views of the program. Individual interviews were also employed to explore in more depth the ways in which personnel articulated their arguments for and experiences of providing the program, and to delve deeper into the individual experiences of a selected subset of participants.

The study formed part of a Norwegian clinical investigation on the effectiveness of different interventions (weight-loss program versus surgery) for persons with obesity. The study was registered in Clinicaltrials.gov with registration number NCT00239850. An effect study from this program is published elsewhere [6].

Setting

The study took place at a Danish residential weight-loss centre (Ebeltoft kurcenter), which offers an 18-week on-site program. A total of 80 participants were at the centre at the time of the data collection. This study focused on 30 participants who were referred to this program due to taking part in a Norwegian study on the effect of different weight-loss approaches [6].

Most of the activities at the centre were organized and conducted in groups. The program consisted of group-based intensive exercise, diet (individual calorie intake was based on energy calculations for a normal weight person with a sedentary activity level), and an educational program. Structured group exercise and educational sessions took place from Monday to Friday every week. Exercise was compulsory three times per day. Typical exercise activities were badminton, ball games, aerobics, and swimming. In addition, participants were encouraged to be physically active on their own. Program participants were advised to start on an individually suitable exercise level and to focus on the activities that they liked the most.

Participants lived together in groups in small houses with sleeping and cooking facilities, and they cooked their own dinners with provided groceries. Breakfast, lunch, and in-between meals were served in a buffet. The different types of food were marked with labels of energy content. The diet contained 55–60% carbohydrates, 15% protein, and less than 30% fat. At the beginning of the program, each participant learned to calculate their energy intake by counting calories and weighing the food. If necessary, diet plans were adjusted individually according to weight measurements during the program.

The educational program comprised lessons about nutrition, monitoring of food intake and instruction in behavioral techniques from cognitive therapy. The personal development component included a minimum of two individual conversations with one of the psychotherapists, motivational meetings for all participants, and "family meetings" (for all residents in each house) to discuss and resolve problems regarding the living arrangements. In addition, mandatory group therapy classes in personal development were held one hour per week. Classes in personal development were led by a psychotherapist. The goals of this component of the program were to gain insight into personal problems and to consider how these problems may be associated with the obesity. The overall aim of the personal development component was to help the participants live a mentally healthy life as an important part of the process of lifestyle change.

Participants and Personnel

The participants in this qualitative study also participated in a clinical study investigating the effect of different types of weight-loss interventions, including a stay at the residential weight-loss centre [6]. A total of 30 (21 female and 9 male) Norwegian participants took part and stayed at the centre. These participants were between 22 and 56 years old, their body mass index was between 40 and 63, and the group's mean body weight was 144 kg. All those who were referred to the program from Norway also participated in a Norwegian effect study [6] in addition to this qualitative study. The inclusion criteria for the effect study were ages between 18 and 60 years old and a BMI over 40 kg/m^2 or a BMI over 35 kg/m^2 including comorbidities. There were no additional inclusion criteria for the qualitative study.

The personnel were recruited among the staff at the centre. Since an important aim was to explore the rationale for the choice of program content and structure, key personnel with responsibility for providing the program were recruited to individual interviews.

Data Collection

For the purpose of achieving immersion in the field of inquiry, the first author (UD) took part in the program over the three week period and observed and interviewed personnel and participants. A data collection period was predefined to two periods totaling to three weeks and was conducted during the 1st, 2nd, and 14th weeks of the 18-week program. These weeks were chosen to ensure that the initial expectations and experiences (the first two weeks), as well as the experiences from most of the stay (week 14), were captured. During this period, thematic saturation was reached, and no new experiences, discussions, or perceptions emerged in the last interviews or observation.

Observation. The first author (UD) participated in all activities as an ordinary participant during the three weeks of data collection, that is, she took part in exercise, classes, and social life. The participants were informed before arrival at the weight-loss camp that the researcher would be present during some of the program, and that this was a part of the larger research project. This period of overt participant observation provided an opportunity to observe real life settings and the interaction between personnel and participants. It also provided insights that were used to inform the development of the question guide that was subsequently used in the focus groups and interviews. The observation also involved informal conversations between the researcher and those personnel and participants who did not later participate in interviews. Field notes were written during observation and at the end of every day. Summaries were developed at several points during the three weeks of data collection. Video and audio recording was not used during observation. The field notes were used to record facts about the program and to write reflections during observations and interviews. Field notes were also used as background for focus group interviews and were later compared with the interview transcripts.

Interviews. Those who took part in interviews were asked directly by the first author (UD) and all agreed to participate. A total of 10 Norwegian participants took part in interviews (8 in focus groups and 2 individually). The age and weight range for these 10 persons were the same as for the total sample. Six personnel participated in individual interviews. All interviews were conducted by the first author (UD).

Semi-structured interview guides were employed during focus groups and individual interviews. The guides were used as a memo list to ensure that all the topics were covered in all interviews. The interviewer introduced themes from the interview guide if the participants did not spontaneously talk about them. The participants were also free to talk about issues and topics outside the guide. The main opening question in all interviews was "What are your experiences with taking part in/providing this program?" Additionally, the personnel were asked to describe the content of the program and the rationale for it. Other important topics in interviews with personnel were how personnel attitudes might influence the participants, and how they motivated persons with obesity to change their life-styles.

Each interview lasted from 45 to 90 minutes and was audio taped. All interviews were subsequently transcribed verbatim and de-identified. Redundant words and pauses were removed, and local dialect was changed to written Norwegian.

A strategic sample of eight of the participants, three men and five women, were recruited to focus group interviews. The sampling strategy aimed to achieve some variation in experiences, BMI, age and gender. Focus group participants were interviewed two times, in the 1st and the 14th weeks of the program. The topics discussed in the first interview were immediate perceptions and experiences with the program: overall setting, exercise, living arrangements, and food. The second interviews focused more on the positive and negative experiences that participants reported of the program.

Two female participants were interviewed individually in depth in the first and second weeks of the program. The interviews were conducted to elaborate on specific issues discussed in the focus group interviews. The two participants were selected based on observations of their role in the group of Norwegian participants. During interactions and discussions they had expressed clear views and had provided rich descriptions that offered the potential for meaningful further exploration of key topics in a one-on-one interview. Topics discussed in these interviews included how to make sustainable lifestyle changes, overall approach to weight-loss, and collaboration between the personnel.

Six personnel (2 males and 4 females) took part in individual in-depth interviews. These were considered to be key personnel; the director, the administrative executive, and the leaders of the main areas diet, exercise and personal development. The personnel were selected because they strongly influenced the content and form of the program. Five personnel were interviewed during the first and second weeks. In week 14, three of these were interviewed a second time. In addition, one substitute personnel member was added in week 14 since one of the members had a leave of absence. The second interviews were done to elaborate on the topics discussed in the first interview, and to discuss personnel experiences with the current class of participants.

Data Analysis

Interview transcripts and fields notes were analysed together. The focus of the analysis was to explore the connections between the personnel rationale for the program and the participant experiences of the program. The analysis was inductive and thematically based. The interview transcripts and field notes were read and the main themes generated. The first and fourth author (UD and AS) read all transcripts and made separate analyses. These analyses were subsequently discussed in a group including the third author (BK), and finally with the second author to get a different point of view and to look for alternative interpretations. It was observed that the personnel articulated several recurrent descriptions or characterisations of people with obesity. These perceptions were described by personnel as forming the basis of the rationale for the program, and so perceptions were employed as a coding structure to categorise and interpret the findings. The validity of the results was cross-checked by re-reading the interviews. To illustrate the main themes the most illustrative and comprehensive quotes were chosen. Quotes were translated from Norwegian to English by the first (UD) and second author (MR) and controlled by the fourth author (AS). Quotes from participants used in the result presentation are identified by interview number and gender. Quotes from personnel are not identified to ensure anonymity.

Results and Discussion

A total of 10 participants and six personnel participated in interviews. In addition, observation included 30 participants and approximately 10 personnel.

During the analysis of data from interviews and observations, it became clear that there were important differences between how the personnel and participants experienced and perceived the purpose and activities of the weight-loss program. The personnel were very consistent in their perceptions, while the participant experiences and reactions were more diverse. Personnel described very similar rationales for the aims of the program, and for the content and teaching principles used to achieve these aims. Participants had, on the other hand, experiences that ranged from joy and excitement to feeling the program was a struggle and waste of time.

During the analysis, the personnel rationale for the program was categorised into five recurrent themes in their characterization of people with obesity: lack of acknowledgement of resources, lack of structure, a distorted relationship to food, social problems, and lack of personal insight. The program was seen to be intended to change these five characterisations, and the program content and teaching principles were selected deliberately in order to achieve these changes. However, interviews and observation both revealed that participants viewed the program content very differently to the personnel. All participants were very enthusiastic about the exercise program and complied with the strict structures and routines involved in the exercise and diet program. The participants were, however, divided on the usefulness of the parts of the program concerning social and personal development. Division was particularly strong on the usefulness and necessity of the personal development classes.

In the following section, the five characterisations of participants that personnel believed had to be changed in order for participants to achieve and maintain weight-loss, are discussed in more detail. Quotes that describe participant experiences and reactions to the program content are woven into these sections in order to be contrasted with personnel views. Quotes from personnel and participants are used to illustrate and support the findings.

1. Lack of Acknowledgement of Resources

The first characterisation ascribed to participants by personnel was the perceived lack of acknowledgement of their own resources in being physically active and sticking to a diet. This was said to be due mainly to low self-efficacy beliefs and low self-confidence. Personnel emphasised the importance of strengthening participants' self-confidence by ensuring experiences of success in their efforts to exercise and eat properly during the program. Personnel thus highlighted the importance of compulsory and regular exercise and of setting achievable goals.

The participants shall first and foremost get confirmation that they can do a hundred times more than they believe they can. [...] It is about transferring the joy of success from one situation to another. (Personnel).

Participants confirmed this practice by reporting that even small efforts were praised regularly by personnel, and that personnel also frequently encouraged participants individually during group exercise. This was also confirmed during observation. In the interviews, participants confirmed that personnel supported and respected them. This attitude was described as strengthening their motivation, self-confidence, and sense of self-worth.

That is how they [the personnel] *build our self-confidence. We are all of a sudden humans again and not just the chubby clown.* (Female, first focus group interview).

When it came to the exercise part of the program participants confirmed personnel claims that success experiences encouraged participants to keep working towards weight-loss. The positive experience of exercise was highlighted both in the observation and interviews from the beginning, and was in fact the most striking finding of the research. The participants were enthusiastic about learning and managing new activities, or activities that they had not engaged in for years or had never tried before. They were also very enthusiastic about improving their physical condition and achieving new goals.

My first victory [...] *was that I managed to ride a bike for 15 kilometres. I think I walked on air for four days after that experience.* (Female, second focus group interview).

This attitude towards exercise remained stable through the program. In the 14th week of the program, participants unanimously agreed that exercise was fun, increased their energy, and instilled in them a desire to be active in the future.

2. Lack of Structure

The second characterisation that personnel attributed to persons with obesity was that they generally lead unstructured lives. Personnel reported a belief that many things in obese persons' lives were seen to happen by chance, such as when and what to eat, and what activities to engage in.

Our experience is that many obese people are incredibly unstructured and live extremely unstructured lives. Such as keeping appointments and being stringent and sharp in the things you do. (Personnel).

According to the personnel, the program schedule had to be very strict. They described that their strictness was important to provide the participants with a daily structure that they could then seek to maintain at home.

And those who succeed, those we know maintain their weight-loss… they are those who have a structured life afterwards. Those who eat at fixed times and who get out of bed at approximately the same time each day, and those who know what they want with themselves. (Personnel).

This strict structure was evident across all program activities. All participants had to be present at the morning meetings, group exercises, and at six meals every day (at fixed hours). Other activities, like teaching classes and the weekly weighting, were also scheduled. According to the participants, adhering to the structure was an effort they felt they had to do to comply with the program. Although there were several who compared the program to military training and being back at school with very little freedom, both observation and interviews showed that the participants all agreed that a strict schedule was appropriate to initiate lifestyle change.

You are supposed to learn some things while you are here. It is like school… subjects are mandatory. And when you know that you have to be ticked off on a list to be registered [by an employee] you show up. There's no alternative. And if it hadn't been like that the activity would not have been very great. If you are not there at eight o'clock in the morning they call you and ask: "Where are you?" (Female, Individual interview).

3. Distorted Relationship to Food

The third characterisation articulated by personnel proposed that people with obesity have a distorted relationship to food.

Those with a really high BMI have a behaviour that fully can be compared with an alcoholic or a drug addict. They have no control over food. They always have their form of drug lying around in small storages on secret places so they always know that they are close to the food and they become insecure if it is not in the house. (Personnel)

According to the personnel, most persons with obesity were addicted to food and would have to live with an eating disorder for the rest of their lives. The program was designed to give the participants the practical experience of eating like a standard weight person and, through that, to change their relationship towards food.

I tell the participants that they all have a normal weight person inside. They shall hold on to this person and from now on leave the decisions to the standard weight person. […] This is not a treatment; it is four months of practice in living as a standard weight person. (Personnel).

In the beginning of the program, it was observed that participants talked a lot about how to manage the new diet chart, frequency of eating and the amount of food, and the result of their latest weighing. Although there was a consensus among participants about the necessity of managing new eating habits, the experience of enthusiasm and success were less apparent in discussions about diet than about the exercise program. The participants expressed the belief that changing eating habits and adhering to the diet regime was laborious. At the end of the program this was, however, a theme the participants talked less about. When asked, the participants said that following a diet plan had become an automatic behaviour.

In the beginning I was strictly adhering to the written diet chart, but now I have noticed that I am composing the diet based on my energy target. (Female, second focus group interview).

4. Social Problems

The fourth characterisation of people with obesity highlighted by personnel was a belief that they often have poor social skills. Personnel expressed that persons with obesity often had felt like an outsider and therefore had adopted unfavourable approaches to handle this. Examples of approaches mentioned by the personnel were taking on the comedian role, social isolation, avoiding conflict, and not recognizing or acknowledging their own needs. The housing structure during the program, in which people share apartments and house work and cooking takes place in groups, was deliberately chosen to give the participants practice in socializing with other people.

Quite a few experience social isolation. So groups are a way to get out of isolation and learn to associate with others. (Personnel).

The main approach to managing conflict between participants was in line with this focus on social training. Personnel facilitated problem-solving techniques by asking participants what they believed they could do themselves. According to the personnel, this was intended to assist participants with focusing their skills and abilities.

If two persons share a flat and have different habits… If one of them stays up at night to watch TV and the other person wants to sleep but can't. Then one is forced to say: "Do you know what, I need to sleep. Could you be so kind as to turn off the TV?" […] Then there are two people learning. One is the person who gets the message and the other is the person giving the message. They both learn how it is to enter a conflict. (Personnel).

The participants had different experiences with the social development component of the program. Some participants described that living closely together was a very positive experience and that they had made friends for life. Others described it as challenging and difficult, and said that they hardly talked with the other participants in their group.

They have strong emphasis on the mental part, and it is very important… and I feel that I have come a long way […] and at the same time… I don't want to participate so much because… well, there is a lot of gossip here. And if you open up with sensitive stuff you cannot be sure whether it is on every corner afterwards. So I have… I told them [the personnel] that this was my decision, and it was fine. This part of the program should have been different. I don't think all will benefit from it. (Female, second focus group interview).

5. Lack of Personal Insight

The fifth characterisation described by personnel asserted that persons with obesity lack insight into their personal problems. Personnel expressed the belief that lifestyle change require a development in personal insight, and that participants have to reflect on their own behaviour to achieve this. Therefore, the program included classes in personal development. During these group classes, participants were urged to identify underlying factors that may have influenced their obesity problem, such as family relations, working conditions, or economic contexts. The personnel believed such personal development was crucial to enable participants to implement and maintain lifestyle changes.

It is important that they become aware of what in their life makes a difference in being obese or not. They have some enemies out there and some accomplices […] Not only persons… it might also be their work. But to get control over them you first have to be conscious of what they are. (Personnel).

The participants were clearly divided in their perceptions of the usefulness of the personal development program and views on whether personal development classes should be compulsory. Some strongly disagreed, while others shared personnel views about the importance of personal development. Some participants even asked for more individual sessions with the psychotherapists.

I think we get tools to gain self-insight and retrieve things we have forgotten… also things that you might not want to touch. […] And suddenly […] you have changed as a person. And you experience things and perceive things quite differently and you get a different outlook in life. (Female, second focus group interview).

Other participants saw losing weight, not as a matter of personal development and psychological factors, but rather as a matter of changing bad habits - to eat less and exercise more. These participants were more likely to avoid or refuse to engage in the process of identifying any underlying factors, either by not attending the personal development classes, or by not participating in the discussions during class.

It is about the personal [development] part… I cannot benefit from it. I will never open up in that room and talk among the others. I will not do that. (Male, second focus group interview).

According to the personnel, those who withdrew from classes or discussions were the persons most in need of attending. Personnel expressed the belief that they were prepared for the type of conflicts that arose, e.g., participants not showing up to classes due to feeling alienated from the content. Personnel explanations of this type of reaction drew on their understandings of some of the potential mechanisms underlying obesity, such as having an eating disorder or a lack of self-insight.

There are some who do not want to attend our classes. There are some who do anything to avoid them. But they are mandatory. And this is because sometimes you have to hear things more than once before it sticks […]. Those who do not bother to attend the lessons are often those who need it the most. (Personnel).

Personnel perceptions of persons with obesity as persons in need of insight and personal development were noticed by the participants. Some of the participants expressed the view that these generalised characterisations were demeaning, and explained that this could make them feel misunderstood and not treated as individuals.

…they talk to us like we are all the same. […] As if we are flawed. […] And as if we all have the same problems." (Male, second focus group interview).

Discussion of Results

According to the participants, the best part of the program was the strict and intensive exercise regime, which was overwhelmingly embraced. Here, the participants and personnel also interacted without any noteworthy conflicts. The participants were very enthusiastic, and the major reason for this was in accordance with the personnel rationale for the program: the experience of success. Experiencing success to strengthen self-confidence and self-efficacy is a well-known approach [26], and such experiences are also recommended to give persons with obesity faith in their own capacity to make healthier choices [27]. Experiencing self-efficacy regarding exercise has also been shown to increase success when trying to lose weight [28]. Experiences of receiving support from personnel were also highlighted by the participants. This is in line with previous findings in which support and encouragement from personnel have been described by persons with obesity as important facilitators for change [29,30].

The rationale of the program was explained and justified in personnel accounts by articulating a series of claims about the most common characteristics of persons with obesity. The personnel viewed this population as typically resourceful, even if people with obesity didn't always find this easy to acknowledge themselves, and also lacking in structure, social skills, and personal insight. It was a clear finding that the stereotyping helped the personnel build a joint understanding among themselves regarding both the objectives of and activities in the program. Stereotypes also seemed to help the personnel to remain focused and firm during the program. There are some similarities between the personnel views and the negative attitudes and stigmas that persons with obesity are likely to experience both in society and among health care professionals [31,32]. Some of the more typical obesity-related stigmas are laziness, non-compliance, and being more overindulgent and less successful than persons without obesity [31,32]. Persons with obesity have reported experiences of stigmatization on a regular basis, and even more so for those most obese [33]. Inappropriate comments from doctors are in fact among the most common stigmatizing experiences for persons with obesity [34]. Health personnel also attribute obesity to negative characteristics, such as a lack of willpower, sloppiness and laziness [35], lack of discipline, lack of motivation, denial of eating wrongly, and psychological problems [36].

Personnel attitudes observed in the present study might be in conflict with the ideal of building a relationship of trust and acceptance [22]. It would be problematic if the very common negative attitudes towards persons with obesity [33,34,37,38] also exist among the professionals who are supposed to help and support. It is also reasonable to believe that if the personnel perceptions border on stigmatizing, the result for participants will be poor. Obesity stigma has, for instance, been found to negatively influence the motivation to exercise [37] and to increase consumption of calories [34,39]. Therefore, stigma will most likely not be a facilitator for weight-loss. Personnel in this study are thus likely to walk on a very fine line in perpetuating these ideas for the purpose of supporting participants. One the one hand, it was found that their rationale for the program resembles the same topics as is noticed in the literature on stigma. On the other hand, it was found both in the observation and interviews that the participants for the most part felt that personnel respected them. Therefore, a very clear awareness among personnel also, or especially, at residential weight-loss centres is needed.

According to the personnel in this study, characteristics of persons with obesity were also described as the reasons for being obese, and changing these characteristics was described as a potential way towards a healthier lifestyle. Some studies have shown that persons with obesity agree that determination, commitment, and discipline are necessary to change diet and exercise [29]. To maintain weight-loss, attitudes have been described as important [40], for instance, taking responsibility for one's life, caring about appearance, self-confidence, and believing that success is possible. Maintaining weight-loss has also been linked to determination, commitment, and patience [41]. Persons with obesity have also confirmed that eating habits can be linked to sadness, lack of motivation, lack of control [29], and negative emotions [42], and obesity has been connected to depression and mood disorders [40]. The mandatory social training in the housing arrangement and the personal development classes in the program were meant to help strengthen participants' social skills and increase their personal insight. The results showed, however, that some of the participants strongly disagreed with personnel on the usefulness and appropriateness of this approach, especially the personal development classes. Personnel did not see the resistance and objections from some of the participants as an indication that they should rethink and revise the program. Neither did they see the reactions as relating to their own attitudes, but rather as a confirmation that personal development was needed.

Most agree that making changes in exercise and diet is essential to losing weight. The idea that personal development is important to achieve and maintain weight-loss is less well recognised or understood. Claiming that a person with obesity has to change internal factors such as self-knowledge and self-acceptance is also a more sensitive topic than claiming that more exercise and healthy food is necessary to lose weight. Yet personnel argued consistently that personal development was especially important to maintaining lifestyle changes long-term. Studies have also confirmed that knowledge about how to eat healthy is not always sufficient to be able to change eating behaviours long-term [42]. A possible explanation for the participant responses to the different parts of the program could be the length of time between taking part in an activity to experiencing results [43]. While the exercise gave almost immediate results and experiences of success, the area of personal development, with its focus on increasing self-insight to prepare for a lasting lifestyle change, can be seen as more abstract and remote. A focus on the positive outcomes from lifestyle changes are important [44], and the reward for strengthening social skills and gaining personal insight can appear too distant and indistinct.

It could also be important to discuss the criteria of success in weight-loss programs. While successful obesity treatment can be related to absolute weight-loss, another main focus could be to reduce the risk of cardiovascular and metabolic diseases [7]. Although successful long-term weight-loss has been defined as losing 10% of initial weight and maintaining this for one year or more [45], studies have shown that even modest weight-loss can

improve psychosocial problems [40]. In addition, moderate changes and the experience of success can motivate persons with obesity to maintain lifestyle changes [46]. This was confirmed by the participants in the present study. Research has, however, also shown that persons with obesity tend to choose immediate rewards over long term benefits [47], and they tend to overestimate the reward from the immediate fulfilment of a need [48]. To wait for long-term reward can be extra challenging, and the connection between the input (personal development classes) and output (weight-loss) would be more challenging, yet imperative, to communicate.

Strengths and Limitations

The strength of this study is that it is the first to explore and contrast personnel and participant perspectives in a residential weight-loss program that has been reported to be successful [12]. The program has been developed and modified over time according to experiences made in a real-world setting [16]. There are, however, some limitations. The observation took place in 3 out of 18 weeks. A somewhat different picture might have emerged if the whole length of the program had been observed.

The interviews partially revealed situations that were not observed, and in our judgement, we achieved a good overview of the main picture through quite extensive observation and long interviews. However, this study only collected data from a limited group of participants in this program. Although the personnel confirmed that the participants taking part in this study did not differ substantially from other groups, the results could have been different if a different group was studied.

Conclusion

Personnel characterisation of persons with obesity as a rationale for the design and delivery of a weight-loss program could lead to stigmatizing and discrimination, and thereby less adherence to the program. Participants embraced and adapted to the exercise part of the program. However, the personnel claim that social training and personal development is necessary to lose weight and to maintain weight was not supported by all participants. The social and personal development part of the program is therefore likely to be the most challenging and controversial part of residential weight-loss programs.

Acknowledgments

We thank the personnel and participants at Ebeltoft Kurcenter for participating in the study.

Author Contributions

Analyzed the data: UD MBR BK AS. Wrote the paper: UD MBR BK AS.

References

1. Jeffery RW, Harnack LJ (2007) Evidence implicating eating as a primary driver for the obesity epidemic. Diabetes 56: 2673–2676. db07-1029 [pii];10.2337/db07-1029 [doi].

2. Must A, Spadano J, Coakley EH, Field AE, Colditz G et al. (1999) The disease burden associated with overweight and obesity. JAMA 282: 1523–1529. joc81719 [pii].

3. Calle EE, Thun MJ, Petrelli JM, Rodriguez C, Heath CW Jr (1999) Body-mass index and mortality in a prospective cohort of U.S. adults. N Engl J Med 341: 1097–1105. MJBA-411501 [pii];10.1056/NEJM199910073411501 [doi].

4. Chang SH, Stoll CR, Song J, Varela JE, Eagon CJ et al. (2014) The Effectiveness and Risks of Bariatric Surgery: An Updated Systematic Review and Meta-analysis, 2003–2012. JAMA Surg 149: 275–287. 1790378 [pii];10.1001/jamasurg.2013.3654 [doi].

5. Kwok CS, Pradhan A, Khan MA, Anderson SG, Keavney BD et al. (2014) Bariatric surgery and its impact on cardiovascular disease and mortality: A

systematic review and meta-analysis. Int J Cardiol. S0167-5273(14)00379-9 [pii];10.1016/j.ijcard.2014.02.026 [doi].

6. Martins C, Strommen M, Stavne OA, Nossum R, Marvik R et al. (2011) Bariatric surgery versus lifestyle interventions for morbid obesity–changes in body weight, risk factors and comorbidities at 1 year. Obes Surg 21: 841–849. 10.1007/s11695-010-0131-1 [doi].

7. Rossner S, Hammarstrand M, Hemmingsson E, Neovius M, Johansson K (2008) Long-term weight loss and weight-loss maintenance strategies. Obes Rev 9: 624–630. OBR516 [pii];10.1111/j.1467-789X.2008.00516.x [doi].

8. Sodlerlund A, Fischer A, Johansson T (2009) Physical activity, diet and behaviour modification in the treatment of overweight and obese adults: a systematic review. Perspect Public Health 129: 132–142.

9. Nicholson F, Rolland C, Broom J, Love J (2010) Effectiveness of long-term (twelve months) nonsurgical weight loss interventions for obese women with

polycystic ovary syndrome: a systematic review. Int J Womens Health 2: 393–399. 10.2147/IJWH.S13456 [doi].

10. Romanova M, Liang LJ, Deng ML, Li Z, Heber D (2013) Effectiveness of the MOVE! Multidisciplinary weight loss program for veterans in Los Angeles. Prev Chronic Dis 10: E112. 10.5888/pcd10.120325 [doi];E112 [pii].

11. Foster-Schubert KE, Alfano CM, Duggan CR, Xiao L, Campbell KL et al. (2012) Effect of diet and exercise, alone or combined, on weight and body composition in overweight-to-obese postmenopausal women. Obesity (Silver Spring) 20: 1628–1638. oby201176 [pii];10.1038/oby.2011.76 [doi].

12. Christiansen T, Bruun JM, Madsen EL, Richelsen B (2007) Weight loss maintenance in severely obese adults after an intensive lifestyle intervention: 2- to 4-year follow-up. Obesity (Silver Spring) 15: 413–420.

13. Powell LH, Calvin JE, III, Calvin JE, Jr. (2007) Effective obesity treatments. Am Psychol 62: 234–246. 2007-04834-009 [pii];10.1037/0003-066X.62.3.234 [doi].

14. Avenell A, Broom J, Brown TJ, Poobalan A, Aucott L et al. (2004) Systematic review of the long-term effects and economic consequences of treatments for obesity and implications for health improvement. Health Technol Assess 8: iii–182. 99-02-02 [pii].

15. Douketis JD, Macie C, Thabane L, Williamson DF (2005) Systematic review of long-term weight loss studies in obese adults: clinical significance and applicability to clinical practice. Int J Obes (Lond) 29: 1153–1167.

16. Foster GD, Makris AP, Bailer BA (2005) Behavioral treatment of obesity. Am J Clin Nutr 82: 230S–235S.

17. Middleton KM, Patidar SM, Perri MG (2012) The impact of extended care on the long-term maintenance of weight loss: a systematic review and meta-analysis. Obes Rev 13: 509–517. 10.1111/j.1467-789X.2011.00972.x [doi].

18. Foster GD, Makris AP, Bailer BA (2005) Behavioral treatment of obesity. Am J Clin Nutr 82: 230S–235S. 82/1/230S [pii].

19. Grol R, Wensing M (2004) What drives change? Barriers to and incentives for achieving evidence-based practice. Med J Aust 180: S57–S60. gro10753_fm [pii].

20. Bleich SN, Huizinga MM, Beach MC, Cooper LA (2010) Patient use of weight-management activities: A comparison of patient and physician assessments. Patient Education and Counseling 79: 344–350.

21. Shay LE (2008) A concept analysis: adherence and weight loss. Nurs Forum 43: 42–52.

22. Anderson RM, Funnell MM (2008) The art and science of diabetes education: a culture out of balance. Diabetes Educ 34: 109–117.

23. Maldonato A, Piana N, Bloise D, Baldelli A (2010) Optimizing patient education for people with obesity: possible use of the autobiographical approach. Patient Educ Couns 79: 287–290.

24. Beach MC, Duggan PS, Cassel CK, Geller G (2007) What does 'respect' mean? Exploring the moral obligation of health professionals to respect patients. J Gen Intern Med 22: 692–695. 10.1007/s11606-006-0054-7 [doi].

25. May CR, Mair F, Finch T, MacFarlane A, Dowrick C et al. (2009) Development of a theory of implementation and integration: Normalization Process Theory. Implement Sci 4: 29. 1748-5908-4-29 [pii];10.1186/1748-5908-4-29 [doi].

26. Bandura A (2004) Health promotion by social cognitive means. Health Educ Behav 31: 143–164.

27. Cochrane G (2008) Role for a sense of self-worth in weight-loss treatments: helping patients develop self-efficacy. Can Fam Physician 54: 543–547.

28. Byrne S, Barry D, Petry NM (2012) Predictors of weight loss success. Exercise vs. dietary self-efficacy and treatment attendance. Appetite 58: 695–698. S0195-6663(12)00006-2 [pii];10.1016/j.appet.2012.01.005 [doi].

29. Jones N, Furlanetto DL, Jackson JA, Kinn S (2007) An investigation of obese adults' views of the outcomes of dietary treatment. J Hum Nutr Diet 20: 486–494. JHN810 [pii];10.1111/j.1365-277X.2007.00810.x [doi].

30. Chan RS, Lok KY, Sea MM, Woo J (2009) Clients' experiences of a community based lifestyle modification program: a qualitative study. Int J Environ Res Public Health 6: 2608–2622. 10.3390/ijerph6102608 [doi].

31. Fabricatore AN, Wadden TA (2003) Psychological functioning of obese individuals. Diabetes spectrum 16: 245–252.

32. Puhl RM, Heuer CA (2009) The Stigma of Obesity: A Review and Update. Obesity 17: 941–964.

33. Myers A, Rosen JC (1999) Obesity stigmatization and coping: relation to mental health symptoms, body image, and self-esteem. Int J Obes Relat Metab Disord 23: 221–230.

34. Puhl RM, Brownell KD (2006) Confronting and coping with weight stigma: an investigation of overweight and obese adults. Obesity (Silver Spring) 14: 1802–1815. 14/10/1802 [pii];10.1038/oby.2006.208 [doi].

35. Foster GD, Wadden TA, Makris AP, Davidson D, Sanderson RS et al. (2003) Primary care physicians' attitudes about obesity and its treatment. Obes Res 11: 1168–1177. 10.1038/oby.2003.161 [doi].

36. Ferrante JM, Fyffe DC, Vega ML, Piasecki AK, Ohman-Strickland PA et al. (2010) Family physicians' barriers to cancer screening in extremely obese patients. Obesity (Silver Spring) 18: 1153–1159. oby2009481 [pii];10.1038/oby.2009.481 [doi].

37. Vartanian LR, Novak SA (2011) Internalized societal attitudes moderate the impact of weight stigma on avoidance of exercise. Obesity (Silver Spring) 19: 757–762. oby2010234 [pii];10.1038/oby.2010.234 [doi].

38. Sikorski C, Luppa M, Kaiser M, Glaesmer H, Schomerus G et al. (2011) The stigma of obesity in the general public and its implications for public health - a systematic review. BMC Public Health 11: 661. 1471-2458-11-661 [pii];10.1186/1471-2458-11-661 [doi].

39. Schvey NA, Puhl RM, Brownell KD (2011) The impact of weight stigma on caloric consumption. Obesity (Silver Spring) 19: 1957–1962. oby2011204 [pii];10.1038/oby.2011.204 [doi].

40. Elfhag K, Rossner S (2005) Who succeeds in maintaining weight loss? A conceptual review of factors associated with weight loss maintenance and weight regain. Obes Rev 6: 67–85. OBR170 [pii];10.1111/j.1467-789X.2005.00170.x [doi].

41. Byrne SM (2002) Psychological aspects of weight maintenance and relapse in obesity. J Psychosom Res 53: 1029–1036. S0022399902004877 [pii].

42. Christiansen B, Borge L, Fagermoen MS (2012) Understanding everyday life of morbidly obese adults-habits and body image. Int J Qual Stud Health Well - being 7: 17255. 10.3402/qhw.v7i0.17255 [doi];QHW-7-17255 [pii].

43. Reach G (2009) Obstacles to patient education in chronic diseases: a trans-theoretical analysis. Patient Educ Couns 77: 192–196.

44. Bandura A (2004) Health promotion by social cognitive means. Health Educ Behav 31: 143–164. 10.1177/1090198104263660 [doi].

45. Wing RR, Hill JO (2001) Successful weight loss maintenance. Annu Rev Nutr 21: 323–341. 10.1146/annurev.nutr.21.1.323 [doi];21/1/323 [pii].

46. Penn L, Moffatt SM, White M (2008) Participants' perspective on maintaining behaviour change: a qualitative study within the European Diabetes Prevention Study. BMC Public Health 8: 235. 1471-2458-8-235 [pii];10.1186/1471-2458-8-235 [doi].

47. Mobbs O, Crepin C, Thiery C, Golay A, Van der Linden M (2010) Obesity and the four facets of impulsivity. Patient Education and Counseling 79: 372–377.

48. Mobbs O, Crepin C, Thiery C, Golay A, Van der Linden M (2010) Obesity and the four facets of impulsivity. Patient Educ Couns 79: 372–377. S0738-3991(10)00119-9 [pii];10.1016/j.pec.2010.03.003 [doi].

5

PTPRT Regulates High-Fat Diet-Induced Obesity and Insulin Resistance

Xiujing Feng[1,2⑨], Anthony Scott[1,2⑨], Yong Wang[1,2], Lan Wang[3], Yiqing Zhao[1,2], Stephanie Doerner[1], Masanobu Satake[4], Colleen M. Croniger[3], Zhenghe Wang[1,2,5]*

1 Department of Genetics and Genome Sciences, Case Western Reserve University, Cleveland, Ohio, United States of America, 2 Case Comprehensive Cancer Center, Case Western Reserve University, Cleveland, Ohio, United States of America, 3 Department of Nutrition, Case Western Reserve University, Cleveland, Ohio, United States of America, 4 Department of Molecular Immunology, Institute of Development, Aging and Cancer, Tohoku University, Sendai, Japan, 5 Genomic Medicine Institute, Cleveland Clinic Foundation, Cleveland, Ohio, United States of America

Abstract

Obesity is a risk factor for many human diseases. However, the underlying molecular causes of obesity are not well understood. Here, we report that protein tyrosine phosphatase receptor T (PTPRT) knockout mice are resistant to high-fat diet-induced obesity. Those mice avoid many deleterious side effects of high-fat diet-induced obesity, displaying improved peripheral insulin sensitivity, lower blood glucose and insulin levels. Compared to wild type littermates, PTPRT knockout mice show reduced food intake. Consistently, STAT3 phosphorylation is up-regulated in the hypothalamus of PTPRT knockout mice. These studies implicate PTPRT-modulated STAT3 signaling in the regulation of high-fat diet-induced obesity.

Editor: Xiaoli Chen, University of Minnesota-Twin Cities, United States of America

Funding: Funding came from the National Institutes of Health R01CA127590, P50CA150964, U24 DK059630 and U24-DK76174. The funders had no role in study design, data collection and analysis, decision to publish, or preparation of the manuscript.

Competing Interests: The authors have declared that no competing interests exist.

* E-mail: zxw22@case.edu

⑨ These authors contributed equally to this work.

Introduction

Numerous studies have shown the deleterious effects of obesity on health, increasing all-cause mortality [1] and predisposing individuals to cardiovascular disease, diabetes and cancer [2]. Diet plays a crucial role in obesity, specifically those high in fats and sugar that increase body fat [3,4]. Adipocytes, which increase in size and number during obesity, can dramatically influence a variety of metabolic processes by disturbing normal homeostatic signals [5]. Chief among these disturbances is insulin resistance, leading to hyperglycemia and diabetes [6–9].

Energy imbalance – essentially a combination of increased food intake with decreased energy expenditure – causes obesity [3,10]. Circulating hormones, such as insulin and leptin, are readouts of the body's energy state and act at the hypothalamus to affect food intake [3,11–15]. Ideally, energy intake is equal to energy expenditure, leading to weight homeostasis. However, if not enough energy is released proportional to calories consumed, the excess energy is stored as lipid in adipocytes and weight gain ensues [13]. For example, dietary fat consumption affects both sides of the energy imbalance equation. Since it releases less satiety signals in comparison to protein and carbohydrate, it leads to increased food intake [10]. Conversely, since fats are an efficient form of energy and because they are stored instead of used as an energy source after feeding, dietary lipids also contribute to decreased energy expenditure [10,13]. Therefore, from both biochemical and physiologic perspectives of energy homeostasis, an excess of food intake over what is expended leads to weight gain.

Protein tyrosine phosphatases (PTPs) modulate signaling pathways that regulate a variety of metabolic processes through dephosphorylating tyrosine residues on proteins [16]. Increasing evidence suggests that PTPs play a crucial role in obesity and metabolic disease [16]. It has long been known that PTP1B is implicated in obesity, insulin resistance and type-2 diabetes mellitus by regulating insulin signaling [17]. A recent study showed that TCPTP is also involved in obesity through modulating leptin signaling [18]. TCPTP dephosphorylates STAT3 at the tyrosine 705 (Y705) residue. STAT3 Y705 phosphorylation is a key mediator of leptin signaling in the hypothalamus [19,20]. Leptin-STAT3 signaling suppresses the drive for food intake by increasing the expression of anorectic neuropeptides and repress those favoring orexigenic responses [11,17,18,21,22].

Because we previously showed that STAT3 is a substrate of protein tyrosine phosphatase receptor T (PTPRT) [23], we investigate here whether PTPRT regulates food intake and obesity in mice.

Results

Ptprt$^{-/-}$ Mice are Resistant to High-fat Diet-induced Obesity

As described previously [24], we bred the Ptprt knockout allele into the C57BL/6 strain for over 15 generations. When mouse body weights were measured from 8-week-old to 36-week-old mice, we observed that the body weight of Ptprt$^{-/-}$ mice were slightly and consistently lower than those of Ptprt$^{+/+}$ littermates on

A

B

C

D

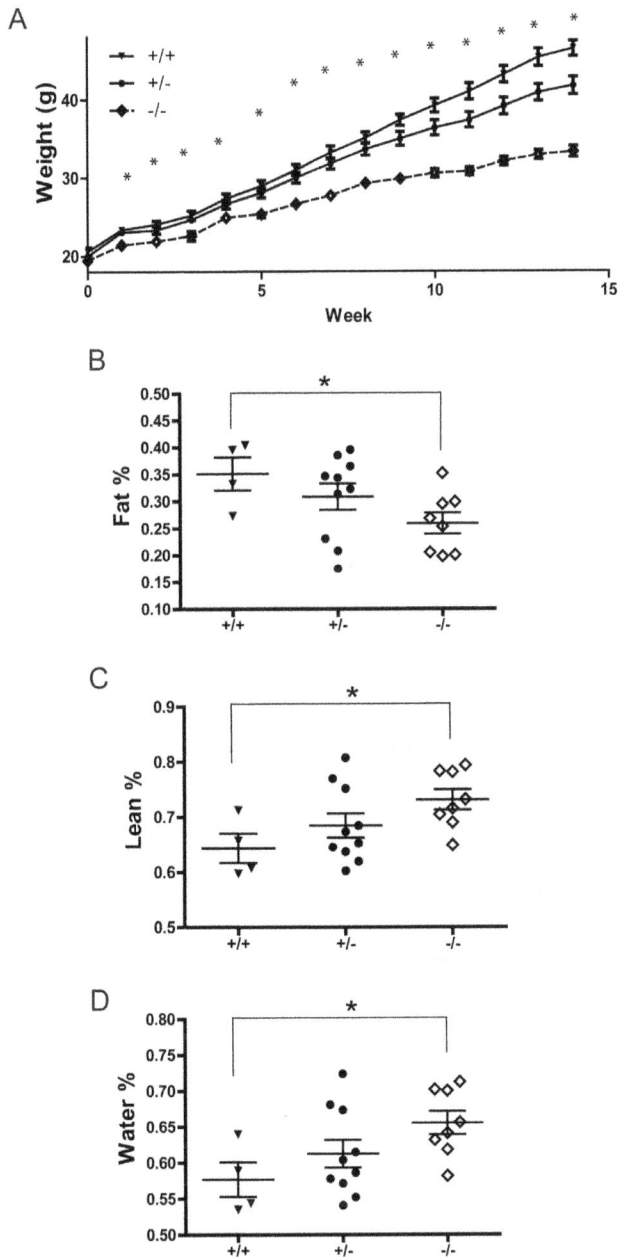

Figure 1. PTPRT KO mice are resistant to high-fat diet-induced body composition changes. A) Five week-old male mice of Ptprt⁺/⁺ (n = 8), Ptprt⁺/⁻ (n = 14) and Ptprt⁻/⁻ (n = 11) genotypes were fed with high-fat diet for 14 weeks. Body weight of the three genotypes was assessed weekly. (*p<0.05; t-test comparing Ptprt⁺/⁺ and Ptprt⁻/⁻ genotypes). B–D) Body composition was analyzed by quantitative magnetic resonance after 14 weeks on a high-fat diet (Fat % – B, Lean % – C, Water % – D; *p<0.05; t-test comparing Ptprt⁺/⁺ and Ptprt⁻/⁻ genotypes; t-test comparing Ptprt⁺/⁻ and Ptprt⁻/⁻ genotypes was not significant).

chow diet (Figure S1). However, these mice were not obese. It is well documented that high-fat diet induces obesity and insulin resistance in C57BL/6 male mice [25]. To interrogate if PTPRT plays a role in obesity, five-week-old male Ptprt⁺/⁺, Ptprt⁺/⁻ and Ptprt⁻/⁻ mice were fed with a high-fat diet for 14 weeks. Although Ptprt⁻/⁻ mice were largely indistinguishable from their Ptprt⁺/⁺ and Ptprt⁺/⁻ littermates in terms of body weight on a normal diet

at a baseline of five weeks (Figure 1A, Time 0 weeks), the body weights of Ptprt⁻/⁻ mice are significantly lower than those of Ptprt⁺/⁺ littermates through the course of the high-fat diet (Figure 1A). Consistent with previous reports [25], the Ptprt⁺/⁺ male mice were obese at the end of 14 weeks (average body weight = 46.5 g). In contrast, the Ptprt⁻/⁻ male mice remained lean (average body weight = 33.3 g) after being fed with a high-fat diet for 14 weeks, suggesting that knocking out of Ptprt renders male mice resistant to high-fat diet-induced obesity.

Ptprt⁻/⁻ Mice have Less Body Fat by Percentage than Wild Type Littermates

Obesity and its associated co-morbidities are caused by excess amounts of body fat. Therefore, we set out to determine body composition of the high-fat diet fed mice using quantitative magnetic resonance to ensure that the difference in weight gain can be attributed to increased obesity [26]. As expected, the percentages of body fat of Ptprt⁺/⁺ and Ptprt⁺/⁻ mice were significantly higher than that of Ptprt⁻/⁻ mice (Figure 1B). Consistently, the lean body mass and water in Ptprt⁺/⁺ mice were lower than the Ptprt⁻/⁻ mice (Figure 1 C and D). Since the body weight and body fat percentage of Ptprt⁺/⁻ mice were not significantly different from those of Ptprt⁺/⁺ mice (Figure 1), we focused on Ptprt⁺/⁺ versus Ptprt⁻/⁻ mice for in depth analyses in this study.

Ptprt⁻/⁻ Reduces Food Intake

Next, we set out to interrogate the variety of mechanisms by which Ptprt⁻/⁻ mice are resistant to high-fat diet-induced obesity. Given that food intake is one of the major factors that impact body weight, we measured food intake of Ptprt⁺/⁺ and Ptprt⁻/⁻ mice both at the beginning and at the end of the high-fat diet. For a ten-day period, food intake was measured daily and average values were calculated. As shown in Figure 2A, Ptprt⁻/⁻ mice ate significantly less than their Ptprt⁺/⁺ counterparts at the beginning of the high-fat diet period. However, at the end of the high-fat diet

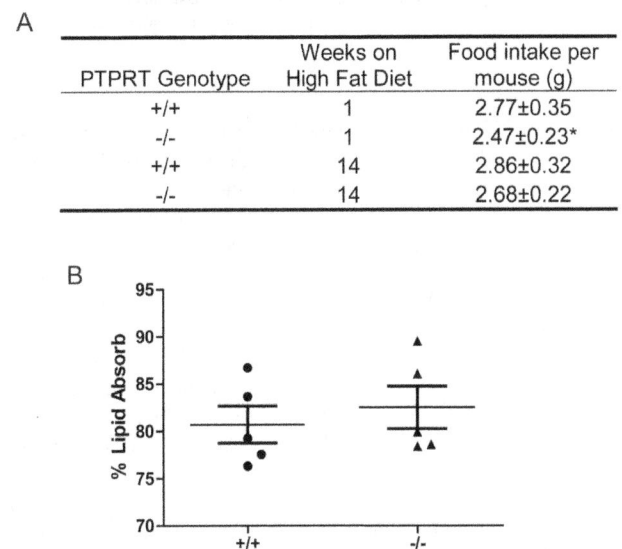

A

PTPRT Genotype	Weeks on High Fat Diet	Food intake per mouse (g)
+/+	1	2.77±0.35
-/-	1	2.47±0.23*
+/+	14	2.86±0.32
-/-	14	2.68±0.22

B

Figure 2. PTPRT KO mice eat less but do not absorb dietary fats differently. A) Average daily food intake of Ptprt⁺/⁺ (n = 8), and Ptprt⁻/⁻ (n = 12) mice was assessed at beginning and end of the high-fat diet (*p<0.05; t-test). B) Dietary lipid absorption was assessed using fecal samples of Ptprt⁺/⁺ and Ptprt⁻/⁻ mice after being fed a butter oil and sucrose behenate diet.

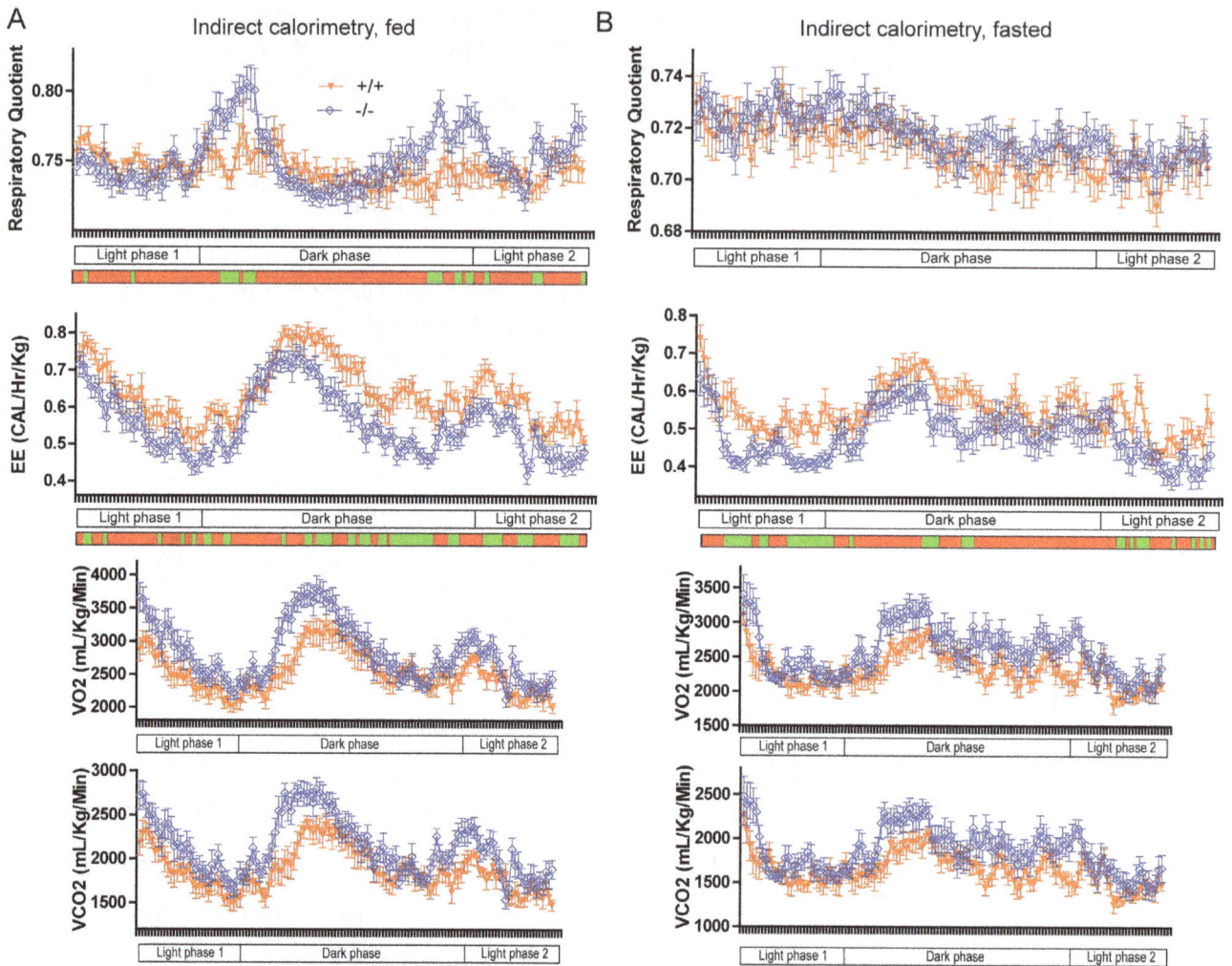

Figure 3. PTPRT KO mice utilize more glucose and expend less energy than wild type mice. Energy expenditure (Heat Released and Respiratory Quotient) was assessed through indirect calorimetry for mice in fed (A) and fasted (B) state through 24 hours (green segments: $p < 0.05$; t-test). $Ptprt^{+/+}$ (n = 8); $Ptprt^{-/-}$ (n = 11).

period, $Ptprt^{-/-}$ mice again trended toward lower food intake, but the difference did not reach statistical significance (p = 0.17).

To determine the reason behind the increased food intake in $Ptprt^{+/+}$ versus $Ptprt^{-/-}$ mice, we inferred that neurohormonal signals may play a role. Chief among these signals are leptin and neuropeptide Y, serving in an anorectic or orexigenic fashion, respectively. While plasma levels of leptin did not show significant difference between $Ptprt^{+/+}$ versus $Ptprt^{-/-}$ mice (Figure S2), $Ptprt^{+/+}$ mice had a significantly higher plasma level of neuropeptide Y versus $Ptprt^{-/-}$ mice at the beginning of the high-fat diet period (Figure S3A). This difference disappeared by the end of the high-fat diet period (Figure S3B). The plasma NPY appears to correlate with the food intake patterns at the beginning and end of the high-fat diet period.

Since dietary lipid absorption also impacts body weight and because PTPRT is expressed in the small intestine and colon [24], we set out to assess the lipid absorption capacity of $Ptprt^{+/+}$ and $Ptprt^{-/-}$ mice using non-invasive fecal analysis [27]. However, these two cohorts did not show any difference in absorbing dietary fats (Figure 2B). Taken together, our data suggest that the reduced body weight of $Ptprt^{-/-}$ mice may be due to less food consumption.

$Ptprt^{-/-}$ Mice have Reduced Energy Expenditure than Wild Type Mice

Since reduced energy expenditure and differences in nutrient utilization could also contribute to obesity, we assessed the energy expenditure and respiratory quotient of $Ptprt^{+/+}$ and $Ptprt^{-/-}$ mice via indirect calorimetry in both the fed (Figure 3A) and fasted (Figure 3B) state. According to the energy expenditure values, $Ptprt^{-/-}$ mice had reduced energy expenditure than $Ptprt^{+/+}$ mice in both the fed and fasted state (Figure 3, second row). However, reduced energy expenditure in $Ptprt^{-/-}$ mice does not explain the weight difference between $Ptprt^{+/+}$ and $Ptprt^{-/-}$ mice. The respiratory quotient of $Ptprt^{+/+}$ and $Ptprt^{-/-}$ mice was indistinguishable in the fasted state (Figure 3B, top panel). In the fed state, $Ptprt^{-/-}$ mice had a higher respiratory quotient than $Ptprt^{+/+}$ mice during the dark phase (Figure 3A, top panel), indicating they preferentially use glucose, but this difference was not sustained through the whole 24-hour period of testing. Taken as a whole, indirect calorimetry demonstrates that the metabolic phenotype of $Ptprt^{-/-}$ mice does not explain their reduced body weight versus their $Ptprt^{+/+}$ littermates.

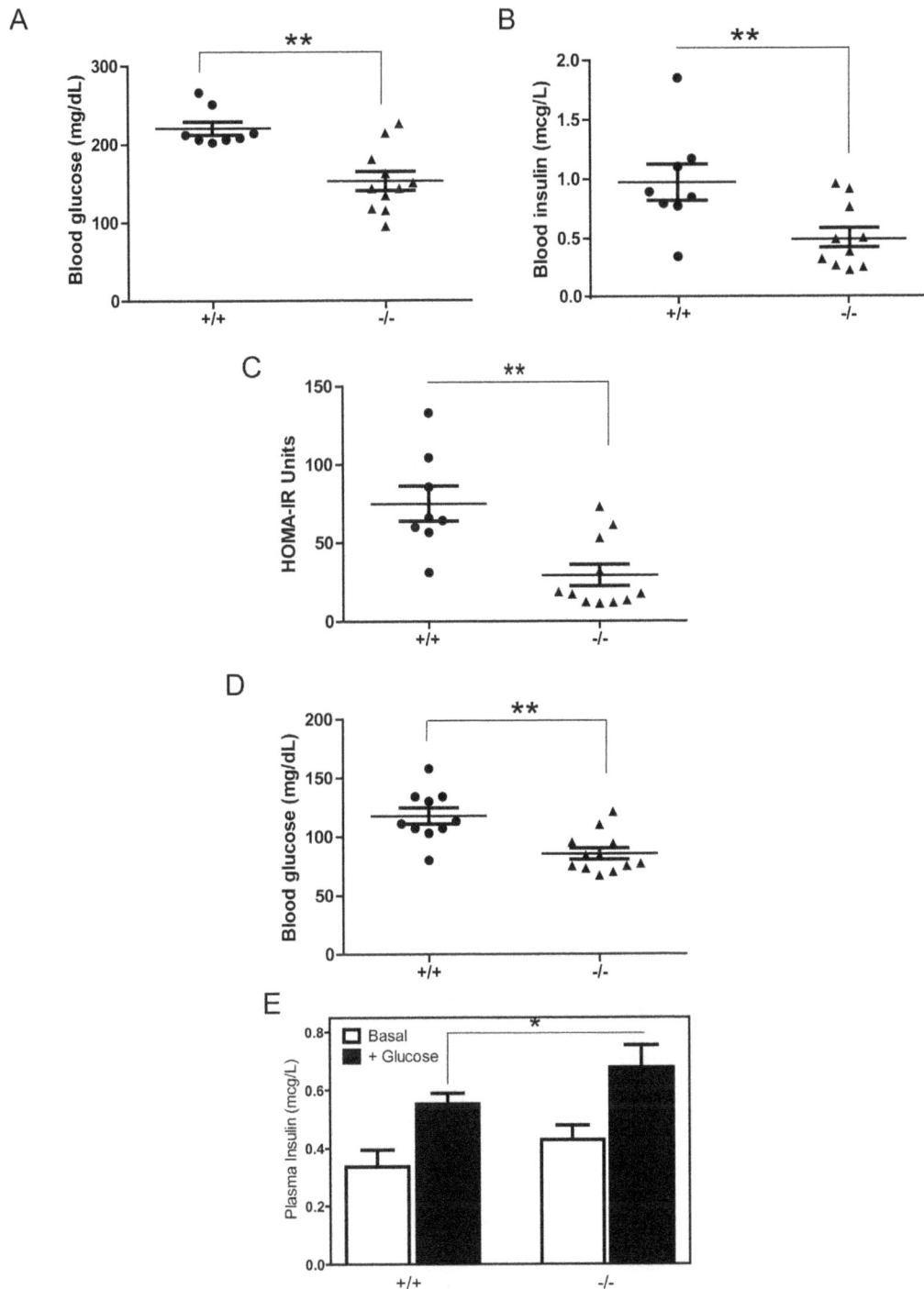

Figure 4. PTPRT KO mice have less insulin resistance than wild type mice after high-fat diet. A) Fasting blood glucose levels were assessed after 14 weeks on a high-fat diet (**$p<0.01$; t-test). B) Fasting insulin levels were assessed after 14 weeks on a high-fat diet (**$p<0.01$; t-test). C) HOMA-IR calculations of $Ptprt^{+/+}$ and $Ptprt^{-/-}$ mice after high-fat diet (**$p<0.01$; t-test). D) Fasting blood glucose levels were assessed before high-fat diet (*$p<0.05$). E) Glucose stimulated insulin secretion test between $Ptprt^{+/+}$ (n = 5) and $Ptprt^{-/-}$ (n = 5) mice before high-fat diet (*$p<0.05$; t-test).

$Ptprt^{-/-}$ Mice Resist High-fat Diet-induced Hyperglycemia and Insulin Resistance

Given that obesity often causes metabolic syndrome, such as hyperglycemia and peripheral insulin resistance, we measured fasting glucose and insulin levels in $Ptprt^{+/+}$ and $Ptprt^{-/-}$ littermates. At the end of the high-fat diet treatment, the average blood glucose levels in $Ptprt^{+/+}$ mice reached 220.6 mg/dL;

therefore, these mice were hyperglycemic. In contrast, the blood glucose levels in $Ptprt^{-/-}$ were within normal range at 152.6 mg/dL (Figure 4A). Accordingly, $Ptprt^{+/+}$ mice also had higher insulin levels than the $Ptprt^{-/-}$ counterparts after the high-fat diet (Figure 4B). These blood glucose and blood insulin values can be used to estimate peripheral insulin resistance using the HOMA-IR model [28]. By this calculation, $Ptprt^{-/-}$ had much lower HOMA-

A

B

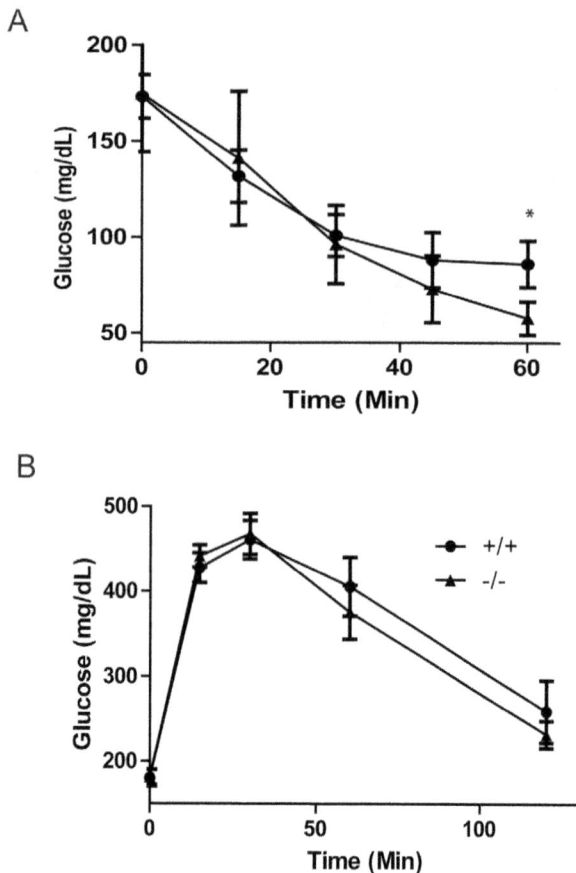

Figure 5. PTPRT KO mice demonstrate better insulin regulation than wild type mice. A) Insulin tolerance test of $Ptprt^{+/+}$ (n = 8) and $Ptprt^{-/-}$ (n = 11) mice after high-fat diet (*$p<0.05$; t-test). B) Glucose tolerance test of $Ptprt^{+/+}$ (n = 8) and $Ptprt^{-/-}$ (n = 11) mice after high-fat diet.

IR values than their $Ptprt^{+/+}$ littermates (Figure 4C), indicating a prediction of insulin sensitivity versus $Ptprt^{+/+}$ mice. It is worth noting that the $Ptprt^{+/+}$ mice were neither hyperglycemic nor insulin resistant before high-fat diet treatment, although the blood glucose levels in the $Ptprt^{+/+}$ mice were slightly higher than that of $Ptprt^{-/-}$ littermates (Figure 4D). No blood insulin difference was observed before high-fat diet treatment among the $Ptprt^{+/+}$ and $Ptprt^{-/-}$ mice (data not shown). Before the high-fat diet, $Ptprt^{-/-}$ mice also secreted more insulin in response to a glucose bolus (Figure 4E).

We then further measured peripheral insulin resistance in mice fed the high-fat diet via an insulin tolerance test. Consistent with the HOMA-IR calculation, $Ptprt^{-/-}$ mice had a lower glucose value in response to an insulin bolus after 60 minutes (Figure 5A). However, $Ptprt^{-/-}$ and $Ptprt^{+/+}$ mice did not have different glucose clearance in response to a glucose tolerance test (Figure 5B). Taken together, our data suggest that loss of PTPRT function attenuates the development of peripheral insulin resistance after a high-fat diet.

Metabolic Differences between $Ptprt^{+/+}$ and $Ptprt^{-/-}$ Littermates

Given the deviation seen between $Ptprt^{+/+}$ and $Ptprt^{-/-}$ mice as it relates to glucose and insulin metabolism, we decided to interrogate their blood plasma for differences in other nutrients.

These blood metabolites will shed additional light onto the metabolic disturbances that $Ptprt^{+/+}$ mice are experiencing. Interestingly, $Ptprt^{-/-}$ mice had lower cholesterol and higher free-fatty acids than $Ptprt^{+/+}$ mice, but we did not observe an increase in triglycerides or β-hydroxybutyrate in these mice (Figure 6). Once fatty acids are oxidized, the acetyl CoA produced is used to generate ketone bodies such as β-hydroxybutyrate. As such, PTPRT may also regulate the utilization and storage of dietary fats.

Phospho-STAT3 Increased in the Hypothalamus of $Ptprt^{-/-}$ Mice

To elucidate the molecular mechanisms by which $Ptprt^{-/-}$ mice resist high-fat diet-induced obesity, we assessed PTPRT expression in tissues implicated in metabolic regulation. Consistent with a previous report that PTPRT is expressed in the hypothalamus [29], we detected PTPRT protein in the hypothalamus using Western blot analyses (Figure 7A). However, it is not expressed in the liver, adipose or muscle. Since STAT3 is a substrate of PTPRT, we reasoned that phospho-STAT3 levels may be increased in the hypothalamus of $Ptprt^{-/-}$ mice. As expected, we found that pY705 STAT3 is up-regulated in hypothalamus of $Ptprt^{-/-}$ mice compared to $Ptprt^{+/+}$ mice (Figure 7B), suggesting that PTPRT modulates food intake by affecting phospho-STAT3 levels in the hypothalamus (Figure 7C).

Discussion

Our study reveals that PTPRT regulates metabolism and body weight. Our data suggest that PTPRT could be a drug target for obesity, because $Ptprt^{-/-}$ mice resist many key effects of a high-fat diet, including increased body mass, hyperglycemia, hypercholesterolemia, insulin resistance and increased adiposity. Consistent with this notion, several recent human genetic studies linked obesity to chromosome 20q12–13 [30–32], the genomic region in which PTPRT is located.

The decreased food intake in $Ptprt^{-/-}$ mice suggests a behavioral mechanism as to why they weigh less than their $Ptprt^{+/+}$ littermates. Food intake is primarily decreased by leptin signaling pathway [11,21,22] and increased by neuropeptide Y [14,15,33]. Leptin suppresses food intake by activating STAT3 phosphorylation in the hypothalamus [19,20]; our previous study shows that PTPRT dephosphorylates STAT3 in colorectal cancers [23]. The decreased food intake in $Ptprt^{-/-}$ mice in the absence of increased circulating leptin levels suggests that $Ptprt^{-/-}$ mice have increased phospho-STAT3 independent of leptin activity in the hypothalamus (Figure 7C). Our data indicate that STAT3 hyperphosphorylation in the hypothalamus represses food intake in $Ptprt^{-/-}$ mice. As such, we propose that the central nervous system plays a dominant role in the phenotype of $Ptprt^{-/-}$ mice.

$Ptprt^{-/-}$ mice demonstrate decreased peripheral insulin resistance as well as lower levels of blood insulin and glucose. Although a human study shows that PTPRT expression levels in adipose tissue are much higher in insulin-resistant individuals compared to insulin-sensitive individuals [34], we failed to detect PTPRT protein in mouse adipose (Figure 7A). Neither could we detect PTPRT protein in liver or muscle (Figure 7A). Our data indicate that PTPRT does not directly modulate insulin sensitivity in peripheral tissues. Instead, PTPRT may indirectly impact peripheral insulin resistance through affecting the nervous system control of energy homeostasis. It is well documented that increased plasma NPY from autonomic nervous system sources is associated with greater adiposity and increased insulin resistance [35–40]. Consistently, our data show increased NPY secretion in $Ptprt^{+/+}$

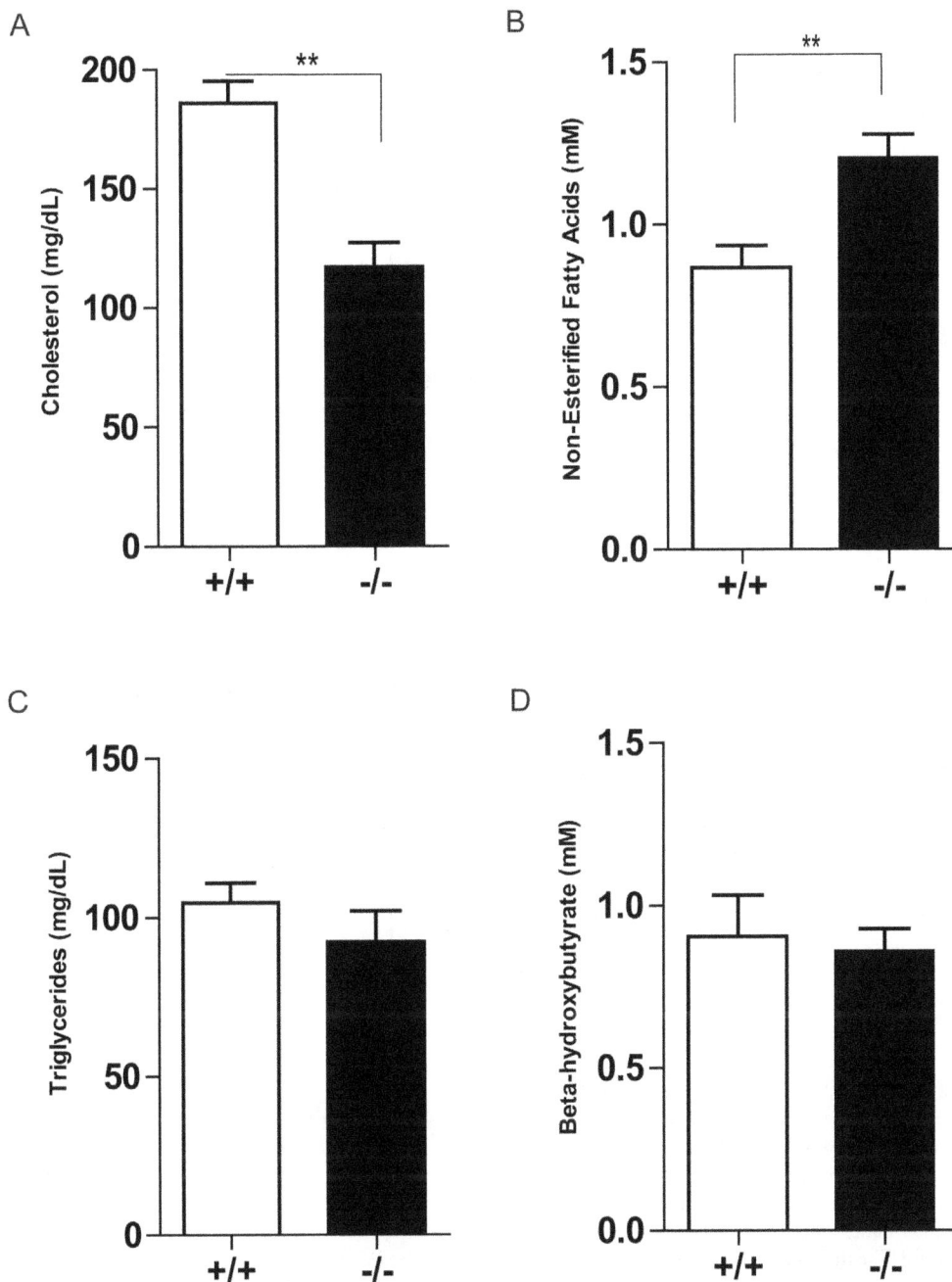

Figure 6. PTPRT KO mice have different blood chemistry values after high-fat diet. Fasted plasma concentrations of cholesterol (A), non-esterified fatty acids (B), triglycerides (C) and beta-hydroxybutyrate (D) of $Ptprt^{+/+}$ (n = 8) and $Ptprt^{-/-}$ (n = 11) mice after high-fat diet (**$p<0.01$; t-test).

mice that go on to develop obesity and insulin resistance. As such, the decrease in NPY in $Ptprt^{-/-}$ mice further suggests the role of PTPRT in nervous system regulating obesity and peripheral insulin resistance [41–46].

Materials and Methods

Animals and Diet

Treatment of experimental mice and related protocols were done in accordance with the Institutional Animal Care and Use Committee at Case Western Reserve University (CWRU). The protocol (Number 2010-0125) was approved by the IACUC Committee at CWRU. Male and female PTPRT heterozygous and homozygous knockout mice in a C57BL/6 background were generated as described previously [24]; referred to as $Ptprt^{+/-}$ and $Ptprt^{-/-}$, respectively. Colonies of these mice were maintained on a normal chow diet (#5010 (4.5% fat by weight, 12.7% fat by calorie), LabDiet St. Louis, MO). Five weeks after birth, male $Ptprt^{+/+}$, $Ptprt^{+/-}$ and $Ptprt^{-/-}$ mice were put on a high-fat diet (#D12331: 33% Hydrogenated Coconut Oil (35% fat by weight; 58% fat by calorie), Research Diets, Inc. New Brunswick, NJ) for

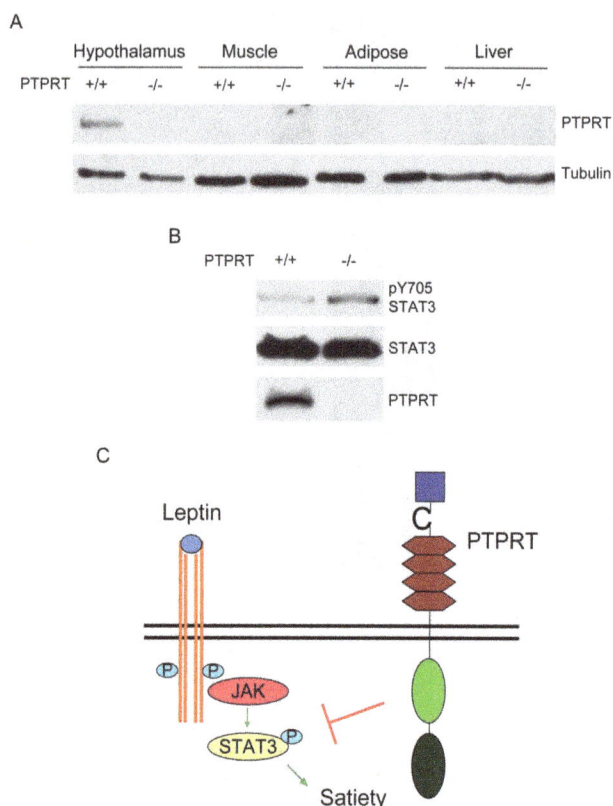

Figure 7. PTPRT regulates STAT3 phosphorylation in mouse hypothalamus. A) Tissue lysates from *Ptprt*^+/+ and *Ptprt*^−/− mice were blotted with the indicated antibodies. B) Hypothalamic lysates from *Ptprt*^+/+ and *Ptprt*^+/− mice were blotted with the indicated antibodies. C) Proposed model for the effect of PTPRT on food intake.

14 weeks. Body composition was analyzed by quantitative magnetic resonance as described previously [26].

Glucose and Insulin Tolerance Test

Tests were done as described previously [47,48]. Mice were deprived of food overnight and then injected intraperitoneally with glucose (2 g/Kg) or insulin (0.9 g/Kg). Tail vein blood was sampled for glucose levels at 0, 15, 30 and 60 minutes after insulin or 0, 15, 30, 60 and 120 minutes after glucose using an UltraTouch meter. GTT was performed at the Mouse Metabolic Phenotyping Center of CWRU.

HOMA-IR

Insulin resistance was estimated for *Ptprt*^+/+ and *Ptprt*^−/− mice after high-fat diet using the homeostatic model assessment [28]. Formula = (Insulin (mcU/L)×Glucose (mg/dl)/405).

Insulin, Neuropeptide Y and Blood Chemistry Measurements

Tests were done as described previously [48]. Mouse blood plasma was isolated using Microtainer plasma separator tubes (BD Biosciences). Insulin was measured using an insulin enzyme-linked immunosorbent assay (Mercodia, Inc., Uppsala, SWE). Neuro-peptide Y was measured using neuropeptide Y insulin enzyme-linked immunosorbent assay (EMD Millipore, Billerica, Massa-chusetts, USA). Mouse plasma was sent to Marshfield Laboratories

to assess β-hydroxybutyrate, triglycerides, non-esterified fatty acids and total cholesterol.

Lipid Absorption

Tests were done as described previously [49]. Mice were fed a diet consisting of butter oil and 5% sucrose polybehenate for three days. Fecal pellets from these mice were collected and analyzed via gas chromatography of fatty acid methyl esters. The ratio of behenic acid to other fatty acids then was used to determine intestinal lipid absorption [27]. The lipid absorption studies were performed at the Cincinnati Mouse Metabolic Phenotyping Center.

Glucose-stimulated Insulin Secretion Test

Tests were done as described previously [50]. For the glucose-stimulated insulin secretion test, mice were starved overnight and 2 g/kg of glucose was injected intraperitoneally into the mice. Tail vein blood was collected and plasma insulin concentrations were measured at 0 and 30 minutes after glucose injection.

Indirect Calorimetry

Tests were done as described previously [51]. Indirect calorimetry (IDC) was performed using the Oxymax system (Columbus Instruments' Comprehensive Lab Animal Monitoring System (CLAMS), Columbus, OH). VO2, VCO2, respiratory quotient (RQ) and heat (energy expenditure – EE) were determined. Energy expenditure was normalized to body mass. IDC was performed on mice either after an overnight fast and water (Fasted) or with ad lib food and water (Fed). The experiments ran for 22 hours on a 12 hour dark cycle (6 pm to 6 am).

Food Intake

Mice with ad lib access to food and water were placed in a clean cage and the food was weighed. The remaining food after 24 hours was weighed and the average food intake per mouse was calculated [47].

Western Blot

Tissues were lysed as described previously [52]. Total brain, hypothalamus, hind leg muscle, perigonadal adipose or liver tissue were lysed in RIPA lysis buffer (150 mM NaCl, 10 mM Tris-HCl (pH 7.5), 0.1% SDS, 1% Triton X-100, 1% Deoxycholate, 0.5 M 5 mM EDTA) supplemented with protease (Roche, Penzberg, GER) and phosphatase inhibitors (1 mM NaVO4, 50 mM NaF). Western blots were performed as described previously [24]. Antibodies for pSTAT3^Y705 and STAT3 were from Cell Signaling (Danvers, MA, USA), tubulin from Sigma-Aldrich (St. Louis, MO, USA) and PTPRT from Biovendor (Asheville, NC, USA).

Statistical Analysis

Results were assessed using two-tailed unpaired Student's *t*-test with significance set at $p < 0.05$.

Supporting Information

Figure S1 PTPRT KO mice demonstrate slightly lower body weight than wild type littermates on normal chow diet. Eight week-old male mice of *Ptprt*^+/+ (n = 13), *Ptprt*^+/− (n = 13) and *Ptprt*^−/− (n = 13) genotypes were maintained on a normal chow diet for 29 weeks. Body weight of the three genotypes was assessed weekly. (*$p < 0.05$; t-test comparing *Ptprt*^+/+ and *Ptprt*^−/− genotypes).

Figure S2 PTPRT KO mice do not have different circulating levels of leptin. A) Fasting plasma leptin levels of $Ptprt^{+/+}$ and $Ptprt^{-/-}$ mice were assessed before high-fat diet. B) Fasting plasma leptin levels of $Ptprt^{+/+}$ and $Ptprt^{-/-}$ mice were assessed after 14 weeks on a high-fat diet.

Figure S3 PTPRT KO mice have decreased NPY levels before high-fat diet. A) Fasting plasma neuropeptide Y levels of $Ptprt^{+/+}$ and $Ptprt^{-/-}$ mice were assessed before high-fat diet (**$p < 0.01$; t-test). B) Fasting plasma neuropeptide Y levels of $Ptprt^{+/+}$ and $Ptprt^{-/-}$ mice were assessed after 14 weeks on a high-fat diet.

Acknowledgments

The authors would like to thank Drs. Brian Bai and Zhi Huang for their technical assistance.

Author Contributions

Conceived and designed the experiments: XF AS SD CMC ZW. Performed the experiments: XF AS YW LW YZ. Analyzed the data: XF AS SD CMC ZW. Contributed reagents/materials/analysis tools: MS. Wrote the paper: XF AS CMC ZW.

References

1. Berrington de Gonzalez A, Hartge P, Cerhan JR, Flint AJ, Hannan L, et al. (2010) Body-mass index and mortality among 1.46 million white adults. N Engl J Med 363: 2211–2219.
2. Flegal KM, Graubard BI, Williamson DF, Gail MH (2007) Cause-specific excess deaths associated with underweight, overweight, and obesity. JAMA 298: 2028–2037.
3. Ahima RS (2011) Digging deeper into obesity. J Clin Invest 121: 2076–2079.
4. Bray GA (2010) Soft drink consumption and obesity: it is all about fructose. Curr Opin Lipidol 21: 51–57.
5. Haslam DW, James WP (2005) Obesity. Lancet 366: 1197–1209.
6. Rossmeisl M, Rim JS, Koza RA, Kozak LP (2003) Variation in type 2 diabetes-related traits in mouse strains susceptible to diet-induced obesity. Diabetes 52: 1958–1966.
7. Schreyer SA, Wilson DL, LeBoeuf RC (1998) C57BL/6 mice fed high fat diets as models for diabetes-accelerated atherosclerosis. Atherosclerosis 136: 17–24.
8. Surwit RS, Seldin MF, Kuhn CM, Cochrane C, Feinglos MN (1991) Control of expression of insulin resistance and hyperglycemia by different genetic factors in diabetic C57BL/6J mice. Diabetes 40: 82–87.
9. Hussain AH, M.Z.I.; Claussen, B.; Asghar, S. (2010) Type 2 Diabetes and obesity: A review. Journal of Diabetology 2.
10. Jequier E (2002) Pathways to obesity. Int J Obes Relat Metab Disord 26 Suppl 2: S12–17.
11. Schwartz MW, Woods SC, Porte D, Jr., Seeley RJ, Baskin DG (2000) Central nervous system control of food intake. Nature 404: 661–671.
12. Williams KW, Elmquist JK (2012) From neuroanatomy to behavior: central integration of peripheral signals regulating feeding behavior. Nat Neurosci 15: 1350–1355.
13. Hariri N, Thibault L (2010) High-fat diet-induced obesity in animal models. Nutr Res Rev 23: 270–299.
14. Beck B (2006) Neuropeptide Y in normal eating and in genetic and dietary-induced obesity. Philos Trans R Soc Lond B Biol Sci 361: 1159–1185.
15. Bi S, Kim YJ, Zheng F (2012) Dorsomedial hypothalamic NPY and energy balance control. Neuropeptides 46: 309–314.
16. Xu E, Schwab M, Marette A (2013) Role of protein tyrosine phosphatases in the modulation of insulin signaling and their implication in the pathogenesis of obesity-linked insulin resistance. Rev Endocr Metab Disord.
17. Bence KK, Delibegovic M, Xue B, Gorgun CZ, Hotamisligil GS, et al. (2006) Neuronal PTP1B regulates body weight, adiposity and leptin action. Nat Med 12: 917–924.
18. Loh K, Fukushima A, Zhang X, Galic S, Briggs D, et al. (2011) Elevated hypothalamic TCPTP in obesity contributes to cellular leptin resistance. Cell Metab 14: 684–699.
19. Vaisse C, Halaas JL, Horvath CM, Darnell JE, Jr., Stoffel M, et al. (1996) Leptin activation of Stat3 in the hypothalamus of wild-type and ob/ob mice but not db/db mice. Nat Genet 14: 95–97.
20. Bates SH, Stearns WH, Dundon TA, Schubert M, Tso AW, et al. (2003) STAT3 signalling is required for leptin regulation of energy balance but not reproduction. Nature 421: 856–859.
21. Munzberg H, Bjornholm M, Bates SH, Myers MG, Jr. (2005) Leptin receptor action and mechanisms of leptin resistance. Cell Mol Life Sci 62: 642–652.
22. Elmquist JK, Maratos-Flier E, Saper CB, Flier JS (1998) Unraveling the central nervous system pathways underlying responses to leptin. Nat Neurosci 1: 445–450.
23. Zhang X, Guo A, Yu J, Possemato A, Chen Y, et al. (2007) Identification of STAT3 as a substrate of receptor protein tyrosine phosphatase T. Proc Natl Acad Sci U S A 104: 4060–4064.
24. Zhao Y, Zhang X, Guda K, Lawrence E, Sun Q, et al. (2010) Identification and functional characterization of paxillin as a target of protein tyrosine phosphatase receptor T. Proc Natl Acad Sci U S A 107: 2592–2597.
25. West DB, Boozer CN, Moody DL, Atkinson RL (1992) Dietary obesity in nine inbred mouse strains. Am J Physiol 262: R1025–1032.
26. Tinsley FC, Taicher GZ, Heiman ML (2004) Evaluation of a quantitative magnetic resonance method for mouse whole body composition analysis. Obes Res 12: 150–160.
27. Jandacek RJ, Heubi JE, Tso P (2004) A novel, noninvasive method for the measurement of intestinal fat absorption. Gastroenterology 127: 139–144.
28. Matthews DR, Hosker JP, Rudenski AS, Naylor BA, Treacher DF, et al. (1985) Homeostasis model assessment: insulin resistance and beta-cell function from fasting plasma glucose and insulin concentrations in man. Diabetologia 28: 412–419.
29. Visel A, Carson J, Oldekamp J, Warnecke M, Jakubcakova V, et al. (2007) Regulatory pathway analysis by high-throughput in situ hybridization. PLoS Genet 3: 1867–1883.
30. Lembertas AV, Perusse L, Chagnon YC, Fisler JS, Warden CH, et al. (1997) Identification of an obesity quantitative trait locus on mouse chromosome 2 and evidence of linkage to body fat and insulin on the human homologous region 20q. J Clin Invest 100: 1240–1247.
31. Lee JH, Reed DR, Li WD, Xu W, Joo EJ, et al. (1999) Genome scan for human obesity and linkage to markers in 20q13. Am J Hum Genet 64: 196–209.
32. Borecki IB, Rice T, Perusse L, Bouchard C, Rao DC (1994) An exploratory investigation of genetic linkage with body composition and fatness phenotypes: the Quebec Family Study. Obes Res 2: 213–219.
33. Brothers SP, Wahlestedt C (2010) Therapeutic potential of neuropeptide Y (NPY) receptor ligands. EMBO Mol Med 2: 429–439.
34. Elbein SC, Kern PA, Rasouli N, Yao-Borengasser A, Sharma NK, et al. (2011) Global gene expression profiles of subcutaneous adipose and muscle from glucose-tolerant, insulin-sensitive, and insulin-resistant individuals matched for BMI. Diabetes 60: 1019–1029.
35. Ruohonen ST, Pesonen U, Moritz N, Kaipio K, Roytta M, et al. (2008) Transgenic mice overexpressing neuropeptide Y in noradrenergic neurons: a novel model of increased adiposity and impaired glucose tolerance. Diabetes 57: 1517–1525.
36. Han R, Li A, Li L, Kitlinska JB, Zukowska Z (2012) Maternal low-protein diet up-regulates the neuropeptide Y system in visceral fat and leads to abdominal obesity and glucose intolerance in a sex- and time-specific manner. FASEB J 26: 3528–3536.
37. Kuo LE, Kitlinska JB, Tilan JU, Li L, Baker SB, et al. (2007) Neuropeptide Y acts directly in the periphery on fat tissue and mediates stress-induced obesity and metabolic syndrome. Nat Med 13: 803–811.
38. Ruohonen ST, Vahatalo LH, Savontaus E (2012) Diet-induced obesity in mice overexpressing neuropeptide y in noradrenergic neurons. Int J Pept 2012: 452524.
39. Warne JP, Dallman MF (2007) Stress, diet and abdominal obesity: Y? Nat Med 13: 781–783.
40. Bagherian A, Kalhori KA, Sadeghi M, Mirhosseini F, Parisay I (2010) An in vitro study of root and canal morphology of human deciduous molars in an Iranian population. J Oral Sci 52: 397–403.
41. Koch L, Wunderlich FT, Seibler J, Konner AC, Hampel B, et al. (2008) Central insulin action regulates peripheral glucose and fat metabolism in mice. J Clin Invest 118: 2132–2147.
42. Lam CK, Chari M, Rutter GA, Lam TK (2011) Hypothalamic nutrient sensing activates a forebrain-hindbrain neuronal circuit to regulate glucose production in vivo. Diabetes 60: 107–113.
43. Pagotto U (2009) Where does insulin resistance start? The brain. Diabetes Care 32 Suppl 2: S174–177.
44. Gerozissis K (2008) Brain insulin, energy and glucose homeostasis; genes, environment and metabolic pathologies. Eur J Pharmacol 585: 38–49.
45. Obici S, Zhang BB, Karkanias G, Rossetti L (2002) Hypothalamic insulin signaling is required for inhibition of glucose production. Nat Med 8: 1376–1382.
46. Carey M, Kehlenbrink S, Hawkins M (2013) Evidence for central regulation of glucose metabolism. J Biol Chem 288: 34981–34988.
47. Marwarha G, Berry DC, Croniger CM, Noy N (2013) The retinol esterifying enzyme LRAT supports cell signaling by retinol-binding protein and its receptor STRA6. FASEB J.
48. Berry DC, Jacobs H, Marwarha G, Gely-Pernot A, O'Byrne SM, et al. (2013) The STRA6 receptor is essential for retinol-binding protein-induced insulin

resistance but not for maintaining vitamin A homeostasis in tissues other than the eye. J Biol Chem 288: 24528–24539.

49. Buchner DA, Geisinger JM, Glazebrook PA, Morgan MG, Spiezio SH, et al. (2012) The juxtaparanodal proteins CNTNAP2 and TAG1 regulate diet-induced obesity. Mamm Genome 23: 431–442.

50. Millward CA, Desantis D, Hsieh CW, Heaney JD, Pisano S, et al. (2010) Phosphoenolpyruvate carboxykinase (Pck1) helps regulate the triglyceride/fatty acid cycle and development of insulin resistance in mice. J Lipid Res 51: 1452–1463.

51. Prince A, Zhang Y, Croniger C, Puchowicz M (2013) Oxidative metabolism: glucose versus ketones. Adv Exp Med Biol 789: 323–328.

52. Marwarha G, Berry DC, Croniger CM, Noy N (2014) The retinol esterifying enzyme LRAT supports cell signaling by retinol-binding protein and its receptor STRA6. FASEB J 28: 26–34.

6

Single Rapamycin Administration Induces Prolonged Downward Shift in Defended Body Weight in Rats

Mark Hebert[1], Maria Licursi[2], Brittany Jensen[1], Ashley Baker[1], Steve Milway[1], Charles Malsbury[1], Virginia L. Grant[1], Robert Adamec[1], Michiru Hirasawa[2]*, Jacqueline Blundell[1]*

1 Department of Psychology, Memorial University of Newfoundland, St. John's, Newfoundland, Canada, 2 Division of Biomedical Sciences, Memorial University of Newfoundland, St. John's, Newfoundland, Canada

Abstract

Manipulation of body weight set point may be an effective weight loss and maintenance strategy as the homeostatic mechanism governing energy balance remains intact even in obese conditions and counters the effort to lose weight. However, how the set point is determined is not well understood. We show that a single injection of rapamycin (RAP), an mTOR inhibitor, is sufficient to shift the set point in rats. Intraperitoneal RAP decreased food intake and daily weight gain for several days, but surprisingly, there was also a long-term reduction in body weight which lasted at least 10 weeks without additional RAP injection. These effects were not due to malaise or glucose intolerance. Two RAP administrations with a two-week interval had additive effects on body weight without desensitization and significantly reduced the white adipose tissue weight. When challenged with food deprivation, vehicle and RAP-treated rats responded with rebound hyperphagia, suggesting that RAP was not inhibiting compensatory responses to weight loss. Instead, RAP animals defended a lower body weight achieved after RAP treatment. Decreased food intake and body weight were also seen with intracerebroventricular injection of RAP, indicating that the RAP effect is at least partially mediated by the brain. In summary, we found a novel effect of RAP that maintains lower body weight by shifting the set point long-term. Thus, RAP and related compounds may be unique tools to investigate the mechanisms by which the defended level of body weight is determined; such compounds may also be used to complement weight loss strategy.

Editor: Thierry Alquier, CRCHUM-Montreal Diabetes Research Center, Canada

Funding: This work was supported by the National Alliance for Research on Schizophrenia and Depression Young Investigator Award (JB), Natural Sciences and Engineering Research Council of Canada (JB), Canadian Institute for Health Research (ROP 91548, REA; MOP 84409, MH), and Memorial University psychology departmental funding. ML is a CIHR/RDC Fellow. The funders had no role in study design, data collection and analysis, decision to publish, or preparation of the manuscript.

Competing Interests: The authors have declared that no competing interests exist.

* E-mail: jblundell@mun.ca (JD); michiru@mun.ca (M. Hirasawa)

Introduction

The most common weight loss strategy is caloric restriction and exercise, as obesity is typically due to chronic excess in caloric intake over energy expenditure [1]. However, weight loss is strongly countered by physiological compensatory responses that often defeat attempts to stay on a diet regimen and maintain weight loss [2,3]. It has been proposed that obesity is not a state where energy homeostasis is dysregulated, but where the defended body weight level, or set point, is shifted upwards [4]. This is a major obstacle that needs to be overcome if obesity and overeating are to be contained.

Rapamycin (RAP) is an inhibitor of the mammalian target of rapamycin (mTOR). mTOR is a highly conserved serine/threonine kinase that is inhibited by energy deficiency but activated by energy and nutrient signals to promote cell growth through well described pathways (see [5] for recent review). Inhibition of mTOR by daily RAP administration reduces both food intake and body weight gain in free-feeding animals and provides resistance to diet-induced obesity [6,7,8]. In hypothalamic neurons that regulate energy balance and food intake, mTOR has been shown to mediate the anorexic and orexigenic effects of leptin and ghrelin, respectively. These effects can be blocked by direct injections of RAP into these areas [9,10,11].

Thus, peripherally administered RAP could exert actions either peripherally or centrally, or both.

In the present study, we examined the effect of a single injection of RAP (peripheral or central) on eating and body weight. Consistent with chronic administration [6,7,8], acute RAP produced a dose-dependent reduction in both food intake and body weight gain. Unexpectedly, however, RAP treated animals voluntarily maintained a lower body weight for weeks and months in the absence of additional RAP administration. The persistent lowered body weight by RAP could be explained by a sustained downward shift in body weight set point or a disruption of compensatory mechanisms for regaining body weight. Thus, the goal of the current study was to test the hypothesis that acute RAP causes a downward shift in body weight set point. Overall, our findings suggest a novel role of mTOR in *establishing* a homeostatic set point and that RAP may be a unique tool for probing the determinants of body weight set point.

Methods

Animals

Male Sprague Dawley rats were obtained from the Vivarium at Memorial University of Newfoundland at 7 weeks of age. The rats were housed individually in a temperature- and humidity-controlled environment with a 12-h light/12-h dark cycle (lights on at 7:00 am). Rats were given free access to a standard rodent diet (Prolab RMH 3000: PMI Nutrition International LLC, Brentwood, MO, USA) and water, unless otherwise stated. Body weight and food intake were measured every 1–2 days unless indicated otherwise at the same time each day (9:00–11:00 am).

Ethics Statement

All procedures involving animals adhered to the guidelines of the Canadian Council on Animal Care, and were approved by the Institutional Animal Care Committee of Memorial University.

Rapamycin injection

For intraperitoneal (i.p.) administration, rats received either vehicle (VEH: 5% ethanol in 5% Tween 80 and 5% PEG 400 in distilled water) or RAP (LC Laboratories, Woburn, MA, USA) in vehicle at 0.1, 1, or 10 mg/kg, similar to [12,13].

For intracerebroventricular (i.c.v.) injection, rats were initially implanted stereotaxically with guide cannulae aimed at the left lateral ventricle under 4% chloral hydrate (400 mg/kg i.p.). After 16–19 days of recovery, rats received an i.c.v. injection of 1 μL of DMSO as vehicle (VEH-ICV) or 50 μg of RAP in 1 μL DMSO (RAP-ICV), similar to [9]. Upon completion of the experiment, rats were anesthetized with 15% urethane and brains were collected. To verify location of cannula tips, brains were sectioned and stained with cresyl violet and examined microscopically. Tracks formed by the guide cannulae reached the lateral ventricle in all subjects.

Visceral Fat

Visceral fat was assessed in a sub-set of subjects that received 2 injections a week apart of VEH or RAP (RAP-RAP and VEH-VEH groups). Rats were killed by CO_2 inhalation approximately 2 weeks following the second injection. Retroperitoneal and epididymal fat pads were dissected and weighed immediately to determine total visceral fat mass.

Glucose Tolerance Test

Two groups of rats matched by weight were injected i.p. with RAP (10 mg/kg) or VEH. Two weeks later, the rats were fasted overnight for 16 hours. To establish basal values of blood glucose (fasted), a drop of blood was drawn by nicking the tail vein with a razor blade and glucose level in whole blood was measured with Blood Glucose Monitoring System (Free Style Lite, Abbott). Then the rats were injected with glucose solution (in H_2O, 2 g/kg i.p., Time 0). Blood glucose levels were measured at 15, 30, 60, 90 and 120 minutes post-injection.

Conditioned Taste Aversion Test

All rats had unrestricted access to rodent chow and restricted access to water (one hour each day, 9:00–10:00 am) for one week (Days 1–7) prior to injection. Body weights were measured approximately 2 hours later each day. On injection day (Day 8), all rats were presented with one bottle containing 0.1% saccharin in water for one hour (between 9:00–10:00 am). Immediately following saccharin consumption, rats received an i.p. injection of RAP (10 mg/kg), VEH, LiCl (as a positive control, dose of 127.17 mg/kg), or saline (vehicle for LiCl). The next day (Day 9), the rats had 1-hour access to water. On Day 10, the rats were given a 1-hour two-bottle preference test during which they had access to a bottle containing 0.1% saccharin solution and another

bottle containing water. At 30 minutes into the test, the places of bottles were exchanged to control for side preference effects.

Saccharin preference was calculated as a ratio of the total amount of saccharin consumed during the one hour period to the total amount of fluid (water + saccharin solution) consumed. A a lower saccharin preference measure from controls indicates whether the drug has produced a conditioned aversion to the associated saccharin.

Yoke Procedure

Rats were divided into three groups having approximately equal baseline food intake (differed by less than 1 gram). Rats in two of the groups were ranked by food intake from highest to lowest. Pairs were formed by taking the two subjects with the highest food intake, the two with the next highest food intake, and so forth. Within each pair one animal was randomly assigned to the RAP and the other to the yoked condition (YOKE). The third group formed the VEH group. Following the five day baseline period, rats were injected i.p. on Day 0 with RAP (10 mg/kg) or vehicle (VEH and YOKE groups). During the yoked period, daily food intake was determined for each rat in the RAP group and expressed as a percentage of its averaged daily food intake during the baseline period. Each day of the 5-day yoked period, yoked rats were given a percentage of their daily baseline food intake amount which corresponded to that of their RAP counterpart for the preceding 24 hour period. Animals were then placed back on ad lib food; daily food intake and body weight were measured for an additional week.

Food deprivation

Rats were given either a RAP (10 mg/kg) or VEH injection i.p., and then 24 hours or 2 weeks later, food was restricted to 5 grams for 24 hours. All rats were returned to *ad libitum* feeding following deprivation for the remainder of the experiment. Water was available *ad libitum* at all times.

Data analysis

All body weights were expressed as a percentage of injection day (Day 0) body weight in order to adjust for individual differences in absolute body weight. Food efficiency (FE) was calculated by dividing body weight gain by food intake (FI) (both in grams) for each 24 hour period. There were no pre-treatment differences among the groups in weight gain, FI or FE in any of the experiments. One-way ANOVAs with Tukey's post hoc tests were used to test for differences among three or more groups, whereas unpaired t-test was used to compare two groups, as appropriate. Two-way Mixed ANOVA was used to compare two or more groups that were repeatedly measured to follow the time course. Paired t-test was used for within-group comparisons. Data are expressed as mean ± S.E.M. $p < 0.05$ was considered significant.

Results

Single injection of rapamycin inhibits food intake and body weight gain

To examine the effect of acute RAP treatment on energy balance, 40 male Sprague Dawley rats were given a single i.p. injection of either 0 (vehicle, VEH), 0.1, 1.0 or 10 mg/kg of RAP (n = 10 for each group). We found that food intake (FI) and food efficiency (FE) were significantly reduced during the first 3–5 days post-injection in a dose-dependent manner (Fig. 1A–D). The effect of 10 mg/kg RAP on FI was observed as early as Day 1 post-injection, while the response was delayed to the second day at 1 mg/kg. This was accompanied by a transient (2–3 days)

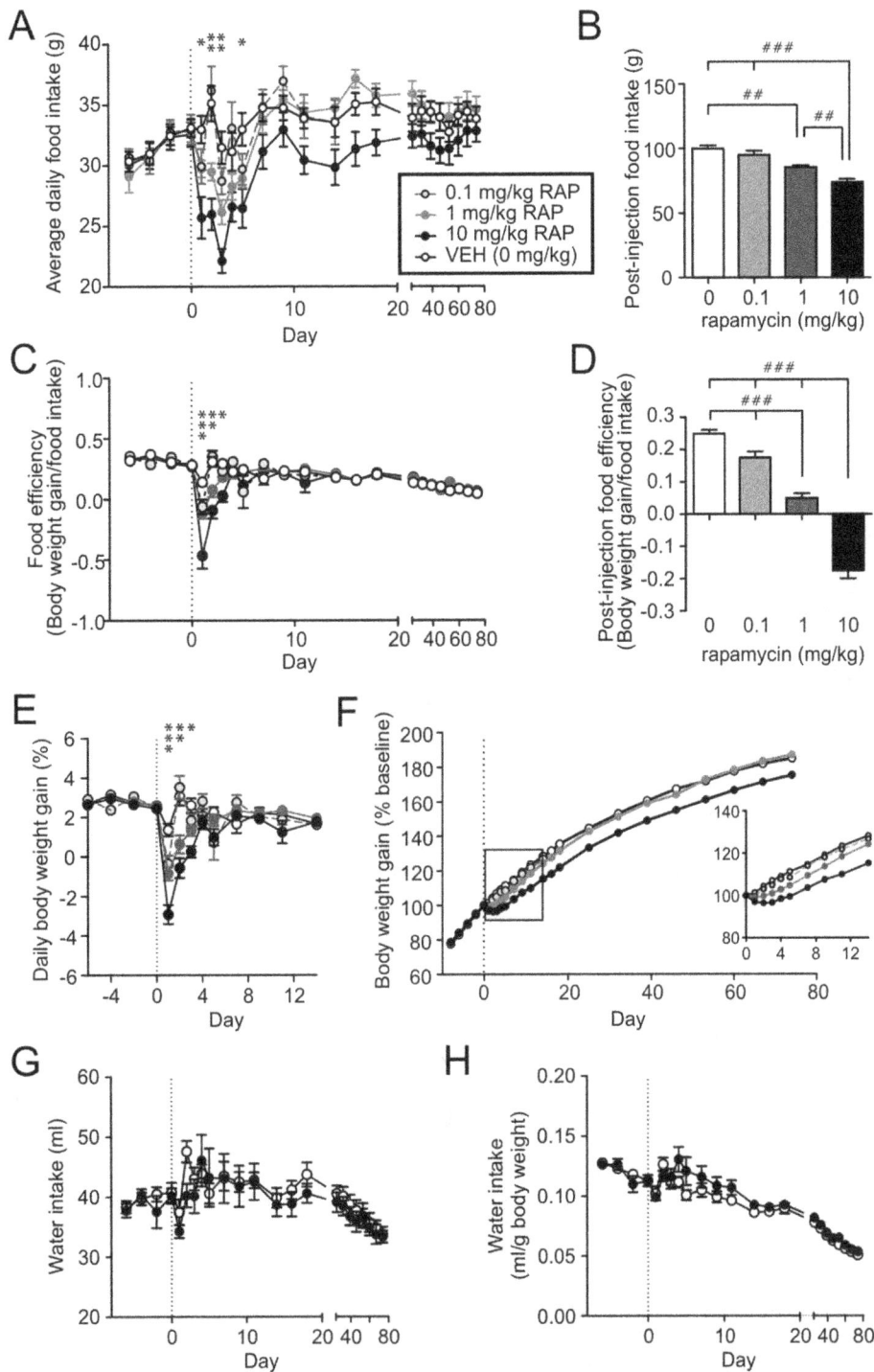

Figure 1. Single systemic injection of rapamycin induces prolonged decrease in body weight gain. A: Rapamycin (RAP) i.p. injection on Day 0 (vertical broken line) induces a transient decrease in daily food intake. B: Three-day cumulative food intake (Day 1–3 post-injection) shows a dose-dependent inhibition. C: RAP induces a transient decrease in food efficiency. D: Three-day cumulative food efficiency (Day 1–3) shows a dose-dependent suppression. E: RAP induces a transient decrease in daily weight gain. For panel A, C and E, *p<0.001 for 10 mg/kg vs.VEH; **p<0.05 for 1 and 10 mg/kg vs. VEH; ***p<0.05 for all RAP doses vs.VEH (two-way Mixed-ANOVA). For panels B and D, ##p<0.01, ###p<0.001 (one-way ANOVA with Tukey's test). F: Cumulative body weight gain curve depicting that RAP injection results in a downward shift in body weight. The first 2 weeks (box) is expanded and shown in the inset. The effect is dose-dependent. VEH vs.10 mg/kg, p<0.01 on Day 3–74; VEH vs. 1 mg/kg, p<0.01 on Day 2–11, p<0.05 on Day 14 and 18; VEH vs. 0.1 mg/kg, not significant (two-way Mixed-ANOVA). G, H: Averaged daily water intake (H) and water intake normalized to/body weight (G) shows no difference between RAP (10 mg/kg)-treated animals compared to VEH (two-way Mixed-ANOVA).

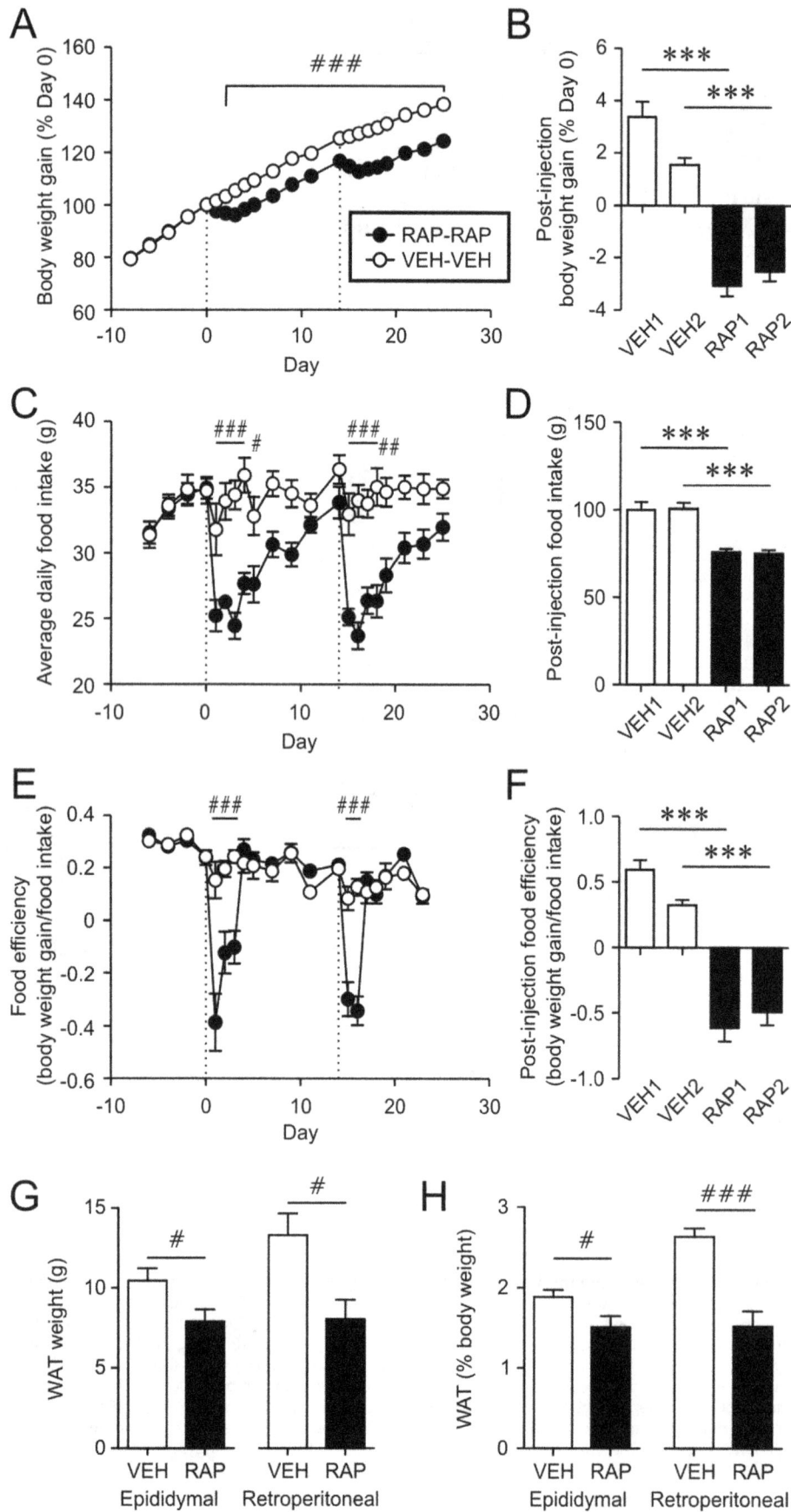

Figure 2. Spaced injections of rapamycin have additive effect on body weight gain. A–C: A, C and E: Body weight gain (A), daily food intake (C) and food efficiency (E) of rats given two i.p. injections (broken lines, Day 0 and 14) of RAP (10 mg/kg each) or VEH with a 2-week interval. # p<0.05, ## p<0.01, ### p<0.001, VEH vs. RAP (two-way Mixed-ANOVA). Horizontal bars indicate the days when significance was seen. B, D and F: Cumulative body weight gain, food intake and food efficiency during the first three days post-injection. ***p<0.001, VEH vs. RAP. There was no statistical difference between two injections within the group (VEH1 vs. VEH2, RAP1 vs. RAP2) (two-way Mixed-ANOVA). G: The weight of white adipose tissues (WAT), epididymal and retroperitoneal pads, in rats treated twice with VEH or RAP. H: WAT weight normalized to the body weight of individual rat. # p<0.05, ### p<0.001 (unpaired t-test).

decrease in daily body weight gain, which subsequently returned to baseline levels in all groups (Fig.1E). As a result, the difference in cumulative body weight gain persisted for up to 14 and 74 days for 1 mg/kg and 10 mg/kg, respectively (Fig.1F). Despite the persistent decrease in body weight, there was no difference in fluid intake across days (Fig. 1G, H) between VEH and RAP-treated animals. These results indicate that single RAP injection dose dependently induces sustained reduction in body weight.

Double rapamycin injections has additive effects on energy balance and reduces fat mass

There are reports of resistance to RAP in other experimental contexts [14]. Thus, we examined whether a RAP administration (10 mg/kg, i.p.) would affect responses to a subsequent RAP treatment. Rats received a pair of RAP or VEH i.p. injections (n = 10 each) with a 2-week interval between injections. We found that the two injections of RAP were equally effective in reducing FI, FE and weight gain (Fig.2A–F). The VEH group showed a tendency of lower weight gain and FE after the second injection compared to the first, which may be due to age-dependent slowing of the rate of weight gain (Fig.2B). In RAP treated animals (n = 6), the white adipose tissues were significantly smaller than those of VEH controls (n = 5) (Fig.2G, H), consistent with previous reports showing decreased adiposity following chronic RAP administration [6,7]. These data suggest that the RAP effect does not desensitize with intermittent injections, at least when injections are separated by 2 weeks, and it effectively reduces adiposity.

Possible side effects of single RAP injection

Since chronic RAP administration is known to induce glucose intolerance [6,15,16,17], we conducted a glucose tolerance test two weeks after single injection of RAP (10 mg/kg i.p., n = 8) or VEH (n = 7). There was no difference in the fasting blood glucose or response to glucose challenge between the two groups (Fig.3A). Furthermore, in a separate cohort of animals (n = 5 each), we

found no difference in non-fasting blood glucose levels at 2-week post-injection (Fig.3B). Therefore, a single RAP injection does not appear to influence glucose homeostasis long-term, unlike the glucose tolerance that develops with daily administrations of RAP over a 2-week period, as shown previously [6,15,16,17].

There is a possibility that the reduced eating from RAP could be due, at least in part, to sickness induced by the drug. The typical conditioned taste aversion (CTA) procedure provides a robust and sensitive test of drug-induced sickness, where animals are allowed to drink a novel-flavored solution following which they are injected with a drug. If the drug induced sickness, the animals would show a CTA later for that flavored solution. Rats were given 1 hour access to a novel saccharin solution (0.1% in water) followed immediately by either RAP (10 mg/kg i.p., n = 9) or VEH (n = 9). A two-bottle choice test was administered 2 days later when the rats were given simultaneous access to water and 0.1% saccharin solution. The total fluid intake from saccharin solution and water was greater in VEH than RAP rats (Fig. 4A). This is in contrast to the lack of effect of RAP on water intake (Fig. 1G, H), which may be due to the experimental condition of the CTA test, involving restricted fluid access and a choice of water and saccharin solution. Saccharin preference was calculated as the proportion of saccharin solution intake over total fluid intake. This test indicated no differences in saccharin preference between RAP and VEH treated rats (Fig.4B). In contrast, LiCl (i.p., n = 5) induced a robust decrease in saccharin preference compared to saline injection (n = 5; Fig. 4D), as expected [9]. Thus there was no evidence that RAP induced CTA, which suggests that RAP-induced anorexia is not due to illness.

Rapamycin lowers the defended level of body weight

Normally, caloric restriction and/or weight loss are followed by rebound hyperphagia and increased efficiency in food storage. However, RAP-treated animals did not display hyperphagia and body weight remained lower than controls following the acute

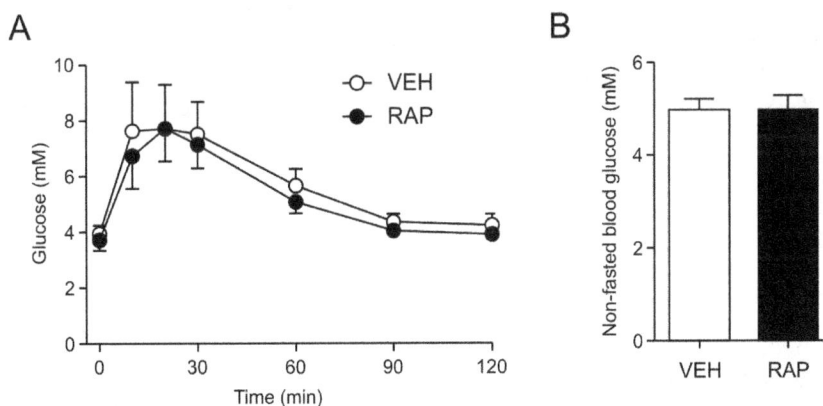

Figure 3. Rapamycin does not affect glucose tolerance. A: There were no differences in blood glucose levels measured at 15, 30, 60, 90 and 120 minutes post-glucose injection in RAP and VEH animals (two-way Mixed-ANOVA). B: There is no difference in non-fasted blood glucose in rats administered with VEH or RAP (10 mg/kg i.p.) at 2 weeks post-injection (unpaired t-test).

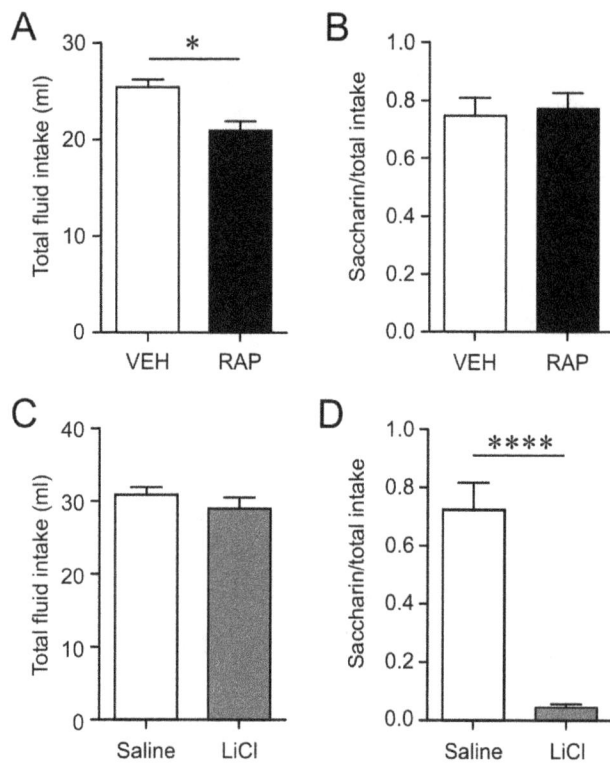

Figure 4. Rapamycin does not induce malaise or illness. A: During the two-bottle test, RAP group ingested significantly less fluid (sum of water and 0.1% Saccharin solution). *p<0.05 (unpaired t-test). B: There was no difference in saccharin preference (unpaired t-test). C: During the two-bottle test, there was no difference in total fluid intake in LiCl- and VEH-treated rats (sum of water and 0.1% Saccharin solution). D: LiCl-treated rats showed a significantly lower saccharin preference compared to VEH-treated rats. ****p<0.0001 (unpaired t-test)

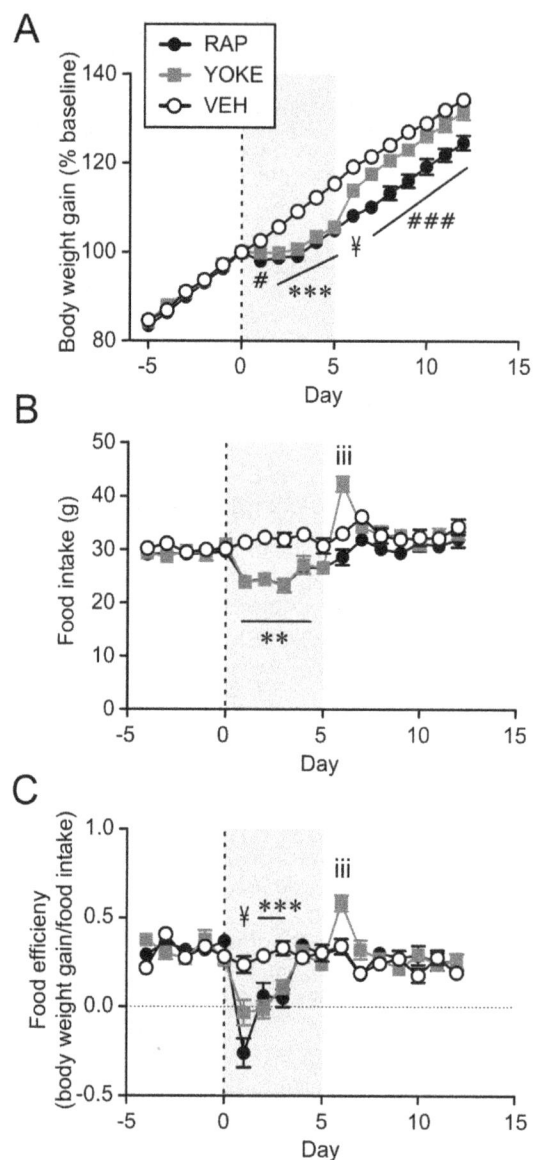

Figure 5. Rats pair-fed with RAP-treated animals show compensatory overfeeding and weight rebound. A: During the pair-feeding (shaded area), YOKE and RAP groups had lower weight gain compared to VEH group. YOKE group regained weight upon returning to *ad libitum* feeding. B and C: YOKE group show a transient increase in food intake (B) and efficiency (C) following pair-fed period. *p<0.05, ***p<0.001 VEH vs. YOKE and RAP; iii p<0.001 YOKE vs. VEH and RAP; ###p<0.001 RAP vs. VEH and YOKE; #p<0.05 VEH vs.RAP; ¥ p<0.05 all three groups are different from each other (two-way Mixed-ANOVA).

anorexic phase. This may be explained by the RAP-induced reduction in FI and body weight not being sufficient to engage counter-regulatory responses. To test this idea, a group of rats (YOKE) were pair-fed to match the daily FI of RAP-treated rats for the 5-day period following injection, but otherwise fed *ad libitum*. As expected, the RAP and YOKE groups had lower body weight gain and FI during the pair-feeding period compared to free-feeding VEH treated rats (n = 8 each, Fig.5A, B). RAP and YOKE groups did not differ in body weight during this period (p> 0.05). However, the YOKE group displayed an immediate rebound in FI and FE during the first day upon returning to free-feeding following the yoked period (Fig.5B, C). This suggests that the degree of anorexia and weight loss induced by RAP is sufficient to activate a counter-regulatory response in non-RAP treated animals.

Next, we sought to determine whether RAP prevented the development of compensatory responses to transient reduction in FI and weight. To do this, rats were treated i.p. with either RAP (10 mg/kg) or VEH, and then challenged with 24 h-food deprivation (FD), beginning at either 24 h or 2 weeks post-injection (immediate or late FD, respectively). At 24 h, acute effects of RAP are present, whereas at 2 weeks the acute effects would have subsided and only long-term effect on body weight remains. Immediate FD induced a transient drop in body weight in both RAP and VEH groups (n = 10 each, Fig.6A). Upon re-feeding, rats recovered their body weight. However, RAP rats

settled to a lower weight level than the VEH controls. Both groups showed a significant increase in FI post-FD compared to their respective pre-FD levels (p<0.001, paired t-test, Fig.6B). The magnitude of the increase was similar between groups (VEH 12.2±1.1 g, RAP 10.6±1.7 g; p>0.05, unpaired t-test). FE was also increased, although the RAP group showed a greater change in FE compared to the VEH group due to reduced FE on Day 1 post-injection (i.e. the day before FD; Fig.6C).

When FD was imposed 2 weeks after injection (late FD), both RAP and VEH groups showed a transient decrease in body weight (n = 10 each, Fig.6D), which recovered at an identical rate (Fig.6D

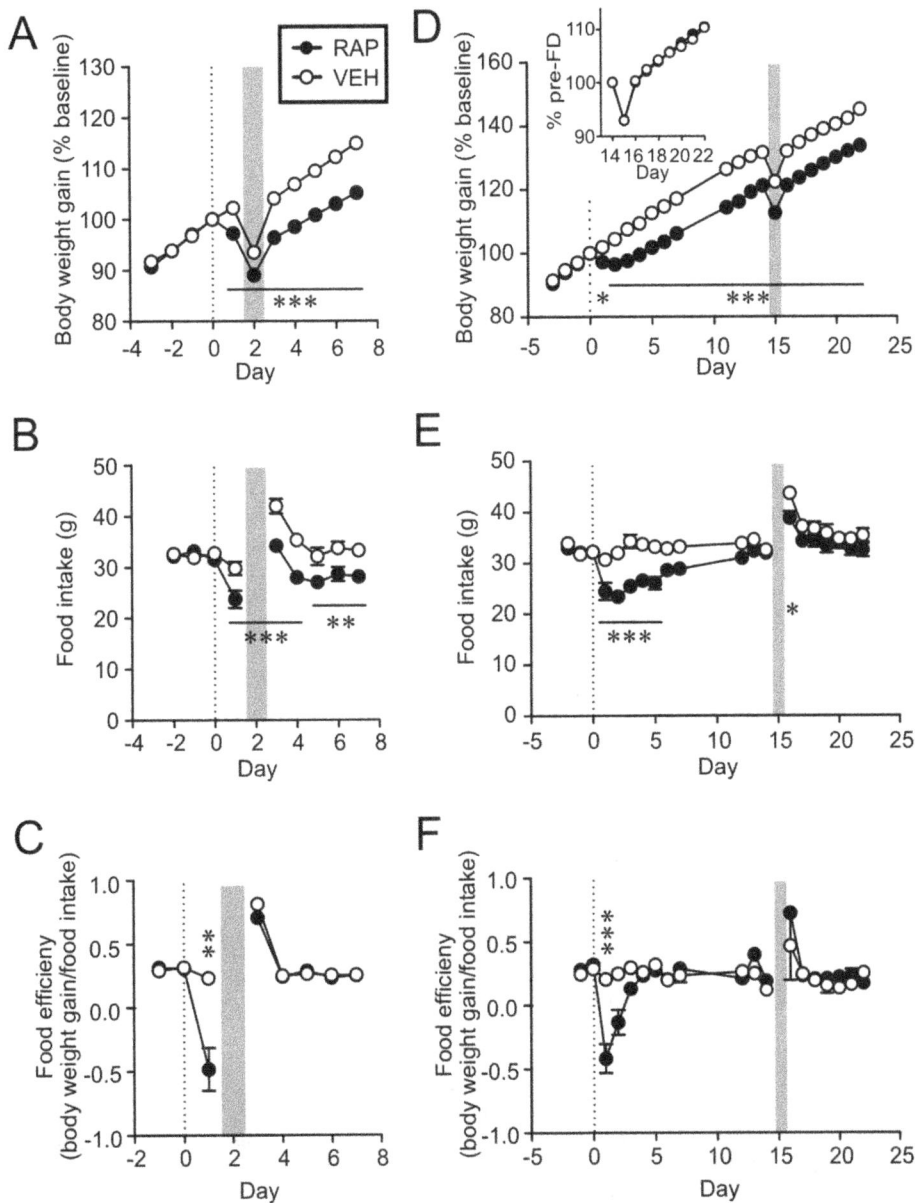

Figure 6. Rapamycin-treated animals defend lower body weight in response to acute perturbation in energy balance. A, B and C: Rats were injected i.p. with RAP or VEH on Day 0 (broken line), and then 24 h later food deprived (FD) as indicated by shaded area. In both groups, FD resulted in an immediate decline in body weight (A), followed by a transient increase in food intake (B) and food efficiency (C) upon refeeding. D, E and F: Rats were injected i.p. with RAP or VEH, then 2 weeks later challenged with FD (shaded area). D inset: Recovery rate of body weight following FD is identical. *$p < 0.05$, **$p < 0.01$, ***$p < 0.001$ (two-way Mixed-ANOVA). Horizontal bars in panel A, B, D and E indicate statistical significance at all time points labeled.

inset). There was a transient increase in FI and FE during re-feeding in both groups (Fig.6E, F). The increase in FI (pre- vs. post-FD) was less pronounced in the RAP group (VEH 11.1±1.0 g, RAP 6.8±0.7 g, p<0.005, unpaired t-test), which is likely due to the difference in absolute body weight, as there was no difference in FI normalized to body weight (VEH 0.100±0.002, RAP 0.098±0.003, p>0.05, unpaired t-test). These results strongly suggest that RAP-treated animals are capable of activating compensatory mechanisms to defend their body weight in response to acute perturbations, even during the early phase post-injection when rats do not attempt to recover the weight loss

by RAP. Therefore, it appears that RAP does not simply inhibit FI and FE, but rather lowers the defended level of body weight.

Intracerebroventricular rapamycin produces prolonged weight reduction

To determine whether the effect of systemic RAP that we observed was due to a central action, we conducted a central injection study. RAP i.c.v. (n = 9) reduced the daily weight gain, FI and FE transiently (Fig.7A,C,D) and cumulative weight gain for up to 15 days compared to those that received equal volume (1 µL) of VEH i.c.v. (n = 11) (Fig.7B). There was some delay in the effect; suppression of FI and cumulative weight gain became significant

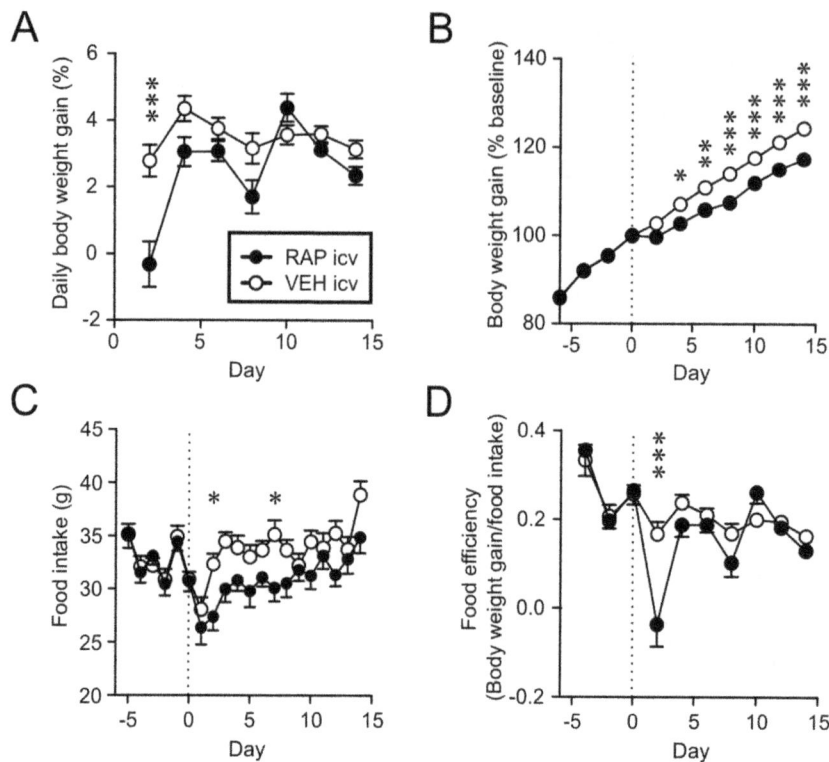

Figure 7. Rapamycin acts in the brain to decrease body weight gain. A and B: RAP i.c.v. injection (Day 0, broken line) induces a transient decrease in body weight gain (A), which results in prolonged shift in body weight (B). C: Food intake is inhibited by RAP i.c.v. D: Food efficiency is inhibited by RAP i.c.v. *p<0.05, **p<0.01, ***p<0.001 (two-way Mixed-ANOVA).

on Day 2 and Day 4 post-injection, respectively. This is similar to the delayed response seen after 1 mg/kg i.p. injection. These results suggest that the effect of RAP is at least partially mediated by the brain.

Discussion

The present study shows that single injection of RAP induces a transient decrease in FI, FE and daily weight gain lasting for several days. Surprisingly, the lowered body weight persists for at least 74 days. These effects are not likely due to malaise or illness, as there was no evidence for conditioned taste aversion to RAP. Once the transient effects subside and lower body weight is attained, RAP treated animals do not compensate for the lost weight, unlike pair-fed controls who overeat upon resuming *ad-libitum* feeding and regain weight. Instead, the rate of weight gain, FI and FE of RAP rats is restored after 3–5 days to the levels of vehicle controls, suggesting that energy homeostasis is re-established. Subsequently, the RAP-treated animals defend the newly established lower body weight upon acute perturbations in energy balance by adjusting FI and FE. Specifically, acute food deprivation induces a similar rebound hyperphagia upon re-feeding and an identical rate of body weight recovery in RAP and vehicle-treated groups. When food deprivation is applied within the first few days following RAP injection during the transient suppression of FI, rats are still able to respond with rebound hyperphagia. This indicates that the lack of hyperphagia following RAP is not due to a RAP-induced failure to activate counter-regulatory responses. Also this supports our contention that anorexia did not result from illness, as these animals are capable of eating as much as pre-injection levels. Taken together, our data

indicate that RAP induces a downward shift in body weight set point, and that the initial transient decrease in FI and FE following injection is a compensatory response to the disparity between the actual body weight and lowered set point. Therefore, our study shows a novel effect of RAP on body weight regulation.

While peripheral action cannot be ruled out, our findings strongly suggest that RAP acts centrally to exert its effect on energy homeostasis. It is known that peripherally administered RAP readily enters the brain [18] and inhibits p70S6K phosphorylation, a downstream substrate of mTOR signaling [19]. While the highest dose tested in our study (10 mg/kg) is higher than those of previous food and body weight studies (0.2–5 mg/kg/day) [6,7,8,10,20], 1 mg/kg in our study also produced a significant effect on long-term weight gain. Also, the highest dose used here is well below the RAP doses (e.g. 40 mg/kg) that are effective in other experimental contexts such as fear conditioning [12]. The elimination half-life of RAP is relatively long (approximately 30 hours in rats [21]), however, this would not appear to be of sufficient duration to suppress its target molecule for the entire 10-week observation period in the present study. Therefore, it is likely that a transient action of RAP is enough to induce a long-lasting change in the neural circuitry for body weight regulation. This may involve cap-dependent translation regulated by mTOR, which has been implicated in synaptic plasticity [22].

It is well established that mTOR plays an important role in the control of FI. Metabolic signals such as leptin and branched-chain amino acids (e.g., leucine) activate mTOR1 to inhibit FI [9]. Accordingly, a single central injection of RAP has been shown to increase FI transiently in sated rats by acting in the arcuate nucleus and nucleus of the solitary tract in the hypothalamus and brainstem, respectively [9,23]. This hyperphagic effect is short-lived; it gradually diminishes with time within the first day post-injection. We did not observe any increase in FI following systemic or central injection of RAP, which may be because our earliest time point was 24 h post-injection. The response we observed was delayed by one day in rats treated with 1 mg/kg i.p. or 50 µg i.c.v., which may be explained by hyperphagic and hypophagic responses balancing out during the first 24 h. Alternatively, the discrepancy may arise from the differences in feeding protocol. Previous studies induced satiation prior to RAP injection by overnight fast followed by re-feeding [23] or exposure to palatable food [9]. Under such conditions, neurons that mediate satiety may be fully activated and cannot be stimulated further. Another possibility is an involvement of a mechanism recruited by orexigenic factors such as ghrelin and thyroid hormone, which activates mTOR pathway and agouti related protein/neuropeptide Y neurons in the hypothalamus to induce FI [11,24]. Single central injection of RAP is sufficient to block the orexigenic effect of ghrelin [11]. Overall, these contrasting effects of mTOR and RAP on FI may involve different brain regions and/or neuronal populations [11,25].

Chronic RAP administration (daily injections) also has mixed effects on FI, which seems to depend on animal species, age, diet and duration of RAP treatment, although weight gain is consistently inhibited [6,7,8,10,20]. In aged mice, mTOR promotes positive energy balance by negatively regulating the activity of POMC neurons in the arcuate nucleus, which is reversed by chronic RAP, suggesting an important role of POMC neurons in the central RAP effect, at least with chronic administration [10]. It is likely that at least some of the known

effects of chronic RAP treatment overlap with those seen in our acute injection study, and reduced body weight may not be readily reversible even when the RAP treatment is terminated.

Manipulation of set point would be an effective weight loss and maintenance strategy, as deviating from the set point normally results in strong activation of physiological compensatory mechanisms. We showed that acute injection of RAP, which presumably induces a transient suppression of mTOR, can have a long-lasting effect on the set point for body weight, suggesting a novel role of mTOR in body weight regulation. Moreover, a single injection has distinct advantage as it can avoid side effects of chronic RAP administration such as glucose intolerance. We propose that RAP and related compounds could be used as tools to investigate how the defended level (apparent set point) of body weight is determined and to complement other weight loss strategies.

Acknowledgments

Special thanks to Dr. Malcolm Grant for help with statistical analysis and Kimberly Williams for her technical assistance, and Gene Hertzberg for his consideration of and enthusiasm for these experimental findings.

Author Contributions

Conceived and designed the experiments: JB M. Hebert M. Hirasawa VG CM. Performed the experiments: M. Hebert ML BJ AB SM. Analyzed the data: M. Hebert M. Hirasawa. Contributed reagents/materials/analysis tools: JB M. Hirasawa VG RA. Wrote the paper: M. Hebert M. Hirasawa JB.

References

1. Bray GA, Tartaglia LA (2000) Medicinal strategies in the treatment of obesity. Nature 404: 672–677.
2. Leibel RL, Rosenbaum M, Hirsch J (1995) Changes in energy expenditure resulting from altered body weight. N Engl J Med 332: 621–628.
3. Wadden TA, Sternberg JA, Letizia KA, Stunkard AJ, Foster GD (1989) Treatment of obesity by very low calorie diet, behavior therapy, and their combination: a five-year perspective. Int J Obes 13 Suppl 2: 39–46.
4. Ryan KK, Woods SC, Seeley RJ (2012) Central nervous system mechanisms linking the consumption of palatable high-fat diets to the defense of greater adiposity. Cell Metab 15: 137–149.
5. Zhou H, Huang S The complexes of mammalian target of rapamycin. Curr Protein Pept Sci 11: 409–424.
6. Fang Y, Westbrook R, Hill C, Boparai RK, Arum O, et al. (2013) Duration of rapamycin treatment has differential effects on metabolism in mice. Cell Metab 17: 456–462.
7. Deblon N, Bourgoin L, Veyrat-Durebex C, Peyrou M, Vinciguerra M, et al. (2012) Chronic mTOR inhibition by rapamycin induces muscle insulin resistance despite weight loss in rats. Br J Pharmacol 165: 2325–2340.
8. Chang GR, Chiu YS, Wu YY, Chen WY, Liao JW, et al. (2009) Rapamycin protects against high fat diet-induced obesity in C57BL/6J mice. J Pharmacol Sci 109: 496–503.
9. Cota D, Proulx K, Smith KA, Kozma SC, Thomas G, et al. (2006) Hypothalamic mTOR signaling regulates food intake. Science 312: 927–930.
10. Yang SB, Tien AC, Boddupalli G, Xu AW, Jan YN, et al. (2012) Rapamycin ameliorates age-dependent obesity associated with increased mTOR signaling in hypothalamic POMC neurons. Neuron 75: 425–436.
11. Martins L, Fernandez-Mallo D, Novelle MG, Vazquez MJ, Tena-Sempere M, et al. (2012) Hypothalamic mTOR signaling mediates the orexigenic action of ghrelin. PLoS One 7: e46923.
12. Blundell J, Kouser M, Powell CM (2008) Systemic inhibition of mammalian target of rapamycin inhibits fear memory reconsolidation. Neurobiol Learn Mem 90: 28–35.
13. Fifield K, Hebert M, Angel R, Adamec R, Blundell J (2013) Inhibition of mTOR kinase via rapamycin blocks persistent predator stress-induced hyperarousal. Behav Brain Res 256: 457–463.
14. Huang S, Houghton PJ (2001) Resistance to rapamycin: a novel anticancer drug. Cancer Metastasis Rev 20: 69–78.
15. Fraenkel M, Ketzinel-Gilad M, Ariav Y, Pappo O, Karaca M, et al. (2008) mTOR inhibition by rapamycin prevents beta-cell adaptation to hyperglycemia and exacerbates the metabolic state in type 2 diabetes. Diabetes 57: 945–957.
16. Houde VP, Brule S, Festuccia WT, Blanchard PG, Bellmann K, et al. (2010) Chronic rapamycin treatment causes glucose intolerance and hyperlipidemia by upregulating hepatic gluconeogenesis and impairing lipid deposition in adipose tissue. Diabetes 59: 1338–1348.
17. Lamming DW, Ye L, Katajisto P, Goncalves MD, Saitoh M, et al. (2012) Rapamycin-induced insulin resistance is mediated by mTORC2 loss and uncoupled from longevity. Science 335: 1638–1643.
18. Yanez JA, Forrest ML, Ohgami Y, Kwon GS, Davies NM (2008) Pharmacometrics and delivery of novel nanoformulated PEG-b-poly(epsilon-caprolactone) micelles of rapamycin. Cancer Chemother Pharmacol 61: 133–144.
19. Erlich S, Alexandrovich A, Shohami E, Pinkas-Kramarski R (2007) Rapamycin is a neuroprotective treatment for traumatic brain injury. Neurobiol Dis 26: 86–93.
20. Wang CY, Kim HH, Hiroi Y, Sawada N, Salomone S, et al. (2009) Obesity increases vascular senescence and susceptibility to ischemic injury through chronic activation of Akt and mTOR. Sci Signal 2: ra11.
21. Yatscoff RW, Wang P, Chan K, Hicks D, Zimmerman J (1995) Rapamycin: distribution, pharmacokinetics, and therapeutic range investigations. Ther Drug Monit 17: 666–671.
22. Hoeffer CA, Klann E (2010) mTOR signaling: at the crossroads of plasticity, memory and disease. Trends Neurosci 33: 67–75.
23. Blouet C, Schwartz GJ (2012) Brainstem nutrient sensing in the nucleus of the solitary tract inhibits feeding. Cell Metab 16: 579–587.
24. Varela L, Martinez-Sanchez N, Gallego R, Vazquez MJ, Roa J, et al. (2012) Hypothalamic mTOR pathway mediates thyroid hormone-induced hyperphagia in hyperthyroidism. J Pathol 227: 209–222.
25. Zhang W, Zhang C, Fritze D, Chai B, Li J, et al. (2013) Modulation of food intake by mTOR signalling in the dorsal motor nucleus of the vagus in male rats: focus on ghrelin and nesfatin-1. Exp Physiol 98: 1696–1704.

7

Tracking of a Dietary Pattern and Its Components over 10-Years in the Severely Obese

David J. Johns[1]*, **Anna Karin Lindroos**[3], **Susan A. Jebb**[1], **Lars Sjöström**[2], **Lena M. S. Carlsson**[2], **Gina L. Ambrosini**[1]

1 Medical Research Council Human Nutrition Research, Elsie Widdowson Laboratory, Cambridge, United Kingdom, **2** Institute of Medicine, University of Gothenburg, Gothenburg, Sweden, **3** The National Food Administration, Uppsala, Sweden

Abstract

Understanding how dietary intake changes over time is important for studies of diet and disease and may inform interventions to improve dietary intakes. We investigated how a dietary pattern (DP) tracked over 10-years in the Swedish Obese Subjects (SOS) study control group. Dietary intake was assessed at multiple time-points in 2037 severely obese individuals (BMI 41 ± 4 kg/m^2). Reduced rank regression was used to derive a dietary pattern using dietary energy density (kJ/g), saturated fat (%) and fibre density (mg/kJ) as response variables and score respondents at each follow-up. Tracking coefficients for the DP, its key foods and macronutrient response variables and corrected for time-dependent and time-independent covariates were calculated using generalised estimating equations to take into account all available data. The DP tracking coefficient was moderate for women (0.40; 95% CI: 0.38–0.42) and men (0.38; 95% CI: 0.35–0.41). Of the eleven foods key to this DP, fruit and vegetable intakes had the strongest tracking coefficient for both sexes. Fast food and candy had the lowest tracking coefficients for women and men respectively. Scores for an energy dense, high saturated fat, low fibre density DP appear moderately stable over a 10-year period in this severely obese population. Furthermore, some food groups appear more amenable to change while others, often the most healthful, appear more stable and may require intervention before adulthood.

Editor: Michael Müller, University of East Anglia, United Kingdom

Funding: This work was funded by a programme grant from the UK Medical Research Council (U105960389 Nutrition and Health) and a MRC PhD studentship. The funders had no role in study design, data collection and analysis, decision to publish, or preparation of the manuscript.

Competing Interests: The authors have declared that no competing interests exist.

* E-mail: david.johns@mrc-hnr.cam.ac.uk

Introduction

In epidemiological literature tracking is used to describe the stability of the longitudinal development of a certain outcome variable [1]. Understanding how dietary intake changes over time i.e. its stability or tracking, can help to in identify targets for interventions to improve diet quality. Individual food groups with strong tracking of dietary intake are likely less amenable to change than those with weak tracking. Tracking of dietary intake over time in different populations also has methodological implications for cohort design and dietary assessment.

Little is known about longitudinal tracking of dietary intake in adults. Some work has previously been conducted looking at the tracking of food intake in children and adolescents [2,3,4,5,6] and their transition into adulthood [7,8]. Few studies have considered changes in dietary patterns over time [9,10,11].

To our knowledge, no studies examining changes in dietary intake over time have been conducted in obese populations. This is despite weight-loss attempts being more prevalent and frequent in this group [12,13]. Understanding the role of tracking in a severely obese population will provide an insight into the opportunities for lifestyle interventions in this population.

Dietary patterns allow us to account for total dietary intake and the various combinations of food and nutrients typically consumed [14]. However, few studies have considered how dietary patterns change over time in adults using more than two time points [15,16,17] and even fewer have made use of all the available data in longitudinal analysis [16].

Reduced rank regression (RRR) is a dietary pattern method that balances hypothesis and exploratory driven aspects [18,19]. It creates patterns in food groups that optimise the amount of variation in a set of variables chosen by the researcher using a priori knowledge.

The aim of this study was to investigate tracking of dietary intake in a severely obese population over a 10-year period, in particular, a hypothesis driven dietary pattern identified using RRR. The tracking of intakes of selected food groups and nutrients was also examined.

Materials and Methods

Ethics statement: Written Informed consent was obtained for all study participants. The Swedish Obese Subjects (SOS) study has been conducted according to the principles expressed in the Declaration of Helsinki. The SOS study protocol was approved by the Research Ethics Committee of University of Gothenburg and seven other Swedish regional ethics review boards, each harbouring one or several of the involved study sites. The SOS trial has been registered in the ClinicalTrials.gov registry (NCT01479452, http://clinicaltrials.gov/ct2/show/NCT01479452?term).

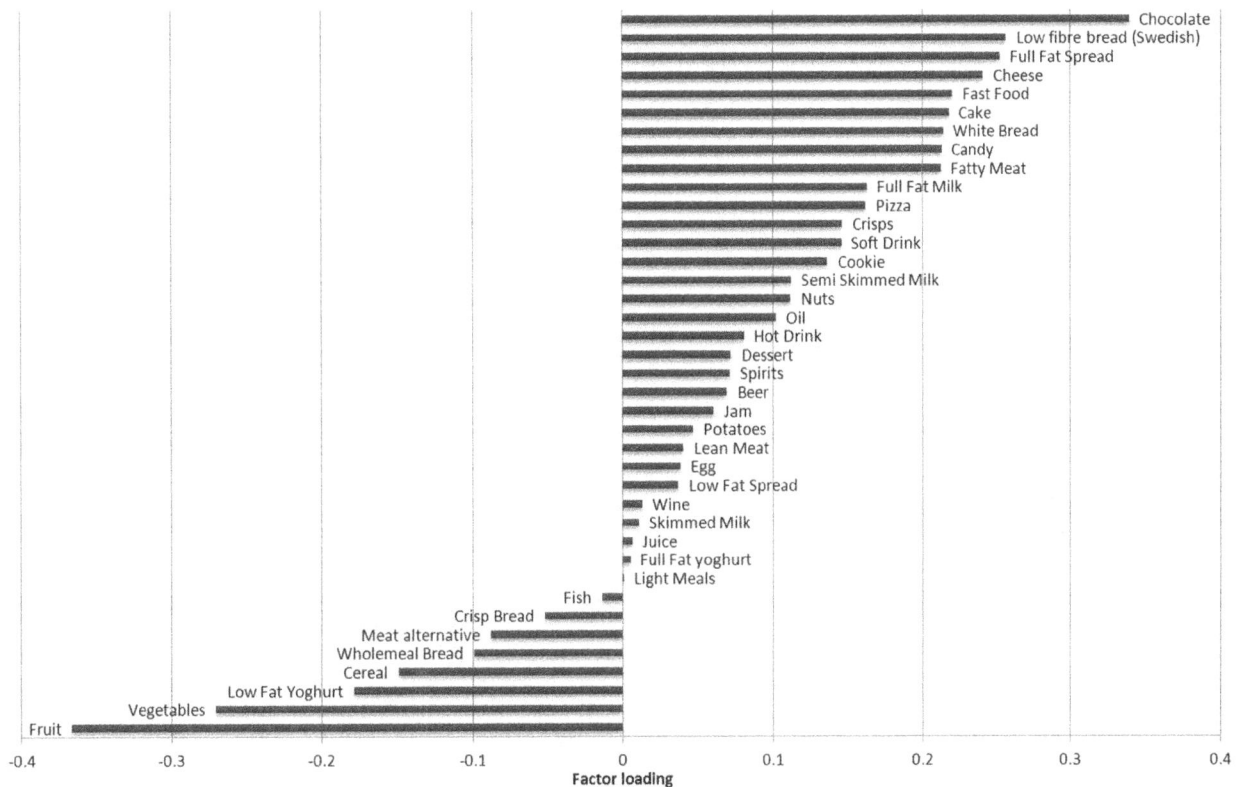

Figure 1. Factor loadings for the first dietary pattern high in energy density, high in percentage saturated fat and low in fibre density.

Data were from the Swedish Obese Subjects Study (SOS) study. Full details of the study have been reported elsewhere [20]. In brief, SOS is a prospective intervention study investigating the impact of bariatric surgery on morbidity and mortality in severely obese adults. SOS recruited 6905 severely obese individuals between 1987 and 2001 in Sweden, all of whom underwent health examinations and completed multiple questionnaires including diet. Of this registry population, 2010 individuals chose to have bariatric surgery and were matched with 2037 control who received no standardised treatment [20].

The current analysis uses data from the control group only (n = 2037). Participants completed medical, dietary and health related quality-of-life questionnaires as well as health examinations at registration, baseline (t = 0), 0.5, 1, 2, 3, 4, 6, 8, and 10-years. Data collected up to 16th June 2009 is included.

Dietary Assessment

Dietary data were collected using a validated, semi-quantitative diet questionnaire. This questionnaire was adapted from a diet history interview developed for the general population in Sweden, based on clinical experience of the problematic eating characteristics of obese individuals. The questionnaire included 50 questions with additional sub-questions intended to add detail and clarification. The questionnaire, described in more detail elsewhere [21], has been validated in both obese and non-obese groups [21,22]. The correlations between the questionnaire and food records for reported intakes of dietary fibre, saturated fat and total energy were 0.59, 0.62 and 0.45 respectively [21]. The reproducibility of dietary fibre, saturated fat and energy intake was also assessed with correlation coefficients of 0.61, 0.72 and 0.73

respectively [21]. The dietary questionnaire was completed at registration for the study (t = R), at commencement of the study (on average 13 months later, t = 0) and then at 0.5, 1, 2, 3, 4, 6, 8, and 10 years later.

All completed food questionnaires were linked to Swedish food composition tables to provide nutrient information. For this analysis, 39 food groups were defined based on nutrient content and usual culinary usage (Table S1). Average intakes of each food group (g/day) were calculated at each time-point and standardised relative to data collected at study registration ((Intake − mean (t = R))/SD(t = R)).

Dietary Patterns

Dietary patterns were firstly derived using available data collected at registration (n = 6869). The RRR model (PROC PLS in Statistical Analysis Software (version 9.3, SAS Institute Inc)) included dietary energy density (DED), fibre density (FD) and saturated fat intake (% of total energy) as response variables. These variables are the focus of current recommendations from healthcare bodies and governments for weight-loss and improving diet quality conducive to cardiovascular health [23,24,25]. The 39 food groups were used as predictors.

Energy density (kJ/g) was calculated as food energy (kJ) divided by food weight (g). Food was defined as solid food and liquids consumed as food (for example, soups and yoghurt). All beverages, both energy-containing (alcoholic drinks, milk, sweetened drinks and fruit juices) and non-energy-containing (water, coffee and diet beverages), were excluded from this calculation [26]. The percentage energy from saturated fat was calculated by dividing daily energy from saturated fat (kJ) by total daily energy intake (kJ)

Table 1. Mean (±SD) measures and dietary characteristics of the 2037 individuals at their registration to the study (t = R).

Measure	Male	Female
	n = 590	*n = 1447*
Age (yrs)	46.8±5.8	47.6±6.2
Smoking (%)	23.7	18.8
Weight (kg)	126.8±14.4	112.9±13.9
Height (m)	1.80±0.07	1.65±0.06
BMI (kg/m²)	39.2±4.1	41.6±4.2
Total Energy intake (kJ)	13240±4802	11334±4697
Saturated fat intake (%)	16.0±3.1	16.5±3.1
Fibre density (mg/kJ)	1.93±0.60	2.24±0.62
Dietary energy density (kJ/g)	7.63±1.55	7.12±1.55

then multiplying by 100. FD (mg/kJ) was calculated by dividing dietary fibre intake (mg) by total energy intake (kJ).

Three dietary patterns were extracted at t = R. Every subject received a z-score for the dietary pattern to quantify how their reported dietary intake reflected the pattern. Dietary patterns were similarly derived at each follow-up to ensure no new patterns appeared during the 10-year follow-up; no new patterns were identified. Only the dietary pattern explaining most variation in response variables was of interest.

To investigate tracking of the same dietary pattern over the 10-year follow up, an applied dietary pattern z-score was calculated for t = 0, 0.5, 1, 2, 3, 4, 6, 8, and 10 years using the scoring

Table 2. Tracking coefficient and 95% CI of a standardised dietary pattern score, reported food intake and macronutrient intake of 2037 severely obese Swedish men and women.

	Men		Women	
	Tracking coefficient	**95% CI**	**Tracking coefficient**	**95% CI**
Dietary pattern score (time points)				
Dietary Pattern score (R-10)	0.38*	0.35–0.41	0.40*	0.38–0.42
Dietary pattern score (0–10)	0.39*	0.36–0.41	0.45*	0.43–0.47
Dietary Pattern score (2–10)	0.45*	0.31–0.49	0.50*	0.44–0.55
Dietary Pattern score (6–10)	0.47*	0.38–0.57	0.53*	0.46–0.61
Food groups (t = R to 10)				
Vegetables	0.52	0.48–0.56	0.37	0.35–0.39
Fruit	0.46	0.43–0.49	0.36	0.34–0.38
Chocolate	0.23	0.21–0.25	0.25	0.24–0.27
Swedish sweet bread (low-fibre)	0.25	0.23–0.27	0.28	0.27–0.29
Full Fat spread (butter)	0.19	0.09–0.29	0.25	0.21–0.30
Cheese	0.28	0.25–0.30	0.20	0.18–0.21
Fast Food	0.25	0.22–0.27	0.14	0.13–0.15
Cake	0.21	0.19–0.23	0.24	0.23–0.25
White Bread	0.24	0.19–0.30	0.26	0.22–0.30
Fatty Meat	0.23	0.20–0.25	0.27	0.25–0.28
Candy	0.10	0.08–0.12	0.18	0.17–0.20
DP response variables (t = R to 10)				
Total Energy	0.38*	0.36–0.41	0.38*	0.36–0.39
Saturated fat %	0.37	0.35–0.40	0.35	0.33–0.37
Fibre density	0.45	0.41–0.48	0.47	0.45–0.49
Dietary Energy Density	0.35	0.33–0.38	0.38	0.36–0.39

*Adjusted for age and smoking. All other coefficients are adjusted for age, smoking and total energy intake.

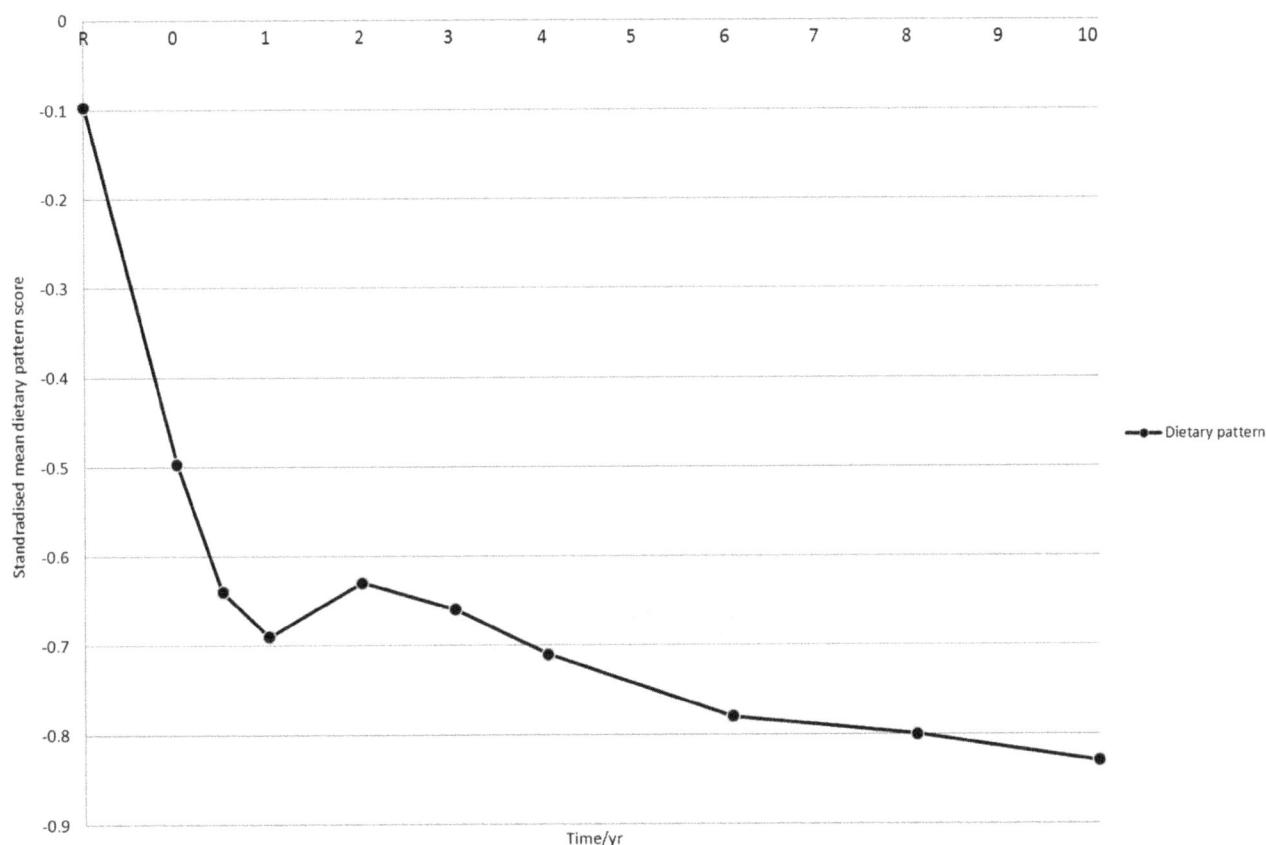

Figure 2. Mean dietary pattern z-score* over 10 years. * z-scores from t0 – t10 were standardised to intakes at t = R.

coefficients produced by the RRR analysis of registration data [6,27]. The applied z-score reflected adherence to the dietary pattern observed at registration. This consistency in measurement over time provides a better variable for longitudinal measurement than using z-scores from similar dietary patterns derived from an exploratory analysis conducted at each time point.

Statistical Analyses

Tracking (or stability) coefficients were calculated for dietary pattern z-scores using a generalised estimating equation (GEE) model [28].

The model regressed repeated measurements of the DP z-score against the DP z-score at registration, adjusting for time between each measurement. As the relationship between the initial DP score and subsequent DP scores is tested simultaneously, the standardised regression coefficient (β_1) can be interpreted as a tracking coefficient [28]. Adjustments were included for time-dependent covariates and time-independent covariates.

GEE was carried out using the GENMOD procedure in SAS v9.3. As dietary intake is included as z-scores, the standardised tracking coefficient ranges from 0 to 1, with 1 indicating perfect tracking and 0 indicating no tracking. There are no universally accepted cut offs to classify good or poor tracking, as the magnitude of the tracking coefficient can depend on the length of follow up and measurement error in the variable being tracked [28]. However, it is possible to contrast tracking coefficients observed within the same study, and we considered weak tracking coefficients as ≤0.3; moderate tracking 0.3–0.6; and high tracking ≥0.6 [29,30].

Interactions between the tracking coefficient and gender were observed (p<0.0001) and so tracking coefficients were estimated separately for men and women. To test the effect of different durations of follow-up, tracking coefficients were estimated between registry (t = R) and 10-years, baseline (t = 0) and 10-years, from 2 to 10-years and from 6 to 10-years. Age, and smoking were examined as time-varying covariates. A similar GEE model was applied to estimate tracking coefficients for key food group intakes and the macronutrients chosen as response variables, which were included in the models as standardised intakes (z-scores). Only those food groups with a factor loading ≥ 0.2 or ≤ −0.2 were examined (Figure 1).

Because of potential differences in dietary habits and dietary reporting, tracking analyses were conducted separately for men and women. Only those respondents who completed a food diary on at least two follow ups were included in the analyses.

Results

A total of 2037 participants had dietary data from at least two follow-ups with data from 2037, 1771, 1766, 1667, 1544, 1477, 1364, 1219 and 1131 individuals at t = 0, 0.5, 1, 2, 3, 4, 6, 8, 10 respectively. At registration, the study population had a greater proportion of women (71%) than men (29%) and a mean age of 47.6±6.2 and 46.8±5.8 years respectively for women and men (Table 1). BMI was higher in women (41.6±4.2 kg/m²) than men (39.2±4.12 kg/m²) due to the SOS inclusion criteria [31].

The first of the three derived dietary patterns explained most of the variance in the response variables (54%). The remaining

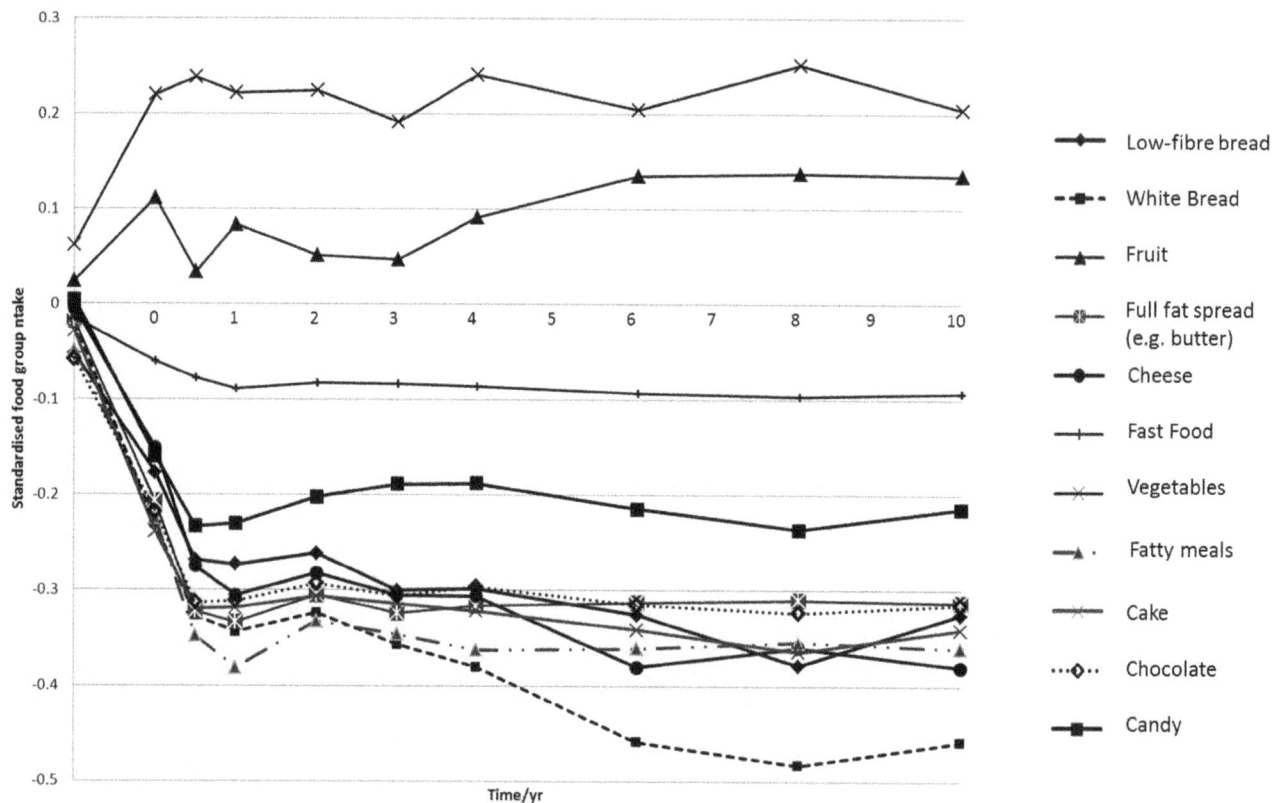

Figure 3. Mean intakes of standardised food group z-scores* over 10 years. *z-scores from t0– t10 were standardised to intakes at t = R.

two patterns combined explained only 26% of the total variation in response variables and were therefore not included in the tracking analysis. The first pattern explained 31% of the variance in percentage energy from saturated fat, 60% of the variance in fibre density and 71% of the variance in dietary energy density. This pattern was negatively correlated with fibre density $(r = -0.61)$ and positively correlated with percentage energy from saturated fat $(r = 0.44)$ and dietary energy density $(r = 0.66)$. It was therefore labelled an 'energy-dense, high saturated fat, low fibre dietary pattern'. A high dietary pattern z-score was characterised by higher intakes of chocolate, low-fibre bread, cheese, fast food and cake and lower intakes of fruit and vegetables (Figure 1).

The tracking coefficient (β_1) for the energy-dense, high saturated fat, low fibre dietary pattern over the entire study period was moderate, at 0.40 (95% CI: 0.38–0.42) for women and 0.38 (95% CI: 0.35–0.41) for men. The DP tracking coefficients were stronger when follow up time was shorter in women (Table 2). Tracking coefficients were consistently stronger in women than men.

Tracking coefficients for the food group intakes were generally lower (Table 2). The highest coefficients for men and women were for vegetable $(\beta_1 = 0.52$ [95% CI: 0.48–0.56] and $\beta_1 = 0.37$ [95% CI: 0.35–0.39] respectively) and fruit $(\beta_1 = 0.46$ [95% CI: 0.43–0.49] and $(\beta_1 = 0.36$ [95% CI: 0.34–0.38]) intake. The lowest coefficients were for intake of fast food in women $(\beta_1 = 0.14$ [95% CI: 0.13–0.15]) and candy in men $(\beta_1 = 0.10$ [95% CI: 0.08–0.12]). Tracking coefficients for macronutrient response variables were similar to the DP ranging from 0.35 to 0.47. The tracking coefficients for macronutrients and foods did not alter with different follow-up times (not shown).

Population mean z-scores for the dietary pattern over 10-years are displayed in Figure 2. Between study registration and one year after baseline, there was a marked decrease in z-score. Population mean z-scores for food groups (Figure 3) and macronutrients (Figure 4) were also plotted. The observed changes in mean intakes correspond with the changes observed in dietary pattern scores.

Discussion

In this cohort of severely obese adults, an energy dense, high saturated fat, low fibre dietary pattern showed moderate levels of tracking over a 10-year period. However, intakes of food groups important to this dietary pattern showed varying degrees of tracking. This suggests that dietary intake among severely obese adults is not fixed and may be subject to change.

The differences in tracking among food groups indicate that some food intakes are less stable over time, than others. The weaker tracking coefficients for fast food and candy highlight that these foods are more susceptible to change than others and this weaker tracking may be in part due to repeated attempts to lose weight and eliminate these 'treat' foods. While food groups with weaker levels of tracking are likely appropriate, modifiable targets for interventions, they may also represent food groups where changes are short term and difficult to maintain. Fruit and vegetable intakes had stronger tracking coefficients, a possible result of strong habit formation and food preference. This is consistent with another study that showed while men made a number of positive dietary changes as part of a weight-loss intervention, increased vegetable intake was least likely [32].

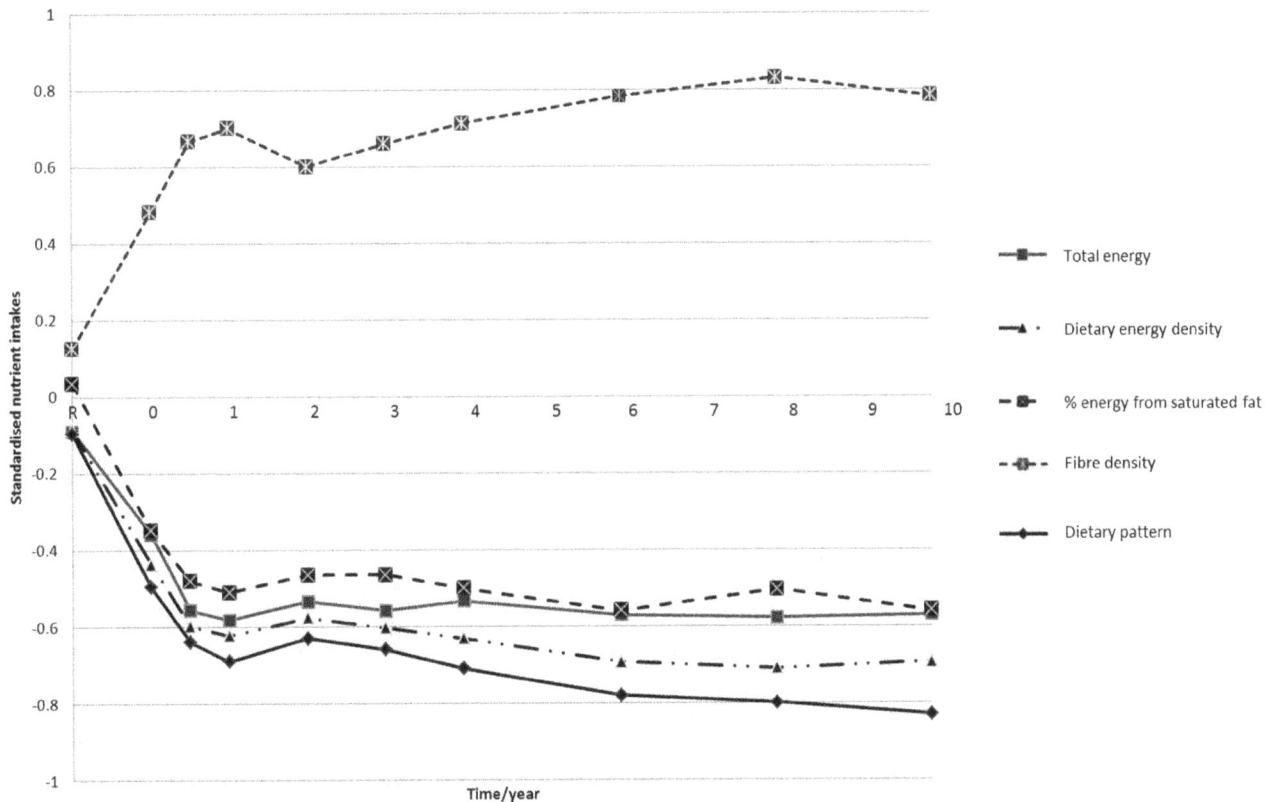

Figure 4. Mean dietary pattern z-score and mean total energy intake, percentage energy from saturated fat, dietary energy density and fibre density z-scores* over 10 years. *z-scores from t0– t10 were standardised to intakes at t = R.

The changes observed in food intakes in this cohort may be explained by a number of factors. For instance, in the waiting period between registration and baseline, participants showed improvements in cardiovascular disease risk factors [33]. This is likely to be a 'study effect' and may also be responsible for initial dietary changes as individuals' have demonstrated a willingness to change by expressing interest in the study. Secondly, it may be that diagnosis of disease may be a precursor to dietary changes in this high risk population. For example, despite receiving no intervention, the control group has demonstrated a rate of recovery from diabetes at 2 years of 21% and at 10 years of 13% [20]. It is likely these recoveries are influenced by dietary and lifestyle changes.

To our knowledge, no other study has looked at the tracking of dietary patterns over time in severely obese adults. Two studies have investigated the tracking of nutrient and food intakes in the Amsterdam Growth and Health Study (AGHS) using similar methods. Post *et al.* [7] analysed repeated dietary measures in 200 individuals over a period of 20 years spanning adolescence and adulthood. The tracking coefficients for total energy (β1 = 0.33; 0.22–0.4) and percentage energy from saturated fat (β1 = 0.41; 0.32–0.50) were similar to that observed in our analysis. In another analysis of AGHS data, Velde *et al.* tracked fruit and vegetable intake in 168 men and women between the ages of 12 and 36 years [34]. Their tracking coefficient for fruit was 0.33 (95% CI: 0.25–0.41); similar to the tracking coefficient for SOS women but weaker than SOS men. Velde's tracking coefficient for vegetables was 0.27 (95% CI: 0.19–0.36), much lower than that of the current study [34]. However, these studies are not directly comparable as partic-

ipant numbers, population type and length between follow up measurements may influence tracking coefficients [1]. Despite being over a longer follow-up, the studies in the AGHS, had fewer participants and measurements. This may contribute to the differences observed between the two studies.

This is one of very few studies to track a dietary pattern that was identified using reduced rank regression. The SOS study is one of the largest cohorts of severely obese individuals with repeated dietary measures collected using consistent data collection methods. Despite this, some study limitations should be noted. Firstly, it is possible the categorisation of food groups has influenced the degree to which foods appear to track. For example, it may be easier to swap intake of foods in the 'Candy' group with foods in another food group than it is for groups such as fruits or vegetables. We cannot rule out recall bias or dietary misreporting, which is common, to some degree, to all dietary assessments, and this may have influenced some tracking coefficients. However, the dietary assessment in the SOS study was adapted for the severely obese population to minimise biases due to the nature of the population. The levels of under-reporting in the SOS dietary assessment tool are no greater than those seen in the general population [21]. As the study occurred over ten years, a potential limitation is loss to follow up as is frequently observed in longitudinal studies. The loss to follow-up was greatest at 10-years with a 56% loss. However, this remains a large cohort of severely obese individuals at 10-years with 1130 responses. Furthermore, the application of longitudinal models to estimate the tracking coefficients utilised all available data at each follow up, thus avoiding some of the potential bias associated with 'completers only' analyses. By comparison, most studies reporting dietary tracking have been limited to only two measurements over time and the use of more simple statistical methods e.g. Pearson's correlation and measures of agreement (e.g. Kappa statistics) as indicators of dietary tracking [35]. Finally, while we considered the influence of gender, age and smoking, it is possible that there are other important influences on dietary tracking that were not considered in this study.

The varying degrees of tracking of diet in this population, highlights the importance of repeated measures in cohort studies. A single measure of diet at baseline is likely to be a poor proxy for total diet over later years or decades. Further work is needed in a variety of cohorts to better understand the stability of diet in different adult populations.

In conclusion, an energy dense, high saturated fat, low fibre density dietary pattern and selected food and nutrient intakes track moderately over a 10-year period in this cohort of severely obese adults. The intakes of some food groups appear more

subject to change than others while some, often the most healthful, appear more stable and may require earlier intervention. While there is continued scepticism about the long-term effectiveness of lifestyle interventions in the severely obese [36,37,38,39], these results suggest that while changes in some foods are likely to be more successful than others, there are opportunities to change dietary habits in the severely obese, which deserve further investigation.

Author Contributions

Conceived and designed the experiments: DJJ AKL SAJ GLA. Analyzed the data: DJJ GLA AKL. Wrote the paper: DJJ GLA SAJ AKL LMSC LS. Design of original SOS study and data collection: AKL LS LMSC.

References

1. Twisk JWR (2009) Applied longitudinal data analysis for epidemiology: a practical guide. Cambridge: Cambridge University Press.
2. Totland TH, Gebremariam MK, Lien N, Bjelland M, Grydeland M, et al. (2013) Does tracking of dietary behaviours differ by parental education in children during the transition into adolescence? Public Health Nutrition 16: 673–682.
3. Demory-Luce D, Morales M, Nicklas T, Baranowski T, Zakeri I, et al. (2004) Changes in food group consumption patterns from childhood to young adulthood: The Bogalusa Heart Study. Journal of the American Dietetic Association 104: 1684–1691.
4. Wang YF, Bentley ME, Zhai FY, Popkin BM (2002) Tracking of dietary intake patterns of Chinese from childhood to adolescence over a six-year follow-up period. J Nutr 132: 430–438.
5. Li J, Wang YF (2008) Tracking of dietary intake patterns is associated with baseline characteristics of urban low-income African-American adolescents. Journal of Nutrition 138: 94–100.
6. Ambrosini GL, Emmett PM, Northstone K, Jebb SA (2013) Tracking a dietary pattern associated with increased adiposity in childhood and adolescence. Obesity: n/a-n/a.
7. Post GB, de Vente W, Kemper HCG, Twisk JWR (2001) Longitudinal trends in and tracking of energy and nutrient intake over 20 years in a Dutch cohort of men and women between 13 and 33 years of age: The Amsterdam growth and health longitudinal study. Br J Nutr 85: 375–385.
8. Lien N, Lytle LA, Klepp K-I (2001) Stability in Consumption of Fruit, Vegetables, and Sugary Foods in a Cohort from Age 14 to Age 21. Preventive Medicine 33: 217–226.
9. Borland SE, Robinson SM, Crozier SR, Inskip HM (2007) Stability of dietary patterns in young women over a 2-year period. European Journal of Clinical Nutrition 62: 119–126.
10. Mikkilae V, Rasnan L, Raitakari OT, Marniemi J, Pietinen P, et al. (2007) Major dietary patterns and cardiovascular risk factors from childhood to adulthood. The Cardiovascular Risk in Young Finns Study. British Journal of Nutrition 98: 218–225.
11. Prevost AT, Whichelow MJ, Cox BD (1997) Longitudinal dietary changes between 1984–5 and 1991–2 in British adults: associations with socio-demographic, lifestyle and health factors. British Journal of Nutrition 78: 873–888.
12. Mokdad AH, Bowman BA, Ford ES, Vinicor F, Marks JS, et al. (2001) The continuing epidemics of obesity and diabetes in the United States. Jama-Journal of the American Medical Association 286: 1195–1200.
13. Hjartaker A, Laake P, Lund E (2001) Body mass index and weight change attempts among adult women - The Norwegian Women and Cancer Study. European Journal of Public Health 11: 141–146.
14. Ashima K K (2004) Dietary patterns and health outcomes. Journal of the American Dietetic Association 104: 615–635.
15. Crozier SR, Robinson SM, Godfrey KM, Cooper C, Inskip HM (2009) Women's Dietary Patterns Change Little from Before to During Pregnancy. Journal of Nutrition 139: 1956–1963.
16. Mishra GD, McNaughton SA, Bramwell GD, Wadsworth MEJ (2006) Longitudinal changes in dietary patterns during adult life. British Journal of Nutrition 96: 735–744.
17. Weismayer C, Anderson JG, Wolk A (2006) Changes in the stability of dietary patterns in a study of middle-aged Swedish women. Journal of Nutrition 136: 1582–1587.
18. Hoffmann K, Schulze MB, Schienkiewitz A, Nothlings U, Boeing H (2004) Application of a new statistical method to derive dietary patterns in nutritional epidemiology. Am J Epidemiol 159: 935–944.
19. Hoffmann K, Schulze MB, Schienkiewitz A, Nothlings U, Boeing H (2004) Application of a new statistical method to derive dietary patterns in nutritional epidemiology. American Journal of Epidemiology 159: 935–944.
20. Sjostrom L, Lindroos AK, Peltonen M, Torgerson J, Bouchard C, et al. (2004) Lifestyle, diabetes, and cardiovascular risk factors 10 years after bariatric surgery. New England Journal of Medicine 351: 2683–2693.
21. Lindroos AK, Lissner L, Sjostrom L (1993) Validity and Reproducibility of a Self-Administered Dietary Questionnaire in Obese and Nonobese Subjects. European Journal of Clinical Nutrition 47: 461–481.
22. Lindroos AK, Lissner L, Sjostrom L (1999) Does degree of obesity influence the validity of reported energy and protein intake? Results from the SOS Dietary Questionnaire. European Journal of Clinical Nutrition 53: 375–378.
23. U.S. Department of Agriculture and U.S. Department of Health and Human Services (2010) Dietary Guidelines for Americans, 2010. Washington, DC: U.S. Government Printing Office.
24. NHS Choices Available: http://www.nhs.uk/livewell/goodfood/Pages/Goodfoodhome.aspx Accessed 2014 Apr 25.
25. American Dietetic Association Available: http://www.eatright.org/public/default.aspx Accessed 2014 Apr 25.
26. Johnson L, Mander AP, Jones LR, Emmett PM, Jebb SA (2008) Energy-dense, low-fiber, high-fat dietary pattern is associated with increased fatness in childhood. American Journal of Clinical Nutrition 87: 846–854.
27. Imamura F, Lichtenstein AH, Dallal GE, Meigs JB, Jacques PF (2009) Generalizability of dietary patterns associated with incidence of type 2 diabetes mellitus. The American Journal of Clinical Nutrition 90: 1075–1083.
28. Twisk JWR, Kemper HCG, Mellenbergh GJ, van Mechelen W, Post GB (1996) Relation between the longitudinal development of lipoprotein levels and lifestyle parameters during adolescence and young adulthood. Annals of Epidemiology 6: 246–256.
29. Pearson N, Salmon J, Campbell K, Crawford D, Timperio A (2011) Tracking of children's body-mass index, television viewing and dietary intake over five-years. Preventive Medicine 53: 268–270.
30. Twisk JWR, Kemper HCG, van Mechelen W, Post GB (1997) Tracking of Risk Factors for Coronary Heart Disease over a 14-Year Period: A Comparison between Lifestyle and Biologic Risk Factors with Data from the Amsterdam Growth and Health Study. American Journal of Epidemiology 145: 888–898.
31. Sjostrom L, Backman L, Bengtsson C, Dahlgren S, Jonsson E, et al. (1987) Announcement of the Multicenter Project Swedish Obese Subjects-Sos. International Journal of Obesity 11: 87–87.
32. Collins CE, Morgan PJ, Warren JM, Lubans DR, Callister R (2010) Men participating in a weight-loss intervention are able to implement key dietary messages, but not those relating to vegetables or alcohol: the Self-Help, Exercise and Diet using Internet Technology (SHED-IT) study. Public Health Nutrition 14: 168–175.
33. Sjostrom L, Lindroos AK, Peltonen M, Torgerson J, Bouchard C, et al. (2004) Lifestyle, diabetes, and cardiovascular risk factors 10 years after bariatric surgery. N Engl J Med 351: 2683–2693.
34. Velde SJ, Twisk JWR, Brug J (2007) Tracking of fruit and vegetable consumption from adolescence into adulthood and its longitudinal association with overweight. British Journal of Nutrition 98: 431–438.
35. Northstone K, Emmett PM (2008) A comparison of methods to assess changes in dietary patterns from pregnancy to 4 years post-partum obtained using principal components analysis. British Journal of Nutrition 99: 1099–1106.
36. Gondoni LA, Liuzzi A (2011) Diet and Physical Activity Interventions in Severely Obese Adults. Journal of the American Medical Association 305: 563.

37. Goodpaster BH, DeLany JP, Jakcic JM (2011) Diet and Physical Activity Interventions in Severely Obese Adults Reply. Journal of the American Medical Association 305: 564–564.

38. Goodpaster BH, DeLany JP, Otto AD, Kuller L, Vockley J, et al. (2010) Effects of Diet and Physical Activity Interventions on Weight Loss and Cardiometabolic Risk Factors in Severely Obese Adults A Randomized Trial. Journal of the American Medical Association 304: 1795–1802.

39. Hemmingsson E, Udden J, Rossner S (2011) Diet and Physical Activity Interventions in Severely Obese Adults. Journal of the American Medical Association 305: 563–564.

8

Fast Food Consumption and Gestational Diabetes Incidence in the SUN Project

Ligia J. Dominguez[1]*, Miguel A. Martínez-González[2], Francisco Javier Basterra-Gortari[2,3], Alfredo Gea[2], Mario Barbagallo[1], Maira Bes-Rastrollo[2]*

1 Geriatric Unit - Department of Internal Medicine and Specialties, University of Palermo, Palermo, Sicily, Italy, 2 Department of Preventive Medicine and Public Health, University of Navarra, Pamplona, Navarra, Spain, and CIBER Fisiopatologia de la Obesidad y Nutricion (CIBERobn), Instituto de Salud Carlos III, Madrid, Spain, 3 Department of Internal Medicine (Endocrinology), Hospital Reina Sofia, Tudela, Navarra, Spain

Abstract

Background: Gestational diabetes prevalence is increasing, mostly because obesity among women of reproductive age is continuously escalating. We aimed to investigate the incidence of gestational diabetes according to the consumption of fast food in a cohort of university graduates.

Methods: The prospective dynamic "Seguimiento Universidad de Navarra" (SUN) cohort included data of 3,048 women initially free of diabetes or previous gestational diabetes who reported at least one pregnancy between December 1999 and March 2011. Fast food consumption was assessed through a validated 136-item semi-quantitative food frequency questionnaire. Fast food was defined as the consumption of hamburgers, sausages, and pizza. Three categories of fast food were established: low (0–3 servings/month), intermediate (>3 servings/month and ≤2 servings/week) and high (>2 servings/week). Non-conditional logistic regression models were used to adjust for potential confounders.

Results: We identified 159 incident cases of gestational diabetes during follow-up. After adjusting for age, baseline body mass index, total energy intake, smoking, physical activity, family history of diabetes, cardiovascular disease/hypertension at baseline, parity, adherence to Mediterranean dietary pattern, alcohol intake, fiber intake, and sugar-sweetened soft drinks consumption, fast food consumption was significantly associated with a higher risk of incident gestational diabetes, with multivariate adjusted OR of 1.31 (95% confence interval [CI]:0.81–2.13) and 1.86 (95% CI: 1.13–3.06) for the intermediate and high categories, respectively, versus the lowest category of baseline fast food consumption (p for linear trend: 0.007).

Conclusion: Our results suggest that pre-pregnancy higher consumption of fast food is an independent risk factor for gestational diabetes.

Editor: Rudolf Kirchmair, Medical University Innsbruck, Austria

Funding: The SUN Study has received funding from the Spanish Ministry of Health Government (current grants PI10/02658, PI10/02293, RD06/0045, G03/140 and PI13/00615), the Navarra Regional Government (45/2011) and the University of Navarra. AG is supported by a FPU fellowship from the Spanish Government. Funding sources had no role in the design, collection, analysis, and interpretation of the data; in the writing, and in the decision to submit the paper for publication.

Competing Interests: The authors have declared that no competing interests exist.

* Email: ligia.dominguez@unipa.it (LJD); mbes@unav.es (MBR)

Introduction

Gestational diabetes mellitus (GDM), traditionally defined as carbohydrate intolerance first diagnosed during pregnancy [1], has long been recognized as a risk factor for a number of unfavorable outcomes. These include short- and long-term complications for mothers (i.e., preeclampsia or eclampsia, and type 2 diabetes after delivery), and for offspring (i.e., macrosomia, increased likelihood of trauma at birth, cesarean delivery, and neonatal metabolic abnormalities, such as hypoglycemia or hyperbilirubinemia) [1,2]. GDM prevalence is increasing, mostly because obesity among women of reproductive age is continuously escalating [3–5]. It has been reported that gestational diabetes affects 1–14% of all pregnancies in the US [1], and about 2–6% of pregnancies in Europe [5]. A recent meta-analysis of RCTs suggested that interventions on glucose control/monitoring, diet, or pharmacological treatment including insulin, did not significantly reduce the risk of adverse outcomes (i.e. cesarean delivery and perinatal or neonatal death) [6]. Hence, the identification of modifiable risk factors, such as excess adiposity, decreased physical activity, and unhealthy diet for the prevention of gestational diabetes is criticalin order to avoid associated harmful outcomes [7].

Several observational studies have related some pre-pregnancy dietary factors with GDM risk. An inverse association between pre-pregnancy adherence to healthful dietary patterns and gestational diabetes incidence has been reported [8]. Low fiber intakes or high dietary glycemic index [9], high red meat/ processed meat consumptions [10], and high intakes of animal fat and cholesterol [11] have all been associated with an elevated risk of GDM.

Previous studies have suggested that Western-style fast food, which is calorically dense and usually served in large portions, is a determinant of weight gain, insulin resistance, and type 2 diabetes incidence [12,13]. However, the association of fast food consumption with GDM risk remains unknown. Therefore, we conducted the present analyses to appraise whether fast food consumption (hamburgers, sausages, and pizza) was associated with GDM risk in the SUN project ("Seguimiento Universidad de Navarra", University of Navarra Follow-up).

Materials and Methods

Study population

The SUN project is a prospective dynamic cohort study entirely composed of university graduates. The recruitment started in 1999 and it is permanently open. The design and methods utilized in the SUN study have been previously described in detail [14,15]. In brief, graduates from the University of Navarra and other Spanish universities, registered nurses and other health professionals from different Spanish provinces were invited to participate by a mailed questionnaire. The study protocol was approved by the Institutional Review Board of the University of Navarra and the initial response to the questionnaire was considered as informed consent to participate. After the baseline assessment, participants receive a follow-up questionnaire every two years on diet, lifestyle, risk factors, and medical conditions.

For the present analyses we examined the last available database as of the 1st of December 2013. From a total 13,231 women, we included 12,456 women who had answered the baseline questionnaire before the 1st of March 2011, to have enough time to answer the first follow-up questionnaire. Up to that date, 3,137 pregnant women were identified among them. Women were excluded from the analyses if they reported extremely low (below percentile 1) or high (above percentile 99) values for total energy intake (n = 67), had prevalent or previous gestational diabetes (n = 10), or had a previous diagnosis of diabetes (n = 17). Women who reported gestational diabetes in a previous pregnancy were not included in the analyses because they were thought to be more likely to have changed their diet and lifestyle during the next pregnancy to prevent recurrent gestational diabetes. The final analytic population included 3,048 pregnant women (**Fig. 1**).

Since we have excluded those pregnant women with implausible levels of total energy intake, missing data on fast food consumption (n = 40) was considered as no consumption. Those with missing values in smoking were treated as another category (current, former smoker, never smokers, and missing [n = 67]). For missing anthropometrics data we used the last value carried forward, although in this sample of the SUN Project there were no missing data for these variables.

Dietary assessment

Dietary habits at baseline were assessed by a semi-quantitative food frequency questionnaire (FFQ) with 136 items previously validated and described in detail [16,17]. The validity [18] and reproducibility [19] of this questionnaire have been also recently assessed. Updated Spanish food composition tables were used to assess nutrient intake. Nutrient scores were computed using an *ad hoc* computer software specifically developed for this aim. A trained dietitian updated the nutrient data bank using the latest available information included in food composition tables for

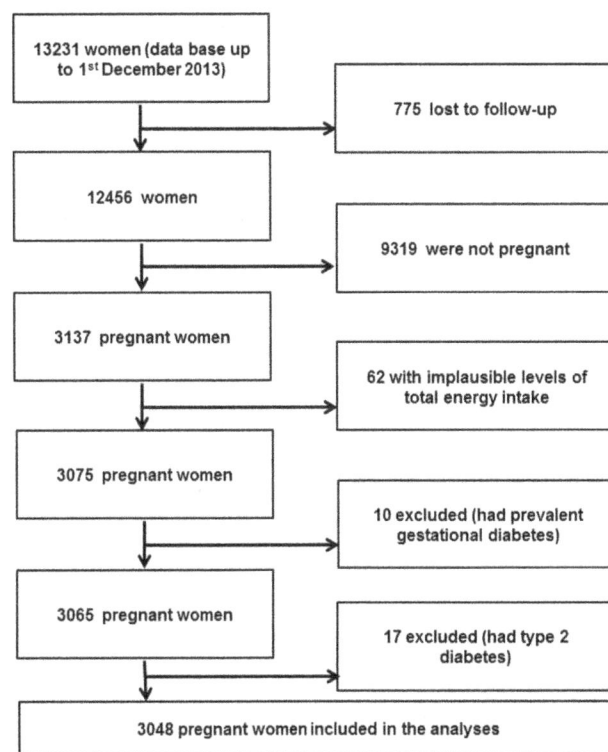

Figure 1. Flow chart depicting the selection process among participants of the SUN project to be included in the present analyses.

Spain [20,21]. Adherence to the Mediterranean food pattern was appraised using the sample-specific score proposed by Trichopoulou [22]. Fast food was defined as the consumption of hamburgers, sausages, and pizza, as previously reported [23,24]. We estimated energy adjusted fast food consumption through the residuals method [25]. Three categories of fast food (serving = 100 g) consumption were also established: low (0–3 servings/month), intermediate (>3 servings/month and ≤2 servings/week) and high (>2 servings/week).

Ascertainment of gestational diabetes

The outcome of interest was the incidence of gestational diabetes. Pregnant women identified in the SUN project (**Fig. 1**) who reported a diagnosis of GDM made by a physician in the biennial questionnaire and did not have diabetes at baseline were considered as incident cases of new-onset GDM.

We sent an additional questionnaire to participants who reported probable new onset GDM, requesting their written confirmation and date of the diagnosis of either GDM, a previous diagnosis of diabetes, their highest fasting glucose value, their first glycated haemoglobin during pregnancy, whether they had ever undergone an oral glucose tolerance test and its results, and the use during pregnancy of insulin. We also asked them to send us the medical report detailing their diagnoses. A panel of medical doctors, blinded to the information about dietary habits, used the information provided by participants (additional questions and medical reports) to classify the diagnosis or each candidate woman as an incident case of GDM or not. For the present analyses we considered only those cases of confirmed GDM. From the potential cases of GDM we had avalaible information in 98% of them. Among those with information, 80% of them were confirmed as incident cases of GDM. There is a lack of universally accepted diagnostic criteria for GDM [5]. Several different protocols are in regular use internationally, each with its own recommendations on which pregnant women should be selected for biochemical testing, how the test should be performed and what glycemic thresholds should be considered diagnostic [26]. The usual diagnosis criteria for GDM in Spain were those with 100-g oral glucose tolerance test with the cut-off points of Carpenter and Coustan [27] or the cut-off points from the National Diabetes Data Group [28] after a positive 50-g glucose challenge test.

Other covariates

Information on other covariates was assessed in the baseline questionnaire. This included socio-demographic parameters (age), anthropometric measurements (weight, body mass index [BMI]), health related habits (smoking status, physical activity, sedentary lifestyle), and clinical variables (use of medication, self-reported pregnancy, family history of diabetes, cardiovascular disease/ hypertension, parity). Age was calculated from the date of birth to the date of the questionnaire's return. Self-reported weight and BMI have previously shown a high validity in a specific study conducted in a sub-sample of this cohort [29 27]. Physical activity was assessed using a previously validated questionnaire with a Spearman correlation coefficient of 0.51 ($p<0.001$) between questionnaire information and objectively obtained measurements [30]. Physical activity was expressed in metabolic equivalent tasks (METs/weeks) as calculated from the time spent at each activity in hours per week multiplied by its typical energy expenditure [31].

Statistical analysis

Consumption frequencies of the fast food items considered (hamburgers, sausages, and pizza) were standardized, summed, and divided into categories that allowed logical cut points with a sufficient number of subjects. Means with SDs for continuous baseline characteristics and proportions for categorical characteristics were calculated by categories of fast food frequency consumption. Since the risk of gestational diabetes depends on the fact to be pregnant and not on a timely basis, we estimated the odds ratios (and 95% CIs) for each of the three categories of fast food consumption using non-conditional regression models taking as the reference category those women with the lowest fast food consumption (0–3 servings per month). For each exposure category, we fitted a crude (univariate) model, an age-adjusted model (Model 1), and multivariate models with additional adjustments (Models 2 and 3) for a priori-selected pre-pregnancy dietary and non-dietary covariables (see below). Non-conditional logistic regression models were adjusted for age, BMI (Kg/m^2), total energy intake (Kcal/day), smoking status (never, former and current smokers), physical activity (expressed in METs-h/w), family history of diabetes (yes or no), presence at baseline of cardiovascular disease or hypertension (yes or no), parity (first pregnancy/1 or 2 pregnancies before/3 or more pregnancies before), score of adherence to the Mediterranean dietary pattern (0–9 points score), and other potential dietary confounders, such as fiber intake (g/day), and sugar-sweetened soft drink consumption (ml/day). The p for trend was calculated taking the median consumption of fast food for each category and introducing this new variable as a continuous variable in the models. We evaluated the interaction between fast food consumption and sugar-sweetened soft drink consumption, physical activity, and family history of diabetes on the risk of GDM through likelihood ratio tests for each of the product-terms introduced (each at one time) in fully-adjusted models.

The analyses were performed with Stata software package version 12 (Stata Corp). All tests were two sided and statistical significance was set at the conventional cut-off of $p<0.05$.

Results

Baseline characteristics of the studied population of pregnant women in the SUN project, according to categories of fast food consumption (low to high), are shown in **Table 1.** Almost half of the participants had an intermediate consumption of fast food (>3 servings/month and ≤2 servings/week; 48%) whereas lower values (0–3 servings/week; 20%) were less frequently observed. Higher frequency of fast food consumption (>2 servings/week; 32%) was present in nearly a third of participants. The participants with a higher consumption of fast food were on average younger, more likely to be current smokers, multiparous, less physically active, less adherent to the Mediterranean dietary pattern, and also more likely to have a lower fiber intake and a higher fat intake.

During 28,064 person-year follow-up (mean follow-up: 10.2, sd: 2.9 years) 159 women reported a first diagnosis of GDM among 3,048 pregnant women of the SUN project, corresponding to 5.2% of pregnant participants. When we analyzed the association between the consumption of fast food and incident gestational diabetes during follow-up with adjustments for a set of potential dietary and non-dietary confounders (age, baseline BMI, total energy intake, smoking, physical activity, family history of diabetes, cardiovascular disease/hypertension at baseline, parity, adherence to Mediterranean dietary pattern score, fiber intake, and sugar-sweetened soft drinks consumption), we found that fast food consumption was strongly and positively associated with incident gestational diabetes risk. In fully-adjusted models, those women in the highest category of fast food consumption presented almost a twofold risk of developing GDM (adjusted OR: 1.86; 95% CI: 1.13–3.06) compared to pregnant women with the lowest fast food consumption (0–3 servings per month) (**Table 2 and Fig. 2**).The risk of incident gestational diabetes was 3.9% for those in the lowest category of fast food consumption, 4.8% for those with intermediate fast food consumption, and 6.7% for the highest category. These associations remained statistically significant after further adjustment for the presence of polycystic ovary syndrome (**Table 2**).

When we used tertiles of energy-adjusted fast food consumption the results were similar: we observed an adjusted OR of 1.82 (95% CI 1.18–2.78) for the highest versus the lowest tertile (p for trend: 0.004).

No significant interaction was observed between fast food consumption and sugar-sweetened soft drinks consumption (p for interaction: 0.645) or between fast food consumption and physical activity (p for interaction: 0.280).

When we assessed fast food consumption as a continuous variable, each additional serving (100 g) per day of fast food consumption was associated with an increased risk, although this association was not statistically significant (adjusted OR: 1.42; 95% CI: 0.76–2.63).

Several sensitivity analyses were carried out in order to appraise the robustness of our findings. This included restricting our analysis to first births from nulliparous women to reduce possible confounding by experiences from previous pregnancies. Excluding multiparous women, the results of the fully adjusted model did not change materially (highest consumption: OR 1.70 (95% CI: 1.00–2.89 for highest vs. lowest consumption; p for linear trend: 0.014). The exclusion of obese women did not either change the results of

Table 1. Characteristics of 3,048 pregnant women in the SUN cohort according to their frequency of fast food consumption.

	Fastfood consumption		
	0–3 servings*/month	>3 servings/month and ≤2 servings/week	>2 servings/week
N (%)	616 (20.2)	1461 (47.9)	971 (31.9)
Age (years)	29.3 (5.5)	28.9 (4.6)	28.6 (4.4)
Gestational diabetes (%)	3.9	4.8	6.7
Family history of diabetes (%)	10.4	11.2	9.8
Current smoking (%)	22.2	25.0	26.6
Body Mass Index (kg/m^2)	21.3 (2.6)	21.4 (2.6)	21.6 (2.7)
Nulliparous (%)	81.0	81.1	77.9
Physical activity (METs-h/week)	22.4 (23.8)	17.9 (18.2)	16.7 (19.1)
Prevalence of hypertension (%)	2.6	2.4	1.7
Prevalence of CVD (%)	0.7	0.7	0.5
Mediterranean diet score	4.6 (1.7)	4.5 (1.7)	3.9 (1.7)
Alcohol intake (g/d)	3.3 (4.9)	4.0 (5.2)	3.6 (4.7)
Fiber intake (g/d)	32.9 (17.0)	28.2 (11.7)	25.4 (10.9)
Total energy intake (kcal/d)	2479 (842)	2357 (685)	2395 (689)
Carbohydrate intake (% energy)	43.6 (8.8)	42.5 (7.2)	41.4 (6.7)
Protein intake (% energy)	18.7 (3.7)	18.6 (3.0)	18.4 (2.9)
Fat intake (% energy)	36.7 (7.9)	37.6 (6.2)	39.1 (6.1)
SFA intake (% energy)	12.0 (4.0)	12.8 (3.1)	13.5 (2.9)
MUFA intake (% energy)	16.0 (4.5)	16.2 (3.8)	16.4 (3.4)
PUFA intake (% energy)	5.4 (1.9)	5.4 (1.6)	5.7 (1.7)
Trans fatty acid intake (% energy)	0.3 (0.2)	0.4 (0.2)	0.4 (0.2)
Soft drinks (servings/week)	1.3 (2.9)	1.5 (2.4)	1.9 (2.9)

*1 serving = 100 g.
METs: metabolic equivalent tasks; CVD: cardiovascular disease; SFA: saturated fatty acid; MUFA: monounsaturated fatty acid; PUFA: polyunsaturated fatty acid.
Values are means (SD) for age, BMI, physical activity, Mediterranean diet score, alcohol, fiber, and total energy intake.

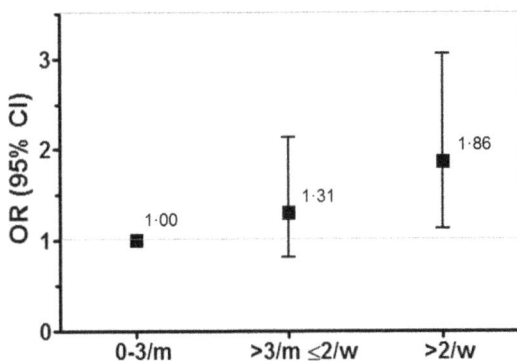

Figure 2. Odds ratio (OR) and 95% confidence interval (CI) for incident gestational diabetes according to frequency of fast food consumption in the SUN project (n = 3,048 pregnant women). Respective numbers (gestational diabetes incidence) for fast food intake of 0–3 times per month (low), >3 times a month and ≤2 times per week (intermediate), and >2 times per week (high) were 616 (24), 1,461 (70), 971 (65). Results represent fully adjusted model (age, baseline BMI, total energy intake, smoking, physical activity, family history of diabetes, cardiovascular disease/hypertension at baseline, parity, adherence to Mediterranean dietary pattern, fiber intake, alcohol intake, and sugar-sweetened soft drinks consumption).

the fully adjusted model (highest consumption: OR 1.81 (95% CI: 1.10–2.99 vs. low consumption; p for linear trend: 0.010).

To test the existence of a potential selection bias due to lost during follow-up we compared baseline characteristics between those participants included in the analyses and those lost during follow-up. There were no statistical significant differences between both groups.

Discussion

The results of the present analyses, using data from a large, well-characterized, prospective cohort of Spanish university graduates, found that among pregnant women in the cohort, a higher pre-pregnancy consumption of fast food was associated with a significantly higher risk of developing gestational diabetes. This association remained significant after several adjustments for potential confounders. Our findings are relevant in the context of a global epidemic of diabetes, which is likely driven, at least in part, by an unhealthy Westernized diet and lifestyle [32]. Fast food is a hallmark of such unhealthy diet; hence, the present results help to reinforce its harmful consequences in women of reproductive age.

It is widely accepted that one of the characteristics of normal pregnancy is a state of insulin resistance with a relative intolerance to dietary carbohydrates and a compensatory increase in insulin secretion from the beta cells in the pancreas [1,33]. The pregnancy-related insulin resistance mainly occurs after the second trimester when insulin requirements are higher. Pregnant women

Table 2. OR and 95% confidence interval of incident gestational diabetes according to fast food consumption.

| | Fastfood consumption | | | |
	0–3 servings*/month	>3 servings/month and ≤2 servings/week	>2 servings/week	p for trend
Cases, n/N	24/616	70/1461	65/971	
Rate	$4.11*10^{-3}$	$5.16*10^{-3}$	$7.41*10^{-3}$	
Model 1: OR (95% CI)	1.0 (Ref.)	1.26 (0.78–2.02)	1.80 (1.11–2.91)	0.009
Model 2: OR (95% CI)	1.0 (Ref.)	1.31 (0.81–2.12)	1.90 (1.15–3.12)	0.005
Model 3: OR (95% CI)	1.0 (Ref.)	1.31 (0.81–2.13)	1.86 (1.13–3.06)	0.007
Model 2: OR (95% CI) and PCO	1.0 (Ref.)	1.31 (0.81–2.13)	1.90 (1.16–3.13)	0.005
Model 3: OR (95% CI) and PCO	1.0 (Ref.)	1.31 (0.81–2.13)	1.86 (1.13–3.07)	0.007
Model 2: OR (95% CI) w/o multiparous	1.0 (Ref.)	1.07 (0.64–1.79)	1.71 (1.01–2.90)	0.013
Model 3: OR (95% CI) w/o multiparous	1.0 (Ref.)	1.07 (0.64–1.79)	1.70 (1.00–2.89)	0.014
Model 2: OR (95% CI) w/o obese	1.0 (Ref.)	1.25 (0.77–2.04)	1.83 (1.11–3.03)	0.008
Model 3: OR (95% CI) w/o obese	1.0 (Ref.)	1.25 (0.77–2.04)	1.81 (1.10–2.99)	0.010

The SUN project 1999–2012.
*1 serving = 100 g.
OR: odd ratio; CI: confidence interval; Rate: crude incident gestational diabetes rate per 10,000 person-years; PCO: polycystic ovary syndrome; w/o: without.
Model 1: adjusted for age.
Model 2: model 1 plus adjustment for total energy intake, smoking, physical activity, family history of diabetes, cardiovascular disease/hypertension at baseline, parity, adherence to Mediterranean dietary pattern score, alcohol intake, fiber intake, and sugar-sweetened soft drinks consumption.
Model 3: model 2 plus adjustment for baseline BMI.

who develop gestational diabetes are assumed to have a compromised beta cell capacity unable to adapt to the increased demand of insulin due to target-organ insulin resistance. This resembles closely the pathophysiology of type 2 diabetes [34], thus, pregnancy-related metabolic challenges may unmask a predisposition to glucose metabolic disorders, which may attain hyperglycemia in the range of diabetic disease. Furthermore, gestational diabetes is a recognized risk factor for future development of type 2 diabetes [35].This and other unfavorable outcomes of gestational diabetes, from delivery complications to perinatal mortality, call for efforts to identify modifiable factors, such as an unhealthy diet, predisposing to the condition.

We found an incidence of gestational diabetes (5.8%) that was similar to that previously reported in the Nurses' Health Study (6.4%) [8]. Attempts to identify women at risk of developing gestational diabetes have been conventionally directed to sociodemographic characteristics, such as maternal weight/BMI, and family history of type 2 diabetes [1,2]. However, in the past decade, several investigations have been focused on the pregravid dietary and lifestyle factors that may contribute to such risk, in parallel with studies pointing to these factors as main drivers of type 2 diabetes epidemic. Even if former studies showed inconsistentresults on the role of total dietary fat intake and gestational diabetes risk, two recent studies taking in consideration specific types of fat concluded that higher pre-pregnancy consumptions of animal fat and cholesterol [11], and of saturated fat [36] were associated with an elevated risk of gestational diabetes. Likewise, a very recent study reported that a higher intake of animal protein, particularly red meat, was significantly associated with a greater risk of gestational diabetes. Conversely, a greater intake of vegetable protein, in particular nuts, was associated with a significantly lower risk [37]. Our findings are consistent with these studies, since the components of fast food considered in our study (hamburgers, sausages, and pizza) mainly include animal proteins and saturated fat. Our results also showed that pregnant women with a higher consumption of fast food had a

lower dietary fiber intake and were less adherent to the Mediterranean dietary pattern. In accordance with our observations, previous study have reported inverse and significant associations for fiber intake [9] and adherence to healthful dietary patterns, including Mediterranean diet [8], and the risk of developing gestational diabetes.

Even if the precise molecular mechanisms are unclear, the observed associations between pre-pregnancy fast food consumption and gestational diabetes risk are biologically plausible. First, maternal pregravid BMI has been linked to gestational diabetes risk [1,2] and Western-style fast food has been identified among the greatest dietary contributors to weight gain [38]. However, our results were significant even when the data were adjusted for baseline BMI. Second, there is evidence that frequent consumption of Western-style fast food contributes to insulin resistance [12], a mechanism that is central, as mentioned above, in the development of gestational diabetes. Third, saturated fat and cholesterol, which are components of red and processed meats (part of fast food in our study), have shown to adversely affect not only insulin sensitivity but also beta cell function [39], relevant in the pathophysiology of gestational diabetes. Other components of red and processed meats, including heme iron, and nitrosamines may add to oxidative stress and beta cell damage [9].

It has been previously shown that a 'fast food' dietary pattern was associated with weight gain rate during pregnancy in a dose-dependent manner in a study conducted in Finland [40]. The result of that study is indirectly supportive of our findings. The independent association between pregravid fast food consumption and GDM incidence that we observed was present even after adjustment for baseline BMI. Indeed, we may wonder whether this association was driven by gestational weight gain among fast food eaters or by other mechanisms. Unfortunately, we do not have specific available information to track weight changes in detail throughout the gravid period of our participants and we cannot provide further specific analyses to assess whether gestational weight changes are the most important mediators of this association. However, overadjustment for intermediate mechanisms through which fast food consumption can increase the risk of GDM would be inappropriate if the aim of our analyses is to control for confounding. It is recognized that in an attempt to estimate the total effect of an exposure on some outcome, control for intermediate factors in the causal chain (that is, overadjustment) will generally bias estimates of the total effect of the exposure on the outcome [41,42].

The global epidemic of diabetes, including increasing rates of gestational diabetes [43], parallels obesity trends worldwide. These "twin epidemics" have been linked to Westernization of diet and lifestyle [32,44]. Fast food advertising and broad accessibility may have considerable influence on food choices, particularly among young people. This includes women of childbearing age, increasing their risk for gestational diabetes. In our study, women with the highest consumption of fast food were younger and more likely to have other unhealthy behaviors (i.e. sedentary, smoking, and lower adherence to Mediterranean dietary pattern). Thus, fast food consumption could be a marker of an unhealthy diet/lifestyle and not a cause of increased gestational diabetes risk. However, the multivariate analyses we present here accounted for other unhealthy behaviors, and after multiple adjustments and sensitivity analyses the consumption of fast food remained strongly significantly associated with gestational diabetes risk. Emphasis is needed to develop strategies that persuade women to optimize their diet and lifestyle before conceiving.

The present study has a number of strengths, including a large sample size with a high retention rate, prospective design, a lengthy follow-up and use of a FFQ that has been specifically and repeatedly validated in Spain [17–19]. Importantly, we were able to obtain a high degree of control for confounding, including potential lifestyle and demographic confounders, and we also controlled for overall dietary patterns. In addition, we conducted several sensitvity analyses and the results were robust. The study also has some potential limitations. Information that was self-reported and this may lead to some degree of misclassification, which most often would drive associations toward the null value. However, parameters such as self-reported weight and BMI have been previously validated in a sub-sample of this cohort [29]. The cohort is composed mostly of middle-aged, Spanish highly educated persons.Therefore the generalizability of our results must be based on common biological mechanisms instead of on statistical representativeness, and we used restriction to reduce potential confounding by disease, education, socioeconomic status, and presumed access to health care. Future studies are needed in order to test the applicability of our results in pregnant women from other populations.

Since our all participants are Spanish university graduates, we can guarantee that the vast majority of them are Caucasians and therefore, unfortunately we were not able to assess the effect of ethnicity in the association between fast food consumption and GDM.

Finally, possible concerns may arise from the use of dietary data derived from an FFQ, which may be subject to information bias. However, the FFQ used has been repeatedly validated [17–19]. Moreover, it is difficult to find a better alternative than a FFQ as an advantageous method to characterize the food habits of large samples of individuals' that need to be followed-up over long periods of time in order to rank them and assess associations with incident clinical end-points [45].

In conclusion, findings from this prospective study suggest that pre-pregnancy higher consumption of fast food (i.e. hamburgers, sausages, and pizza) is an independent risk factor for gestational diabetes. Further research to confirm these findings in other populations and to decipher underlying molecular mechanisms is warranted. Nevertheless, our results emphasize the potential importance of considering pregravid dietary recommendations for the prevention of gestational diabetes. The information on the increased gestational risk associated with fast food intake might be disseminated to women of reproductive age.

Acknowledgments

The authors thank the participants of the SUN Project for their enthusiastic collaboration.

Author Contributions

Conceived and designed the experiments: LJD MBR MAMG. Performed the experiments: MAMG FJBG MBR. Analyzed the data: MBR LJD MAMG. Contributed to the writing of the manuscript: LJD MBR MAMG MB FJBG AG. Critical revision and approval of the final version to be published: LJD MBR MAMG MB FJBG AG.

References

1. ADA (2004) Gestational diabetes mellitus. Diabetes Care (suppl 1):S88–90.
2. Bellamy L, Casas JP, Hingorani AD, Williams D (2009) Type 2 diabetes mellitus after gestational diabetes: a systematic review and meta-analysis. Lancet 373: 1773–1779.
3. Albrecht SS, Kuklina EV, Bansil P, Jamieson DJ, Whiteman MK, et al. (2010) Diabetes trends among delivery hospitalizations in the United States, 1994-2004. Diabetes Care 33: 768–773.
4. Dabelea D, Snell-Bergeon JK, Hartsfield CL, Bischoff KJ, Hamman RF, et al. (2005) Increasing prevalence of gestational diabetes mellitus(GDM) over time and by birth cohort: Kaiser Permanente of Colorado GDM Screening Program. Diabetes Care 28: 579–584.
5. Buckley BS, Harreiter J DammP, Corcoy R, Chico A, et al, on behalf of the DALI Core Investigator Group (2012) Gestational diabetes mellitus in Europe: prevalence,current screening practice and barriers to screening. A review.Diabet Med29: 844–854.
6. Horvath K, Koch K, Jeitler K, Matyas E, Bender R, et al. (2010) Effects of treatment in women with gestational diabetes mellitus: systematic review and meta-analysis. BMJ 340:c1395.
7. Zhang C, Ning Y (2011) Effect of dietary and lifestyle factors on the risk of gestational diabetes: review of epidemiologic evidence. Am J Clin Nutr 94(suppl):1975S–1979S.
8. Tobias DK, Zhang C, Chavarro J, Bowers K, Rich-Edwards J, et al. (2012) Prepregnancy adherence to dietary patterns and lower risk of gestational diabetes mellitus. Am J ClinNutr 96: 289–295.
9. Zhang C, Liu S, Solomon CG, Hu FB (2006) Dietary fiber intake, dietary glycemic load, and the risk for gestational diabetes mellitus. DiabetesCare 29: 2223–2230.
10. Zhang C, Schulze MB, Solomon CG, Hu FB (2006) A prospective study of dietary patterns, meat intake and the risk of gestational diabetes mellitus. Diabetologia 49: 2604–2613.
11. Bowers K, Tobias DK, Yeung E, Hu FB, Zhang C (2012) A prospective study of prepregnancy dietary fat intake and risk of gestational diabetes. Am J ClinNutr 95: 446–453.
12. Pereira MA, Kartashov AI, Ebbeling CB, Van Horn L, Slattery ML, et al. (2005) Fast-food habits, weight gain, and insulinresistance (the CARDIA study): 15-year prospective analysis. Lancet 365: 36–42.
13. Odegaard AO, Koh WP, Yuan J-M, Gross MD, Pereira MA (2012) Western-Style Fast Food Intake and Cardiometabolic Risk in an Eastern Country Circulation 126: 182–188.
14. Martinez-Gonzalez MA, Sanchez-Villegas A, De Irala J, Marti A, Martinez JA (2002) Mediterranean diet and stroke: objectives and design of the SUN project. SeguimientoUniversidad de Navarra. NutrNeurosci 5: 65–73.

15. Segui-Gomez M, de la Fuente C, Vazquez Z, de Irala J, Martinez-Gonzalez MA (2006) Cohort profile: the 'Seguimiento Universidad de Navarra' (SUN) study. Int J Epidemiol 35: 1417–1422.

16. Martinez-Gonzalez MA, de la Fuente-Arrillaga C, Nunez-Cordoba JM, Basterra-Gortari FJ, Beunza JJ, et al. (2008) Adherence to Mediterranean diet and risk of developing diabetes: prospective cohort study. BMJ 336: 1348–1351.

17. Martin-Moreno JM, Boyle P, Gorgojo L, Maisonneuve P, Fernandez-Rodriguez JC, et al. (1993) Development and validation of a food frequency questionnaire in Spain. Int J Epidemiol 22: 512–519.

18. Fernandez-Ballart JD, Pinol JL, Zazpe I, Corella D, Carrasco P, et al. (2010) Relative validity of a semi-quantitative food-frequency questionnaire in an elderly Mediterranean population of Spain. Br J Nutr 103: 1808–1816.

19. de la Fuente-Arrillaga C, VazquezRuiz Z, Bes-Rastrollo M, Sampson L, Martinez-Gonzalez MA (2010) Reproducibility of an FFQ validated in Spain. Public Health Nutr 13: 1364–1372.

20. Moreiras O, Carbajal A, Cabrera L, Cuadrado C (2013) Tablas de composición de alimentos [Food composition tables]. 16th ed. Madrid: Piramide.

21. Mataix Verdu J (2009) Tabla de composición de alimentos [Food composition tables]. 5th ed. Granada: Universidad de Granada.

22. Trichopoulou A, Costacou T, Bamia C, Trichopoulos D (2003) Adherence to a Mediterranean diet and survival in a Greek population. N Engl J Med 348: 2599–2608.

23. Bes-Rastrollo M, Sánchez-Villegas A, Gómez-Gracia E, Martínez JA, Pajares RM, et al. (2006) Predictors of weight gain in a Mediterranean cohort: the Seguimiento Universidad de Navarra Study. Am J Clin Nutr 83: 362–370.

24. Bes-Rastrollo M, Basterra-Gortari FJ, Sánchez-Villegas A, Marti A, Martínez JA, et al. (2010) A prospective study of eating away-from-home meals and weight gain in a Mediterranean population: the SUN (Seguimiento Universidad de Navarra) color. Public Health Nutrition 13: 1356–1363.

25. Hu FB, Stampfer MJ, Rimm E, Ascherio A, Rosner BA, et al. (1999) Dietary fat and coronary heart disease: a comparison of approaches for adjusting for total energy intake and modeling repeated dietary measurements. Am J Epidemiol 149: 531–540.

26. NICE (2008) Diabetes in Pregnancy: Management of Diabetes and its Complications from Pre-Conception to the Post-Natal Period. Clinical guideline no. 63. London: National Institute for Health and Clinical Excellence / National Collaborating Centre for Women's and Children's Health.

27. ADA (2010) Position statement. Diagnosis and classification of diabetes mellitus. Diabetes Care 33:S62–S69.

28. National Diabetes Data Group (1979) Classification and diagnosis of diabetes mellitus and other categories of glucose intolerance. Diabetes 28: 1039–1057.

29. Bes-Rastrollo M, Perez Valdivieso JR, Sanchez-Villegas A, Alonso A, Martinez-Gonzalez MA (2005) Validación del peso e índice de masa corporal auto-declarados de los participantes de una cohorte de graduados universitarios.[Validation of self-reported weight and body mass index of participants in a cohort of university graduates] Rev EspObes 3: 183–189.

30. Martinez-Gonzalez MA, Lopez-Fontana C, Varo JJ, Sanchez-Villegas A, Martinez JA (2005) Validation of the Spanish version of the physical activity questionnaire used in the Nurses' Health Study and the Health Professionals' Follow-up Study. Public Health Nutr 8: 920–927.

31. Ainsworth BE, Haskell WL, Whitt MC, Irwin ML, Swartz AM, et al. (2000) Compendium of physical activities: an update of activity codes and MET intensities. Med Sci Sports Exerc 32:S498–S504.

32. Smyth S, Heron A (2006) Diabetes and obesity: the twin epidemics. Nature Medicine 12: 75–80.

33. Buchanan TA, Xiang AH (2005) Gestational diabetes mellitus. J Clin Invest 115: 485–491.

34. Nolan CJ, Damm P, Prentki M (2011) Type 2 diabetes across generations: from pathophysiology to prevention and management. Lancet 378: 169–181.

35. ADA (2014) Standards of Medical Care in Diabetes – 2014. Diabetes Care (Suppl 1): S14–S80.

36. Park S, Kim MY, Baik SH, Woo JT, Kwon YJ, et al. (2013) Gestational diabetes is associated with high energy and saturated fat intakes and with low plasma visfatin and adiponectin levels independent of prepregnancy BMI. Eur J ClinNutr 67: 196–201.

37. Bao W, Bowers K, Tobias DK, Hu FB, Zhang C (2013)Prepregnancy dietary protein intake, major dietary protein sources, and the risk of gestational diabetes mellitus. A prospective cohort study. Diabetes Care 36: 2001–2008.

38. Mozaffarian D, Hao T, Rimm EB, Willett WC, Hu FB (2011) Changes in diet and lifestyle and long-term weight gain in women and men. N Engl J Med 364: 2392–2404.

39. Lopez S, Bermudez B, Ortega A, Varela LM, Pacheco YM, et al. (2011) Effects of meals rich in either monounsaturated or saturated fat on lipid concentrations and on insulin secretion and action in subjects with high fasting triglyceride concentrations. Am J ClinNutr 93: 494–499.

40. Uusitalo U, Arkkola T, Ovaskainen ML, Kronberg-Kippila C, Kenward MG, et al. (2009) Unhealthy dietary patterns are associated with weight gain during pregnancy among Finnish women. Public Health Nutrition 12: 2392–2399.

41. Vander Weele TJ (2009) On the relative nature of overadjustment and unnecessary adjustment. Epidemiology 20: 496–499.

42. Greenland S, Neutra R. (1980) Control of confounding in the assessment of medical technology. Int J Epidemiol 9: 361–367.

43. Lawrence JM, Contreras R, Chen W, Sacks DA (2008) Trends in the prevalence of preexisting diabetes and gestational diabetes mellitus among a racially/ethnically diverse population of pregnant women, 1999–2005. Diabetes Care 31: 899–904.

44. Pan A, Malik VS, Hu FB (2012) Exporting diabetes mellitus to Asia. The impact of Western-style fast food. Circulation 126: 163–165.

45. Willett W (1998) Nutritional epidemiology. 2nd ed. New York: Oxford University Press.

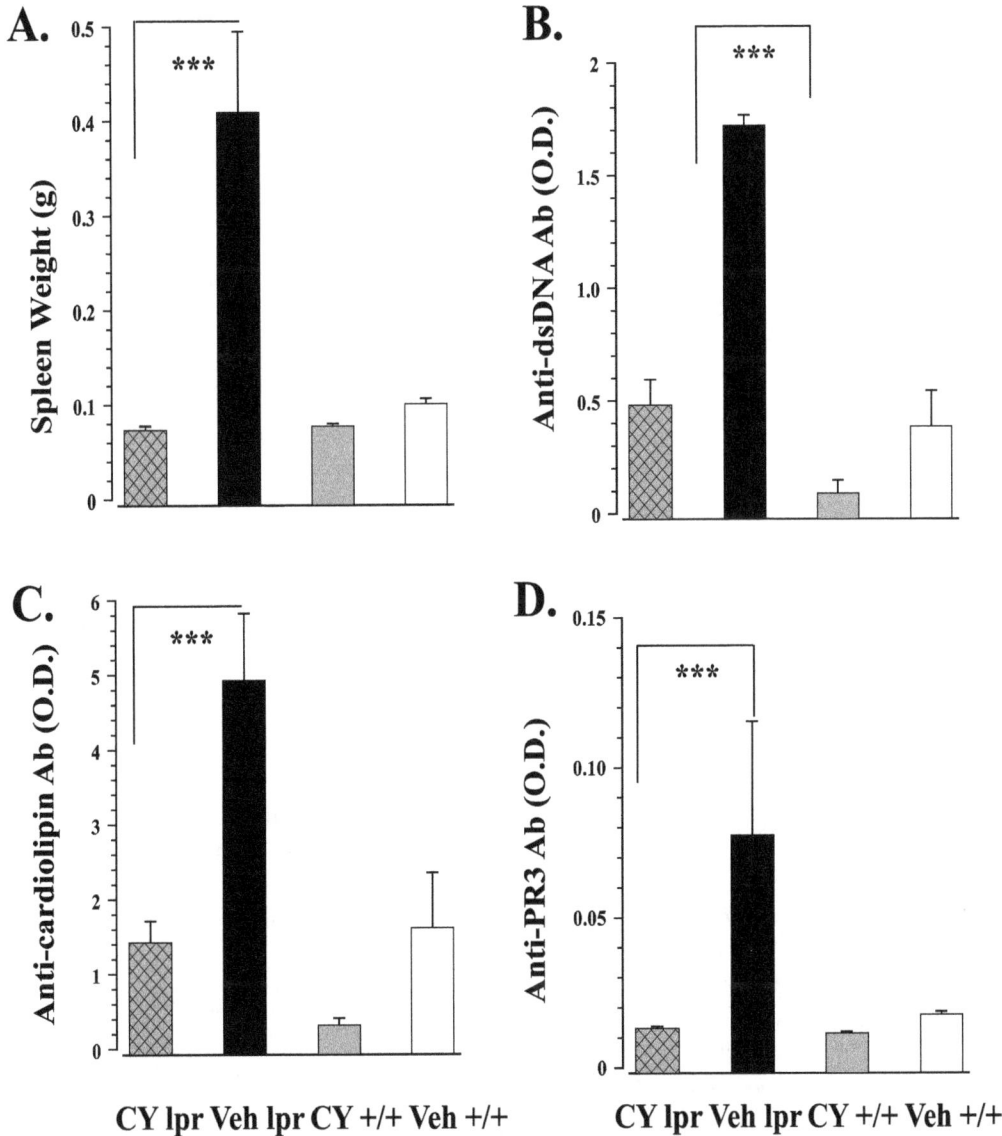

Figure 6. Immunosuppressive effects of continuous exposure to CY/sucrose solution. Generalized immunosuppression in CY-exposed MRL/lpr mice evidenced by normalized spleen weight (A), as well as significant reductions in serum levels of autoantibodies directed towards dsDNA (B), cardiolipin (C), and PR3 (D) antigens.

findings are consistent with previous observations that brief, 1-hr access to CY-laced chocolate milk over 3–4 weeks results in significant reductions in lymphadenopathy and serum anti-ssDNA antibody levels [17,18], as well as with generalized immunosuppression after repeated 100 mg/kg/week intraperitoneal CY injections [42]. Along the same line, *ad lib* access to the immunosuppressive solution abolished several behavioural deficits, as shown previously for anxiety-like and motivated behaviours following CY treatment [24,42]. Conversely, the lack of effectiveness in preventing excessive floating in the forced swim test and brain atrophy suggest a role for non-immunological factors, such as genetic lesions and/or neuroendocrine changes [43].

The second concept that emerged relates to the non-invasive administration route of a noxious drug in experimental mice. Indeed, the present results suggest that lacing it with a palatable ingredient can produce a therapeutic dose comparable to the injection route. As shown in Figure 3, the weekly dose voluntarily ingested was relatively constant in each substrain and can likely be adjusted by increasing the number of exposure days and/or concentration of sucrose or the drug itself. By avoiding repeated exposure to handling, oral gavage, and injections, such an approach can minimize the confounding effects of stress and reduce variance, thus increasing consistency and precision across behavioural and physiological studies.

Author Contributions

Conceived and designed the experiments: MK BS. Performed the experiments: MK HZ RH MM. Analyzed the data: MK BS. Contributed reagents/materials/analysis tools: DM BS. Contributed to the writing of the manuscript: MK BS.

References

1. Attardo C, Sartori F (2003) Pharmacologically active plant metabolites as survival strategy products. Boll Chim Farm 142: 54–65.
2. Davies AG, Baillie IC (1988) Soil-Eating by Red Leaf Monkeys (Presbytis rubicunda) in Sabah, Northern Borneo. Biotropica 20: 252–258.
3. Oates JF (1978) Water-Plant and Soil Consumption by Guereza Monkeys (Colobus guereza): A Relationship with Minerals and Toxins in the Diet? Biotropica 10: 241–253.
4. Huffman MA, Caton JM (2001) Self-induced Increase of Gut Motility and the Control of Parasitic Infections in Wild Chimpanzees. International Journal of Primatology 22: 329–346.
5. Huffman MA (2003) Animal self-medication and ethno-medicine: exploration and exploitation of the medicinal properties of plants. Proceedings of the Nutrition Society 62: 371–382.
6. Wrangham RW (1995) Relationship of chimpanzee leaf-swallowing to a tapeworm infection. American Journal of Primatology 37: 297–303.
7. Villalba JJ, Landau SY (2012) Host behavior, environment and ability to self-medicate. Small Ruminant Research 103: 50–59.
8. Kreulen DA (1985) Lick use by large herbivores: a review of benefits and banes of soil consumption. Mammal Review 15: 107–123.
9. Diamond JM (1999) Evolutionary biology. Dirty eating for healthy living. Nature 400: 120–121.
10. Raman R, Kandula S (2008) Zoopharmacognosy. Resonance 13: 245–253.
11. Andrews BS, Eisenberg RA, Theofilopoulos AN, Izui S, Wilson CB, et al. (1978) Spontaneous murine lupus-like syndromes. Clinical and immunopathological manifestations in several strains. J Exp Med 148: 1198–1215.
12. Watanabe-Fukunaga R, Brannan CI, Copeland NG, Jenkins NA, Nagata S (1992) Lymphoproliferation disorder in mice explained by defects in Fas antigen that mediates apoptosis. Nature 356: 314–317.
13. Watanabe-Fukunaga R, Brannan CI, Itoh N, Yonehara S, Copeland NG, et al. (1992) The cDNA structure, expression, and chromosomal assignment of the mouse Fas antigen. J Immunol 148: 1274–1279.
14. Dixon FJ, Andrews BS, Eisenberg RA, McConahey PJ, Theofilopoulos AN, et al. (1978) Etiology and pathogenesis of a spontaneous lupus-like syndrome in mice. Arthritis Rheum 21: S64–S67.
15. Theofilopoulos AN (1992) Murine models of lupus. In: Lahita RG, editors. Systemic lupus erythematosus. New York: Churchill Livingstone. 121–194.
16. Grota LJ, Ader R, Cohen N (1987) Taste aversion learning in autoimmune Mrl-lpr/lpr and Mrl +/+ mice. Brain Behav Immun 1: 238–250.
17. Grota LJ, Schachtman TR, Moynihan JA, Cohen N, Ader R (1989) Voluntary consumption of cyclophosphamide by Mrl mice. Brain Behav Immun 3: 263–273.
18. Grota LJ, Ader R, Moynihan JA, Cohen N (1990) Voluntary consumption of cyclophosphamide by nondeprived Mrl-lpr/lpr and Mrl +/+ mice. Pharm Biochem Behav 37: 527–530.
19. Zahorik DM, Maier SF, Pies RW (1974) Preferences for tastes paired with recovery from thiamine deficiency in rats: appetitive conditioning or learned safety? J Comp Physiol Psychol 87: 1083–1091.
20. Garcia J, Hankins WG, Rusiniak KW (1974) Behavioral regulation of the milieu interne in man and rat. Science 185: 824–831.
21. Kim A, Feng P, Ohkuri T, Sauers D, Cohn ZJ, et al. (2012) Defects in the Peripheral Taste Structure and Function in the MRL/lpr Mouse Model of Autoimmune Disease. PLoS ONE 7: e35588.
22. Sakic B, Szechtman H, Denburg SD, Carbotte RM, Denburg JA (1993) Spatial learning during the course of autoimmune disease in MRL mice. Behav Brain Res 54: 57–66.
23. Sakic B, Szechtman H, Talangbayan H, Denburg SD, Carbotte RM, et al. (1994) Disturbed emotionality in autoimmune MRL-lpr mice. Physiol Behav 56: 609–617.
24. Sakic B, Denburg JA, Denburg SD, Szechtman H (1996) Blunted sensitivity to sucrose in autoimmune MRL-lpr mice: a curve-shift study. Brain Res Bull 41: 305–311.
25. Willner P, Muscat R, Papp M (1992) Chronic mild stress-induced anhedonia: a realistic animal model of depression. Neurosci Biobehav Rev 16: 525–534.
26. Monleon S, D'Aquila P, Parra A, Simon VM, Brain PF, et al. (1995) Attenuation of sucrose consumption in mice by chronic mild stress and its restoration by imipramine. Psychopharmacology (Berl) 117: 453–457.
27. Muscat R, Willner P (1992) Suppression of sucrose drinking by chronic mild unpredictable stress: a methodological analysis. Neurosci Biobehav Rev 16: 507–517.
28. Deacon RM, Rawlins JN (2006) T-maze alternation in the rodent. Nat Protoc 1: 7–12.
29. Ballok DA, Woulfe J, Sur M, Cyr.M., Sakic B (2004) Hippocampal damage in mouse and human forms of systemic autoimmune disease. Hippocampus 14: 649–661.
30. Sakic B, Szechtman H, Denburg JA, Gorny G, Kolb B, et al. (1998) Progressive atrophy of pyramidal neuron dendrites in autoimmune MRL-lpr mice. J Neuroimmunol 87: 162–170.
31. Sled JG, Spring S, van Eede M, Lerch JP, Ullal S, et al. (2009) Time course and nature of brain atrophy in the MRL mouse model of central nervous system lupus. Arthritis Rheum 60: 1764–1774.
32. Galef BG (1991) A contrarian view of the wisdom of the body as it relates to dietary self-selection. Psychological Review 98: 218–223.
33. Ballok DA, Szechtman H, Sakic B (2003) Taste responsiveness and diet preference in autoimmune MRL mice. Behav Brain Res 140: 119–130.
34. Sakic B, Szechtman H, Keffer M, Talangbayan H, Stead R, et al. (1992) A behavioral profile of autoimmune lupus-prone MRL mice. Brain Behav Immun 6: 265–285.
35. Mukherjee N, Delay ER (2011) Cyclophosphamide-induced disruption of umami taste functions and taste epithelium. Neuroscience 192: 732–745.
36. Mukherjee N, Carroll BL, Spees JL, Delay ER (2013) Pre-Treatment with Amifostine Protects against Cyclophosphamide-Induced Disruption of Taste in Mice. PLoS ONE 8: e61607.
37. Bellush LL, Rowland NE (1986) Dietary self-selection in diabetic rats: an overview. Brain Res Bull 17: 653–661.
38. Jimenez JA, Hughes KA, Alaks G, Graham L, Lacy RC (1994) An experimental study of inbreeding depression in a natural habitat. Science 266: 271–273.
39. Meagher S, Penn DJ, Potts WK (2000) Male-male competition magnifies inbreeding depression in wild house mice. Proc Natl Acad Sci U S A 97: 3324–3329.
40. Williams S, Sakic B, Hoffman SA (2010) Circulating brain-reactive autoantibodies and behavioral deficits in the MRL model of CNS lupus. J Neuroimmunol 218: 73–82.
41. Loheswaran G, Stanojcic M, Xu L, Sakic B (2010) Autoimmunity as a principal pathogenic factor in the refined model of neuropsychiatric lupus. Clin Exp Neuroimmunol 1: 141–152.
42. Sakic B, Szechtman H, Denburg SD, Denburg JA (1995) Immunosuppressive treatment prevents behavioral deficit in autoimmune MRL-lpr mice. Physiol Behav 58: 797–802.
43. Loheswaran G, Kapadia M, Gladman M, Pulapaka S, Xu L, et al. (2013) Altered neuroendocrine status at the onset of CNS lupus-like disease. Brain Behav Immun 32: 86–93.

9

Learning to Eat Vegetables in Early Life: The Role of Timing, Age and Individual Eating Traits

Samantha J. Caton[1,2]**, Pam Blundell**[1]**, Sara M. Ahern**[1]**, Chandani Nekitsing**[1]**, Annemarie Olsen**[3]**,
Per Møller[3]**, Helene Hausner**[3]**, Eloïse Remy**[4,5,6]**, Sophie Nicklaus**[4,5,6]**, Claire Chabanet**[4,5,6]**,
Sylvie Issanchou[4,5,6]**, Marion M. Hetherington**[1]*

1 Institute of Psychological Sciences, University of Leeds, Leeds, United Kingdom, **2** School of Health and Related Research, University of Sheffield, United Kingdom, **3** Department of Food Science, University of Copenhagen, Copenhagen, Denmark, **4** CNRS, UMR6265, Centre des Sciences du Goût et de l'Alimentation, Dijon, France, **5** INRA, UMR1324, Centre des Sciences du Goût et de l'Alimentation, Dijon, France, **6** Université de Bourgogne, Centre des Sciences du Goût et de l'Alimentation, Dijon, France

Abstract

Vegetable intake is generally low among children, who appear to be especially fussy during the pre-school years. Repeated exposure is known to enhance intake of a novel vegetable in early life but individual differences in response to familiarisation have emerged from recent studies. In order to understand the factors which predict different responses to repeated exposure, data from the same experiment conducted in three groups of children from three countries (n = 332) aged 4–38 m (18.9±9.9 m) were combined and modelled. During the intervention period each child was given between 5 and 10 exposures to a novel vegetable (artichoke puree) in one of three versions (basic, sweet or added energy). Intake of basic artichoke puree was measured both before and after the exposure period. Overall, younger children consumed more artichoke than older children. Four distinct patterns of eating behaviour during the exposure period were defined. Most children were "learners" (40%) who increased intake over time. 21% consumed more than 75% of what was offered each time and were labelled "plate-clearers". 16% were considered "non-eaters" eating less than 10 g by the 5th exposure and the remainder were classified as "others" (23%) since their pattern was highly variable. Age was a significant predictor of eating pattern, with older pre-school children more likely to be non-eaters. Plate-clearers had higher enjoyment of food and lower satiety responsiveness than non-eaters who scored highest on food fussiness. Children in the added energy condition showed the smallest change in intake over time, compared to those in the basic or sweetened artichoke condition. Clearly whilst repeated exposure familiarises children with a novel food, alternative strategies that focus on encouraging initial tastes of the target food might be needed for the fussier and older pre-school children.

Editor: Mihai Covasa, INRA, France

Funding: The research leading to these results has received funding from the European Community's Seventh Framework Program (FP7/2007–2013) under the grant agreement n°FP7-245012-HabEat. The French part was also supported by the Regional Council of Burgundy. The authors have no conflict of interest to disclose. The funders had no role in study design, data collection and analysis, decision to publish, or preparation of the manuscript.

Competing Interests: The authors have declared that no competing interests exist.

* E-mail: m.hetherington@leeds.ac.uk

Introduction

Despite current recommendations and the apparent health related benefits, vegetable consumption is below the recommended level in both adults and children [1,2]. Children dislike vegetables [3] and when given the option pre-schoolers avoid vegetables when allowed to choose lunch [4]. Children prefer foods which are high in energy density [5–7] and appear to accept sweet taste more than bitter taste from birth [8]. Therefore, lower energy density and bitter taste might inhibit intake of vegetables among children. Nevertheless, exposure to the taste of vegetables promotes acceptance. The mere exposure phenomenon first described by Zajonc [9] predicts that familiarisation to a stimulus results in a positive attitude to that particular stimulus. Thus, applying this to food acceptance, it is predicted that repeated experience will be effective in increasing liking and intake of novel vegetables [10–15].

Whilst there is extensive evidence regarding the effectiveness of repeated exposure on promoting vegetable liking and intake, whether this is equally effective across children remains unclear. Food preferences have been shown to occur through pre-natal experience and breastfeeding [16–18]. For example, flavours experienced in amniotic fluid or breast milk might be sufficient to promote the intake of those specific or associated flavours later in life [19,20]. Breastfed babies are more likely to accept novel foods including vegetables compared to those who were not breastfed [20,21], breastfeeding also affects the healthfulness of the habitual diet later in life [22,23].

In contrast to weanlings, as children get older they become more reluctant to consume novel foods and by 2–3 years of age many develop neophobia [24]. During this stage even previously liked foods might be refused [25]. Many different techniques have been tested to promote vegetable preference and intake. Such techniques range from the relatively subtle and covert such as observational learning and social modelling [26–28], availability [3], hiding vegetables [29,30], to the more overt and direct such as using social praise or tangible rewards [31]. Attempts to improve the acceptability of vegetables by offering dips and condiments [28,32,33] or by adding energy (flavour-nutrient learning) and/or an already liked flavour (flavour-flavour learning) yields variable results [15,34,35].

Repeated exposure is the simplest and most convenient method to enhance vegetable intake in children, and is ecologically valid since it mimics what mothers generally do at home when introducing new vegetables. However, mothers often give up after only 5 exposures [36] yet current recommendations suggest at least 8–10 exposures [37]. Anecdotally, mothers adopt different strategies for encouraging vegetable intake in their children [30] and these different strategies may be more or less successful depending on a general child's temperament, prior experience with vegetables and eating traits. For example, fussy eaters are likely to refuse to try new foods whereas children who are less fussy might be more receptive to any new foods including vegetables. Measuring eating traits [38] which are stable over time [39] such as fussy eating and enjoyment of food offers a means to predict eating patterns including those children who respond well to repeated exposure and those who do not.

The current study was designed to investigate three questions; what individual characteristics predict initial acceptance of a novel vegetable, what individual characteristics predict patterns of acceptance (intake) over time, for example which characteristics predict those children who consume everything offered compared to those who do not, and what individual characteristics predict the effectiveness of repeated exposure in promoting vegetable intake. Data from three investigations using the same target vegetable products, and following the same procedure in the UK [40], Denmark [41] and France [42] were combined to examine the impact of characteristics of the child such as age, BMI, eating traits and diet history (breastfeeding duration, age of introduction of solid food) on initial vegetable acceptance, pattern of intake during repeated exposure, and effectiveness of repeated exposure on learning to like this novel vegetable. Children were also grouped into eating categories, using the pattern of their intake during the exposure period, and logistic regressions were conducted to investigate predictors of category membership.

Materials and Methods

Participants

Managers of private day care nurseries and parents of preschool children were invited to take part in the investigation. A total of 403 preschool children from the UK (n = 108, aged 6–36 m), France (n = 123 aged 4–8 m) and Denmark (n = 172, aged 6–36 m) were recruited for the study between Jan and May 2011. Children were enrolled if they were aged between 4 and 38 months at the beginning of the study and for the French, younger cohort they were included if the introduction of complementary foods was started more than 2 weeks and less than 2 months prior to the start of the study. Children were excluded from taking part in the investigation if they had any known food allergies.

Ethics Statement

The studies were approved by the University of Leeds, Institute of Psychological Science ethics committee (UK), Comité de Protection de Personnes Est I Bourgogne (France) and after reviewing the study protocol the study was found not to require formal approval by the Copenhagen Regional Research ethics committee. The study procedures complied with the Helsinki Declaration. Written parental consent was given for the participating children in the three countries.

Experimental Design and Measurements of Intake

Pre-intervention levels of intake were initially evaluated. Children were given up to 200 g (2×100 g pots) of basic artichoke puree, and their intake was weighed. For the intervention, children were randomly assigned to one of three groups: repeated exposure (basic artichoke puree, n = 112), flavour-flavour learning (basic artichoke puree with added sweetness, n = 112) or flavour-nutrient learning (basic artichoke puree with added energy n = 108). Each child received 5–10 exposures to one of the purees (variation due to unplanned absences from nursery) during a state of hunger, either before a main meal or as an afternoon snack (UK and Denmark) or at the beginning of a meal (France). In UK children were offered 100 g per exposure and in Denmark and France children were offered up to 200 g. Finally, the effectiveness of the intervention was assessed by offering 200 g of basic artichoke, and the intake was weighed. Detailed descriptions of the study have been previously published elsewhere [40–42]. Intake was measured before and after the intervention and throughout the exposure period. Change in intake following the intervention was calculated by subtracting the baseline from the post-intervention intake.

In order to characterise the patterns of intake for each child during the intervention period, linear regressions for each child were calculated, with weight of food consumed on each trial as the outcome variable, and exposure number as the predictor variable. This provided a value for each child of the intercept corresponding to the predicted intake on exposure one and a value of the slope corresponding to the rate of change in consumption over successive exposures that were not overly affected by behaviour on each individual day. However, simple regressions could underestimate the rate of change of consumption for children who rapidly learned to like the puree. Therefore a quadratic regression was also calculated for each child. In the case where the quadratic term was a significant negative predictor, indicating that the rate of change was decreasing overtime, we used only the first five trials to create our linear model (n = 41). We then classified the children into one of four categories, according to the following algorithm. If the slope was a significant predictor at $\alpha = 0.1$ level, and greater than 2 g/exposure, a child was classified as a "learner". If their predicted intake at exposures 1 and 5 were greater than 75 g, and the slope was greater than - 2.5 g, they were classified as "plate-clearers". If their predicted intake at exposure 5 was less than 10 g, they were classified as non-eaters. All other children were assigned to the category 'others'.

Study Foods

For the investigation baby-food grade ingredients were used, in order to meet the European regulation (Directive 2006/125/CE), because the study was conducted with children younger than 3 years old. One recipe was developed for each condition. The recipe, used for the repeated exposure condition and for the pre and post-intervention measurement, was a basic artichoke puree (48 kcal/100 g). For the flavour-flavour condition the chosen unconditioned stimulus was sweetness (sucrose, 51 kcal/100 g) and for the flavour-nutrient condition, the chosen unconditioned stimulus was a higher energy density (addition of sunflower oil, 144 kcal/100 g). The ingredients selected were baby food-grade frozen artichoke heart (France Recherche & Développement FRDP, Avignon, France), water, sucrose (Vermandoise, Peronne Cedex, France), sunflower oil (Huileries de Lapalisse, Lapalisse, France) and salt. A full description of the study foods used can be obtained elsewhere [40–42].

Measurements of Individual Characteristics

Demographic and anthropometric information. Parents and caregivers were asked to report their child's age and sex. Height and weight were self-reported by mothers in France and Denmark based on measurements taken by a medical doctor and recorded in the health notebook and in the UK measurements were recorded the experimenters using digital scales (Seca) and a portable stadiometer (LeicesterSMSSE-0260; Seca Model 416 infantometer). Using the WHO anthropometric calculator, weights and lengths or heights were entered (http://www.who.int/childgrowth/software/en/), for children over 12 months of age, weight-for-height z-scores were calculated, and for children 12 months of age and younger, weight-for-length z-scores were calculated.

Early feeding practices and child eating behavior. Parents and caregivers provided information related to early feeding practices. Thus, they answered the following questions "Did you breast feed your child, if so for how long?", and "How old was your child when you introduced formula-milk?" to determine the duration of total breastfeeding. Parents and caregivers were asked "How old was your child when they were introduced to solid foods?" to determine age of introduction of solid food. Individual eating traits of the child were reported by parents using the Child Eating Behaviour Questionnaire [38] adapted for 15 month-old infants. The CEBQ is a validated psychometric tool that measures eating behaviour styles and individual appetitive traits. In the present study, four out of seven dimensions were evaluated: food fussiness, enjoyment of food, satiety responsiveness and food responsiveness. Items were scored on a 5-point Likert scale ranging from "never" to "always". The Danish CEBQ was completed using a 7-point scale and so this was rescaled for comparison with the other countries.

Output Variables and Statistical Analysis

Different output variables characterising children's eating behaviour were considered. First, the initial intake of the basic artichoke at pre-intervention was considered. Second, the change in intake of basic artichoke puree from pre- to post-intervention was calculated. Third, the intercept and slope, which characterise the pattern of intake during the exposure phase, were calculated by individual regressions.

Correlations between individual characteristics were investigated using Kendall's tau.

Multiple linear regressions were used to identify individual characteristics which predicted the initial intake before the intervention, the change in intake from pre- to post-intervention, the predicted intake on exposure one during intervention (intercept), and the rate of change of consumption over successive exposures (slope). Z-scores for all the scalar predictor variables (age, total breastfeeding duration, BMI z-score, enjoyment of food, food fussiness, food responsiveness, satiety responsiveness and age of introduction of solid food) were calculated and entered into the model as predictors, along with experimental condition, simultaneously. The normality of the distribution of the residuals was tested with the Shapiro-Wilk test. Where this assumption was not met, robust regression was carried out using iterated re-weighted least squares. We report only the significant predictors, that is those for which the gradient co-efficient was significantly different from zero (alpha <0.05) as assessed by t-test.

One-way ANOVA was used to investigate differences in individual characteristics between eating categories (learners, non-eaters, plate-clearers and others) and Bonferroni corrected pairwise comparisons were used to interpret significant differences. Chi-square tests were used to investigate frequency of eating category in each country and frequency of eating category across experimental conditions. Logistic regressions were used to predict which variables characterised eating category. Multicollinearity was assessed using the variance inflation factor (VIF), and models with VIF greater than 10 were disregarded [43]. All data are presented as means ± SD and the alpha value was set at 0.05 except for individual regressions.

Results

Participant Characteristics

403 children were recruited to take part and data from 71 children were excluded due to one of the following reasons: not meeting the inclusion criteria, withdrawal from the study or insufficient exposures during the intervention period. Data for 332 pre-school children aged between 4 and 38 months (mean age 18.9±9.95 months) are presented. Table 1 displays the demographic characteristics overall and for each country.

There were significant differences in the age of children between each country ($F_{(2,329)} = 268.7$, $p<0.001$). This was due to differences between France and both Denmark and UK (both $p<0.001$). ANOVA demonstrated a main effect of country for length of time of exclusive breastfeeding ($F_{(2,219)} = 13.07$, $p<0.001$) with differences existing between Denmark and France ($p<0.001$). Differences between the UK and Denmark were borderline ($p = 0.051$). ANOVA also demonstrated significant differences between the countries for total breastfeeding duration ($F_{(2,237)} = 12.04$, $p<0.001$). This was due to differences between France and both Denmark ($p = 0.004$) and UK ($p<0.001$). The difference between Denmark and UK was borderline significant ($p = 0.051$). There were also differences between all countries in the age of introduction of solid food ($F_{(2,245)} = 13.3$, $p<0.001$), with differences existing between France and both Denmark ($p<0.001$) and UK ($p = 0.01$). Additionally there were differences between all countries for the BMI z-score ($F_{(2,225)} = 34.23$, $p<0.001$), satiety responsiveness ($F_{(2,245)} = 342.1$, $p<0.001$), and food fussiness ($F_{(2,244)} = 103.0$, $p<0.001$). There was no difference between the countries for food responsiveness or enjoyment of food.

Correlations between Measured Individual Characteristics

Table 2 shows the pattern of correlations between the different measures of individual characteristics. The Shapiro-Wilk test of normality revealed significant deviations from normality for most variables. Therefore, Kendall's tau was calculated to indicate associations between the variables. Age was significantly positively correlated with BMI z-score, satiety responsiveness, food fussiness, and duration of breastfeeding. Age was negatively correlated with enjoyment of food and age of introduction of solid food. BMI z-score had a positive correlation with duration of breastfeeding, satiety responsiveness and food fussiness; this last correlation is to be expected since age correlated with BMI z-score and with food fussiness. Enjoyment of food had a significant negative correlation with satiety responsiveness. Satiety responsiveness also positively correlated with food fussiness and duration of breastfeeding and had a significant negative correlation with age of introduction of solid food. Food fussiness had a significant positive correlation with duration of breastfeeding, which is to be expected given that older children had been breastfed for longer and were also more food fussy. Food fussiness also had a significant negative correlation with age of introduction of solid food.

Table 1. Participant characteristics of pre-school children who took part in the intervention (Means ± SD) overall and split by country (DK = Denmark, UK = United Kingdom, FR = France)*.

	All		DK		UK		FR	
	N =		N =		N =		N =	
Number of participants	332		165		72		95	
Age (months)	332	18.92 (9.95)	165	24.01 (7.03) a	72	23.56 (7.75) a	95	6.57 (0.92) b
Sex	332	M = 175 F = 157	165	M = 86 F = 79	72	M = 32 F = 40	95	M = 57 F = 38
BMI z-score	228	0.36 (1.21)	103	0.57 (1.12) b	47	1.14 (0.74) a	78	−0.39 (1.15) c
Duration of exclusive breastfeeding (weeks)	222	13.56 (8.61)	98	16.35 (6.92) a	35	14.17 (8.95) a, b	89	10.25 (9.09) b
Duration of total breastfeeding (weeks)	240	21.27 (17.94)	114	23.1 (13.07) a	35	31.09 (30.9) a	91	15.21 (14.16) b
Age of introduction of solid food (weeks)	248	21.24 (4.48)	118	20.02 (4.79) b	35	20.57 (4.16) b	95	23 (3.57) a
Enjoyment of food	247	4.11 (0.61)	119	4.09 (0.59)	35	3.94 (0.63)	93	4.21 (0.61)
Satiety responsiveness	248	2.63 (0.91)	119	3.34 (0.37) a	35	2.84 (0.47) b	94	1.65 (0.58) c
Food fussiness	247	2.48 (0.79)	119	3.02 (0.32) a	35	2.3 (0.78) b	93	1.87 (0.75) c
Food responsiveness	248	2.49 (0.74)	119	2.54 (0.75)	35	2.25 (0.63)	94	2.52 (0.77)

*not all parents answered all questions.
Means with a different letter are significantly different (a, b, c).

Table 2. Patterns of correlations between all measures taken in pre-school children.

	BMI z-score	Enjoyment of food	Satiety responsiveness	Food fussiness	Food responsiveness	Duration of total breastfeeding (weeks)	Age of introduction of solid food (weeks)
Age (months)	.24***	-.13**	.48***	.42***	-.01	.22***	-.15**
BMI z-score		.03	.16***	.12**	.06	.11*	-.06
Enjoyment of food			-.11*	.001	.06	-.02	.03
Satiety responsiveness				.45***	.08	.17***	-.24***
Food fussiness					.003	.13**	-.18***
Food responsiveness						-.02	.05
Duration of total breastfeeding (weeks)							.06

*p<0.05, **p<0.01, ***p<0.001.

Predictors of Pre-intervention Intake

We calculated regression models to investigate the predictors of the actual intake of the novel target vegetable in the pre-intervention phase. The only significant predictor of the amount of artichoke consumed during the pre-intervention phase was age (b = −24.25 (6.28), β = −0.40 (0.10), p<0.001, overall model adj R^2 = 0.3, p<0.001). With robust regression, age (b = −16.8 (4.9), β = −0.28 (0.08), p<0.001) and satiety responsiveness (b = −16.1 (4.7), β = −0.26 (0.08), p<0.001) were both significant predictors, with younger and less satiety responsive children consuming more of the novel vegetable in the pre-intervention period.

Predictors of Change in Intake from Pre- to Post-intervention

Overall the model was weak, with adj R^2 = 0.05, p = 0.04. The predictors of change in artichoke intake were enjoyment of food (b = 12.9 (5.9), β = 0.17 (0.08), p = 0.03) and being in the flavour-nutrient learning condition (b = −35.5 (14.9), β = −0.20 (0.09), p = 0.02). With robust regression these two variables remained significant predictors; enjoyment of food (b = 13.1 (6.0), β = 0.17 (0.08), p = 0.03) and being assigned to the flavour-nutrient learning condition (b = −31.8 (15.1), β = −0.18 (0.09), p = 0.04). The change in intake from pre- to post-intervention could suffer from ceiling effects for children who ate most during both pre- and post-intervention tests. Therefore, we repeated this analysis using only children who ate less than 50 g at pre-intervention. This revealed a highly significant model (adj R^2 = 0.21, p<0.001) with age (b = −32.56 (9.90), β = −0.43 (0.13), p = 0.001) and enjoyment of food (b = 13.88 (6.34), β = 0.21 (0.09), p = 0.03) as significant predictors of the change in intake. With robust regression age (b = −37.5 (10.1), β = −0.49 (0.13). p<0.001) and enjoyment of food (b = 14.3 (6.5), β = 0.21 (0.10), p = 0.03) were still the only significant predictors.

Predictors of Initial Intake (Intercept) during the Intervention

The model was a good predictor of the intercept (adj R^2 = 0.49, p<.001), with age (b = −27.5 (4.88), β = −0.48 (0.08), p<0.001), enjoyment of food (b = 6.76 (3.07), β = 0.12 (0.05), p = 0.0) and satiety responsiveness (b = −9.87 (4.88), β = −0.17 (0.08), p = 0.04) as the significant predictors. Younger children consumed more during the initial exposures than older children and children who scored higher on enjoyment of food and lower on satiety responsiveness consumed more during the initial exposures than those with lower scores on these variables.

Predictors of Slope during the Intervention

During the intervention period the children were exposed to one of three versions of the artichoke puree. In order to investigate whether the impact of the factors in the model differed by group, linear regressions were constructed in which the interaction terms of the variables with the experimental condition were entered as predictors. For the slope model, no interaction terms were significant predictors. However, for the intercept, the interaction terms were significant for the predictors of age, enjoyment of food, and satiety responsiveness. Therefore, the individual linear regressions were conducted for each experimental condition separately to further understand the impact of individual differences within the three groups. This found that in the repeated exposure group (adj R^2 = 0.40, p<.001), age (b = −27.12 (8.50), β = −0.51 (0.16), p = 0.002) and enjoyment of food (b = 11.80 (5.65), β = 0.22 (0.10), p = 0.04) were significant predictors of intercept. However, in the flavour-flavour learning

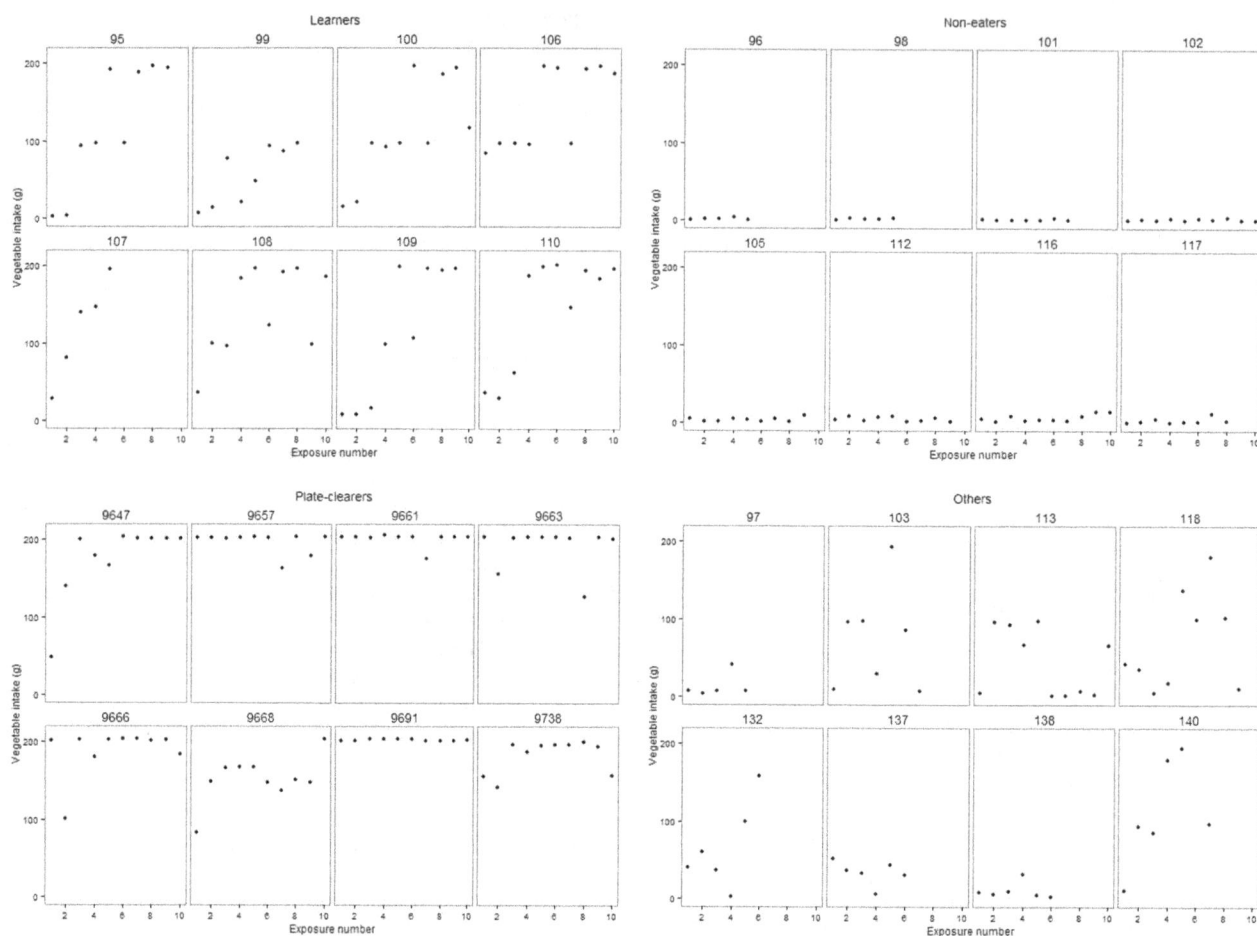

Figure 1. Eating categories: examples of individual profiles of intake (g) of artichoke over the intervention period. (Numbers for each case represent participant ID).

group (adj $R^2 = 0.62$, p<0.001) only age was a significant predictor (b = −46.52 (9.25), β = −0.66 (0.13), p<0.001). In the flavour-nutrient learning group (adj $R^2 = 0.48$, p<0.0001), only satiety responsiveness (b = −24.53 (8.99), β = −0.40 (0.15), p = 0.009) predicted initial intake.

As there were no interactions with group for predicting slope, a simpler regression model was constructed with the experimental condition, and z-scored age, total breastfeeding duration, BMI z-score, enjoyment of food, food fussiness, food responsiveness, satiety responsiveness, and age of introduction of solid food as predictors. This produced a significant model (adj $R^2 = 0.05$, p = 0.03), with experimental condition (being assigned to the flavour-nutrient learning condition) as the only significant predictor of slope (b = −4.31(2.04), β = −0.17 (0.08), p = 0.04). That is, those children in the flavour-nutrient learning condition had a flatter slope than those in either of the two other conditions. With robust regression, being in the flavour-nutrient group was the only predictor of slope (b = −4.8 (1.8), β = −0.19 (0.07), p = 0.009).

Eating Categories (Plate-clearers, Non-eaters, Learners and Others) and Their Predictors

Most children, 40%, in our sample were characterised as "learners" (n = 133), 16% (n = 53) of the children were classified as "non-eaters", 21% (n = 70) as "plate-clearers", and 23% (n = 76)

were classified as "others" since their pattern did not fit any of the other categories in a systematic way. Figure 1 shows typical intake profiles of a sample of children from each of the eating categories (with child ID presented above each profile).

Significant differences in the age of children in each eating category were observed (F(3,328) = 33.8, p<0.001). Differences were significant between all eating categories except between the learners and "others". ANOVA also demonstrated significant differences between the eating categories for satiety responsiveness (F(3,244) = 18.22, p<0.001) this was due to plate-clearers differing from both learners (p<0.001) and others (p = 0.008), and non-eaters differing from both plate-clearers (p<0.001) and others (p = 0.008). Food fussiness also differed significantly between the categories of eaters, (F(3,243) = 19.7, p<0.001). Non-eaters were significantly more food fussy than others and plate-clearers (both p<0.001) and learners (p = 0.02). Learners were more food fussy than others (p = 0.01) and plate-clearers (p<0.001). There was no difference between the categories of BMI z-score, enjoyment of food, food responsiveness, total breastfeeding duration, or age of introduction of solid food (Table 3).

Chi-square tests revealed significant differences between distribution of eating categories across the different countries (Chi-sq (6) = 70.8, p<0.001) (Table 4). There were more plate-clearers and fewer non-eaters in the French sample than in the other two countries. This is likely to be because the French children were significantly younger than the other children in the study. There

Table 3. Eating category individual characteristics of pre-school children (Means ± SD).

	Learners	Non-eaters	Plate-clearers	Others
N	133	53	70	76
Mean consumption during intervention (g)	94.0 (45.1) b	3.9 (2.0) d	118.3 (39.5) a	65.4 (39.5) c
Predicted initial intake (Intercept, g)	35.4 (41.5) c	3.6 (3.6) d	118.9 (36.4) a	63.0 (39.5) b
Rate of change in intake (Slope, g/exp)	17.2 (11.8) a	0.0 (0.8) b	3.0 (5.2) b	1.5 (9.5) b
Age (months)	20.0 (9.4) b	27.8 (5.7) a	12.4 (8.0) c	16.8 (9.9) b
BMI z-score	0.44 (1.23)	0.52 (0.96)	0.38 (1.32)	0.08 (1.14)
Enjoyment of food	4.1 (0.6)	3.9 (0.7)	4.2 (0.6)	4.1 (0.6)
Satiety responsiveness	2.8 (0.9) b	3.2 (0.5) a	2.0 (0.8) c	2.5 (0.9) b
Food fussiness	2.7 (0.7) b	3.1 (0.5) a	2.0 (0.8) c	2.3 (0.8) c
Food responsiveness	2.5 (0.7)	2.4 (0.9)	2.0 (0.8)	2.5 (0.8)
Duration of total breastfeeding (weeks)	20.8 (14.4)	23.0 (14.9)	19.7 (16.8)	22.7 (25.2)
Age of introduction of solid food (weeks)	20.7 (4.6)	21.0 (5.6)	21.7 (4.0)	21.8 (3.9)

Means with a different letter are significantly different (a, b, c, d).

were fewer plate-clearers in the Danish sample, more learners and non-eaters compared to the other countries.

Chi-square tests also revealed significant differences between distribution of eating categories across experimental conditions (Chi-sq = (6) = 20.3, p = 0.002) (Table 5). Children who were in the flavour-nutrient group were more likely to be "others", and less likely to be learners.

Predictors of Eating Category

Two logistic regression models were constructed to discriminate non-eaters from learners, and non-eaters from plate-clearers. The first model was successful at discriminating learners from non-eaters (Chi-sq (10) = 20.8, p = 0.02). The only significant predictors of eating category was food fussiness (b = −1.04 (0.50), Z = −2.09, p = 0.04), although age approached significance (b = −0.72 (0.39) Z = −1.85, p = 0.06). That is, younger children were more likely to be learners than non-eaters, and children who scored higher on food fussiness were more likely to be non-eaters than learners. The second model (plate-clearers vs. non-eaters) was a highly predictive model (Chi-sq (10) = 73.8, p<0.001), with age (b = 3.90 (1.32), Z = −2.95, p<0.001) as the only significant predictor, although food fussiness approached significance (b = 2.02 (1.16), Z = −1.917, p = 0.055). Younger children were more likely to be plate-clearers than older children and children with higher food fussiness scores were more likely to be non-eaters.

Discussion

The aim of the current study was to investigate how individual characteristics influence initial acceptance and effectiveness of repeated exposure to a novel vegetable. Our results demonstrated that the younger children were less fussy, enjoyed food more and had lower satiety responsiveness, representing a profile of characteristics that together contributed to increased acceptance of a novel food. Change in target vegetable intake (post - pre intervention) was predicted by age and enjoyment of food, with younger children and those with higher enjoyment of food scores being more responsive to repeated exposure. Age, enjoyment of food and satiety responsiveness were significant predictors of initial intake in the intervention period. Younger children consumed more as did those children who scored higher on enjoyment of food and lower on satiety responsiveness. Recipe was also important since experimental condition was the only significant predictor of the rate of change in intake over time, with those children in the flavour-nutrient learning condition (receiving a more energy-dense purée) demonstrating a flatter slope. The present investigation demonstrated that children respond differently to an intervention aimed at enhancing vegetable intake. Four categories of eaters were identified, plate-clearers, non-eaters, learners and others. Children who demonstrated a gradual increase over time were classified as learners and constituted the largest group. Age, food fussiness and being in the flavour-nutrient learning condition were predictors of eating category membership.

Table 4. Distribution (frequency and percentage of sample) of eating categories across countries.

	France	Denmark	UK
Learners	28 (29%)	82 (50%)*	23 (32%)
Non-eaters	0 (0%)**	38 (23%)**	15 (21%)
Others	26 (27%)	35 (21%)	15 (21%)
Plate-clearers	41 (43%)**	10 (6%)**	19 (26%)

(*chi-sq p<0.05, **contributes to chi-sq p<0.01).
Those in bold represent categories where there were more pre-school children than expected, those in italic represent those categories where there were less than expected.

Table 5. Distribution (frequency and percentage of sample) of eating categories across conditions.

	Flavour-flavour learning	Flavour-nutrient learning	Repeated exposure
Learners	51(46%)	30 (28%)*	52 (46%)
Non-eaters	6 (14%)	23 (21%)	14 (13%)
Others	17 (15%)	**37(34%)***	22 (20%)
Plate-clearers	28 (25%)	18 (17%)	24 (21%)

(*chi-sq p<0.05).
Those in bold represent categories where there were more pre-school children than expected, those in italic represent those categories where there were less than expected.

Younger children were more likely to be learners than non-eaters and those children scoring higher on food fussiness were more likely to be non-eaters than learners. Similarly, younger children were more likely to be plate-clearers than older pre-school children.

The age of child is key when introducing novel foods [44]. Age predicted initial intake of the novel vegetable both pre-intervention and during the initial exposure of the intervention period, with younger children consuming more compared to older children. Plate-clearers were also younger and less fussy whilst older children ate less consistently, were more likely to be non-eaters and were fussier compared to younger children. Successful repeated exposure is dependent upon tasting even small amounts of the target food [45,46]. Thus, repeated exposure is more likely to be effective at a time when most tastes are easily accepted, namely the weaning period. The first year of life presents a window of opportunity before the onset of food neophobia, which then peaks around 2–6 years [44,47,48], thus introducing novel foods such as different vegetables is optimal earlier rather than later. Repeated exposure has been reported to reduce neophobia [10], however, in pre-school children sufficient taste exposures are required to establish learned safety of novel foods [49]. Therefore, alternative methods that focus on encouraging initial tastes of the target food might be needed for the fussier and older pre-school children.

Learners were approximately 6 m younger than the non-eaters and this might reflect the period before neophobia develops. Learners in this sample scored higher on satiety responsiveness, and were relatively fussy compared to the plate-clearers and others. Food fussiness has been reported to correlate positively with satiety responsiveness [38,50] and negatively with enjoyment of food and food responsiveness [38,51]. In the current study food fussiness and satiety responsiveness, both "food avoidant" appetitive traits, positively correlated with age [50]. Age was also negatively correlated with enjoyment of food with younger children scoring higher on this scale.

Children in the "other" eating category had high initial intake, which dropped off or failed to increase over time. These children have low levels of food fussiness (same as plate-clearers) but higher levels of satiety responsiveness (same as learners). It is possible that these children were more subject to learned satiety, as there were more than the expected number of children in the flavour-nutrient condition within this group.

Children in the flavour-nutrient learning condition consumed less artichoke over the intervention period confirming previous investigations where energy was added to vegetables from carbohydrates [35] or a mixture of carbohydrates and fat [15]. It appears that flavour-nutrient learning is not effective in promoting intake of vegetables in young children. The current

investigation demonstrated the benefit of repeated exposure to a novel food [10–14,32,52] and to a limited extent some advantage of flavour-flavour learning [34,53] on promoting its acceptance. Although children prefer energy dense foods [6,7,54], adding oil directly to a novel vegetable changes both taste and texture and might reduce liking. However, de Wild et al. [15] demonstrated that pre-school children reported liking a high energy dense soup more than a matched low energy dense soup, but this preference did not affect intake. One possible explanation for this is expected satiation and learned satiety [55]. In support of this, satiety responsiveness was the only significant predictor of intake in the flavour-nutrient learning condition, suggesting detection of extra energy such as oil might limit intake. Alternatively children might simply have preferred the sweeter lower-fat version [34,56,57] or the pure, unadulterated taste of vegetables in the basic artichoke puree. Ahern et al. [58] reported that caregivers use fats such as butter in small amounts, however, little is known about how fat content and palatability might be manipulated to optimise vegetable intake in young children [56].

The findings of this study should be interpreted in light of the potential limitations. For example self-report measures were used to investigate height and weight in the Danish sample and for the French sample this was based on the most recent paediatrician's record of height/length and weight. Similarly age of introduction of solid food, duration of breastfeeding were self-reported and these might not be completely accurate, especially in the older children. This might partially explain the inverse relationship observed between age and age of introduction of solid food. The purees were served cold to the children in the UK and Denmark and warm to the French children and this might have influenced intake. However, this is unlikely since a novel food was used and the children would not have developed any learned expectations that the food should be served warm. Therefore, it is unlikely that warming the food would have had a significant impact on the taste to enhance palatability and intake. During the exposures the Danish and French children were given access up to 200 g during the intervention, therefore increasing the possibility of these children consuming more than UK sample who were offered 100 g. However, there was no evidence that the Danish children consumed more artichoke during the intervention. Additionally, the younger cohort included in this study were exclusively French and so we were not able to include country as a predictor in the models. To eliminate the confounding effect of this and the fact that the UK children and Danish children were offered different amounts the regressions were performed without the French data (data not shown), including the factor of country as a predictor. Interestingly age remained a significant predictor of pre-intervention artichoke intake, change in artichoke intake and of the intercept. This demonstrated indirectly that the portion size offered had no impact on the age effects observed in the current study. Age was no longer significant predictor of the rate of change as indicated by the slope. All other significant predictors became non-significant. For the analysis discriminating learners from non-eaters, no predictors were found. Yet, for the test discriminating plate-clearers from non-eaters, age remained a significant predictor. It is worth noting that the removal of the French data resulted in a dramatic reduction in number of participants and potentially power. Finally, whilst a positive effect of repeated exposure was observed in the younger pre-school children, it remains to be determined how long this lasts, although durable effects were observed at three [42] and six months post intervention [41] it is not clear if the effects would be stable beyond this time.

This is the first study, to our knowledge, to investigate the role of individual differences in response to novel vegetable exposure. It is clear that children respond differently to repeated exposure. This suggests that recommendations to improve vegetable intake in children should take account of individual differences. Novel vegetables are best introduced when children are young during a period when novel foods are readily accepted and before the onset of neophobia. Food preferences are formed early on, tend to be

fixed from early childhood and track into adulthood [4] it is therefore critical that a healthful diet is established early. Whilst repeated exposure might be effective for most children, for older and fussier children, other approaches are needed to improve acceptance. Alternative techniques such as the use of dips and sauces might be an effective way of encouraging these fussy eaters to try the target food [28,32,33]. Alternatively providing vegetables by stealth would ensure that these children are gaining the nutritional benefits of consuming vegetables [29,30]. Offering tangible non-food rewards [31] may be more effective in these children. In future it will be of interest to compare the efficiency of these different strategies to improve vegetable intake in older and fussier pre-school children.

Author Contributions

Conceived and designed the experiments: SJC SMA MMH ER SN SI HH PM. Performed the experiments: SJC SMA ER AO HH. Analyzed the data: PB CC SJC MMH SN SI. Wrote the paper: SJC PB MMH CN. Final approval of the version to be published: SJC PB SMA CN AO HH PM ER SN CC SI MMH.

References

1. Anderson AS, Porteous LE, Foster E, Higgins C, Stead M, et al. (2005) The impact of a school-based nutrition education intervention on dietary intake and cognitive and attitudinal variables relating to fruits and vegetables. Public Health Nutrition 8: 650–656.

2. Ransley JK, Greenwood DC, Cade JE, Blenkinsop S, Schagen I, et al. (2007) Does the school fruit and vegetable scheme improve children's diet? A non-randomised controlled trial. Journal of Epidemiology and Community Health 61: 699–703.

3. Cooke L, Wardle J (2005) Age and gender differences in children's food preferences. British Journal of Nutrition 93: 741–746.

4. Nicklaus S, Boggio V, Chabanet C, Issanchou S (2005) A prospective study of food variety seeking in childhood, adolescence and early adult life. Appetite 44: 289–297.

5. Birch LL (1999) Development of food preferences. Annual Review of Nutrition 19: 41–62.

6. Johnson SL, McPhee L, Birch LL (1991) Conditioned preferences: Young children prefer flavors associated with high dietary fat. Physiology & Behavior 50: 1245–1251.

7. Kern DL, McPhee L, Fisher J, Johnson S, Birch LL (1993) The postingestive consequences of fat condition preferences for flavors associated with high dietary fat. Physiology & Behavior 54: 71–76.

8. Steiner JE (1977) Facial expressions of the neonate infant indicating the hedonics of food-related chemical stimuli; in Weiffenbach JM (ed): Taste and Development: The Genesis of Sweet Preference. Washington, DC: U.S. Government Printing Office. 173–189.

9. Zajonc RB (1968) Attitudinal effects of mere exposure. Journal of Personality and Social Psychology 9: 1–27.

10. Birch LL, Gunder L, Grimm-Thomas K, Laing DG (1998) Infants' Consumption of a New Food Enhances Acceptance of Similar Foods. Appetite 30: 283–295.

11. Gerrish CJ, Mennella JA (2001) Flavor variety enhances food acceptance in formula-fed infants. American Journal of Clinical Nutrition 73: 1080–1085.

12. Sullivan SA, Birch LL (1994) Infant dietary experience and acceptance of solid foods. Pediatrics 93: 271–277.

13. Loewen R, Pliner P (1999) Effects of prior exposure to palatable and unpalatable novel foods on children's willingness to taste other novel foods. Appetite 32: 351–366.

14. Pliner P, Stallberg-White C (2000) "Pass the ketchup, please": familiar flavors increase children's willingness to taste novel foods. Appetite 34: 95–103.

15. de Wild VW, de Graaf C, Jager G (2013) Effectiveness of flavour nutrient learning and mere exposure as mechanisms to increase toddler's intake and preference for green vegetables. Appetite 64: 89–96.

16. Mennella JA (1995) Mother's milk: a medium for early flavor experiences. Journal of Human Lactation 11: 39–45.

17. Mennella JA, Beauchamp GK (1991) Maternal diet alters the sensory qualities of human milk and the nursling's behavior. Pediatrics 88: 737–744.

18. Mennella JA, Beauchamp GK (1999) Experience with a flavor in mother's milk modifies the infant's acceptance of flavored cereal. Developmental Psychobiology 35: 197–203.

19. Mennella JA, Jagnow CP, Beauchamp GK (2001) Prenatal and Postnatal Flavor Learning by Human Infants. Pediatrics 107: e88–.

20. Hausner H, Nicklaus S, Issanchou S, Molgaard C, Moller P (2010) Breastfeeding facilitates acceptance of a novel dietary flavour compound. Clinical Nutrition 29: 141–148.

21. Maier AS, Chabanet C, Schaal B, Leathwood PD, Issanchou SN (2008) Breastfeeding and experience with variety early in weaning increase infants' acceptance of new foods for up to two months. Clinical Nutrition 27: 849–857.

22. Abraham EC, Godwin J, Sherriff A, Armstrong J (2012) Infant feeding in relation to eating patterns in the second year of life and weight status in the fourth year. Public Health Nutrition 15: 1705–1714.

23. Scott JA, Ng SY, Cobiac L (2012) The relationship between breastfeeding and weight status in a national sample of Australian children and adolescents. BMC Public Health 12: 1471–2458.

24. Dovey TM, Staples PA, Gibson EL, Halford JCG (2008) Food neophobia and 'picky/fussy' eating in children: A review. Appetite 50: 181–193.

25. Schwartz C, Scholtens PA, Lalanne A, Weenen H, Nicklaus S (2011) Development of healthy eating habits early in life. Review of recent evidence and selected guidelines. Appetite 57: 796–807.

26. Horne PJ, Greenhalgh J, Erjavec M, Lowe CF, Viktor S, et al. (2011) Increasing pre-school children's consumption of fruit and vegetables. A modelling and rewards intervention. Appetite 56: 375–385.

27. Gregory JE, Paxton SJ, Brozovic AM (2011) Maternal feeding practices predict fruit and vegetable consumption in young children. Results of a 12-month longitudinal study. Appetite 57: 167–172.

28. Savage JS, Peterson J, Marini M, Bordi PL Jr, Birch LL (2013) The addition of a plain or herb-flavored reduced-fat dip Is associated with improved preschoolers' intake of vegetables. Journal of the Academy of Nutrition and Dietetics.

29. Spill MK, Birch LL, Roe LS, Rolls BJ (2011) Hiding vegetables to reduce energy density: an effective strategy to increase children's vegetable intake and reduce energy intake. The American Journal of Clinical Nutrition 94: 735–741.

30. Caton SJ, Ahern SM, Hetherington MM (2011) Vegetables by stealth. An exploratory study investigating the introduction of vegetables in the weaning period. Appetite 57: 816–825.

31. Cooke LJ, Chambers LC, Anez EV, Wardle J (2011) Facilitating or undermining? The effect of reward on food acceptance. A narrative review. Appetite 57: 493–497.

32. Anzman-Frasca S, Savage JS, Marini ME, Fisher JO, Birch LL (2012) Repeated exposure and associative conditioning promote preschool children's liking of vegetables. Appetite 58: 543–553.

33. Johnston CA, Palcic JL, Tyler C, Stansberry S, Reeves RS, et al. (2011) Increasing vegetable intake in Mexican-American youth: a randomized controlled trial. Journal of the American Dietetics Association 111: 716–720.

34. Havermans RC, Jansen A (2007) Increasing children's liking of vegetables through flavour-flavour learning. Appetite 48: 259–262.

35. Zeinstra GG, Koelen MA, Kok FJ, de Graaf C (2009) Children's hard-wired aversion to pure vegetable tastes. A 'failed' flavour-nutrient learning study. Appetite 52: 528–530.

36. Carruth BR, Ziegler PJ, Gordon A, Barr SI (2004) Prevalence of picky eaters among infants and toddlers and their caregivers' decisions about offering a new food. Journal of the American Dietetic Association 104: 57–64.

37. Sullivan SA, Birch LL (1990) Pass the sugar, pass the salt - Experience dictates preference. Developmental Psychology 26: 546–551.

38. Wardle J, Guthrie CA, Sanderson S, Rapoport L (2001) Development of the Children's Eating Behaviour Questionnaire. Journal of Child Psychology and Psychiatry 42: 963–970.

39. Ashcroft J, Semmler C, Carnell S, van Jaarsveld CH, Wardle J (2008) Continuity and stability of eating behaviour traits in children. European Journal of Clinical Nutrition 62: 985–990.

40. Caton SJ, Ahern SM, Remy E, Nicklaus S, Blundell P, et al. (2013) Repetition counts: repeated exposure increases intake of a novel vegetable in UK pre-school children compared to flavour-flavour and flavour-nutrient learning. British Journal of Nutrition 109: 2089–2097.

41. Hausner H, Olsen A, Moller P (2012) Mere exposure and flavour-flavour learning increase 2–3 year-old children's acceptance of a novel vegetable. Appetite 58: 1152–1159.

42. Remy E, Issanchou S, Chabanet C, Nicklaus S (2013) Repeated Exposure of Infants at Complementary Feeding to a Vegetable Puree Increases Acceptance as Effectively as Flavor-Flavor Learning and More Effectively Than Flavor-Nutrient Learning. Journal of Nutrition. 143, 1194–1200.

43. Field A FZ, Miles J (2012) Discovering statistics Using R. London: SAGE publications.

44. Cashdan E (1994) A sensitive period for learning about food. Human Nature 5: 279–291.

45. Birch LL, McPhee L, Shoba BC, Pirok E, Steinberg L (1987) What kind of exposure reduces children's food neophobia?: Looking vs. tasting. Appetite 9: 171–178.

46. Pliner P (1982) The effects of mere exposure on liking for edible substances. Appetite 3: 283–290.

47. Addessi E, Galloway AT, Visalberghi E, Birch LL (2005) Specific social influences on the acceptance of novel foods in 2–5-year-old children. Appetite 45: 264–271.

48. Cooke L, Wardle J, Gibson EL (2003) Relationship between parental report of food neophobia and everyday food consumption in 2–6-year-old children. Appetite 41: 205–206.

49. Kalat JW, Rozin P (1973) "Learned safety" as a mecanism in long-delay taste-aversion learning in rats. Journal of Comparative and Physiological Psychology 83: 198–207.

50. Svensson V, Lundborg L, Cao Y, Nowicka P, Marcus C, et al. (2011) Obesity related eating behaviour patterns in Swedish preschool children and association with age, gender, relative weight and parental weight–factorial validation of the Children's Eating Behaviour Questionnaire. International Journal of Behavavioural Nutrition and Physical Activity 8: 1479–5868.

51. Sleddens EF, Kremers SP, Thijs C (2008) The children's eating behaviour questionnaire: factorial validity and association with Body Mass Index in Dutch children aged 6–7. International Journal of Behavioural Nutrition and Physical Activity 1479–5868.

52. Lakkakula A, Geaghan J, Zanovec M, Pierce S, Tuuri G (2010) Repeated taste exposure increases liking for vegetables by low-income elementary school children. Appetite 55: 226–231.

53. Capaldi ED, Privitera GJ (2008) Decreasing dislike for sour and bitter in children and adults. Appetite 50: 139–145.

54. Birch LL, McPhee L, Steinberg L, Sullivan S (1990) Conditioned flavor preferences in young children. Physiology & Behavior 47: 501–505.

55. Brunstrom JM, Fletcher HZ (2008) Flavour-flavour learning occurs automatically and only in hungry participants. Physiology & Behavior 93: 13–19.

56. Mennella JA, Finkbeiner S, Reed DR (2012) The proof is in the pudding: children prefer lower fat but higher sugar than do mothers. International Journal of Obesity 36: 1285–1291.

57. Steiner JE (1979) Human facial expressions in response to taste and smell stimulation. Advances in Child Development Behaviour 13: 257–295.

58. Ahern SM, Caton SJ, Bouhlal S, Hausner H, Olsen A, et al. (2013) Eating a Rainbow. Introducing vegetables in the first years of life in 3 European countries. Appetite 71: 48–56.

10

Unacylated Ghrelin Suppresses Ghrelin-Induced Neuronal Activity in the Hypothalamus and Brainstem of Male Rats

Darko M. Stevanovic[1,2]*, Aldo Grefhorst[1], Axel P. N. Themmen[1], Vera Popovic[3], Joan Holstege[4], Elize Haasdijk[4], Vladimir Trajkovic[5], Aart-Jan van der Lely[1], Patric J. D. Delhanty[1]*

1 Department of Internal Medicine, Erasmus Medical Center, Rotterdam, The Netherlands, 2 Institute of Medical Physiology, School of Medicine, University of Belgrade, Belgrade, Serbia, 3 Institute of Endocrinology, Diabetes and Diseases of Metabolism, School of Medicine, University of Belgrade, Belgrade, Serbia, 4 Department of Neuroscience, Erasmus Medical Center, Rotterdam, The Netherlands, 5 Institute of Microbiology and Immunology, School of Medicine, University of Belgrade, Belgrade, Serbia

Abstract

Ghrelin, the endogenous growth hormone secretagogue, has an important role in metabolic homeostasis. It exists in two major molecular forms: acylated (AG) and unacylated (UAG). Many studies suggest different roles for these two forms of ghrelin in energy balance regulation. In the present study, we compared the effects of acute intracerebroventricular administration of AG, UAG and their combination (AG+UAG) to young adult Wistar rats on food intake and central melanocortin system modulation. Although UAG did not affect food intake it significantly increased the number of c-Fos positive neurons in the arcuate (ARC), paraventricular (PVN) and solitary tract (NTS) nuclei. In contrast, UAG suppressed AG-induced neuronal activity in PVN and NTS. Central UAG also modulated hypothalamic expression of *Mc4r* and *Bmp8b*, which were increased and *Mc3r*, *Pomc*, *Agrp* and *Ucp2*, which were decreased. Finally, UAG, AG and combination treatments caused activation of c-Fos in POMC expressing neurons in the arcuate, substantiating a physiologic effect of these peptides on the central melanocortin system. Together, these results demonstrate that UAG can act directly to increase neuronal activity in the hypothalamus and is able to counteract AG-induced neuronal activity in the PVN and NTS. UAG also modulates expression of members of the melanocortin signaling system in the hypothalamus. In the absence of an effect on energy intake, these findings indicate that UAG could affect energy homeostasis by modulation of the central melanocortin system.

Editor: Thierry Alquier, CRCHUM-Montreal Diabetes Research Center, Canada

Funding: This work was assisted by the ENDO/ESE International Endocrine Scholars Program Fellowship 2011 (DS) supported by the European Society of Endocrinology, and Serbian Ministry of Science and Technological Development proj. no. III 41025. The funders had no role in study design, data collection and analysis, decision to publish, or preparation of the manuscript.

Competing Interests: The authors have declared that no competing interests exist.

* E-mail: dstevano@bidmc.harvard.edu (DS); p.delhanty@erasmusmc.nl (PJDD)

Introduction

The prevalence of obesity and related diseases worldwide has catalyzed the need for a greater understanding of how physiological signals of energy intake and/or energy expenditure converge within the brain to regulate energy homeostasis. The brain melanocortin system represents a fundamental component of centrally regulated energy balance. It consists of circuits of neurons expressing either anorexigenic pro-opiomelanocortin (POMC)-derived melanocortin 3 (MC3) and 4 (MC4) receptor agonists, as well as MC3R and MC4R expressing cells, which are targets of these neurons. The system also includes orexigenic neurons that express the melanocortin receptor inverse agonist agouti-related peptide (AgRP). Distinct populations of AgRP and POMC expressing neurons are found within the arcuate nucleus of the hypothalamus (ARC) and are co-expressed with neuropeptide Y (NPY) and cocaine- and amphetamine-regulated transcript (CART), respectively [1]. These "first order" neurons are able to receive peripheral signals about current energy balance via a wide range of circulating hormones (e.g. leptin, insulin, ghrelin, peptide YY_{3-36}) and nutrients (e.g. glucose, fatty acids, amino acids), mediate anabolic or catabolic effects on energy balance and

hence modulate food intake and energy expenditure. Melanocortin neurons in the ARC send projections to downstream "secondary" neuronal populations within proximal nuclei of the hypothalamus, especially to the paraventricular nucleus (PVN). The ARC and PVN, which contain neurons that express MC3R and MC4R, serve as branch points for activation of many central melanocortin-induced circuits involved in body weight regulation [2]. POMC-positive neurons and neural projections are also located within the nucleus of the solitary tract (NTS) of the caudal brainstem. This area receives and integrates both vagal afferent satiation and blood born energy status signals, and issues output commands essential to energy balance control [3–5]. The function of POMC neurons within the NTS may differ significantly from those in the ARC. Only a small number of studies address this issue, but they suggest divergent roles for hindbrain and forebrain POMC neurons in energy homeostasis [6–9].

Ghrelin is a 28-amino acid peptide hormone that can be acylated on its third serine residue (acylated ghrelin, AG) by ghrelin *O*-acyl transferase (GOAT), and is produced predominantly by the gastric oxyntic mucosa in mammals [10–12]. Acylation is required for ghrelin to bind to its receptor, the growth hormone secretagogue receptor (GHSR) type 1a [13], located in the hypothalamo-pituitary unit, leading to stimulation of food intake and growth hormone (GH) secretion [10]. Recent studies have revealed that central and peripheral administration of AG results in increased NTS activation, suggesting a role for the NTS in mediating the feed-forward mechanisms of food intake [14–16]. However, Kobelt *et al.* (2008) did not find any change in c-Fos positive neurons in the NTS after peripheral administration of AG [17]. Unacylated ghrelin (UAG) also occurs in the circulation [13]. Although UAG does not activate GHSR1a, it has physiological activity [18–24]. A number of studies report that UAG suppresses food intake in rodents both centrally and peripherally [25–27], and the effect is likely mediated via ARC and PVN neurons [26]. At the level of the NTS UAG has been shown to disrupt motor activity in the gastric antrum under fasting conditions, which could potentially modulate food intake [28]. In contrast, Toshinai *et al.* (2006) reported that centrally applied UAG stimulates food intake, while other reports suggest its peripheral administration has no effect on food intake in rodents and humans [18,29,30].

Because it is currently unclear if central UAG has an effect on food intake, we investigated whether central acute administration of AG, UAG or their combination affect neuronal activity in the ARC, PVN and NTS, and hence food intake. Furthermore, to obtain insight into the ability of the ghrelin system to modulate energy expenditure via central mechanisms, we examined changes in hypothalamic mitochondrial uncoupling protein 2 (*Ucp2*) and *Bmp8b* gene expression, as molecules known to regulate thermogenesis and energy balance [31,32].

Materials and Methods

Animals, animal preparation and treatment

The study was performed with 8 week old male Wistar rats (n = 40, body weight = 230 ± 20 g), bred at the Institute of Biomedical Research "Galenika" in Belgrade, Serbia. They were kept in individual metabolic cages under a 12:12 h light-dark cycle, at $22 \pm 2°C$, and were accustomed to daily handling. Animals received *ad libitum* water and a standard balanced diet (D.D. Veterinarski zavod Subotica, Subotica, Serbia) throughout the experiment.

Animals were anesthetized with intramuscular ketamine (50 mg/kg, Pfizer, New York, NY), xylazine (80 mg/kg, Bayer, Leverkusen, Germany), and surgically equipped with a headset for intracerebroventricular (ICV) injection, consisting of a silastic-sealed 20-gauge cannula positioned in the right lateral cerebral ventricle (1 mm posterior and 1.5 mm lateral to the bregma, and 3 mm below the cortical surface) [33]. A small stainless steel anchor screw was placed at the remote site on the skull. The cannula and screw were cemented to the skull with standard dental acrylic. After surgery, the animals received a single dose of s.c. 0.28 mg/kg buprenorphin (Buprenex; Reckitt Benckiser Health-care, Mannheim, Germany) followed by a recovery period of one week. Only animals demonstrating progressive weight gain during the recovery period were used in subsequent experiments. Proper ICV cannula placement was verified at 48 hours before conducting any experiment by demonstrating short-latency, heart rate, and drinking responses to a bolus injection of Angiotensin II (50 ng/1 ug). Aspiration of CSF from the guide cannula also was used to indicate correct positioning of the cannula in the lateral ventricle. Animals were randomly divided into 4 groups (control, AG, UAG and AG+UAG groups, n = 10). Animals from the control group were treated ICV with 5 μl of phosphate buffered saline (PBS), while those from the AG and UAG groups received ICV 5 μg of peptide (Neosystem, Strasbourg, France) in 5 μl of

PBS. Rats of the combined AG+UAG group were treated ICV with 5 μg of each peptide in a total of 5 μl of PBS. All treatments were administered between 10:00 and 11:00 a.m, and food intake was measured. Differences in food intake were considered statistically significant at p<0.05, and considered trends if the p-value was between 0.05 and 0.1. At 2 hrs after ICV injection four animals from each group were deeply anesthetized with isoflurane and transcardially perfused with sterile PBS, followed by 4% of paraformaldehyde. Whole brains were excised, and later used for immunohistochemical studies. At 5 hrs postinjection the remaining six animals in each group were killed by decapitation under deep anesthesia with isoflurane, and hypothalami were collected and stored at $-20°C$ in RNA*later* stabilisation reagent (Qiagen N.V, Venlo, The Netherlands). The samples were transferred to The Netherlands and analyzed at Erasmus MC, Rotterdam. All experimental procedures were approved by the Ethics Committee of the School of Medicine, University of Belgrade, Serbia. All efforts were made to minimize suffering.

Immunohistochemistry

Expression of the proto-oncogene c-Fos was used as a marker for activation of neurons. After perfusion, post-fixation was performed for 1 h in 4% paraformaldehyde (PFA) followed by overnight incubation in 10% sucrose in 0.1 M PBS at 4°C. Subsequently, the dura mater was removed, the tissue was embedded in 10% sucrose in 10% gelatine, and fixed with 10% PFA in 30% sucrose for 2.5 h at room temperature. This was followed by an overnight incubation in 30% sucrose in 0.1 M PBS at 4°C. Serial coronal sections (40 μm) were made using a sliding microtome with a cryostat modification (Leica, Bensheim, Germany). Free-floating sections were processed for immunohistochemistry.

C-Fos immunohistochemistry. Sections were incubated in 10% heat-inactivated normal horse serum (NHS) with 0.5% Triton-X100 in PBS for 1 h, and then incubated for 48 h with a polyclonal rabbit anti-c-Fos antibody (1:15000, Calbiochem, Billerica, MA; PC38) at 4°C, rinsed in PBS (4×10 min) and incubated in 1:200 biotinylated goat anti-rabbit IgG secondary antibody (Sigma-Aldrich, St. Louis, MO, USA) for 1.5 h at room temperature. After washing in PBS (4×10 min), all sections were treated with avidin-biotin complex (ABC Elite Kit, Vector, Burlingame, CA, USA). After washing (6×10 min) the peroxidase component of the ABC complex was visualized using a solution of 0.05% diaminobenzidine tetrachloride and 0.3% H_2O_2. The sections were mounted, air dried overnight, counterstained with thionine for 5 min, dehydrated through a graded series of ethanol and xylene, and coverslipped. Assessment of c-Fos immunoreactive neurons was obtained by counting the number of c-Fos immunopositive nuclei. Neurons with black or dark brown nuclear staining were considered as c-Fos positive. Coronal sections were counted for c-Fos immunopositive staining bilaterally in the ARC (12 sections per rat; bregma -2.12 mm to -3.24 mm), PVN (8 sections per rat; bregma -1.56 mm to -2.08 mm) and NTS (8 sections per rat; bregma -13.68 mm to -14.20 mm), using a Nikon Eclipse E400 photomicroscope. Anatomic correlations were made according to landmarks given in Paxinos and Watson's stereotaxic atlas [34]. The investigator counting the number of c-Fos immunopositive cells was blinded to treatments received by the animals. The average number of c-Fos immunopositive neurons per section for the brain nuclei mentioned above was calculated for four rats per experimental group. C-Fos data are expressed as mean ± SEM and differences between experimental groups were assessed by ANOVA (Tukey's *post hoc* test), with p<0.05 considered significant.

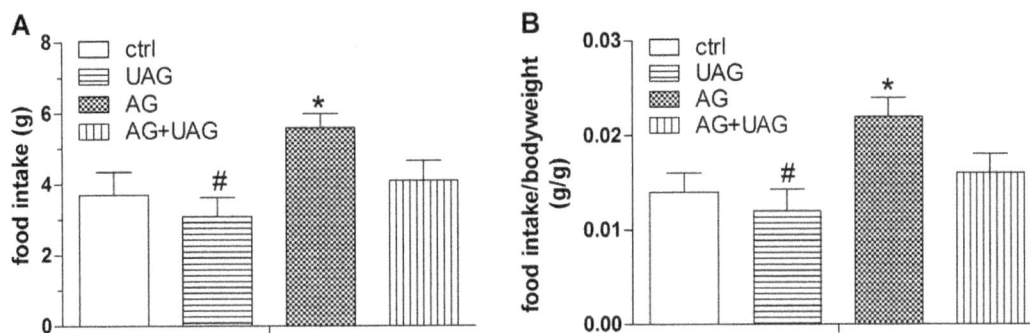

Figure 1. Average food intake (A) and average food intake corrected for body weight (B) 2 hrs after ICV injections of acylated (AG, 5 μg), unacylated (UAG, 5 μg) and combination (AG+UAG, 5 μg of each peptide). Data are mean ± SEM, n = 10 per group, ANOVA, *p<0.05 vs. saline-treated animals, #p<0.05 vs. AG group.

Multi-label immunofluorescence histochemistry and confocal microscopy. Sections were washed (4×15 min) in Tris buffered saline (TBS; 50 mM Tris-Cl, pH 7.5. 150 mM NaCl, pH 7.5), then blocked in TBS, 10% NHS, 0.4% Triton X-100. Sections were then incubated overnight at 4°C in first primary antibody in TBS, 2% NHS, 0.4% Triton X-100 (polyclonal rabbit anti-c-Fos antibody (1:15000; Calbiochem, Billerica, MA, US; PC38)), then washed (4×15 min) in TBS, followed by a 90 min incubation in Cy3-conjugate goat anti-rabbit Fab (1:200; Jackson Immunoresearch Labs Inc., West Grove, PA; 111-167-003). After washing (4×15 min) in TBS, the section were then incubated overnight at 4°C in the second primary antibody in TBS, 2% NHS, 0.4% Triton X-100 (anti-POMC antibody (1:5000; Phoenix Pharmaceuticals Inc. Burlingame, CA; H-029-30)), then after washing in TBS, incubated 90 min in Alexa Fluor 488-conjugated donkey anti-rabbit antibody (1:200; Jackson Immunoresearch Labs Inc.; 711-545-152). Finally, sections were washed 1×10 min in TBS, 1×10 min in PBS, then stained with DAPI. Sections were then mounted in Vectashield on glass slides. Images were acquired with a Zeiss LSM 700 laser-scanning confocal microscope and a 40× oil-immersion objective lens. Levels of gain and laser power were selected to allow optimal visualization of the fluorophores. Each image was saved at a resolution of 1024×1024 pixels.

Quantitative PCR

Six separate hypothalamic samples were used for RNA isolation. Quantitative PCR was performed using a qPCR Core kit for SYBR Green I (Eurogentec, The Netherlands). Gene specific primers were designed to span introns. The sequences forward and reverse were as follows: β-actin, 5'-CCCTGGCT-CCTAGCACCAT and 5'-GAGCCACCAATCCACACAGA, Hprt, 5'-TGGTCAAGCAGTACAGCCCCA and 5'-GGCCT-GTATCCAACACTTCGAGAGG; Mc3r, 5'-GCAACCGGAG-TGGCAGTGGG and 5'-GGGGAGTGCAGGTTGCCGTT; Mc4r, 5'-CTCCCGGGCACGGGTACCAT and 5'-AACGG-GGCCCAGCAGACAAC; Agrp, 5'-AGACAGCAGCAGACC-GAGCAGA and 5'-CACAGCGACGCGGAGAACGA; Pomc, 5'-AGACGTGTGGAGCTGGTGCC and 5'-CTGCAGGCC-CGGATGCAAGC; Ucp2, 5'-ATGAGCTTTGCCTCCGTCC-GC and 5'-GGGCACCTGTGGTGCTACCTG; Bmp8b, 5'-CCACGCCACTATGCAGGCCC and 5'-GGCACTCAGCTT-GGTGGGCA. Gene expression was calculated using the ΔC_t method relative to the mean of 2 housekeeping genes (*Actb* and *Hprt*), and mean values +/- SEM are shown in Table S1.

Statistical analyses

All data were analyzed by ANOVA using Tukey's *post hoc* test, with effects being considered significant at p<0.05. Degrees of freedom, F-values and p-values of the analyses are summarized in Table S2.

Results

Food intake

To evaluate the immediate effect of central AG, UAG or their combination treatment, we assessed food intake in all groups (n = 10 per group) 2 hrs after ICV injection (Fig. 1). In comparison to average food intake in the control group (3.7 g±0.65), AG caused a significant increase in average food intake (5.6 g±0.40), as expected, while there was no significant change in food intake in the UAG (3.1 g±0.53) group compared to controls. Also, a significant difference in food intake was observed between the AG and UAG treated groups, while UAG noticeably reduced the appetitive response to AG when given in combination, although this effect only showed a trend (Fig. 1; AG+UAG treatment, 4.1 g±0.56, p = 0.08).

C-Fos immunoreactivity in the ARC, PVN and NTS

To gain additional insight into the possible interaction between AG and UAG in regulating central neuronal pathways involved in energy homeostasis, we examined whether AG, UAG or combined treatment induce changes in c-Fos immunoreactivity in hypothalamic ARC and PVN as well as in the NTS of the brainstem (see representative micrographs in Figs. 2, 3 and 4A–D). Central AG, UAG and combined treatments all significantly induced c-Fos immunoactivity in all examined brain regions (Figs. 2, 3 and 4, see histograms for quantitative data). Although AG caused a significantly greater induction of c-Fos immunore-activity than UAG, the effects of AG and the combined treatments were not significantly different.

To investigate the biological relevance of the regulation of this neuronal activation by ghrelin peptides in relation to melanocortin signaling, we used multi-label immunofluorescence to discover if the peptides, and UAG treatment in particular, caused c-Fos immunoreactivity in POMC immunopositive neurons. We observed that POMC neurons in the arcuate show co-expression of c-Fos following treatment with these peptides (Fig. 5). We then assessed the relative expression of c-Fos in POMC positive cells, denoting activation of these cells by the different treatments (figure S1A), as well as the relative expression of POMC in c-Fos positive cells (figure S1B). We found that UAG treatment caused a trend to

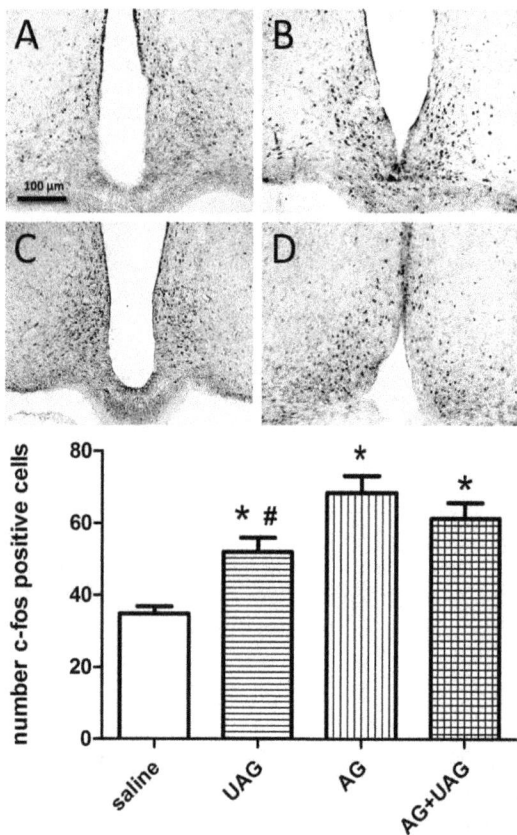

Figure 2. C-Fos immunoreactivity in the arcuate nucleus (ARC) two hrs after ICV injections of vehicle (A) UAG (B), AG (C) or AG+UAG (D). The scale bar applies to all images which are representative of sections from 4 rats. C-Fos positive nuclei were counted in 12 sections from 4 rats and these quantitative data are presented in the histogram (*, $p < 0.01$ v. saline; #, $p < 0.05$ v. AG). Color images were corrected for color balance and contrast before conversion to grayscale.

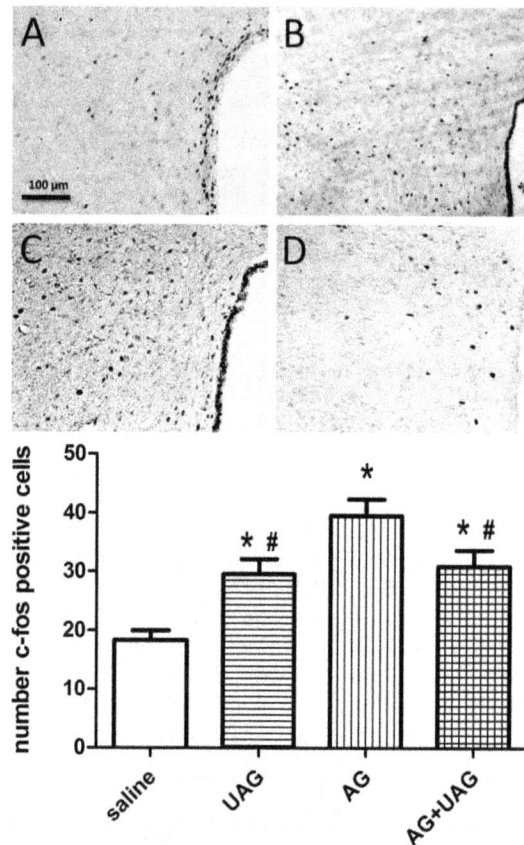

Figure 3. C-Fos immunoreactivity in the paraventricular nucleus (PVN) two hrs after ICV injections of vehicle (A) UAG (B), AG (C) or AG+UAG (D). The scale bar applies to all images. C-Fos positive nuclei were counted in 8 sections from 4 rats and these quantitative data are presented in the histogram (*, $p < 0.01$ v. saline; #, $p < 0.05$ v. AG). Color images were corrected for color balance and contrast before conversion to grayscale.

induce c-Fos immunorectivity in POMC immunoreactive cell bodies, and that this was significantly greater than the levels of c-Fos in AG+UAG treated animals. The distribution of POMC positive c-Fos expressing cells was not affected by treatment.

Hypothalamic *Mc4r, Mc3r, Agrp, Pomc, Ucp2* and *Bmp8b* gene expression

To determine a possible role for UAG in modulating the effect of AG on the hypothalamic melanocortin system, we examined hypothalamic *Mc4r, Mc3r, Agrp* and *Pomc* gene expression following ICV injection of the peptides. *Mc4r* mRNA was significantly ($p < 0.05$) increased in the UAG and AG+UAG treated groups, while there was no change in *Mc4r* mRNA expression in the group treated with AG alone (Fig. 6A–D). In contrast, *Mc3r* mRNA expression was significantly decreased by UAG treatment when compared to controls ($p < 0.05$) and AG treated animals ($p = 0.04$). Both *Agrp* and *Pomc* gene expression were significantly decreased in UAG and AG+UAG groups (*Agrp*-UAG, $p = 0.03$; *Agrp*-AG+UAG = $p < 0.03$; *Pomc*-UAG $p = 0.0004$; *Pomc*-AG+UAG $p = 0.004$). Central AG treatment did not affect expression of these two important components of the melanocortin system. Gene expression of other important players in the central melanocortin system, *Npy* and *Cart*, were not significantly altered by AG and/or UAG treatment (data not shown).

To examine further a possible role for both forms of ghrelin in energy expenditure, we examined changes in hypothalamic *Ucp2* and *Bmp8b* gene expression 5 hrs post-treatment (Fig. 7A and B). Results showed a significant decrease in hypothalamic *Ucp2* mRNA in both UAG and AG+UAG groups compared to control animals ($p < 0.01$) and AG groups alone. UAG treatment alone significantly increased *Bmp8b* mRNA expression ($p < 0.05$) in comparison to control and AG+UAG groups.

Discussion

A large body of evidence shows that AG's most impressive impact on mammalian energy balance appears to be an almost instant induction of food intake when administered in pharmacological doses, even in satiated animals [35,36]. On the other hand, currently available data regarding UAG effects on food intake are inconsistent [37]. A significant anorexigenic effect of UAG was found in fasted and animals fed *ad libitum* during the dark phase and in food-restricted rats throughout the light phase has been described [38]. However, Toshinai and co-workers (2006) did not observe an anorexigenic effect of UAG during the light phase [29]. They also found no significant reduction in food intake in fasted and *ad libitum* fed animals after peripheral UAG administration [30,38]. These differences in UAG's effect on food intake could be due to variability in experimental setup, time of measurement

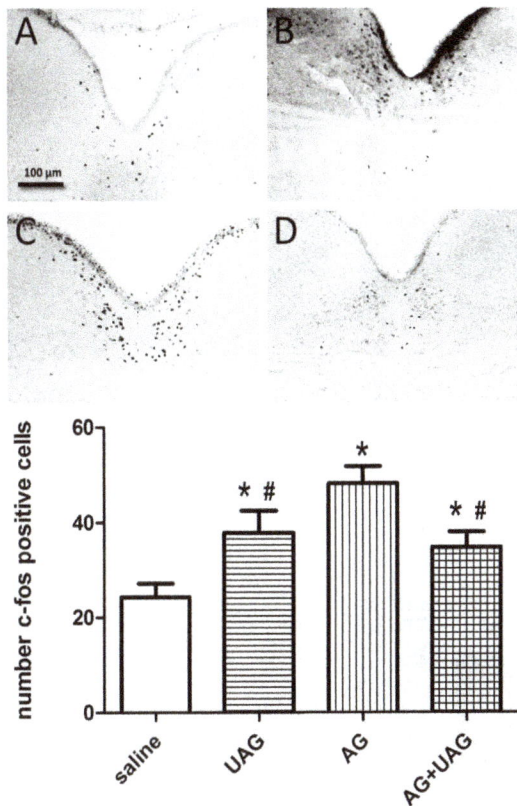

Figure 4. C-Fos immunoreactivity in the nucleus of the solitary tract (NTS) two hrs after ICV injections of vehicle (A) UAG (B), AG (C) or AG+UAG (D). The scale bar applies to all images. C-Fos positive nuclei were counted in 8 sections from 4 rats and these quantitative data are presented in the histogram (*, $p<0.01$ v. saline; #, $p<0.05$ v. AG). Color images were corrected for color balance and contrast before conversion to grayscale.

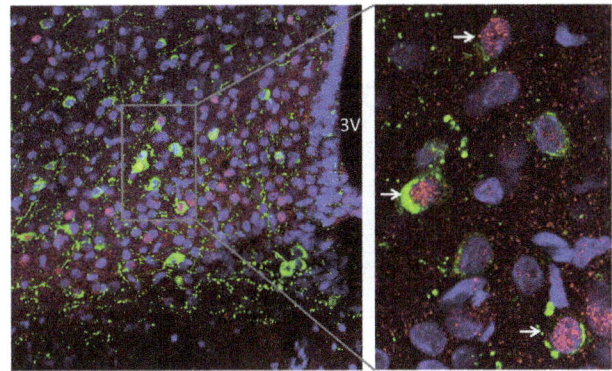

Figure 5. POMC and c-Fos are co-expressed in neurons of the arcuate nucleus of the hypothalamus following UAG treatment. POMC and c-Fos immunoreactivity was identified in sections of the hypothalamus using multi-label immunofluorescence immunohistochemistry. Separate nuclear (DAPI, blue), POMC (green), c-Fos (red) and composite (merge) confocal laser-scanning microscope images are shown from a representative section. The arrowhead indicates a neuron that contains both POMC and c-Fos immunoreactivity. The scale bar represents 20 μm.

(light vs. dark phase) and/or circadian rhythmicity. We found that UAG showed a trend to inhibit AG mediated induction of food intake. This is comparable to the results of Inhoff *et al.* (2008) who found that AG-induced food intake was diminished by i.p. injection of UAG [27]. The inhibitory effect of i.p. UAG on ICV AG-induced increase in food intake was observed at a dose that had no effect on food intake in freely fed rats monitored during the light phase, 5 h after treatment [27]. Since our measurements were performed on rats that had been fed *ad libitum* during the light phase and were likely satiated, it should be noted that it is rather difficult to detect an anorexigenic effect of a satiety hormone under these conditions. This could explain why we failed to observe a significant satiating effect of UAG.

An inhibitory effect of UAG on AG has also been shown in goldfish (*Carrassius auratus*), where UAG administered either ICV or i.p. substantially reduced AG-induced food intake, while having no effect if its own [39].

To gain additional insight into the role of AG-UAG system in central melanocortin neuronal pathways involved in energy homeostasis, we examined whether AG, UAG or their combination treatment induced changes in c-Fos immunoreactivity in the ARC and PVN as well as in the NTS of the brainstem. Our results show that central AG, UAG and combined treatments rapidly induced neuronal activity in all examined brain regions, and UAG reduced AG-induced neuronal activity in the PVN and NTS. Furthermore, we performed c-Fos and POMC double-labeling

immunofluorescence to show that UAG, AG and combination treatments caused the appearance of c-Fos immunoreactivity in POMC-positive neurons. Intriguingly, UAG, but not AG, appeared to induce c-Fos in POMC cell bodies, and this effect of UAG was blocked by combined AG treatment. This fits with activation of anorexigenic circuits in the arcuate by UAG and demonstrate a possible interaction with AG. However, this possible mechanism of action by UAG, and interaction with AG, needs to be verified using POMC-EGFP knock-in mice [40].

Two other studies in rodents also describe increased neuronal activity (c-Fos positive neurons) in the ARC and in the PVN following ICV UAG treatment, while the same treatment had no effect on neuronal activity in the NTS [25,38]. According to these latter results, UAG is involved in the regulation of the synthesis of anorexigenic CART and urocortin 1 in the hypothalamus. However, we did not observe any changes in hypothalamic *Cart* (or *Npy*) gene expression (data not shown) after central AG, UAG and their combined treatment, while hypothalamic *Pomc* and *Agrp* gene expression was suppressed in both UAG and AG+UAG, but not AG, when compared to the control group. Down-regulation of *Pomc* mRNA may indicate the down-regulation of a potent suppressor of food intake (α-melanocyte-stimulating hormone, α-MSH), but also β-endorphin which, on the other hand, stimulates feeding [41]. Thus, an alternative interpretation may be that the down-regulation of *Pomc* mRNA, with subsequent decreased levels of β-endorphin, results in less rewarding signals through the opioid pathway by UAG. Importantly, AG by itself had no effect on these genes, which were only regulated by UAG alone or when in combination with AG, suggesting a specific effect of UAG.

Ucp2 has been shown to be an important negative regulator of reactive oxygen species in the hypothalamus [31]. Interestingly, AgRP (orexigenic) derived action potentials are suppressed by raised ROS, whereas POMC-related (anorexigenic) signals are induced [42]. UAG suppressed *Ucp2* gene expression, therefore we speculate that UAG causes increased levels of ROS in hypothalamic neurons, and if so this would stimulate anorexigenic pathways, in opposition to the effects of AG. Further work is required to confirm this possibility.

To further explore the role of UAG and its interaction with AG in the regulation of the central melanocortin system, we

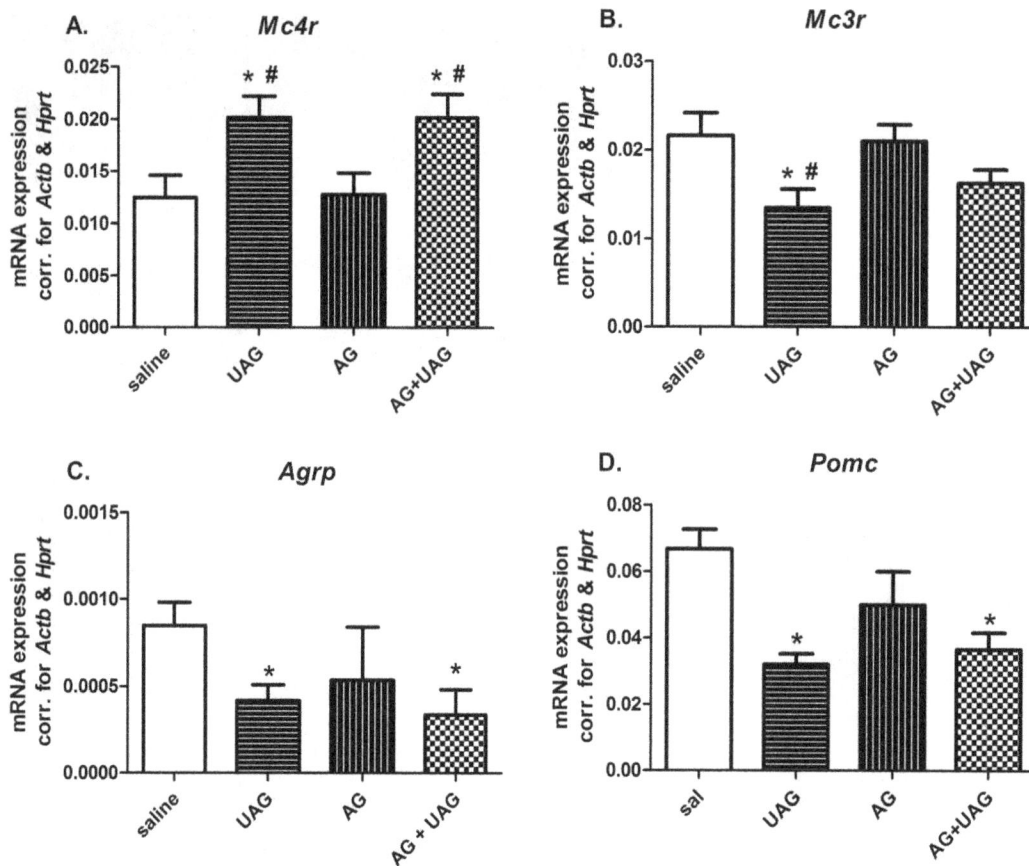

Figure 6. The effects of central AG, UAG and combination of AG and UAG treatment on hypothalamic *Mc4r* **(A),** *Mc3r* **(B),** *Agrp* **(C) and** *Pomc* **(D) gene expression, corrected for** *Hprt* **and** *Actb.* Data are mean ± SEM, n = 6 per group, ANOVA, $^*p<0.05$ vs. saline-treated animals, $^#p<0.05$ vs. AG group.

investigated whether AG, UAG and their combination affect hypothalamic *Mc4r* and *Mc3r* gene expression. Although hypothalamic *Pomc* gene expression was decreased, this is likely compensated for by a significant increase in *Mc4r* after UAG treatment. By increasing MC4R expression, UAG may in effect amplify the POMC driven neuronal signal to induce energy expenditure. Also, possible effects of UAG on complex circuits (e.g. the melanocortin system) regulating energy expenditure in the brain and those regulating appetite may involve other central and peripheral factors/systems, which could separately act on these circuits to modulate overall energy homeostasis. In agreement with this hypothesis, we have recently shown that peripheral UAG infusion can induced expression of genes, such as *Ucp1*, *Pgc1a* and *Bmp8b* in brown adipose tissue of mice on a high fat diet [18].

MC4R is directly activated by α-MSH from POMC neurons, an effect that is inhibited by AgRP from orexigenic neurons in the ARC. MC4R-deficient mice are hyperphagic and obese [43,44]. It has also been shown that stimulation of AgRP-producing neurons involves a melanocortin receptor-independent mechanism to increase food intake, whereas POMC stimulation requires intact melanocortin receptor to reduce food intake [45]. Feeding stimulatory effects of AgRP is instead likely to occur via GABA release/GABA-ergic signaling [46,47]. Taken together, these data point to a distributed control of MC4R signaling within the hypothalamus and between forebrain and hindbrain in regulating energy balance and food intake.

It is known that the melanocortin system modulates energy expenditure via the sympathetic nervous system (SNS). Recently, it was shown that BMP8B has an important role in energy expenditure, by acting both centrally in the hypothalamus and peripherally in brown adipose tissue (BAT) [32]. *Bmp8b* gene expression has been described in the arcuate and ventromedial regions of the hypothalamus [32]. Both regions are important in regulation of BAT activity and energy expenditure. Using *Bmp8b* deficient mice, it was shown that *Bmp8b* is required for the response of BAT to adrenergic stimulation by acting at the hypothalamic level to increase sympathetic output. In this study we observed that hypothalamic *Bmp8b* gene expression is significantly increased by UAG central administration alone. Together with the increase in hypothalamic *Mc4r* gene expression, we could speculate that UAG could affect energy expenditure by upregulating *Mc4r* and *Bmp8b* gene expression in the hypothalamus. Interestingly, however, combined AG+UAG treatment had no effect on *Bmp8b* expression. One possibility is that AG, when it is co-infused, is able to prevent UAG-induced *Bmp8b* mRNA expression in the brain, but by itself has no effect. Like MC4R, MC3R is expressed throughout the brain, especially in the ARC and NTS [48], and endogenous melanocortins activate this receptor [49], suggesting potentially redundant or overlapping actions of both receptors. However, a number of studies suggest that selective MC3R agonists actually stimulate food intake and, unlike MC4R-deleted mice, MC3R-deleted mice are hypophagic [50]. MC3R, but not MC4R, mRNA is expressed in half of the

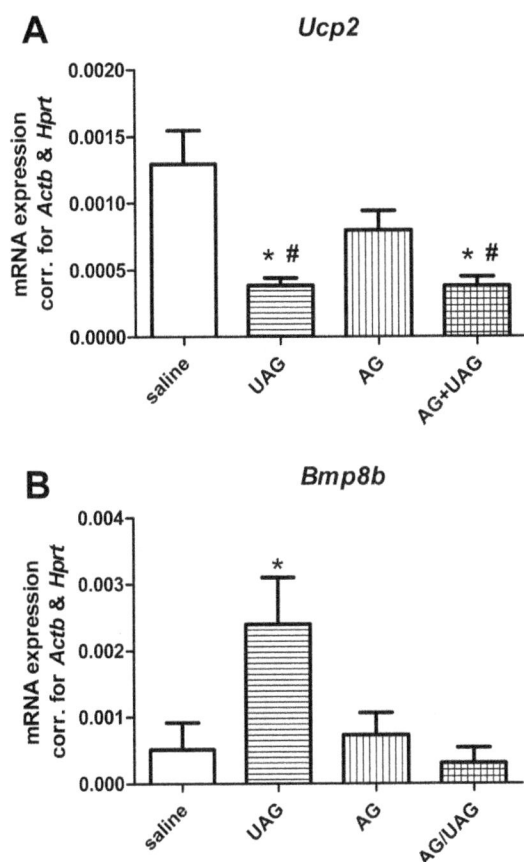

Figure 7. The effects of central AG, UAG and combination of AG and UAG treatment on hypothalamic *Ucp2* (A) and *Bmp8B* (B) gene expression, corrected for *Hprt* and *Actb*. Data are mean ± SEM, n = 6 per group, ANOVA, *p<0.05 vs. saline-treated animals, #p<0.05 vs. AG group.

POMC and AgRP neurons in the ARC [2]. The role of MC3R in these neurons is thought to be auto-inhibitory. They serve as messengers within the ARC and between the ARC and PVN to maintain melanocortin tone, and regulate AgRP/POMC activity via inhibitory GABA-ergic terminals to suppress NPY/AgRP signaling and/or direct activation of POMC signaling [40].

In conclusion, our data indicate a new role for the central ghrelin system in energy balance, in which the unacylated form of ghrelin increases neuronal activity directly and independently of AG. We observed this effect not only in the hypothalamus but also in the brainstem, increasing neuronal activity in the NTS. We also show that UAG can suppress AG-induced neuronal activation in the PVN and NTS, and blunt AG-induced food intake. UAG affected hypothalamic gene expression of several main players in the energy balance regulation, decreasing *Pomc*, *Agrp*, *Mc3r* and *Ucp2*, while increasing *Mc4r* and *Bmp8b* gene expression levels, suggesting a role in induction of energy expenditure via the central melanocortin system. This study also shows that UAG does have biological activity, although its mechanism of action remains to be determined at the cellular level. Further insight is needed to understand the etiology and pathogenesis of human diseases characterized by disturbances of energy balance, such as obesity. Our study further highlights the interplay between AG, UAG and the central melanocortin-ghrelin system in controlling energy homeostasis and, ultimately, body weight.

Supporting Information

Figure S1 A. Co-localization of POMC/c-Fos after saline (A), AG (B), UAG (C) and AG+UAG (D) acute central treatment. POMC (green) and c-Fos (red) immunoreactivity was identified in sections of the hypothalamus using multi-label immunofluorescence immunohistochemistry. Nuclear staining (DAPI) is blue. Composite confocal laser-scanning microscope images are shown from a representative sections. The scale bar represents 20 μm. E. POMC/c-Fos co-localization ratio after saline, AG, UAG and combine treatment. Data are presented and mean ± SEM, * p< 0.05 vs. saline.

Acknowledgments

The authors wish to thank Martin Huisman from the Department of Internal Medicine, Erasmus MC for excellent technical assistance.

Author Contributions

Conceived and designed the experiments: DS APNT VT AJL PJDD. Performed the experiments: DS. Analyzed the data: DS AG APNT VP JH EH VT AJL PJDD. Contributed reagents/materials/analysis tools: AG APNT JH EH VT AJL.

References

1. Cone RD (2005) Anatomy and regulation of the central melanocortin system. Nat Neurosci 8: 571–578.
2. De Jonghe BC, Hayes MR, Bence KK (2011) Melanocortin control of energy balance: evidence from rodent models. Cell Mol Life Sci 68: 2569–2588.
3. Hayes MR, Skibicka KP, Leichner TM, Guarnieri DJ, DiLeone RJ, et al. (2010) Endogenous leptin signaling in the caudal nucleus tractus solitarius and area postrema is required for energy balance regulation. Cell Metab 11: 77–83.
4. Skibicka KP, Grill HJ (2009) Hindbrain leptin stimulation induces anorexia and hyperthermia mediated by hindbrain melanocortin receptors. Endocrinology 150: 1705–1711.
5. Skibicka KP, Grill HJ (2009) Hypothalamic and hindbrain melanocortin receptors contribute to the feeding, thermogenic, and cardiovascular action of melanocortins. Endocrinology 150: 5351–5361.
6. Li G, Zhang Y, Rodrigues E, Zheng D, Matheny M, et al. (2007) Melanocortin activation of nucleus of the solitary tract avoids anorectic tachyphylaxis and induces prolonged weight loss. Am J Physiol Endocrinol Metab 293: E252–258.
7. Ellacott KL, Halatchev IG, Cone RD (2006) Characterization of leptin-responsive neurons in the caudal brainstem. Endocrinology 147: 3190–3195.
8. Huo L, Grill HJ, Bjorbaek C (2006) Divergent regulation of proopiomelanocortin neurons by leptin in the nucleus of the solitary tract and in the arcuate hypothalamic nucleus. Diabetes 55: 567–573.
9. Zhang Y, Rodrigues E, Gao YX, King M, Cheng KY, et al. (2010) Pro-opiomelanocortin gene transfer to the nucleus of the solitary track but not arcuate nucleus ameliorates chronic diet-induced obesity. Neuroscience 169: 1662–1671.
10. Kojima M, Hosoda H, Date Y, Nakazato M, Matsuo H, et al. (1999) Ghrelin is a growth-hormone-releasing acylated peptide from stomach. Nature 402: 656–660.
11. Gutierrez JA, Solenberg PJ, Perkins DR, Willency JA, Knierman MD, et al. (2008) Ghrelin octanoylation mediated by an orphan lipid transferase. Proc Natl Acad Sci U S A 105: 6320–6325.
12. Yang J, Brown MS, Liang G, Grishin NV, Goldstein JL (2008) Identification of the Acyltransferase that Octanoylates Ghrelin, an Appetite-Stimulating Peptide Hormone. Cell 132: 387–396.

13. Delhanty PJ, Neggers SJ, van der Lely AJ (2012) Mechanisms in endocrinology: Ghrelin: the differences between acyl- and des-acyl ghrelin. Eur J Endocrinol 167: 601–608.

14. Faulconbridge LF, Grill HJ, Kaplan JM, Daniels D (2008) Caudal brainstem delivery of ghrelin induces fos expression in the nucleus of the solitary tract, but not in the arcuate or paraventricular nuclei of the hypothalamus. Brain Res 1218: 151–157.

15. Lawrence CB, Snape AC, Baudoin FM, Luckman SM (2002) Acute central ghrelin and GH secretagogues induce feeding and activate brain appetite centers. Endocrinology 143: 155–162.

16. Takayama K, Johno Y, Hayashi K, Yakabi K, Tanaka T, et al. (2007) Expression of c-Fos protein in the brain after intravenous injection of ghrelin in rats. Neurosci Lett 417: 292–296.

17. Kobelt P, Wisser AS, Stengel A, Goebel M, Inhoff T, et al. (2008) Peripheral injection of ghrelin induces Fos expression in the dorsomedial hypothalamic nucleus in rats. Brain Res 1204: 77–86.

18. Delhanty PJ, Huisman M, Baldeon-Rojas LY, van den Berge I, Grefhorst A, et al. (2013) Des-acyl ghrelin analogs prevent high-fat-diet-induced dysregulation of glucose homeostasis. FASEB J 27: 1690–1700.

19. Heijboer AC, Pijl H, Van den Hoek AM, Havekes LM, Romijn JA, et al. (2006) Gut-brain axis: regulation of glucose metabolism. J Neuroendocrinol 18: 883–894.

20. Muccioli G, Baragli A, Granata R, Papotti M, Ghigo E (2007) Heterogeneity of ghrelin/growth hormone secretagogue receptors. Toward the understanding of the molecular identity of novel ghrelin/GHS receptors. Neuroendocrinology 86: 147–164.

21. Favaro E, Granata R, Miceli I, Baragli A, Settanni F, et al. (2012) The ghrelin gene products and exendin-4 promote survival of human pancreatic islet endothelial cells in hyperglycaemic conditions, through phosphoinositide 3-kinase/Akt, extracellular signal-related kinase (ERK)1/2 and cAMP/protein kinase A (PKA) signalling pathways. Diabetologia 55: 1058–1070.

22. Granata R, Settanni F, Julien M, Nano R, Togliatto G, et al. (2012) Des-acyl ghrelin fragments and analogues promote survival of pancreatic beta-cells and human pancreatic islets and prevent diabetes in streptozotocin-treated rats. J Med Chem 55: 2585–2596.

23. Togliatto G, Trombetta A, Dentelli P, Baragli A, Rosso A, et al. (2010) Unacylated Ghrelin Rescues Endothelial Progenitor Cell Function in Individuals with Type 2 Diabetes. Diabetes 59: 1016–1025.

24. Delhanty PJ, Sun Y, Visser JA, van Kerkwijk A, Huisman M, et al. (2010) Unacylated ghrelin rapidly modulates lipogenic and insulin signaling pathway gene expression in metabolically active tissues of GHSR deleted mice. PLoS One 5: e11749.

25. Asakawa A, Inui A, Fujimiya M, Sakamaki R, Shinfuku N, et al. (2005) Stomach regulates energy balance via acylated ghrelin and desacyl ghrelin. Gut 54: 18–24.

26. Chen CY, Chao Y, Chang FY, Chien EJ, Lee SD, et al. (2005) Intracisternal des-acyl ghrelin inhibits food intake and non-nutrient gastric emptying in conscious rats. Int J Mol Med 16: 695–699.

27. Inhoff T, Monnikes H, Noetzel S, Stengel A, Goebel M, et al. (2008) Desacyl ghrelin inhibits the orexigenic effect of peripherally injected ghrelin in rats. Peptides 29: 2159–2168.

28. Fujimiya M, Asakawa A, Ataka K, Kato I, Inui A (2008) Different effects of ghrelin, des-acyl ghrelin and obestatin on gastroduodenal motility in conscious rats. World J Gastroenterol 14: 6318–6326.

29. Toshinai K, Yamaguchi H, Sun Y, Smith RG, Yamanaka A, et al. (2006) Des-acyl Ghrelin Induces Food Intake by a Mechanism Independent of the Growth Hormone Secretagogue Receptor. Endocrinology 147: 2306–2314.

30. Neary NM, Druce MR, Small CJ, Bloom SR (2006) Acylated ghrelin stimulates food intake in the fed and fasted states but desacylated ghrelin has no effect. Gut 55: 135.

31. Diano S, Horvath TL (2012) Mitochondrial uncoupling protein 2 (UCP2) in glucose and lipid metabolism. Trends Mol Med 18: 52–58.

32. Whittle AJ, Carobbio S, Martins L, Slawik M, Hondares E, et al. (2012) BMP8B Increases Brown Adipose Tissue Thermogenesis through Both Central and Peripheral Actions. Cell 149: 871–885.

33. Starcevic VP, Morrow BA, Farner LA, Keil LC, Severs WB (1988) Long-term recording of cerebrospinal fluid pressure in freely behaving rats. Brain Res 462: 112–117.

34. Paxinos G, Watson C (2004) The rat brain in stereotaxic coordinates. San Diego: Academic Press.

35. Tschöp M, Smiley DL, Heiman ML (2000) Ghrelin induces adiposity in rodents. Nature 407: 908–913.

36. Kirchner H, Heppner KM, Tschop MH (2012) The role of ghrelin in the control of energy balance. Handb Exp Pharmacol: 161–184.

37. Inhoff T, Wiedenmann B, Klapp BF, Monnikes H, Kobelt P (2009) Is desacyl ghrelin a modulator of food intake? Peptides 30: 991–994.

38. Chen CY, Inui A, Asakawa A, Fujino K, Kato I, et al. (2005) Des-acyl ghrelin acts by CRF type 2 receptors to disrupt fasted stomach motility in conscious rats. Gastroenterology 129: 8–25.

39. Matsuda K, Miura T, Kaiya H, Maruyama K, Shimakura S, et al. (2006) Regulation of food intake by acyl and des-acyl ghrelins in the goldfish. Peptides 27: 2321–2325.

40. Cowley MA, Smart JL, Rubinstein M, Cerdan MG, Diano S, et al. (2001) Leptin activates anorexigenic POMC neurons through a neural network in the arcuate nucleus. Nature 411: 480–484.

41. Silva RM, Hadjimarkou MM, Rossi GC, Pasternak GW, Bodnar RJ (2001) Beta-endorphin-induced feeding: pharmacological characterization using selective opioid antagonists and antisense probes in rats. J Pharmacol Exp Ther 297: 590–596.

42. Andrews ZB, Liu ZW, Walllingford N, Erion DM, Borok E, et al. (2008) UCP2 mediates ghrelin's action on NPY/AgRP neurons by lowering free radicals. Nature 454: 846–851.

43. Huszar D, Lynch CA, Fairchild-Huntress V, Dunmore JH, Fang Q, et al. (1997) Targeted disruption of the melanocortin-4 receptor results in obesity in mice. Cell 88: 131–141.

44. Pritchard LE, White A (2007) Neuropeptide processing and its impact on melanocortin pathways. Endocrinology 148: 4201–4207.

45. Aponte Y, Atasoy D, Sternson SM (2011) AGRP neurons are sufficient to orchestrate feeding behavior rapidly and without training. Nat Neurosci 14: 351–355.

46. Wu Q, Boyle MP, Palmiter RD (2009) Loss of GABAergic signaling by AgRP neurons to the parabrachial nucleus leads to starvation. Cell 137: 1225–1234.

47. Wu Q, Palmiter RD (2011) GABAergic signaling by AgRP neurons prevents anorexia via a melanocortin-independent mechanism. Eur J Pharmacol 660: 21–27.

48. Roselli-Rehfuss L, Mountjoy KG, Robbins LS, Mortrud MT, Low MJ, et al. (1993) Identification of a receptor for gamma melanotropin and other proopiomelanocortin peptides in the hypothalamus and limbic system. Proc Natl Acad Sci U S A 90: 8856–8860.

49. Fong TM, Mao C, MacNeil T, Kalyani R, Smith T, et al. (1997) ART (protein product of agouti-related transcript) as an antagonist of MC-3 and MC-4 receptors. Biochem Biophys Res Commun 237: 629–631.

50. Chen AS, Marsh DJ, Trumbauer ME, Frazier EG, Guan XM, et al. (2000) Inactivation of the mouse melanocortin-3 receptor results in increased fat mass and reduced lean body mass. Nat Genet 26: 97–102.

11

Behavioral Characterization of the Hyperphagia Synphilin-1 Overexpressing Mice

Xueping Li[1,9,¤], Yada Treesukosol[2,9], Alexander Moghadam[2], Megan Smith[2], Erica Ofeldt[2], Dejun Yang[1], Tianxia Li[1], Kellie Tamashiro[2], Pique Choi[2], Timothy H. Moran[2]*, Wanli W. Smith[1]*

1 Department of Pharmaceutical Sciences, University of Maryland School of Pharmacy, Baltimore, Maryland, United States of America, 2 Department of Psychiatry, Johns Hopkins University School of Medicine, Baltimore, Maryland, United States of America

Abstract

Synphilin-1 is a cytoplasmic protein that has been shown to be involved in the control of energy balance. Previously, we reported on the generation of a human synphilin-1 transgenic mouse model (SP1), in which overexpression of human synphilin-1 resulted in hyperphagia and obesity. Here, behavioral measures in SP1 mice were compared with those of their age-matched controls (NTg) at two time points: when there was not yet a group body weight difference ("pre-obese") and when SP1 mice were heavier ("obese"). At both time points, meal pattern analyses revealed that SP1 mice displayed higher daily chow intake than non-transgenic control mice. Furthermore, there was an increase in meal size in SP1 mice compared with NTg control mice at the obese stage. In contrast, there was no meal number change between SP1 and NTg control mice. In a brief-access taste procedure, both "pre-obese" and "obese" SP1 mice displayed concentration-dependent licking across a sucrose concentration range similar to their NTg controls. However, at the pre-obese stage, SP1 mice initiated significantly more trials to sucrose across the testing sessions and licked more vigorously at the highest concentration presented, than the NTg counterparts. These group differences in responsiveness to sucrose were no longer apparent in obese SP1 mice. These results suggest that at the pre-obese stage, the increased trials to sucrose in the SP1 mice reflects increased appetitive behavior to sucrose that may be indicative of the behavioral changes that may contribute to hyperphagia and development of obesity in SP1 mice. These studies provide new insight into synphilin-1 contributions to energy homeostasis.

Editor: Silvana Gaetani, Sapienza University of Rome, Italy

Funding: This work was supported by National Institutes of Health Grant DK083410 to Wanli W. Smith. The funders had no role in study design, data collection and analysis, decision to publish, or preparation of the manuscript.

* E-mail: wsmith@rx.umaryland.edu (WWS); tmoran@jhmi.edu (THM)

¤ Current address: Xi'an Medical College, Xi'an, China

9 These authors contributed equally to this work.

Introduction

Synphilin-1 (919 aa) is a cellular protein predominantly expressed in the cytosol [1]. Synphilin-1 protein is present in many tissues with enriched expression in neurons [1]. Synphilin-1 has been reported to interact with a number of proteins including alpha-synuclein, parkin and other proteasome/ubiquitin associated proteins [1–5]. Previous reports showed that synphilin-1 enhances the formation of intracellular protein inclusions and may be involved in Parkinson's disease (PD) pathogenesis [1–4,6]. Synphilin-1 can reduce PD-linked mutant alpha-synuclein-, rotenone-, and 6-HODA-induced toxicity in vitro and delays alpha-synucleinopathies in a PD mouse model in vivo [7,8]. Recent studies of human synphilin-1 transgenic *Drosophila* and mouse models have revealed that overexpression of human synphilin-1 results in increases in food intake, body weight and fat deposition, resembling key features of human obesity [9,10]. While these studies suggest a role for synphilin-1 in regulating energy balance, the biological mechanisms underlying synphilin-1-mediated hyperphagia and obesity are unknown.

Hyperphagia is a core feature of many obesity models and alterations in multiple signaling pathways can contribute to hyperphagia [11–17]. Thus, detailed analysis of food intake changes in the SP1 mouse model could provide insights into the underlying mechanisms that are driving the hyperphagia [9]. Increased food intake can be result from increased meal size, meal number, or both [11–17]. The direct controls of meal size can be categorized into positive and negative signals that maintain and terminate eating behavior respectively [18]. Positive feedback is elicited by stimulation of gustatory, olfactory, and somatosensory receptors in the oral cavity whereas negative feedback is produced by contact with receptors in the oral cavity, stomach and small intestine [19,20]. Increased orosensory stimulation and/or reduced sensitivity to postingestive inhibitory signals would alter meal pattern resulting in increases in food intake.

In the present studies, we tested separate cohorts of 6–8-week-old ("pre-obese) and 4-month-old ("obese") SP1 male mice and age-matched NTg controls to assess 1) how meal pattern parameters change to reflect the increased food intake in SP1 mice, 2) whether synphilin-1 expression alters appetitive behaviors or unconditioned licking responses to sucrose that differed from those of NTg control mice, and 3) whether there are developmental feeding behavior changes related to hyperphagia and obesity. The results from these behavioral assessments would provide an increased understanding of how synphilin-1-expression results in hyperphagia and obesity.

Materials and Methods

Subjects

The SP1 mice that expressed human synphilin-1 in neurons under the mouse prion protein promoter were generated as described previously [9]. SP1 mice for behavioral experiments were generated by successive backcrossing with the C57BL6 strain. At 3 weeks of age, SP1 mice and age-matched controls (NTg) were weaned. Then, PCR-genotyping was performed to separate non-transgenic and SP1 mice as described previously [7]. Briefly, a 1 cM tail tip of each mouse was cut and subjected to DNA extraction. Then the resulting DNA was subjected to PCR using primers to detect the synphilin-1 sequences. Male NTg and SP1 mice at 6–10 weeks ("pre-obese") and 4 months ("obese") of age were used as subjects in the behavioral procedures described. All animal experiments were approved by the Johns Hopkins University Institutional Animal Care and Use Committee.

Western Blot Analysis

The brains of Synphilin-1 and NTg control mice were homogenated as described previously [9]. The brain homogenates were subjected to western blot analysis using anti-synphilin-1 antibodies as described previously [9].

Meal Pattern Analysis

Mice for the "pre-obese" comparisons (n = 9–10/group) were single-housed in DietMax System food intake monitoring cages (AccuScan Instruments, Inc., Columbus, OH) (length 32 cm, width 22 cm) with *ad libitum* access to powdered chow (6% fat, 3.1 kcal/g; 2018 Tekland, Harlan) and water. Food intake was monitored continuously over 23-h daily test sessions as previously described [21]. Powdered laboratory chow diet was provided *ad libitum* in a food jar placed on a scale in the feeding compartment of the cage. The animals had access to the food jar via an opening in the wall of the cage. A water bottle was mounted on an adjacent wall of the cage. Water was available *ad libitum*.

Testing began after 7 days of habituation to the experimental environment and maintenance on chow diet. Meal pattern measures to powdered chow were taken for three consecutive days. For the 4–6 weeks old mice, intake of the powdered high fat diet (45% fat, 4.73 kcal/g; D12451, Research Diets) was measured for the next 3 consecutive days A feeding bout was operationally defined as requiring ≥0.02 g food. An interval of ≥10 min without food intake defined the termination of a meal. This bout criteria on average accounted for ~90% of the feeding data in the current study. One mouse produced excessive spillage of the standard chow thus data from that animal were excluded for data analysis.

For the "obese" comparisons (n = 6–7/group), mice were single-housed under similar test conditions but in test cages (Coulbourn Instruments, Allentown PA) (19 cm×19 cm) equipped with a pellet dispenser that delivers 20-mg chow pellets (3.8% fat, 3.35 kcal/g; Bioserve) as previously described [22]. Removal of a pellet from the feeding dish activated the pellet dispenser to deliver another pellet. Meal pattern parameters were measured across four consecutive days. A meal was operationally defined as at least

3 pellets preceded and followed by at least 10 min without food intake.

Brief-access Taste Procedure

This behavior assessment was performed as previously described with slight modification [23]. Additional cohorts of male mice at 6–10 weeks (n = 8/group) or 4 months of age (n = 8/group) were individually housed in a procedure room where humidity, temperature, and a 12 h light-12 h dark cycle were automatically controlled. Behavioral testing started after at least 3 days of adaption to the experimental procedure room. The mice were on a water-restriction schedule during behavioral training. Water was taken away from the home cages for 23 hours before testing. Animals were only allowed access to water during the training sessions. After the last training session, animals were allowed free access water in the home cages. Mice were then tested with different sucrose concentrations under a partial food and water restriction condition in which they were presented ~1 g of chow and ~2 ml of water for ~23 hours before testing. At least one repletion day (free access to water and chow) followed each testing day under food and water restriction.

Training and testing was performed in a lickometer (Davis MS-160, DiLog Instruments, Tallahassee FL) during the light cycle as described previously [23,24]. The animal was put in the testing chamber and had access to a single spout positioned about 5 mm behind a slot. A potential trial was cued with the opening of a shutter exposing the drinking spout. The mouse licked the spout to initiate a trial. The shutter closed after each trial (5 s). During each intertrial interval (8 s), the tube presentation was changed by a motorized block, and then the shutter re-opened for the next trial. Animals were allowed to initiate as many trials as possible during the 25-min sessions. Concentrations were presented in randomized blocks of 7. Water and six concentrations of sucrose (0.03, 0.06, 0.15, 0.3, 0.6 and 1.0 M) that cover the dynamic range of responsiveness for mice were chosen. All solutions were prepared daily using distilled water.

Animals were subjected to a ~23 h water-restriction schedule for the four days of behavioral training during which animals were allowed free access water only during the daily sessions. On days 1 and 2, animals were allowed access a stationary spout of water for 30-min sessions. On days 3 and 4, animals were allowed access to seven spouts of water in 5-s trials across 25-min sessions. After the end of session 4, animals were allowed free access water in their home cages. The next week, animals were subjected to the tests with water and six concentrations of sucrose across three 25-min daily sessions under a partial water and food restriction condition with one repletion day interspersed between testing days.

For each sucrose concentration, the mean number of licks was determined by all trials across the three test sessions. A Licks Relative to Water Value was calculated by subtracting the mean number of licks to water from the mean number of licks at each concentration as described previously [25–29]. This method provided concentration-response curves that were adjusted to a water baseline. The values for each concentration were compared using two-way repeated measures analyses of variance (ANOVA). If an individual animal did not initiate at least 2 trials per concentration collapsed across the three sucrose testing sessions, this animal was excluded from the Licks Relative to Water Value analysis. Data from all animals were included for analysis of number of trials. Two-sample t-tests were used to compare body weight, mean licks to a stationary spout of water, Interlick Interval (ILI) Values, number of trials initiated and chow intake across the two groups. Only ILIs that were between 70 and 200 ms were subjected to analysis given that ILI values less than 70 ms were defined as double licks and values more than 200 ms were considered pauses between licking bursts [25]. A p<0.05 was considered statistical significant for all analyses. The curves were fit to data using a previously described logistic function [23]: $f(x) = a/(1+10^{(x-c)b})$, where x = $\log_{10}$ stimulus concentration, a = asymptotic lick response, b = slope and c = $\log_{10}$ concentration at the inflection point.

Results

Average Meal Size is Increased in SP1 Mice

The cohorts of both pre-obese (6–10 weeks of age) and obese (4 months of age) SP1 transgenic mice were generated as described previously [9]. Human synphilin-1 proteins were expressed in brains at both pre-obese and obese stages as determined by western blot analysis using anti-human synphilin-1 antibodies (Figure 1).

At the pre-obese time point, there was no significant group difference in body weights between SP1 and NTg mice. However, SP1 mice consumed significantly more standard chow than did controls mice {Figure 2A; t(17) = −2.461, p = 0.025}. Meal pattern assessment showed that there was a trend for an increase in meal size (Figure 2B) and a slight decrease in meal number (Figure 2C) although these data did not reach statistical differences in either meal size {t(17) = −1.981, p = 0.064} or meal number {Figure 2C; t(17) = 1.540, p = 0.142}.

When animals were switched to a high-fat diet, a statistical difference in daily intake between the SP-1 and NTg mice was no longer apparent {Figure 2D; t(18) = −0.314, p = 0.757}. Two-sample t-tests did not reveal a significant group difference in meal size {Figure 2E; t(18) = −1.453, p = 0.163} but SP1 mice initiated fewer meals than NTg controls {Figure 2F; (t) = 2.102, p = 0.05}. Paired t-tests revealed that both groups increased meal size when fed high-fat diet, SP1 {t(8) = −2.773, p = 0.024]], NTg {t(9) = −4.534, p = 0.001}. There was also a significant decrease in meal number {SP1 (t(8) = 7.858, p<0.001}, NTg {t(9) = 5.213, p = 0.001} in both groups when switched from chow to the high-fat diet such that this did not result in a significant overall difference in intake for either group {SP1 (t(8) = 0.217, p = 0.833}, NTg {t(9) = −1.782, p = 0.108}.

In the older "obese" cohort, SP1 mice were significantly heavier than NTg controls and SP1 mice consumed significantly more chow than age-matched controls {Fig. 3A,t(11) = −2.302, p = 0.042}. Two-sample t-tests revealed that SP1 mice consumed significantly larger meals compared with the non-transgenic control mice {Figure 3B; (t(11) = −2.628, p = 0.024)}. In contrast,

there was no significant group difference in meal number between SP1 and non-transgenic control mice {Figure 3C; (t) = 0.227, p = 0.824}.

Pre-obese SP1 Mice Initiated Significantly More Trials to Sucrose and Licked More Vigorously to the Higher Sucrose Concentration

Increased intake and larger meal sizes may be driven by elevated orosensory stimulation. Thus, to assess orosensory responsivity, unconditioned licking responses to a concentration array of sucrose was measured in SP1 and NTg controls in a brief-access taste test. The brief-access taste procedure involves presenting a range of taste solution concentrations in short (5-s) trials across one session and thus minimizes the effect of postingestive cues. The procedure also allows for some segregation of the appetitive and consummatory components of ingestive behavior. The mouse's approach behavior to the spout and initiating licking can be considered appetitive behavior. The licking response within a 5-s trial follows contact with the taste stimulus and can be considered consummatory behavior.

At the pre-obese stage, a two-way ANOVA comparing body weights between the two groups across brief assess test days revealed no main effect of group {F(1,14) = 2.403, p = 0.143}, a main effect of day {F(10,140) = 214.831, p<0.001} and a significant interaction {F(10,140) = 4.242, p<0.001}. The SP1 group was significantly lighter on two testing days but this group difference did not reach Bonferonni correction (Figure 4). The groups did not differ in the total number of licks {Figure 5A; t(14) = 0.008, p = 0.994} or interlick interval (ILI) values to a stationary spout of water {Figure 5B; t(14) = −1.148, p = 0.270}. Two-way ANOVAs comparing sucrose licks relative to water values revealed no significant main effect of genotype {F(1,11) = 3.279, p = 0.098}, a main effect of concentration {F(5,55) = 81.148, p<0.001} and no significant interaction {F(5,55) = 1.996, p = 0.094} at the pre-obese stage. Post hoc t-tests revealed SP1 mice displayed significantly more licks to 1.0 M sucrose adjusted for water, compared to the controls {Figure 5C; t(11) = 3.492, p = 0.005 (p = 0.030)} that survived Bonferroni corrections. SP1 mice initiated significantly more trials to sucrose compared to the NTg controls {Figure 5D; t(14) = 2.140, p = 0.050}.

At the obese stage (4 months of age), SP1 mice were significantly heavier than their NTg controls as confirmed by a main effect of group {F(1,14) = 26.586, p<0.001}, a main effect of day {F(10,140) = 88.759, p<0.001} and a significant interaction{(F(10,140) = 7.568, p<0.001}. To a stationary spout of water, total number of licks across a 30-min session was significantly lower in the SP1 group compared to the NTg controls {t(14) = −2.303, p = 0.037; Figure 6A} but there was no significant group difference in ILI values {t(14) = 1.799, p = 0.094; Figure 6B}. Two-way ANOVAs comparing sucrose licks relative to water values between the two groups revealed no main effect of group {F(1,10) = 0.090, p = 0.770}, revealed a main effect of concentration {F(5,50) = 51.297, p<0.001} and a significant interaction {F(5,50) = 3.331, p = 0.011}. Although a significant group × concentration interaction was revealed, two-sample t-tests conducted at each concentration did not reveal significant group differences (Figure 6C). There was no significant difference in the number of trials initiated during the sucrose testing sessions between SP1 and non-transgenic mice {t(14) = 0.274, 0.788; Figure 6D}.

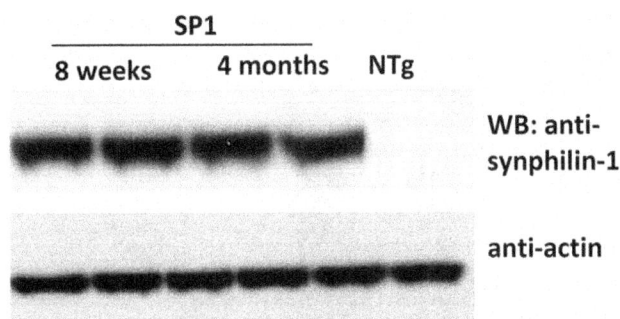

Figure 1. Human synphilin-1 expression in brains of SP1 mice. Brain homogenates from SP1 and non-transgenic control mice were subjected to western blot analysis using anti-synphilin-1 and anti-actin antibodies.

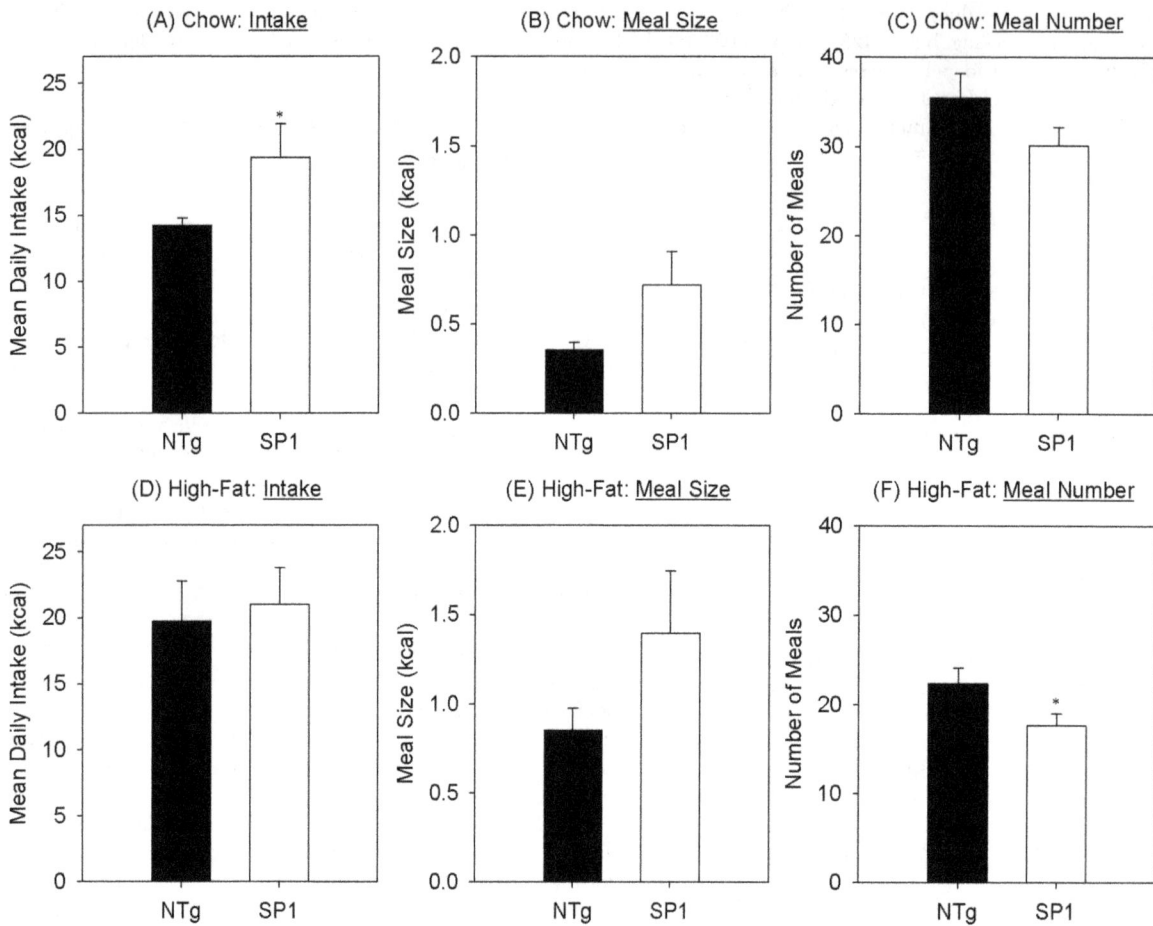

Figure 2. Meal pattern of pre-obese SP1 mice. Mean (A) daily intake, (B) meal size and (C) meal number for standard chow and (D) intake, (E) meal size and (F) meal number for high-fat diet for NTg (black bars) synphilin-1 mice (white bars) at the "pre-obese" time-point. *p<0.05, compared with non-transgenic mice.

Discussion

In this study, we assessed various aspects of feeding behavior to further characterize the hyperphagia in SP1 mice. Meal pattern assessment revealed increases of meal size but no change of meal numbers in SP1 mice compared with non-transgenic controls. SP1 mice at pre-obese stage initiated significantly more trials to sucrose across the testing sessions and licked more vigorously to the highest

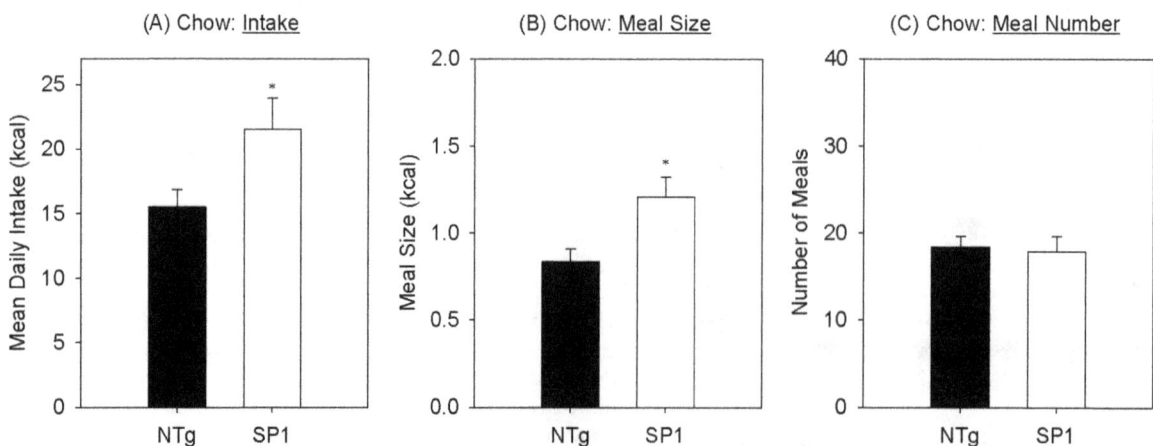

Figure 3. Meal pattern of SP1 obese mice. A. body weight. Mean (B) daily intake, (C) meal size and (D) meal number for standard chow for NTg (black bars) synphilin-1 mice (white bars) at the "obese" time-point. *<0.05, compared with non-transgenic mice.

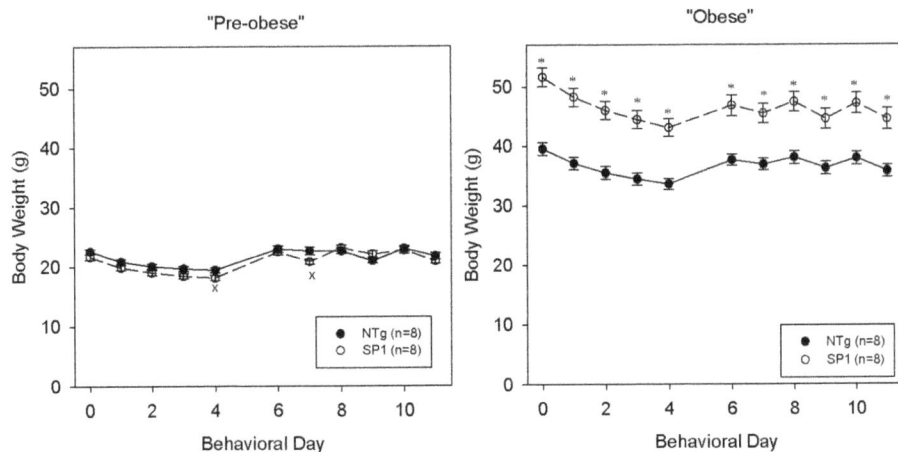

Figure 4. Body weight of pre-obese and obese SP1 mice. Body weight of mice as indicated were measured during testing days in the brief-access taste tests. NTg (black symbols) and synphilin-1 mice (white symbols). *p<0.05 compared with non-transgenic mice.

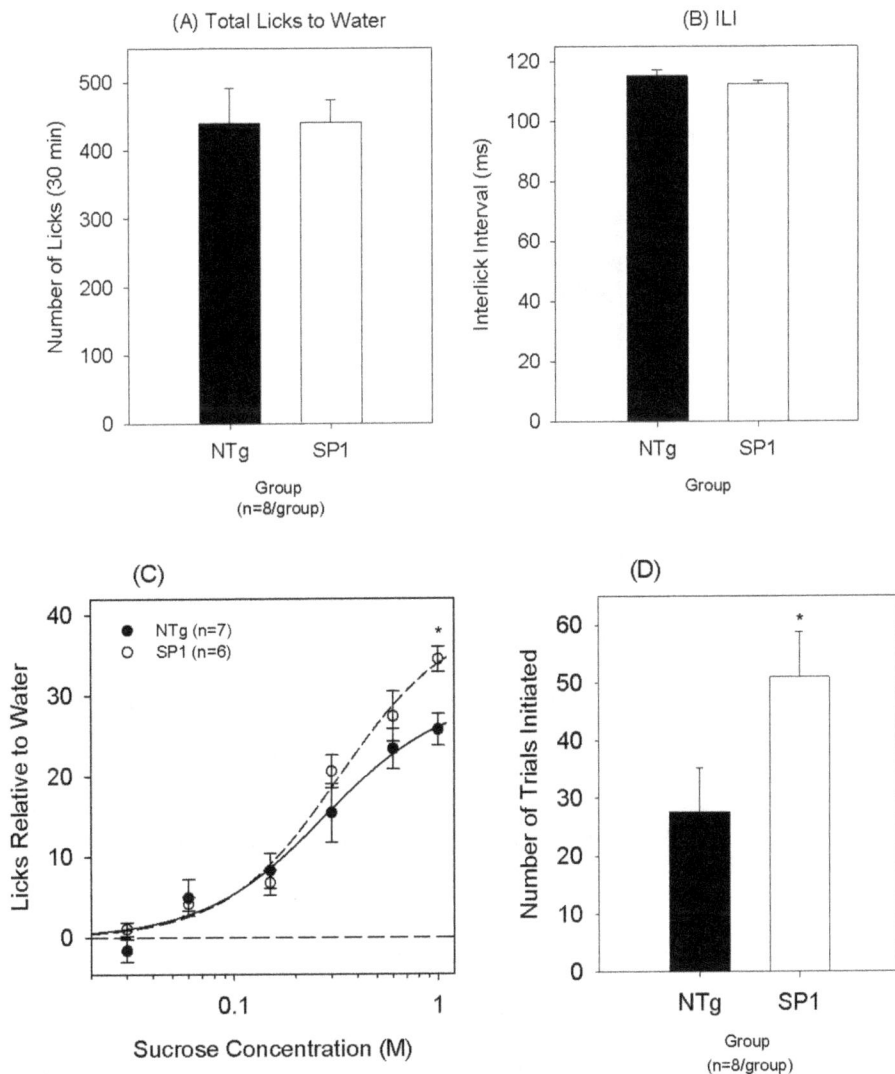

Figure 5. Licking response to sucrose in SP1 pre-obese mice. Mean (A) total licks and (B) interlick interval values to a stationary spout of water. (C) Licks relative to water across a sucrose concentration array and (D) number of trials initiated for NTg (black symbols) and synphilin-1 (white symbols) mice at the "pre-obese" time-point. *p<0.05, compared with non-transgenic mice.

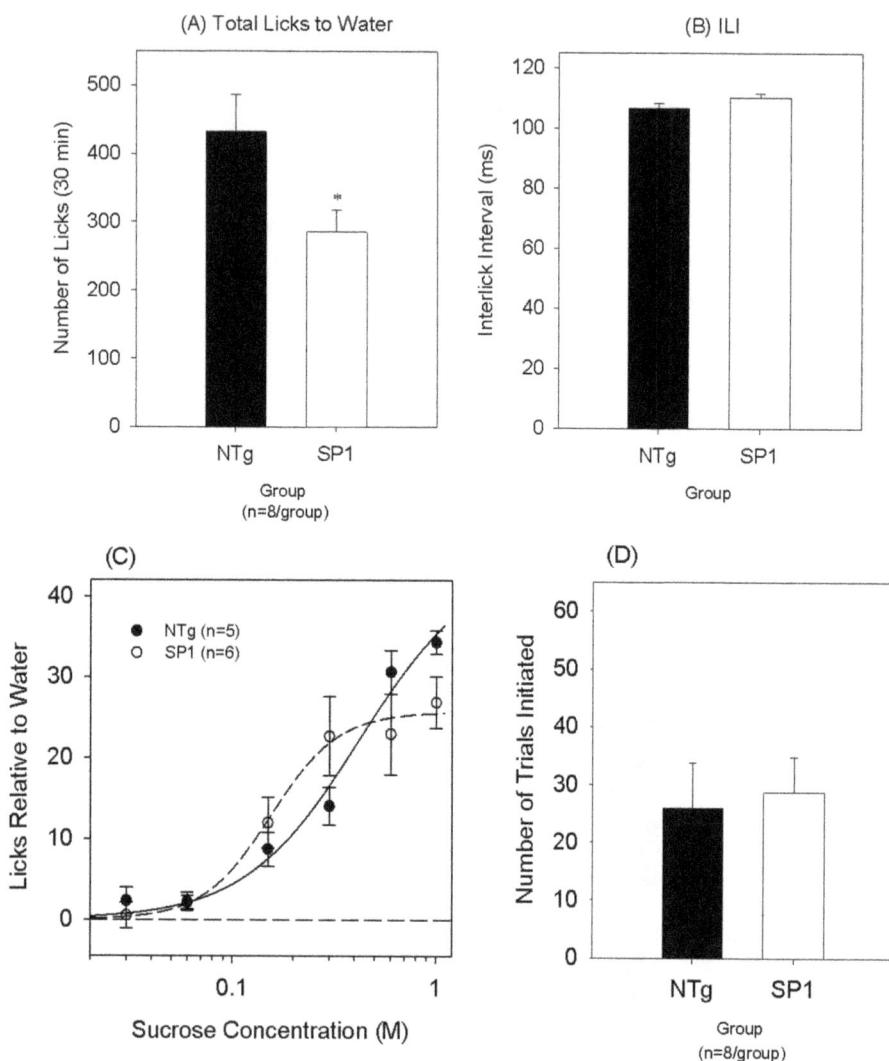

Figure 6. Licking response to sucrose of SP1 obese mice. Mean (A) total licks and (B) interlick interval values to a stationary spout of water. (C) Licks relative to water across a sucrose concentration array and (D) number of trials initiated for NTg (black symbols) and synphilin-1 (white symbols) mice at the "obese" time-point. *p<0.05, compared with non-transgenic mice.

concentration presented, compared to NTg controls. These findings indicate that the hyperphagia in SP1 mice is at least partly due to an increased appetitive behavioral component that may be indicative of enhanced food reward-related mechanisms.

Hyperphagia can be expressed as increases in meal size without significant changes in meal number in some obesity models [11,12]. For instance, leptin and melanocortins that regulate food intake through modulating hypothalamic signaling pathways reduce food intake by altering the size of meal [13–17]. In SP1 mice, synphilin-1 is highly expressed in the arcuate and paraventricular nuclei of the hypothalamus, two central regions that play important roles in the controls of food intake [9]. Moreover, food deprivation significantly increases endogenous synphilin-1 expression in these two regions, suggesting a role for synphillin-1 in deprivation induced feeding [9]. Here, our data demonstrate that SP1 mice, whether tested at the younger "pre-obese" or older "obese" time points, had higher daily chaw intake compared to NTg controls. This increase of food intake was presented by increases in meal size without changing meal number. Increases in meal size of SP1 mice appear to be more

robust in the older mice in which increased body weight was more pronounced. However, even in the "pre-obese" stage, there was a trend increase in meal size although it did not reach statistical significance.

When the diet was switched from standard chow to a 45% high-fat diet, both SP1 mice and NTg controls decreased meal number and increased meal size. Although SP1 mice showed significantly higher intake to standard chow compared to the NTg group, the group difference was no longer apparent when the mice were maintained on the high-fat diet. In the SP1 mice, total caloric intake was already relatively high on chow diet, thus it is plausible that in these mice, although high-fat diet elicits an increase in meal size, there is a ceiling effect preventing further increase in total daily calories.

Increased orosensory stimulation contributes greatly to human obesity and other obesity animal models [19,30]. Our data showed that unconditioned licking responses to sucrose were similar across both groups at the "pre-obese" time point. However, the SP1 mice displayed more vigorous licking at the highest concentration and also initiated significantly more trials than their NTg counterparts giving evidence of increased ingestive behavioral components that likely contribute to increase in food intake in early development of obesity in SP1 mice. At the "obese" time point, these group differences were no longer apparent. Animals approach to the spout can be considered appetitive behavior while licking during a trial is behavior elicited by contact with the stimulus and thus can be considered consummatory. Collectively, these data indicate that consummatory behavior to a palatable liquid measured by unconditioned lick responses to sucrose are relatively normal in SP1 mice. Although measures of sensory threshold and affective responding can be experimentally differentiated [31,32], these assessments are not necessarily mutually exclusive. The SP1 and NTg mice did not significantly differ in affective responding, therefore it suggest that synphilin-1 has little or no effect on orosensory responsivity to sucrose. Thus increased taste responsivity does not likely contribute to the hyperphagia observed in the SP1 mice. At the younger pre-obese time point, SP1 mice display elevated appetitive behavior towards trials to sucrose, suggesting that there may be increased motivation-related mechanisms that may contribute to the hyperphagia that eventually results in increased body weight in SP1 mice. Our data are similar to a previous report of a rat obesity model (OLETF rat), in which that alterations in taste response results in increase of meal size and food intake [33–36]. OLETF rats exhibit an increased response to sweet tastants but unaltered responses to other taste stimuli [34]. At the older obese time point at which SP1 mice are significantly heavier, this difference in appetitive behavior is no longer observed suggesting other underlying mechanism(s) may be in play that remains further studies.

At the older time point (obese stage), SP1 mice displayed significantly fewer licks to a stationary spout of water during a 30-min session when tested under a water-restricted condition. This was not observed at the "pre-obese" time point but only at the "obese" time point when the SP1 mice were significantly heavier. The two groups did not significantly differ in interlick-interval values, a measure of local lick rate [37], nor did the groups differ in licks to water during the 10-s trials thus suggesting the difference in water intake was not likely attributed to oromotor-related alterations or immediate consummatory responses to water. There are reports of increased intake or licks to water in other genetically obese rat models compared to lean controls [38,39] and decreased water intake in diet-induced obesity rat studies [40]. It is unclear how the decreased licks to a stationary spout of water in the obese SP1 mice may be related to the alterations in ingestive behavior observed, which requires further investigation.

In summary, synphilin-1 overexpression induced hyperphagia appears to present as increases in meal size but not meal number. SP1 mice displayed an increased appetitive behavioral response to sucrose at pre-obese stage but no change with the addition of a high-fat diet. These findings indicate that synphilin-1-mediated hyperphagia is associated with meal size contribution factors, such as appetitive response to sucrose. These studies provide a novel insight into synphilin-1 regulating food intake behavior and energy homeostasis.

Author Contributions

Conceived and designed the experiments: XL YT WWS THM. Performed the experiments: XL YT AM MS EO DY TL KT PC. Analyzed the data: XL YT WWS THM. Wrote the paper: XL YT WWS THM.

References

1. Engelender S, Kaminsky Z, Guo X, Sharp AH, Amaravi RK, et al. (1999) Synphilin-1 associates with alpha-synuclein and promotes the formation of cytosolic inclusions. Nat Genet 22: 110–114.
2. Smith WW, Margolis RL, Li X, Troncoso JC, Lee MK, et al. (2005) Alpha-synuclein phosphorylation enhances eosinophilic cytoplasmic inclusion formation in SH-SY5Y cells. J Neurosci 25: 5544–5552.
3. Ribeiro CS, Carneiro K, Ross CA, Menezes JR, Engelender S (2002) Synphilin-1 is developmentally localized to synaptic terminals, and its association with synaptic vesicles is modulated by alpha-synuclein. J Biol Chem 277: 23927–23933.
4. Chung KK, Zhang Y, Lim KL, Tanaka Y, Huang H, et al. (2001) Parkin ubiquitinates the alpha-synuclein-interacting protein, synphilin-1: implications for Lewy-body formation in Parkinson disease. Nat Med 7: 1144–1150.
5. Liani E, Eyal A, Avraham E, Shemer R, Szargel R, et al. (2004) Ubiquitylation of synphilin-1 and alpha-synuclein by SIAH and its presence in cellular inclusions and Lewy bodies imply a role in Parkinson's disease. Proc Natl Acad Sci U S A 101: 5500–5505.
6. Wakabayashi K, Engelender S, Yoshimoto M, Tsuji S, et al. (2000) Synphilin-1 is present in Lewy bodies in Parkinson's disease. Ann Neurol 47: 521–523.
7. Smith WW, Liu Z, Liang Y, Masuda N, Swing DA, et al. (2010) Synphilin-1 attenuates neuronal degeneration in the A53T {alpha}-synuclein transgenic mouse model. Hum Mol Genet 19: 2087–2098.
8. Li X, Liu Z, Tamashiro K, Shi B, Rudnicki DD, et al. (2010) Synphilin-1 exhibits trophic and protective effects against Rotenone toxicity. Neuroscience 165: 455–462.
9. Li X, Tamashiro KL, Liu Z, Bello NT, Wang X, et al. (2012) A novel obesity model: synphilin-1-induced hyperphagia and obesity in mice. Int J Obes 36: 1215–1221.
10. Liu J, Li T, Yang D, Ma R, Moran TH, Smith WW (2012) Synphilin-1 alters metabolic homeostasis in a novel Drosophila obesity model. Int J Obes 36: 1529–1536.
11. Moran TH, Katz LF, Plata-Salaman CR, Schwartz GJ (1998) Disordered food intake and obesity in rats lacking cholecystokinin A receptors. Am J Physiol 274: R618–R625.
12. Strohmayer AJ, Smith GP (1987) The meal pattern of genetically obese (ob/ob) mice. Appetite 8: 111–123.
13. Eckel LA (2004) Estradiol: a rhythmic, inhibitory, indirect control of meal size. Physiol Behav 82: 35–41.
14. Azzara AV, Sokolnicki JP, Schwartz GJ (2002) Central melanocortin receptor agonist reduces spontaneous and scheduled meal size but does not augment duodenal preload-induced feeding inhibition. Physiol Behav 77: 411–416.
15. Lee MD, Kennett GA, Dourish CT, Clifton PG (2002) 5-HT1B receptors modulate components of satiety in the rat: behavioural and pharmacological analyses of the selective serotonin1B agonist CP-94,253. Psychopharmacology (Berl) 164: 49–60.
16. Bi S, Moran TH (2002) Actions of CCK in the controls of food intake and body weight: lessons from the CCK-A receptor deficient OLETF rat. Neuropeptides 36: 171–181.
17. Lee MD, Clifton PG (2002) Meal patterns of free feeding rats treated with clozapine, olanzapine, or haloperidol. Pharmacol Biochem Behav 71: 147–154.
18. Smith GP (1996) The direct and indirect controls of meal size. Neurosci Biobehav Rev 20: 41–46.
19. Davis JD, Smith GP, Miesner J (1993) Postpyloric stimuli are necessary for the normal control of meal size in real feeding and sham feeding rats. Am J Physiol 265: R888–R895.
20. Davis JD, Smith GP (1990) Learning to sham feed: behavioral adjustments to loss of physiological postingestional stimuli. Am J Physiol 259: R1228–R1235.
21. Zhu G, Yan J, Smith WW, Moran TH, Bi S (2012) Roles of dorsomedial hypothalamic cholecystokinin signaling in the controls of meal patterns and glucose homeostasis. Physiol Behav 105: 234–241.
22. Aja S, Bi S, Knipp SB, McFadden JM, Ronnett GV, et al. (2006) Intracerebroventricular C75 decreases meal frequency and reduces AgRP gene expression in rats. Am J Physiol Regul Integr Comp Physiol 291: R148–R154.
23. Treesukosol Y, Bi S, Moran TH (2013) Overexpression of neuropeptide Y in the dorsomedial hypothalamus increases trial initiation but does not significantly alter concentration-dependent licking to sucrose in a brief-access taste test. Physiol Behav 110–111: 109–114.
24. Smith JC (2001) The history of the "Davis Rig". Appetite 36: 93–98.
25. Glendinning JI, Gresack J, Spector AC (2002) A high-throughput screening procedure for identifying mice with aberrant taste and oromotor function. Chem Senses 27: 461–474.
26. Jiang E, Blonde G, Garcea M, Spector AC (2008) Greater superficial petrosal nerve transection in rats does not change unconditioned licking responses to putatively sweet taste stimuli. Chem Senses 33: 709–723.
27. Spector AC, Redman R, Garcea M (1996) The consequences of gustatory nerve transection on taste-guided licking of sucrose and maltose in the rat. Behav Neurosci 110: 1096–1109.

28. Treesukosol Y, Blonde GD, Spector AC (2009) T1R2 and T1R3 subunits are individually unnecessary for normal affective licking responses to Polycose: implications for saccharide taste receptors in mice. Am J Physiol Regul Integr Comp Physiol 296: R855–R865.

29. Treesukosol Y, Smith KR, Spector AC (2011) Behavioral evidence for a glucose polymer taste receptor that is independent of the T1R2+3 heterodimer in a mouse model. J Neurosci 31: 13527–13534.

30. De Jonghe BC, Hajnal A, Covasa M (2005) Increased oral and decreased intestinal sensitivity to sucrose in obese, prediabetic CCK-A receptor-deficient OLETF rats. Am J Physiol Regul Integr Comp Physiol 288: R292–R300.

31. Spector AC (2000) Linking gustatory neurobiology to behavior in vertebrates. Neurosci Biobehav Rev 24: 391–416.

32. Spector AC, Glendinning JI (2009) Linking peripheral taste processes to behavior. Curr Opin Neurobiol 19: 370–377.

33. Moran TH, Bi S (2006) Hyperphagia and obesity in OLETF rats lacking CCK-1 receptors. Philos Trans R Soc Lond B Biol Sci 361: 1211–1218.

34. Hajnal A, Covasa M, Bello NT (2005) Altered taste sensitivity in obese, prediabetic OLETF rats lacking CCK-1 receptors. Am J Physiol Regul Integr Comp Physiol 289: R1675–R1686.

35. Moran TH, Katz LF, Plata-Salaman CR, Schwartz GJ (1998) Disordered food intake and obesity in rats lacking cholecystokinin A receptors. Am J Physiol 274: R618–R625.

36. Schwartz GJ, Whitney A, Skoglund C, Castonguay TW, Moran TH (1999) Decreased responsiveness to dietary fat in Otsuka Long-Evans Tokushima fatty rats lacking CCK-A receptors. Am J Physio 277: R1144–R1151.

37. Travers JB, Dinardo LA, Karimnamazi H (1997) Motor and premotor mechanisms of licking. Neurosci Biobehav Rev 21: 631–647.

38. Drewnowski A, Grinker JA (1978) Food and water intake, meal patterns and activity of obese and lean Zucker rats following chronic and acute treatment with delta9-tetrahydrocannabinol. Pharmacol Biochem Behav 9: 619–630.

39. Hajnal A, Covasa M, Bello NT (2005) Altered taste sensitivity in obese, prediabetic OLETF rats lacking CCK-1 receptors. Am J Physiol Regul Integr Comp Physiol 289: R1675–R1686.

40. Cottone P, Sabino V, Nagy TR, Coscina DV, Levin BE, et al. (2013) Centrally administered urocortin 2 decreases gorging on high-fat diet in both diet-induced obesity-prone and -resistant rats. Int. J. Obes. In press.

12

Multiplex Real-Time PCR for Detection of *Staphylococcus aureus*, *mecA* and Panton-Valentine Leukocidin (PVL) Genes from Selective Enrichments from Animals and Retail Meat

Valeria Velasco[1,2], Julie S. Sherwood[1], Pedro P. Rojas-García[3], Catherine M. Logue[4]*

1 Department of Veterinary and Microbiological Sciences, North Dakota State University, Fargo, North Dakota, United States of America, 2 Department of Animal Sciences, University of Concepción, Chillán, Chile, 3 Laboratory of Animal Physiology and Endocrinology, Veterinary Sciences, University of Concepción, Chillán, Chile, 4 Department of Veterinary Microbiology and Preventive Medicine, College of Veterinary Medicine, Iowa State University, Ames, Iowa, United States of America

Abstract

The aim of this study was to compare a real-time PCR assay, with a conventional culture/PCR method, to detect *S. aureus*, *mecA* and Panton-Valentine Leukocidin (PVL) genes in animals and retail meat, using a two-step selective enrichment protocol. A total of 234 samples were examined (77 animal nasal swabs, 112 retail raw meat, and 45 deli meat). The multiplex real-time PCR targeted the genes: *nuc* (identification of *S. aureus*), *mecA* (associated with methicillin resistance) and PVL (virulence factor), and the primary and secondary enrichment samples were assessed. The conventional culture/PCR method included the two-step selective enrichment, selective plating, biochemical testing, and multiplex PCR for confirmation. The conventional culture/PCR method recovered 95/234 positive *S. aureus* samples. Application of real-time PCR on samples following primary and secondary enrichment detected *S. aureus* in 111/234 and 120/234 samples respectively. For detection of *S. aureus*, the *kappa* statistic was 0.68–0.88 (from substantial to almost perfect agreement) and 0.29–0.77 (from fair to substantial agreement) for primary and secondary enrichments, using real-time PCR. For detection of *mecA* gene, the *kappa* statistic was 0–0.49 (from no agreement beyond that expected by chance to moderate agreement) for primary and secondary enrichment samples. Two pork samples were *mecA* gene positive by all methods. The real-time PCR assay detected the *mecA* gene in samples that were negative for *S. aureus*, but positive for *Staphylococcus* spp. The PVL gene was not detected in any sample by the conventional culture/PCR method or the real-time PCR assay. Among *S. aureus* isolated by conventional culture/PCR method, the sequence type ST398, and multi-drug resistant strains were found in animals and raw meat samples. The real-time PCR assay may be recommended as a rapid method for detection of *S. aureus* and the *mecA* gene, with further confirmation of methicillin-resistant *S. aureus* (MRSA) using the standard culture method.

Editor: Franklin D. Lowy, Columbia University, College of Physicians and Surgeons, United States of America

Funding: This study was funded through the Dean's Office, College of Agriculture, North Dakota State University and The Dean's Office, College of Veterinary Medicine, Iowa State University. The funders had no role in study design, data collection and analysis, decision to publish, or preparation of the manuscript.

Competing Interests: The authors have declared that no competing interests exist.

* E-mail: cmlogue@iastate.edu

Introduction

Staphylococcus aureus is considered as an important cause of a wide variety of diseases in humans such as: food poisoning, pneumonia, wound and nosocomial infections [1,2]. There are many anti-staphylococcal agents; however, the bacterium has developed mechanisms to neutralize them such as the methicillin resistance mechanism [3]. Methicillin-resistant *S. aureus* (MRSA) is an increasing cause of health care-associated (HA-MRSA) [1], community-associated (CA-MRSA) [2], and livestock-associated (LA-MRSA) infections worldwide [4].

The altered penicillin-binding protein (PBP2a) is associated with methicillin resistance. This protein has a reduced affinity for β-lactam antibiotics [5,6], and is encoded by the *mecA* gene, which is carried on the staphylococcal cassette chromosome *mec* (SCC*mec*) [5]. CA-MRSA strains are more likely to encode the Panton–Valentine leukocidin (PVL) toxin, which is a pore-forming toxin considered as a virulence factor [7,8]. The PVL toxin has been

related to life-threatening CA-MRSA infections and deaths, primarily severe skin infections and tissue necrosis [9].

In the United States, approximately 29% (78.9 million people) and 1.5% (4.1 million) of the population were estimated to be nasal carriers of *S. aureus* and MRSA, respectively [10]. An estimated 478,000 hospitalizations corresponded to *S. aureus* infections, of which 278,000 hospitalizations were attributed to MRSA infections in 2005 [11]. In addition, the carriage of MRSA in meat-producing animals [12–14] and the contamination of retail meat with MRSA [15–17] have increased the concern that food may serve as a vehicle to transmit MRSA to the human population [17].

Different culture methods have been used to detect MRSA. Generally, conventional microbiological procedures are laborious, since they require the isolation of *S. aureus* before assessing methicillin resistance. However, culture methods are still considered as standard methods for traditional confirmation of *S. aureus*.

Wertheim *et al.* (2001) [18] developed a selective media containing phenol red and antibiotics (aztreonam and ceftizoxime), increasing the sensitivity of the detection of MRSA after 48 h of incubation, but at the expense of longer time needed for confirmation. The isolation and identification of MRSA, including selective enrichment and plating on selective agars, followed by confirmation using biochemical testing and/or PCR assays, requires 3–7 days approximately [15,16,19]. Therefore, development of a rapid method for detection of MRSA has become an important need in the microbiological analysis of samples especially those where there is a potential risk of exposure for humans.

Real-time PCR technology has been used as an alternative to culture methods for the rapid detection of *S. aureus* and MRSA. Detection using real-time PCR may decrease the time of analysis to 18 h after consecutive broth enrichment in clinical samples [20]; or <2 h in positive blood cultures [21,22]. However, most studies have used real-time PCR to detect MRSA in clinical samples and isolates and a few studies have evaluated the application of this method for the detection of MRSA in animals [23,24] and meat [15,25,26].

Since *S. aureus* and MRSA have been found in food-producing animals and retail meat, increasing the concern about the exposure for humans through the food chain, and there is a need to decrease the time of analysis, we analyzed samples obtained from animals and retail meats using primary and secondary selective enrichments in order to detect *nuc* (identification of *S. aureus*), *mecA* (associated with methicillin resistance) and PVL (virulence factor) genes using a multiplex real-time PCR assay. The results obtained with the real-time PCR assay were compared with the results from a culture method, considered as the standard method, which also included the two-step selective enrichment, followed by selective plating, biochemical testing and conventional multiplex PCR. Positive samples obtained with the culture method were characterized by multilocus sequence typing (MLST) and the antimicrobial resistance profiles were obtained.

Materials and Methods

Samples

A total of 77 nasal swabs (Becton, Dickinson and Company, Sparks, MD, USA) were collected from animals (sheep, n = 35; pigs, n = 28; cows, n = 14) sampled immediately after stunning at the Meat Lab (Department of Animal Sciences); and at the ND Veterinary Diagnostic Lab at North Dakota State University, Fargo, ND. Animal samples were collected during the period May 2010-April 2011. The protocol of sampling was approved by the North Dakota State University Institutional Biosafety Committee (B10014).

In addition, 112 retail raw meat (pork, n = 39; chicken, n = 37; beef, n = 36) and 45 deli meat (ham, n = 20; turkey, n = 16; chicken, n = 9) samples were randomly purchased from four different supermarket chains in Fargo, ND. Sampling visits were made between June 2010 and January 2011. All samples were immediately stored at 4°C and processed within six hours of collection.

Culture method

Staphylococcus aureus were isolated by the two-step selective enrichment procedure according to the method described by de Boer *et al.* (2009) [15] followed by plating steps on selective agar. Briefly, for the primary enrichment, a 25 g sample of retail meat and 225 mL of MHB+6.5%NaCl (Mueller-Hinton broth [Difco, Becton, Dickinson, Sparks, MD, USA] with added 6.5% sodium chloride [VWR International, West Chester, PA, USA]) were

placed in a sterile stomacher bag and homogenized using a stomacher400 circulator (Seaward, England) at 230 rpm for 90 seconds. The suspension was incubated for 18–20 h at 37°C. Following primary enrichment, a secondary enrichment was used by inoculating 1 mL of the primary enrichment broth into 9 mL of PHMB⁺ (D-mannitol in phenol red mannitol broth base [Difco, Becton, Dickinson, Sparks, MD, USA] containing ceftizoxime [5 µg mL^{-1}, US Pharmacopeia, Rockville, MD, USA] and aztreonam [75 µg mL^{-1}, Sigma Chemical CO., Louis, MO, USA] according to Wertheim *et al.* [2001] [18]), followed by incubation for 18–20 h at 37°C. Nasal swabs from animals were placed directly in 9 mL MHB+6.5%NaCl and incubated for 18–20 h at 37°C. Then, the secondary enrichment was used following the procedure described above.

Following incubation of the secondary enrichment broth, all samples were struck directly to BP medium (Baird-Parker medium [Difco, Becton, Dickinson, Sparks, MD, USA]) supplemented with egg yolk tellurite according to manufacturer's recommendations and incubated for 48 h at 37°C. Presumptive *S. aureus* colonies (black colonies surrounded by 2 to 5 mm clear zones) were transferred to TSA II 5%SB plates (Trypticase soy agar with 5% sheep blood [Difco, Becton, Dickinson, Sparks, MD, USA]) and incubated for 18–20 h at 37°C. Suspect *S. aureus* colonies (presence of β-haemolysis) were confirmed using Sensititre Gram Positive ID (GPID) plates (Sensititre, TREK Diagnostic Systems Ltd., Cleveland, OH, USA) according to the manufacturer's instructions.

Conventional multiplex PCR method

Confirmed *S. aureus* strains were recovered from frozen stock to TSA plates (Trypticase soy agar [Difco, Becton, Dickinson, Sparks, MD, USA]) and incubated at 37°C for 18–24 h. DNA extraction was carried out by suspending one colony in 50 µL of DNase/RNase-free distilled water (Gibco Invitrogen, Grand Island, NY, USA), heating (99°C, 10 min) and centrifugation (30,000×g, 1 min) to remove cellular debris. The remaining DNA was transferred to a new tube and stored at −20°C.

A multiplex PCR assay for the detection of 16S rRNA (identification of *S. aureus*), *mecA* (associated with methicillin resistance) and PVL-encoding genes (virulence factor) (Table 1) included 2 µL of the DNA template (described above) added to a 50 µL final reaction mixture containing: 1X Go Taq Reaction Buffer (pH 8.5), 0.025 µL^{-1} of Go Taq DNA polymerase, 200 µM dNTP (Promega, Madison, WI, USA) and 1 µM of primers (16S rRNA, *mecA*, LukS/F-PV) (Integrated DNA Technologies, Inc., Coralville, IA, USA).

Multiplex PCR reactions were carried out in a thermocycler (Eppendorf, Hamburg, Germany), and the PCR conditions were adjusted according to the protocol described by Makgotlho *et al.* (2009) [27] as follows: initial denaturation at 94°C for 10 min, followed by 10 cycles of denaturation at 94°C for 45 s, annealing at 55°C for 45 s and extension at 72°C for 75 s followed by another 25 cycles of 94°C for 45 s, 50°C for 45 s and a final extension step at 72°C for 10 min. An external positive control (DNA extracted from MRSA ATCC 35591, positive for *mecA* and PVL genes) and an external negative control (DNase/RNase-free distilled water) were included with each run.

PCR amplicons (10 µL) were loaded into a 1.5% (wt/vol) agarose gel (Agarose I™) using EzVision One loading dye (Amresco, Solon, OH, USA) and electrophoresis was carried out in 1X TAE buffer at 100 v for 1 h. A molecular weight marker 100-bp ladder (Promega, Madison, WI, USA) were included on each gel. Bands were visualized using an Alpha Innotech UV imager (FluorChem™).

Table 1. Nucleotide sequence of the primers and probes used in conventional multiplex PCR, multiplex real-time PCR.

Primer or probe name	Sequence (5'→3')	5' Reporter dye 3' Quencher
16S rRNA*		
Staph-756F	AAC TCT GTT ATT AGG GAA GAA CA	
Staph-750R	CCA CCT TCC TCC GGT TTG TCA CC	
nuc†		
nuc For	CAA AGC ATC AAA AAG GTG TAG AGA	
nuc Rev	TTC AAT TTT CTT TGC ATT TTC TAC CA	Texas Red
nuc Probe	TTT TCG TAA ATG CAC TTG CTT CAG GAC CA	Iowa Black
mecA		
mecA-1F*	GTA GAA ATG ACT GAA CGT CCG ATA A	
mecA-2F*	CCA ATT CCA CAT TGT TTC GGT CTA A	
mecA For†	GGC AAT ATT ACC GCA CCT CA	
mecA Rev†	GTC TGC CAC TTT CTC CTT GT	FAM†
mecA Probe†	AGA TCT TAT GCA AAC TTA ATT GGC AAA TCC	TAMRA†
PVL		
luk-PV-1F*	ATC ATT AGG TAA AAT GTC TGG ACA TGA TCC A	
luk-PV-2R*	GCA TCA AGT GTA TTG GAT AGC AAA AGC	
PVL For†	ACA CAC TAT GGC AAT AGT TAT TT	
PVL Rev†	AAA GCA ATG CAA TTG ATG TA	Cy5†
PVL Probe†	ATT TGT AAA CAG AAA TTA CAC AGT TAA ATA TGA	Iowa Black

*Conventional multiplex PCR, according to McClure et al. (2006) [43].
†Multiplex real-time PCR, according to McDonald et al. (2005) [35].

Multiplex real-time PCR assay

DNA was extracted from the primary and secondary enrichment broths of the animal and meat samples using the boiling method described previously by de Medici *et al.* (2003) [28]. Five microliters of DNA template extracted was used in the real-time iQ^{TM} Multiplex Powermix (Bio-Rad Laboratories, Hercules, CA, USA), in a final volume of 20 µL per reaction.

The real-time PCR assay targeted the following genes: *nuc* (identification of *S. aureus*), *mecA* (associated with methicillin resistance) and PVL-encoding genes (virulence factor) (Table 1).

The final concentrations in the reaction mixture were: 300 nM of primers (forward and reverse), 200 nM of fluorogenic probes (Applied Biosystems, Foster City, CA, USA), and 1X iQ^{TM} Multiplex Powermix (Bio-Rad Laboratories, Hercules, CA, USA), according to the manufacturer's recommendations.

The thermal cycling conditions were adjusted to an initial denaturation of 3 min at 95°C, followed by 40 PCR cycles of 95°C for 15 s and 55°C for 1 min, using an iCycler IQ^{TM} real time PCR system (Bio-Rad Laboratories, Hercules, CA, USA). An external positive control (DNA extracted from MRSA ATCC 35591, positive for mecA and PVL genes) and an external negative control (DNase/RNase-free distilled water) were included with each plate. Data analysis was carried out using the iCycler software version 3.0 (Bio-Rad Laboratories, Hercules, CA, USA).

Characterization of *S. aureus* strains isolated by culture method

Multilocus Sequence typing (MLST). Briefly, *S. aureus* isolates were struck to TSA plates and incubated at 37°C for 18–24 h. Colonies were picked to 40 µL of single cell lysing buffer (50 µg/mL of Proteinase K, Amresco; in TE buffer [pH = 8]), and then lysed by heating to 80°C for 10 min followed by 55°C for

10 min in a thermocycler. The final suspension was diluted 1:2 in sterile water, centrifuged to remove cellular debris, and transferred to a sterile tube (Marmur, 1961) [29]. The housekeeping genes: *arcC, aroE, glpF, gmk, pta, tpi,* and *yqiL*, were amplified [30]. All PCR reactions were carried out in 50-µL volumes: 1 µL of DNA template, Taq DNA polymerase (Promega) (1.25 U), 1X PCR buffer (Promega), primers (0.1 µM) (Integrated DNA Technologies, Inc.), and dNTPs (200 µM) (Promega). PCR settings were adjusted according to Enright et al. (2000) [30] using a thermocycler (Eppendorf). Ten microliters of the PCR products were loaded into 1% agarose gels in 1X TAE with EzVision One loading dye, and run at 100V in 1X TAE for 1 h. Images were captured using an Alpha Innotech imager. After PCR, each amplicon was purified of amplification primer using the QIAquickPCR Purification Kit (Qiagen, Valencia, CA) as per manufacturer's instructions. Purified DNA was sequenced at Iowa State University's DNA Facility (Ames, IA) using an Applied Biosystems 3730xl DNA Analyzer (Applied Biosystems, Foster City, CA). Sequence data were imported into DNAStar (Lasergene, Madison, WI), trimmed, and aligned to the control sequences (from the MLST site) and interrogated against the MLST database (http://saureus.mlst.net/). Sequence types were added to the strain information for analysis in BioNumerics.

Resistance profiles. The antimicrobial resistance profiles (AR) of *S. aureus* isolates (n = 95) were determined using the broth microdilution method (CMV3AGPF, Sensititre, Trek Diagnostics), according to the manufacturer's and the National Antimicrobial Resistance Monitoring System (NARMS) guidelines for animal isolates [31]. Antimicrobials in the panel and their resistance breakpoints were as follows: erythromycin (≥8 µg/mL), tetracycline (≥16 µg/mL), ciprofloxacin (≥4 µg/mL), chloramphenicol (≥32 µg/mL), penicillin (≥16 µg/mL), daptomycin (no interpretative criteria), vancomycin (≥ 32 μg/mL), nitrofurantoin ($\geq$ 128 μg/mL), gentamicin (>500 μg/mL), quinupristin/dalfopristin (≥ 4 μg/mL), linezolid (≥ 8 μg/mL), kanamycin (≥ 1024 μg/mL), tylosin (≥ 32 μg/mL), tigecycline (no interpretative criteria), streptomycin (>1000 μg/mL), and lincomycin (≥ 8 μg/mL). Resistance to at least three classes of antibiotics was considered as multidrug resistance (MDR) [32].

Statistical analysis

The 95% confidence intervals for prevalence were obtained, using the plus four estimate when positive or negative samples were less than 15. The Chi-square test was used to assess the significance in proportion of positive samples between sample types, only if no more than 20% of the expected counts were less than 5 and all individual expected counts were 1 or greater [33]. On the contrary, Fisher's exact test was used with two-sided p-values. SAS software version 9.2 (SAS Institute Inc., Cary, NC) was used to assess significance with a $P<0.05$.

As there is no true gold standard method for *S. aureus* and MRSA detection, the *kappa* statistic was calculated to compare agreement between real-time PCR assay (using primary and secondary enrichment) and conventional culture/PCR method.

Results

The culture method included a biochemical identification to confirm *S. aureus*, which agreed with the results of the conventional multiplex PCR that detected the gene 16S rRNA. This method detected 95 positive *S. aureus* samples from a total of 234 samples collected (Table 2). The multiplex real-time PCR assay using primary and a secondary enrichment samples, recovered *S. aureus* (detection of *nuc* gene) from 111 and 120 samples of 234 samples respectively.

By the conventional culture/PCR method alone, the rate of positive *S. aureus* samples was found to be 41.6% (CI95%: 30.6–52.6%) in animals and 51.8% (CI95%: 42.5–61.0%) in raw meat samples respectively; a significantly lower rate of 11.1% (CI95%: 4.5–24.1%) was observed in deli meat ($P\leq0.05$). Using the primary enrichment samples and real-time PCR, a significantly higher recovery of *S. aureus* ($P\leq0.05$) was found in animals (55.8%, CI95%: 44.8–66.9%) and raw meat (57.1%, CI95%: 47.9–66.3%) than in deli meat samples (8.9%, CI95%: 3.1–21.4%). However, no significant difference ($P>0.05$) was found between the rate of positive *S. aureus* samples in animals (53.2%, CI95%: 42.1–64.4%), raw meat (53.6%, CI95%: 44.3–62.8%) and deli meat (42.2%, CI95%: 27.8–56.7%), when the secondary enrichment samples were tested by real-time PCR. A significantly higher recovery of *S. aureus* ($P\leq0.05$) was obtained from deli meat when the secondary enrichment samples were assessed by real-time PCR.

The *mecA* gene was detected in two pork samples (5.4%, CI95%: 0.7–18.8%) by the conventional multiplex PCR preceded by the culture method, and by assessing the primary and secondary enrichment samples by real-time PCR. The real-time PCR analysis detected the *mecA* gene using both enrichments in samples that were negative by conventional multiplex PCR in two pork meat and three deli meat samples. Using the primary enrichment, the real-time PCR detected the *mecA* gene in one sample isolated from a sheep, and one from pork meat, which were negative using the secondary enrichment. Using the secondary enrichment, the real-time PCR detected the *mecA* gene from one sample isolated from a pig, one from pork meat, and two from deli meat, which were negative using the primary enrichment.

The PVL gene was not detected in any sample by the conventional culture/PCR method or the real-time PCR assay.

Table 3 shows the results of real-time PCR using primary and secondary enrichments on the detection of *S. aureus* compared with a conventional culture/PCR method. Total agreement and the *kappa* statistic for real-time PCR using the primary enrichment samples were 85.7% ($k=0.72$, CI95%: 0.62–0.82), 83.9% ($k=0.68$, CI95%: 0.59–0.76), and 97.8% ($k=0.88$, CI95%: 0.78–0.97) for animals, raw meat, and deli meat respectively. For real-time PCR using the secondary enrichment samples, the total agreement and the *kappa* statistic were 88.3% ($k=0.77$, CI95%: 0.67–0.86), 87.5% ($k=0.75$, CI95%: 0.67–0.83), and 68.9% ($k=0.29$, CI95%: 0.16–0.43) for animals, raw meat, and deli meat respectively. Positive agreement (sensitivity) was 100% for animal samples using both enrichments. For animals and raw meat, a higher negative agreement (specificity) was obtained for real-time PCR using the secondary enrichment.

Six samples isolated from animals and six from raw meat were deemed *S. aureus* negative by the conventional culture/PCR method, but positive by real-time PCR using the primary and secondary enrichments. Three *S. aureus* samples isolated from raw meat were positive by the conventional culture/PCR method, but negative by the real-time PCR assay.

The real-time PCR method using the primary enrichment failed to detect the presence of *S. aureus* in four samples: three isolated from raw meat (two from beef, one from poultry) and one from deli meat (ham) that were positive by the culture method and by the real-time PCR assay using the secondary enrichment samples. Using the secondary enrichment samples, the real-time PCR assay failed to detect three samples isolated from raw meat (pork) that were *S. aureus* positive by the culture method and using the primary enrichment in real-time PCR.

The results of real-time PCR using primary and secondary enrichment on the detection of the *mecA* gene compared with a conventional culture/PCR method are shown in Table 4. Total agreement for real-time PCR using the primary and secondary enrichment samples ranged from 91.1% to 98.7% and from 86.7 to 98.7%, respectively. The *kappa* statistic was zero when the *mecA* gene was not detected by the conventional culture/PCR method and 0.49 (CI95%: 0.39–0.58) for raw meat. Positive agreement (sensitivity) of 100% was obtained for raw meat samples for both methods.

The real-time PCR assay detected the *mecA* gene in samples that were negative for *S. aureus* by the conventional culture/PCR method (one from a pig, one from a sheep, four from pork, four from deli ham, and one from deli turkey). All of these samples were identified as harboring *S. epidermidis*, *S. saprophyticus* or *S. haemolyticus* using biochemical analysis on isolates recovered. However, three of these samples (one from a pig, two from pork meat) tested positive for the *nuc* gene when primary and secondary enrichments were assessed by real-time PCR.

Table 5 shows the antimicrobial resistance profiles and the sequence types of the ninety-five *S. aureus* strains isolated from animals and retail meat by the conventional culture/PCR method. A total of thirteen antimicrobial resistance profiles were identified among *S. aureus* isolates. Most of the *S. aureus* isolates were resistant to tetracycline and lincomycin, and were of ST9. A total of twenty-two *S. aureus* isolates exhibited multi-drug resistance. Susceptibility to all antimicrobials tested were found in thirty-five *S. aureus* isolates, which were mostly recovered from chicken meat and identified as ST5.

Discussion

In this study, a high recovery of *S. aureus* was found in animals and meat samples by the culture/PCR method and the real-time

Table 2. Detection of *S. aureus*, *mecA* and PVL genes from animals and retail meat using a conventional culture/PCR method and a real-time PCR assay.

Sample type	No. of samples	Culture/PCR method (No. of positives)			Real-time PCR Primary enrichment (No. of positives)			Secondary enrichment (No. of positives)		
		S.aureus	*mecA*	PVL	*S.aureus*	*mecA*	PVL	*S.aureus*	*mecA*	PVL
Animals										
Cow	14	0	0	0	4	0	0	3	0	0
Pig	28	21	0	0	25	0	0	24	1	0
Sheep	35	11	0	0	14	1	0	14	0	0
Total	77	32	0	0	43	1	0	41	1	0
Meat										
Beef	36	9	0	0	10	0	0	12	0	0
Pork	37	25	2	0	26	6	0	27	6	0
Poultry	39	24	0	0	28	0	0	21	0	0
Total	112	58	2	0	64	6	0	60	6	0
Deli meat										
Chicken	9	2	0	0	2	0	0	4	0	0
Ham	20	3	0	0	2	3	0	11	5	0
Turkey	16	0	0	0	0	1	0	4	1	0
Total	45	5	0	0	4	4	0	19	6	0
Total	234	95	2	0	111	11	0	120	13	0

PCR assay (Table 2). The inclusion of selective enrichment steps has been found to increase the rate of detection of *S. aureus* [15]. Waters *et al.* (2011) [26] also found a high prevalence of *S. aureus* in raw meat (47%) using a single step selective enrichment protocol, followed by plating on Baird Parker agar, and confirmation by real-time PCR targeting the *femA* gene.

The *kappa* statistic for detection of *S. aureus* using the primary enrichment in real-time PCR was 0.68–0.88 (Table 3), which indicates a good agreement (substantial to almost perfect agreement) with the conventional culture/PCR method. Using the secondary enrichment and real-time PCR, the *kappa* statistic for detection of *S. aureus* was 0.29–0.77, resulting in a fair agreement when deli meat was tested. This is due to the significantly higher recovery of *S. aureus* from the secondary enrichment samples by real-time PCR (Table 2), and the lower negative agreement (specificity) obtained with this method (Table 3). This observation suggests that small numbers (or levels) of *S. aureus* could be missed when the primary enrichment alone is used in real-time PCR, and that the recovery of potentially injured or non-viable strains appears to be enhanced when a secondary enrichment is applied. The enhanced detection also suggests that the use of a standard culture method or primary enrichment alone could lead to higher false negative results. Therefore, including a secondary selective enrichment step appears to improve the odds of detection of positive *S. aureus* samples.

Multiplex real-time PCR could detect more *S. aureus* positive samples than the conventional culture/PCR method alone. Possible reasons for these discrepant results include: amplification of DNA by the real-time PCR from very low levels of *S. aureus* that were not detectable by the bacteriological methods due to competition or non-viable *S. aureus* in the samples, or false-positive real-time PCR results as a result of cross-reaction rather than

false-negative culture results [23]. However, the possibility that these results are considered as false positives in this study is probably very low, because the gene *nuc*, which was targeted by the real-time PCR assay, has been used for specific detection and identification of *S. aureus* previously [21,22,34,35]. Unfortunately, it was not possible to confirm these results by performing the cultural method as detection was carried out from DNA extracts only, and the cells had already been inactivated. The inability of real-time PCR to detect three *S. aureus* samples isolated from raw meat that were positive by the culture method is somewhat unsatisfactory, and could be considered as false-negative results.

For detection of *mecA* gene, the *kappa* statistic for both enrichments in real-time PCR was 0–0.49 (Table 4). The $k = 0$ indicates no agreement beyond that expected by chance, because the real-time PCR assay detected the *mecA* gene probably from bacteria other than *S. aureus* and the culture/PCR method detected the *mecA* gene from DNA extracted from confirmed *S. aureus* strains. However, a few *mecA* positive samples were obtained from animals and meat in this study (Table 2). Weese *et al.* (2010) [25] detected a low prevalence of MRSA in samples isolated from retail meat (9.6% in pork, 5.6% in beef and 1.2% in chicken), using a single-step selective enrichment protocol, followed of plating and biochemical testing.

The detection of the *mecA* gene by the real-time PCR assay in samples that were negative for *S. aureus* by the conventional culture/PCR method may be due to the fact that either coagulase-negative staphylococci and non *S. aureus* species can also carry the *mecA* gene [21,36–38]. In this study, such samples were identified as *Staphylococcus* spp. positive by biochemical testing. In addition, the *mecA* gene has been found in non-staphylococcal genera, such as: *Proteus vulgaris*, *Morganella morganii*, *Enterococcus faecalis* [39] suggesting that its use in a rapid screening technique would need

Table 3. Raw agreement indices among conventional culture/PCR method and real-time PCR assay, with two-step enrichment procedure for detection of *S. aureus* from animals and retail meat.

Comparison within each sample type	No. of samples	No. positive by culture/PCR method	No. (%) of samples*			*kappa* statistic
			Positive agreement (Sensitivity)	Negative agreement (Specificity)	Total agreement	
Real-time PCR primary enrichment						
Animals	77	32	32 (100.0)	34 (75.6)	66 (85.7)	0.72
Meat	112	58	52 (89.7)	42 (77.8)	94 (83.9)	0.68
Deli meat	45	5	4 (80.0)	40 (100.0)	44 (97.8)	0.88
Real-time PCR secondary enrichment						
Animals	77	32	32 (100.0)	36 (80.0)	68 (88.3)	0.77
Meat	112	58	52 (89.7)	46 (85.2)	98 (87.5)	0.75
Deli meat	45	5	5 (100.0)	26 (65.0)	31 (68.9)	0.29

*Percentages for positive agreement with culture/PCR method number positive as the denominator. Percentages for negative agreement with culture/PCR method number negative as the denominator. Percentage total agreement is obtained from the sum of the positive and negative agreement frequencies divided by the total sample size within each sample type.

further validation to avoid false-positive MRSA data being generated. In this study, the DNA extraction was carried out from selective enrichments, which could contain DNA from coagulase-positive or coagulase-negative staphylococci or non-staphylococcal species that may carry the *mecA* gene, therefore a positive result for the *nuc* and *mecA* genes does not indicate the presence of *S. aureus* carrying the *mecA* gene.

None of the samples obtained from animals and retail meat were positive for the PVL genes using both methods the conventional multiplex PCR and the real-time PCR. A similar observation was reported by Weese *et al.* (2010) [25], who also failed to detect PVL positive samples in raw meat in Canada using the real-time PCR technique. The PVL genes encode the Panton-Valentine leukocidin toxin, which is a virulence factor that have been found in severe cases of CA-MRSA [7,8,9].

Decreasing the time of detection of *S. aureus* and MRSA has become an important goal in microbiological analysis of clinical samples. However, since *S. aureus* ST398, multi-drug resistant *S. aureus* (Table 5), and MRSA are present in animals and meat [12–17], decreasing the time of analysis may allow for prompt action to take place thus reducing the spread of those strains in the food chain. The real-time PCR assay can potentially decrease the total time for detection of *S. aureus* and the presence of the *mecA* gene in animal and meat samples. Using the two-step selective enrichment the total time was <2 days by the real-time PCR method, compared with a total time of 6–7 days using the culture method that includes selective enrichments, plating steps, biochemical

Table 4. Raw agreement indices among conventional culture/PCR method and real-time PCR assay, with two-step enrichment procedure for detection of the *mecA* gene from animals and retail meat.

Comparison within each sample type	No. of samples	No. positive by culture/PCR method	No. (%) of samples*			*kappa* statistic
			Positive agreement (Sensitivity)	Negative agreement (Specificity)	Total agreement	
Real-time PCR primary enrichment						
Animals	77	0	-	76 (98.7)	76 (98.7)	0.00
Meat	112	2	2 (100.0)	106 (96.4)	108 (96.4)	0.49
Deli meat	45	0	-	41 (91.1)	41 (91.1)	0.00
Real-time PCR secondary enrichment						
Animals	77	0	-	76 (98.7)	76 (98.7)	0.00
Meat	112	2	2 (100.0)	106 (96.4)	108 (96.4)	0.49
Deli meat	45	0	-	39 (86.7)	39 (86.7)	0.00

*Percentages for positive agreement with culture/PCR method number positive as the denominator. Percentages for negative agreement with culture/PCR method number negative as the denominator. Percentage total agreement is obtained from the sum of the positive and negative agreement frequencies divided by the total sample size within each sample type.

Table 5. Antimicrobial resistance profiles and sequence types of *S. aureus* isolated by conventional culture/PCR method from animals and retail meat.

Antimicrobial resistance profile*	No. of antimicrobial subclasses	No. of *S. aureus* isolates with the specific profile	Sequence types (n)[†]
PEN-TET-ERY-TYL-LINC-STR-CHL	6	2	Pig-ST9 (2)
PEN-TET-LINC-STR-CHL	5	1	Pig-ST9 (1)
TET-ERY-TYL-LINC	3	7	Pork-ST398 (5) Pork-ST5** Pork-ST9 (1)
PEN-LINC-STR	3	1	Pig-ST9 (1)
TET-ERY-LINC	3	7	Pork-ST9 (4) Pork-ST15 (2) Pork-ST8 (1)
TET-LINC-STR	3	1	Pig-ST9 (1)
ERY-TYL-LINC	2	3	Chicken-ST5 (3)
PEN-ERY	2	3	Pork-ST5 (1) Pork-ST5 (1)** Pork-ST9 (1)
TET-LINC	2	15	Sheep-ST398 (4) Pig-ST9 (11)
ERY-LINC	2	1	Pork-ST9 (1)
TET	1	13	Sheep-ST398 (3) Sheep-ST133 (2) Sheep-ST2111 (1) Pig-ST9 (1) Pork-ST1 (2) Pork-ST5 (2) Pork-ST398 (1) Pork-ST15 (1)
ERY	1	1	Deli chicken Chicken-ST39 (1)
LINC	1	5	Pig-ST9 (3) Sheep-ST133 (1) Deli ham-ST15 (1)
Susceptible to all tested	0	35	Chicken-ST5 (15) Chicken-ST6 (3) Chicken-ST508 (1) Chicken-NT[‡](1) Pork-ST5 (2) Beef-ST1159 (3) Beef-ST2187 (1) Beef-ST188 (1) Beef-ST15 (1) Beef-ST72 (1) Beef-ST5 (1) Beef-ST1 (1) Deli ham-ST146 (1) Deli ham-ST5 (1) Deli chicken-ST5 (1) Pig-ST9 (1)
Total		95	

*Antimicrobial abbreviations are as following: CHL, chloramphenicol; ERY, erythromycin; LINC, lincomycin; PEN, penicillin; STR, streptomycin, TET, tetracycline, TYL, tylosin.
[†]ST, sequence type.
[‡]NT, non-typeable.
**mecA gene positive.

testing and a conventional multiplex PCR for confirmation. However, the presence of MRSA should be confirmed by a culture method if isolates are required for follow on studies. Some real-time PCR assays have been developed for the rapid detection of MRSA from clinical samples [37,40–42]. Danial *et al.* (2011) [42] reported that the real-time PCR assay detected 0.7% more MRSA-positive samples than the routine standard Brilliance Chromogenic MRSA agar culture method in a total time of 8 h. Huletsky *et al.* (2004) [40] detected MRSA directly from clinical specimens containing a mixture of staphylococci in less than 1 h, with a false-positive detection rate of 4.6% for MRSA that was actually MSSA. Paule *et al.* (2005) [41] developed a multiplex real-time PCR that detected the genes *femA* and *mecA* directly from blood culture bottles in 2–3 h, obtaining an indeterminate rate of 0.9% when coagulase-negative staphylococci strains were included.

In conclusion, the application of real-time PCR using selective enrichments appears to improve the detection of *S. aureus* and the *mecA* gene in samples extracted from animals, raw meat and deli meat. The real-time PCR assay may be recommended as a rapid method to detect *S. aureus* and the *mecA* gene in samples obtained from the meat production chain; however, if further confirmation of MRSA should be required (isolate recovery) then the application of the standard culture method in parallel may be warranted.

Acknowledgments

The authors thank the staff of the Meat Lab (Department of Animal Sciences) and the ND Veterinary Diagnostic Lab at North Dakota State University for assistance with the sample collection.

Author Contributions

Conceived and designed the experiments: VV JS CL. Performed the experiments: VV JS CL. Analyzed the data: VV JS PR CL. Contributed reagents/materials/analysis tools: VV CL. Wrote the paper: VV JS PR CL.

References

1. Tiemersma EW, Bronzwaer SL, Lyytikäinen O, Degener JE, Schrijnemakers P, et al. (2004) Methicillin-resistant *Staphylococcus aureus* in Europe, 1999–2002. Emerg Infect Dis 10: 1627–1634.
2. Kennedy AD, Otto M, Braughton KR, Whitney AR, Chen L, et al. (2008) Epidemic community-associated methicillin-resistant *Staphylococcus aureus*: recent clonal expansion and diversification. Proc Natl Acad Sci U S A 105: 1327–1332.
3. Lowy FD (2003) Antimicrobial resistance: the example of *Staphylococcus aureus*. J Clin Invest 111: 1265–1273.
4. Golding GR, Bryden L, Levett PN, McDonald RR, Wong A, et al. (2010) Livestock-associated methicillin-resistant *Staphylococcus aureus* sequence type 398 in humans, Canada. Emerg Infect Dis 16: 587–594.
5. Hartman B, Tomasz A (1981) Altered penicillin-binding proteins in methicillin-resistant strains of *Staphylococcus aureus*. Antimicrob Agents Chemother 19: 726–735.
6. Van De Griend P, Herwaldt LA, Alvis B, DeMartino M, Heilmann K, et al. (2009) Community-associated methicillin-resistant *Staphylococcus aureus*, Iowa, USA. Emerg Infect Dis 15: 1582–1589.
7. Baba T, Takeuchi F, Kuroda M, Yuzawa H, Aoki K, et al. (2002) Genome and virulence determinants of high virulence community-acquired MRSA. Lancet 359: 1819–1827.
8. Dufour P, Gillet Y, Bes M, Lina G, Vandenesch F, et al. (2002) Community-acquired methicillin-resistant *Staphylococcus aureus* infections in France: emergence

of a single clone that produces Panton-Valentine leukocidin. Clin Infect Dis 35: 819–824.

9. Ebert MD, Sheth S, Fishman EK (2009) Necrotizing pneumonia caused by community-acquired methicillin-resistant *Staphylococcus aureus*: an increasing cause of "mayhem in the lung". Emerg Radiol 16: 159–162.

10. Gorwitz RJ, Kruszon-Moran D, McAllister SK, McQuillan G, McDougal LK, et al. (2008) Changes in the prevalence of nasal colonization with *Staphylococcus aureus* in the United States, 2001–2004. J Infect Dis 197: 1226–1234.

11. Klein E, Smith DL, Laxminarayan R (2007) Hospitalizations and deaths caused by methicillin-resistant *Staphylococcus aureus*, United States, 1999–2005. Emerg Infect Dis 13: 1840–1846.

12. van Belkum A, Melles DC, Peeters JK, van Leeuwen WB, van Duijkeren E, et al. (2008) Methicillin-resistant and -susceptible *Staphylococcus aureus* sequence type 398 in pigs and humans. Emerg Infect Dis 14: 479–483.

13. Guardabassi L, O'Donoghue M, Moodley A, Ho J, Boost M (2009) Novel lineage of methicillin-resistant *Staphylococcus aureus*, Hong Kong. Emerg Infect Dis 15: 1998–2000.

14. Persoons D, Van Hoorebeke S, Hermans K, Butaye P, de Kruif A, et al. (2009) Methicillin-resistant *Staphylococcus aureus* in poultry. Emerg Infect Dis 15: 452–453.

15. de Boer E, Zwartkruis-Nahuis JT, Wit B, Huijsdens XW, de Neeling AJ, et al. (2009) Prevalence of methicillin-resistant *Staphylococcus aureus* in meat. Int J Food Microbiol 134: 52–56.

16. Buyukcangaz E, Velasco V, Sherwood JS, Stepan RM, Koslofsky RJ, et al. (2013) Molecular typing of *Staphylococcus aureus* and methicillin-resistant *Staphylococcus aureus* (MRSA) isolated from retail meat and animals in North Dakota, USA. Foodborne Pathog Dis 10: 608–617.

17. O'Brien AM, Hanson BM, Farina SA, Wu JY, Simmering JE, et al. (2012) MRSA in conventional and alternative retail pork products. PLoS One 7(1): e30092.

18. Wertheim H, Verbrugh HA, van Pelt C, de Man P, van Belkum A, et al. (2001) Improved detection of methicillin-resistant *Staphylococcus aureus* using phenyl mannitol broth containing aztreonam and ceftizoxime. J Clin Microbiol 39: 2660–2662.

19. Zhang W, Hao Z, Wang Y, Cao X, Logue CM, et al. (2011) Molecular characterization of methicillin-resistant *Staphylococcus aureus* strains from pet animals and veterinary staff in China. Vet J 190: e125–e129.

20. Söderquist B, Neander M, Dienus O, Zimmermann J, Berglund C, et al. (2012) Real-time multiplex PCR for direct detection of methicillin-resistant *Staphylococcus aureus* (MRSA) in clinical samples enriched by broth culture. Acta Pathol Microbiol Scand 120: 427–432.

21. Thomas LC, Gidding HF, Ginn AN, Olma T, Iredell J (2007) Development of a real-time *Staphylococcus aureus* and MRSA (SAM-) PCR for routine blood culture. J Microbiol Methods 68: 296–302.

22. Kilic A, Muldrew KL, Tang Y-W A, Basustaoglu C (2010) Triplex real-time polymerase chain reaction assay for simultaneous detection of *Staphylococcus aureus* and coagulase-negative staphylococci and determination of methicillin resistance directly from positive blood culture bottles. Diagn Microbiol Infect Dis 66: 349–355.

23. Anderson MEC, Weese JS (2007) Evaluation of a real-time polymerase chain reaction assay for rapid identification of methicillin-resistant *Staphylococcus aureus* directly from nasal swabs in horses. Vet Microbiol 122: 185–189.

24. Morcillo A, Castro B, Rodríguez-Álvarez C, González JC, Sierra A, et al. (2012) Prevalence and Characteristics of methicillin-resistant *Staphylococcus aureus* in pigs and pig workers in Tenerife, Spain. Foodborne Pathog Dis 9: 207–210.

25. Weese JS, Avery BP, Reid-Smith RJ (2010) Detection and quantification of methicillin-resistant *Staphylococcus aureus* (MRSA) clones in retail meat products. Lett Applied Microbiol 51: 338–342.

26. Waters AE, Contente-Cuomo T, Buchhagen J, Liu CM, Watson L, et al. (2011) Multidrug-Resistant *Staphylococcus aureus* in US Meat and Poultry. Clin Infect Dis 52: 1227–1230.

27. Makgotlho PE, Kock MM, Hoosen A, Lekalakala R, Omar S, et al. (2009) Molecular identification and genotyping of MRSA isolates. FEMS Immunol Med Microbiol 57: 104–115.

28. De Medici D, Croci L, Delibato E, Di Pasquale S, Filetici E, et al. (2003) Evaluation of DNA extraction methods for use in combination with SYBR Green I real-time PCR to detect *Salmonella enterica* serotype Enteritidis in poultry. Appl Environ Microbiol 69: 3456–3461.

29. Marmur J (1961) A procedure for the isolation of desoxyribonucleic acid from micro-organisms. J Mol Biol 3: 208–218.

30. Enright MC, Day NP, Davies CE, Peacock SJ, Spratt BG (2000) Multilocus sequence typing for characterization of methicillin-resistant and methicillin-susceptible clones of *Staphylococcus aureus*. J Clin Microbiol 38: 1008–1015.

31. NARMS (2012) National antimicrobial resistance monitoring system animal isolates. Available: http://www.ars.usda.gov/Main/docs.htm?docid = 6750&page = 3. Accessed 5 February 2014.

32. Aydin A, Muratoglu K, Sudagidan M, Bostan K, Okuklu B, et al. (2011) Prevalence and antibiotic resistance of foodborne *Staphylococcus aureus* isolates in Turkey. Foodborne Pathog Dis 8: 63–69.

33. Moore DS (2007) The basic practice of statistics. 4th Edition. W.H. Freeman and Company, New York, USA. 560 p.

34. Costa AM, Kay I, Palladino S (2005) Rapid detection of *mecA* and *nuc* genes in staphylococci by real-time multiplex polymerase chain reaction. Diagn Microbiol Infect Dis 51: 13–17.

35. McDonald RR, Antonishyn NA, Hansen T, Snook LA, Nagle E, et al. (2005) Development of a triplex real-time PCR assay for detection of Panton-Valentine leukocidin toxin genes in clinical isolates of methicillin-resistant *Staphylococcus aureus*. J Clin Microbiol 43: 6147–6149.

36. Ryffel C, Tesch W, Bireh-Machin I, Reynolds PE, Barberis-Maino L, et al. (1990) Sequence comparison of *mecA* genes isolated from methicillin-resistant *Staphylococcus aureus* and *Staphylococcus epidermidis*. Gene 94: 137–138.

37. Hagen RM, Seegmüller I, Navai J, Kappstein I, Lehnc N, et al. (2005) Development of a real-time PCR assay for rapid identification of methicillin-resistant *Staphylococcus aureus* from clinical samples. Int J Med Microbiol 295: 77–86.

38. Higashide M, Kuroda M, Ohkawa S, Ohta T (2006) Evaluation of a cefoxitin disk diffusion test for the detection of *mecA*-positive methicillin-resistant *Staphylococcus saprophyticus*. Int J Antimicrob Agent 27: 500–504.

39. Kassem II, Esseili MA, Sigler V (2008) Occurrence of *mecA* in nonstaphylococcal pathogens in surface waters. J Clin Microbiol 46: 3868–3869.

40. Huletsky A, Giroux R, Rossbach V, Gagnon M, Vaillancourt M, et al. (2004) New real-time PCR assay for rapid detection of methicillin-resistant *Staphylococcus aureus* directly from specimens containing a mixture of staphylococci. J Clin Microbiol 42: 1875–1884.

41. Paule SM, Pasquariello AC, Thomson RB, Kaul KL, Peterson LR (2005) Real-time PCR can rapidly detect methicillin-susceptible and methicillin-resistant *Staphylococcus aureus* directly from positive blood culture bottles. Am J Clin Pathol 124: 404–407.

42. Danial J, Noel M, Templeton KE, Cameron F, Mathewson F, et al. (2011) Real-time evaluation of an optimized real-time PCR assay versus Brilliance chromogenic MRSA agar for the detection of meticillin-resistant *Staphylococcus aureus* from clinical specimens. J Med Microbiol 60: 323–328

43. McClure JA, Conly JM, Lau V, Elsayed S, Louie T, et al. (2006) Novel multiplex PCR assay for detection of the staphylococcal virulence marker Panton-Valentine leukocidin genes and simultaneous discrimination of methicillin-susceptible from -resistant staphylococci. J Clin Microbiol 44: 1141–1144.

13

Irritable Bowel Syndrome Is Positively Related to Metabolic Syndrome: A Population-Based Cross-Sectional Study

Yinting Guo[1,2], Kaijun Niu[1]*, Haruki Momma[3], Yoritoshi Kobayashi[3], Masahiko Chujo[3], Atsushi Otomo[3], Shin Fukudo[2], Ryoichi Nagatomi[3]

1 Nutritional Epidemiology Institute and School of Public Health, Tianjin Medical University, Tianjin, China, 2 Department of Behavioral Medicine, Tohoku University Graduate School of Medicine, Sendai, Japan, 3 Division of Biomedical Engineering for Health & Welfare, Tohoku University Graduate School of Biomedical Engineering, Sendai, Japan

Abstract

Irritable bowel syndrome is a common gastrointestinal disorder that may affect dietary pattern, food digestion, and nutrient absorption. The nutrition-related factors are closely related to metabolic syndrome, implying that irritable bowel syndrome may be a potential risk factor for metabolic syndrome. However, few epidemiological studies are available which are related to this potential link. The purpose of this study is to determine whether irritable bowel syndrome is related to metabolic syndrome among middle-aged people. We designed a cross-sectional study of 1,096 subjects to evaluate the relationship between irritable bowel syndrome and metabolic syndrome and its components. Diagnosis of irritable bowel syndrome was based on the Japanese version of the Rome III Questionnaire. Metabolic syndrome was defined according to the criteria of the American Heart Association scientific statements of 2009. Dietary consumption was assessed via a validated food frequency questionnaire. Principal-components analysis was used to derive 3 major dietary patterns: "Japanese", "sweets-fruits", and "Izakaya (Japanese Pub) "from 39 food groups. The prevalence of irritable bowel syndrome and metabolic syndrome were 19.4% and 14.6%, respectively. No significant relationship was found between the dietary pattern factor score tertiles and irritable bowel syndrome. After adjustment for potential confounders (including dietary pattern), the odds ratio (95% confidence interval) of having metabolic syndrome and elevated triglycerides for subjects with irritable bowel syndrome as compared with non-irritable bowel syndrome are 2.01(1.13–3.55) and 1.50(1.03–2.18), respectively. Irritable bowel syndrome is significantly related to metabolic syndrome and it components. This study is the first to show that irritable bowel syndrome was significantly related to a higher prevalence of metabolic syndrome and elevated triglycerides among an adult population. The findings suggest that the treatment of irritable bowel syndrome may be a potentially beneficial factor for the prevention of metabolic syndrome. Further study is needed to clarify this association.

Editor: Andreas Zirlik, University Heart Center Freiburg, Germany

Funding: This study was supported by a Grant-in-Aid for "Knowledge Cluster Initiative" from the Ministry of Education, Culture, Sports, Science and Technology of Japan. The funders had no role in study design, data collection and analysis, decision to publish, or preparation of the manuscript.

Competing Interests: All the authors have no conflicts of interest exists to disclose.

* Email: nkj0809@163.com

Introduction

Irritable bowel syndrome (IBS) is a common gastrointestinal disorder characterized by episodes of recurrent abdominal pain or discomfort related to disturbed bowel habits [1,2]. The majority of subjects with IBS are conscious that diet may play a role in triggering these episodes and therefore may avoid certain foods and changes in their dietary pattern [3–8]. Furthermore, IBS disrupts the digestion of food, or directly interferes with nutrient absorption [9–11].

Metabolic syndrome (MS) is a well-recognized constellation of risk factors for cardiovascular disease (CVD) [12], which remains a major cause of mortality and morbidity worldwide [13]. Accumulated evidence suggests that dietary factors are the cornerstone for the prevention and treatment of MS [14–16]. Furthermore, with respect to dietary factors studies, researchers have usually focused predominantly on the effects of individual nutrients and sometimes foods, but rarely on dietary patterns. However, daily diets are composed of a wide variety of foods containing complex combinations of nutrients. The surveys that examine a single nutrient in foods, or a single food, may not adequately account for complicated interactions and cumulative effects on human health. Therefore, compared with a single nutrient in foods or a single food, a dietary pattern study may be a more important tool for evaluation of the effects of diet on health [17,18].

Irritable bowel syndrome status may affect the dietary pattern, food digestion, and nutrient absorption, which are important factors for the prevention and treatment of MS and/or its components. Therefore, it is speculated that IBS may be a potential risk factor for MS. However, few epidemiological studies have assessed the relationship between IBS status and MS and its components in an adult population.

At present, we have designed a cross-sectional study to determine whether IBS is related to MS among middle-aged people.

Materials and Methods

Study Population

The current analysis uses data from a population-based longitudinal study designed to investigate the lifestyle risk factors of CVD among Japanese adults. The methods are described in detail elsewhere [19,20].

There were 1,208 subjects who had received a health examination including blood examinations in 2011. Of these, 1,163 subjects agreed to participate and provided informed consent for their data to be analyzed. Subjects were excluded if they did not provide any dietary information (n = 21) or did not answer the Rome III Modular Questionnaire (n = 46). Owing to these exclusions, the final cross-sectional study population comprised 1,096 subjects (mean [standard deviation, SD] age: 46.2 [11.2] years; male, 77.5%). The Institutional Review Board of the Tohoku University Graduate School of Medicine approved the study protocol.

Assessment of IBS

The Japanese version of the Rome III Questionnaire was used to screen for IBS [21]. All subjects were asked to complete self-reported ROME III diagnostic questionnaires. Screening for IBS requires that subjects have abdominal discomfort or pain lasting at least 3 days per month, not necessarily consecutive, during the previous 3 months which is associated with 2 or more of the following: relief by defecation; onset associated with a change in frequency of stool; onset associated with a change in form (appearance) of stool.

Assessment of MS and Other

Waist circumference was measured at the umbilical level with participants standing and breathing normally. Blood pressure (BP) was measured twice from the upper left arm using a YA-MASU605P automatic device (Kenzmedico, Saitama, Japan) after 5 min of rest in the seated position. The mean of these 2 measurements was taken as the BP value. Blood samples were collected in siliconized vacuum glass tubes containing sodium fluoride, for the analysis of fasting blood glucose (FBG), or containing no additives, for the analysis of lipids. Fasting blood glucose was measured by using enzymatic methods (Eerotec, Tokyo, Japan). The concentrations of triglycerides (TG), low-density lipoprotein cholesterol (LDL), and high-density lipoprotein cholesterol (HDL) were measured by enzymatic methods using appropriate kits (Sekisui Medical, Tokyo, Japan). Serum high-sensitive C-reactive protein (hsCRP) levels were determined using N-latex CRP-2 (Siemens Healthcare Japan, Tokyo, Japan). The measurement limit of hsCRP was 0.02 mg/L and an hsCRP value less than the measurement limit was considered to be 0.01 mg/L.

Metabolic syndrome was defined in accordance with the criteria of the American Heart Association scientific statements of 2009 [22]. Participants were considered to have MS when they presented three or more of the following components: 1) elevated waist circumference for Asian individuals (≥90 cm and ≥80 cm in male and female, respectively), 2) elevated TG (≥150 mg/dL), or drug treatment for elevated TG, 3) reduced HDL (<40 mg/dL in male; <50 mg/dL in female) or drug treatment for reduced HDL, 4) elevated blood pressure (SBP ≥130 mm Hg and/or DBP ≥ 85 mm Hg) or antihypertensive drug treatment, 5) elevated fasting glucose (≥100 mg/dL) or drug treatment of elevated glucose.

Assessment of Dietary Intake

The subjects were instructed to complete a brief, self-administered diet history questionnaire (BDHQ) that included questions on 75 food items along with their specified serving sizes, described in terms of natural portions or standard weights and volume measures of the servings, commonly consumed by the study population. For each food item, the subjects indicated their mean frequency of consumption of the food over the past month in terms of the specified serving size by checking 1 of the 7 frequency categories, ranging from "almost never" to "2 or more times/day". The mean daily consumption of nutrients was calculated using an *ad hoc* computer program developed to analyze the questionnaire. The Japanese food composition tables, 5th edition, [23] and other [24] were used as the nutrient database. The reproducibility and validity of the BDHQ have already been described in detail elsewhere [25]. Foods from the BDHQ were categorized into 39 food subgroups, which were used to derive dietary patterns via principal-components analysis.

Factor analysis (principal-components analysis) was used to derive dietary patterns and to determine factor loadings for each of the 39 food subgroups (in g/d) [26]. Factors were rotated with varimax rotation to maintain uncorrelated factors and enhance interpretability [27]. A combined evaluation of the eigenvalues, scree plot test, and factor interpretability was used in determining the number of retained factors. The distinctive dietary patterns of the study population were well described by the 3 factors. Factors were named descriptively according to the food items showing high loading (absolute value) with respect to each dietary pattern as follows: "Japanese" dietary pattern (factor 1), "sweets-fruits" pattern (factor 2), and "Izakaya (Japanese Pub)" pattern (factor 3) (see **Table S1**). For each dietary pattern and each subject, we calculated a factor score by summing the consumption from each food item weighted by its factor loading as follows [27]:

$$\sum[(food\ group_i servings/d) \times (food\ group_i factor\ loading))]$$

where $i =$ food groups 1–39. A higher factor score indicates greater conformity to the dietary pattern. Variables unrelated to a given dietary pattern are weighted close to zero. For further analyses, factor scores were categorized into 3 equal groups by using tertiles cutoffs.

Assessment of Other Variables

Sociodemographic variables, including age, gender, and educational levels were also assessed. The educational level was assessed by determining the last grade level and was divided into 2 categories: <college or ≥college. Body mass index (BMI) was calculated as weight in kilograms divided by squared height in meters (kg/m2). History of physical illness and current medication were noted from "yes" or "no" responses to relevant questions. Information on smoking status, and drinking status were obtained from a questionnaire survey. Levels of daily physical activity (PA) were estimated using the International Physical Activity Questionnaire (IPAQ) (Japanese version) [28]. Total daily PA (metabolic equivalents [METs] ×hours/week) were calculated as follows: (daily hours of walking × days per week with walking×3.3)+(daily hours of moderate-intensity activity×days per week with moderate-intensity activity×4.0)+(daily hours of vigorous activity×days per week with vigorous activity×8.0). The METs values were derived from the IPAQ validity and reliability study [28]. Physical activity was categorized into three groups: no PA, low PA (0<PA<23 METs × hours/week), and high PA (≥23 METs×hours/week) [29]. Depressive symptoms were assessed according to the Japanese version of the Self-Rating Depression Scale (SDS) [30]. An SDS score ≥45 was taken as the cutoff point indicating depressive symptoms [31].

Statistical Analysis

All statistical analyses were performed using the Statistical Analysis System 9.1 edition for Windows (SAS Institute Inc., Cary, NC, USA). The age- and sex-adjusted variable differences according to IBS status were examined by analysis of covariance (ANCOVA) for continuous variables or by the multiple logistic regression analysis for variables of proportion. For main analysis, the MS or its components were used as a dependent variable and IBS status as independent variables. The odds ratio (OR) and 95% confidence interval (CI) of MS and its components compared with IBS status were calculated using multiple logistic regression analysis. We used age, sex, BMI, smoking and drinking status, educational level, PA levels, dietary patterns, total energy intake, depressive symptoms and mutual metabolic syndrome components as covariates for multiple adjustments. Model fit was evaluated using the Hosmer-Lemeshow goodness-of-fit statistic. For all models, the test was not significant ($P \geq 0.31$). Interactions between IBS status and confounders of MS or its components were tested by the addition of cross-product terms to the regression model. All tests were two-tailed and $P < 0.05$ was defined as statistically significant.

Results

Among 1,096 subjects who were available to be analyzed, 213 (19.4%) had self-reported IBS, and 160 (14.6%) had MS.

Age- and sex-adjusted characteristics of study subjects with and without IBS are presented in **Table 1**. Subjects with IBS were significantly younger than the non-IBS subjects ($P < 0.01$) with a mean (95% CI) 43.4 (41.9–45.0) y compared to 46.1 (45.3–47.0) y. Subjects with IBS contained a lower proportion of males, and a higher proportion of ex-smokers and depressive symptoms ($P < 0.05$ for all comparisons). Compared to subjects with IBS, the non-IBS subjects had lower total energy intake, and serum TG levels ($P < 0.05$ for all comparisons).

Because IBS may affect the dietary pattern [32], as an initial step, we evaluated the relationships between dietary patterns and IBS. Three major dietary patterns were identified by factor analysis (**Table S1**). Factor 1, identified as a traditional "Japanese" dietary pattern was characterized by a high consumption of vegetables, seaweeds, soybean products, fish, fruits, miso soup, and green tea. Factor 2 was typified by a greater consumption of cake, ice cream, fruits, bread, dairy products, mayonnaise and lower consumption of alcohol (named the "sweets-fruits" pattern). Factor 3 was typified by a greater consumption of noodles, Squid, octopus, lobster, shellfish, meat, fish, cola, alcohol, coffee, mayonnaise, chicken egg, and bread (named the "Izakaya (Japanese Pub)" pattern). These 3 patterns explained 32.1% of the variance in dietary consumption (18.6% for factor 1, 7.5% for factor 2, and 6.0% for factor 3). Increasing the number of patterns did not materially increase the total proportion of variance in dietary consumption explained by the model. Daily food and nutrient consumption are presented according to tertiles of dietary pattern factor score in **Table S2**. Compared to subjects with factor scores in the lowest tertile for the "Japanese" dietary pattern, those in the highest tertile had a higher consumption of total meats, total fish, seaweeds, total vegetables, soybean products, total fruits, dairy products, green tea, black or oolong tea, total energy intake, animal protein, vegetable protein, animal fat, vegetable fat, carbohydrate, total fiber, calcium, and eicosapentaenoic acid (EPA) + docosahexaenoic acid (DHA), and lower consumptions of cola (P for trend < 0.05). Compared to those in the middle "sweets-fruits" pattern tertile, subjects in the highest tertile had significant higher

consumptions of total fish, total seaweeds, total vegetables, total fruit, dairy products, green tea, cola, total energy intake, animal protein, vegetable protein, animal fat, vegetable fat, carbohydrate, total fiber, calcium, EPA+DHA, and a lower consumption of alcohol ($P < 0.05$). Compared to those in the lowest "Izakaya (Japanese Pub)" pattern tertile, subjects in the highest tertile had a higher consumption of total meats, total fish, seaweeds, coffee, cola, total energy intake, animal protein, vegetable protein, animal fat, vegetable fat, carbohydrate, total fiber, calcium, EPA+DHA, alcohol, lower consumption of total fruits, and dairy products (P for trend < 0.01). The age- and sex-adjusted relationships between tertiles of dietary pattern factor score and IBS status are indicated in **Table 1**. No significant relationships between the tertiles of each dietary pattern and IBS status were observed. These results were unchanged when we adjusted for multiple confounding factors (see Table 2 model 5) ($P > 0.15$ for all comparisons).

We next investigated whether IBS status is related to MS and its components. **Table 2** shows the adjusted relationships between IBS status and MS and its components. In the final multivariate models, the adjusted ORs (95% CI) of MS related to IBS group as compared with the non-IBS group is 2.01 (1.13–3.55). In MS components analysis, IBS status was only positively related to elevated TG in the final model (OR [95% CI]: 1.50 [1.03–2.18]). Although the difference was not statistically significant, the proportion of subjects with elevated waist circumference was higher in IBS group (OR [95% CI]: 1.60 [0.85–2.96]) in the final multivariate models. No significant relationships were observed between IBS status and other MS components in the final multivariate models. The tests for interactions between IBS status and other potential confounders in the final models were also not statistically significant (interaction P values > 0.22).

Discussion

In this cross-sectional study, we investigated the relationships between IBS and MS in an adult population. This study is the first to show that IBS is independently related to a higher prevalence of MS and elevated TG. Further, no significant relationships between IBS and dietary patterns were observed.

In this study, we adjusted for various potential confounders related to IBS and/or MS. First, we considered that age, sex (see Table 1), and body mass index [33] were potential confounders. Second, the effect of lifestyle factors, such as smoking [33] and drinking status [34], physical activity [35], and educational level [36], were adjusted. Moreover, IBS can affect dietary intake and thus affect the dietary pattern. Furthermore, dietary factors are also important for incidence of MS [35]. Accordingly, we made adjustments for total energy intake, and dietary pattern. Third, depressive symptoms are also closely related to IBS (see Table 1) and MS [37]. However, adjustments for these confounding factors did not change the significant positive relationship between IBS and MS. That is, the positive relationship between IBS and MS was independent of these factors.

It is hypothesized that IBS has a potentially adverse effect on MS and its components possibly due to effect on dietary pattern, food digestion, or nutrient absorption. However, no significant relationships between IBS and dietary patterns were observed. The result was similar to a previous study, suggesting that IBS is not related to dietary habits and/or nutritional intake [38]. Furthermore, although several studies have demonstrated self-reported food intolerances in most patients with IBS [3–8], two other studies have indicated that those with IBS appear to have adequate and balanced food and macronutrient intake, with no evidence of inadequate micronutrient intake [39,40]. Therefore,

Table 1. Age- and sex-adjusted characteristics of the subjects in relation to irritable bowel syndrome (n = 1,096)[†]

	Irritable bowel syndrome		P value[‡]
	No (n = 883)	Yes (n = 213)	
Age (y)	46.1 (45.3–47.0)[§]	43.4 (41.9–45.0)	<0.01
Sex (male, %)	79.2	70.4	0.02
BMI (kg/m2)	22.8 (22.5–23.0)	22.6 (22.1–23.1)	0.54
Smoking status (%)			
Current smoking	40.8	36.6	0.46
Ex-smoking	12.5	17.4	0.049
Never-smoking	46.6	46.0	0.53
Drinking status (%)			
Daily	28.3	25.4	0.77
Sometimes	48.5	50.2	0.93
Never-drinking	23.2	24.4	0.89
Educational level (≥ college, %)	33.8	28.2	0.14
PA (%)			
0 METs hours/week	25.4	23.5	0.50
0–23 METs hours/week	39.9	44.6	0.37
≥23 METs hours/week	34.8	31.9	0.76
Total energy intake (kcal/d)	1733.0 (1686.7–1779.4)	1836.6 (1754.4–1918.7)	0.02
"Japanese" dietary pattern			
The lowest tertile of factor score	32.7	35.7	0.85
The middle tertile of factor score	33.9	31.5	0.41
The highest tertile of factor score	33.4	32.9	0.48
"sweets-fruits" dietary pattern			
The lowest tertile of factor score	34.4	28.6	0.37
The middle tertile of factor score	33.6	32.4	0.44
The highest tertile of factor score	31.9	39.0	0.09
"Izakaya (Japanese Pub)" dietary pattern			
The lowest tertile of factor score	33.8	31.5	0.24
The middle tertile of factor score	33.9	31.5	0.55
The highest tertile of factor score	32.4	37.1	0.08
Depressive symptoms (SDS ≥45, %)	30.9	42.7	<0.01
Waist (cm)	80.2 (79.5–81.0)	80.6 (79.3–81.9)	0.61
SBP (mmHg)	122.1 (120.9–123.2)	121.6 (119.6–123.6)	0.67
DBP (mmHg)	75.7 (74.8–76.5)	75.1 (73.6–76.6)	0.49
Log translated TG (mg/dl)[¶]	86.5 (82.7–90.5)	98.4 (90.9–106.5)	<0.01
FBG (mg/dl)	96.7 (94.8–98.6)	96.4 (93.0–99.7)	0.85
HDL (mg/dl)	62.6 (61.5–63.7)	61.5 (59.6–63.4)	0.31
LDL (mg/dl)	114.1 (111.7–116.4)	114.5 (110.2–118.7)	0.87
Log translated hsCRP (mg/L)[¶]	0.32 (0.29–0.35)	0.30 (0.26–0.36)	0.64

[†]BMI, body mass index; PA, physical activity; METs, metabolic equivalents; SDS, Self-rating Depression Scale; SBP, systolic blood pressure; DBP, diastolic blood pressure; TG, triglyceride; FBG, fasting blood glucose; HDL, high-density lipoprotein-cholesterol; LDL, low-density lipoprotein; hsCRP, high-sensitivity C-reactive protein.
[‡]Analysis of covariance or logistic regression analysis adjusted for age and sex where appropriate.
[§]Adjusted least squares mean (95% confidence interval) (all such values).
[¶]Adjusted geometric mean (95% confidence interval).

we consider that while IBS symptoms affect the choice of certain specific food items, it has no essential impact on dietary pattern or the intake of nutrients. In the modern world, especially in developed countries, a variety of different foods are available in daily life. The availability of these choices may well make up for

IBS symptoms-bringing harmful effects from certain food items. Further study is needed to confirm this hypothesis.

Despite underlying causes of pathophysiologic changes still not being completely understood, low grade mucosal inflammation, increased intestinal mucosal permeability, and abnormal intestinal motility are accepted mechanisms which alter gut function and

Table 2. Adjusted odds ratios and 95% confidence interval for the relationship between MS and IBS (n = 1,096) [†]

| | IBS vs non IBS | | | | |
	Model 1[‡]	Model 2[§]	Model 3[¶]	Model 4[ǀ]	Model 5[ǀ]
MS	1.85 (1.05–3.20)	1.94 (1.10–3.37)	2.01 (1.13–3.52)	2.01 (1.13–3.55)	-
MS components					
Waist circumference ≥90 cm for male or ≥80 cm for female	1.61 (0.89–2.92)	1.68 (0.91–3.06)	1.64 (0.88–3.03)	1.60 (0.86–2.97)	1.60 (0.85–2.96)
Triglycerides ≥150 mg/dL	1.52 (1.05–2.18)	1.53 (1.06–2.20)	1.52 (1.04–2.19)	1.51 (1.04–2.19)	1.50 (1.03–2.18)
HDL-cholesterol <40 mg/dL	1.04 (0.56–1.85)	0.99 (0.51–1.80)	1.00 (0.52–1.83)	1.04 (0.53–1.92)	1.00 (0.51–1.85)
SBP ≥130 mmHg or DBP ≥85 mmHg	1.04 (0.74–1.48)	1.03 (0.72–1.47)	1.05 (0.73–1.51)	1.06 (0.74–1.53)	1.04 (0.72–1.50)
High fasting glucose ≥100 mg/dL	1.27 (0.74–2.11)	1.24 (0.72–2.07)	1.26 (0.73–2.12)	1.25 (0.72–2.11)	1.19 (0.68–2.03)

[†]MS, metabolic syndrome; IBS, irritable bowel syndrome; HDL, high-density lipoprotein cholesterol; SBP, systolic blood pressure; DBP, diastolic blood pressure.
[‡]Adjusted for age, sex and body mass index.
[§]Additionally adjusted for smoking and drinking status, educational level, and physical activity.
[¶]Additionally adjusted for dietary patterns, and total energy intake.
[ǀ]Additionally adjusted for depressive symptoms.
[ǀ]Additionally adjusted for mutual metabolic syndrome components.

generate symptoms of IBS [41]. The response of the gastrointestinal tract to ingestion of food is a complex and closely controlled process, which allows optimization of propulsion, digestion, absorption of nutrients, and removal of indigestible remnants. Therefore, it is believed that IBS is an important risk factor in the digestion of food and in nutrient absorption. Many studies have investigated the effects of IBS on the digestion of food or nutrient absorption [9–11]. These studies have consistently demonstrated that increased and discordant absorption of some nutrients, such as mannitol and sorbitol occurs in subjects with IBS compared to healthy controls [9–11]. In the present study, we found that IBS is mainly related to elevated TG, suggesting that IBS may affect the digestion and absorption of fats in the gastrointestinal tract. In fact, several studies have evaluated the relationships between IBS and the digestion and absorption of fats [8,42]. Simren et al. have reported IBS to be related to increased colonic sensitivity and an altered viscerosomatic referral pattern after duodenal lipids infusion [42]. Another study also indicated that gastrointestinal symptoms were frequently reported after intake of fried and fatty foods in IBS patients [8]. Therefore, we consider the relationship between IBS and MS and its components to be possibly due to the disorder of food digestion and nutrient absorption, especially in the fat components. This study, which was designed to investigate the relationships between IBS and MS, is very limited and we therefore cannot determine an exact mechanism to explain our observations. Further studies are needed to make certain the causality and exact mechanisms of IBS in MS.

On the other hand, gut microbiota alterations could also be considered a potential link between IBS and MS and its components. The accumulated evidence has indicated that IBS is related to quantitative and qualitative changes in gut microbiota [43]. Because the gut microbiota is becoming known as a more and more important risk factor for the treatment and prevention of MS [44,45], there is conjecture that IBS may be a potential risk factor for MS due to the effect of IBS on the quantitative and qualitative changes of gut microbiota. Further studies are needed to clarify this hypothesis.

A small-scale case-control study has shown that IBS was significantly related to a higher FBS and higher prevalence of prediabetes than in the control group [46]. In contrast, the present study did not find significant relationships between IBS and FBG

or elevated FBG. Although the reason remains unclear, differences in age (mean age is 46.2 y in our study vs 33.0 y in their study), adjustment factors (all lifestyle factors were adjusted in their study) and population size may partly explain the discrepancy. Further study is needed to investigate this issue.

To the best of our knowledge, no previous study has examined the relationships between dietary pattern and IBS among the general population. The present study first investigated the relationships between dietary patterns and IBS in apparently healthy adults. The results suggest that dietary patterns were not related to the prevalence of IBS. Furthermore, the traditional Japanese diet is a well-known healthy diet pattern [26,47]. Thus, we also evaluated whether the traditional Japanese dietary pattern was significantly related to a lower prevalence of IBS. The results indicated that no significant relationships between traditional Japanese dietary patterns and IBS were observed. Further study is needed to make certain of our observations.

This study had several limitations. First, the Rome III questionnaire is designed for the measurement and screening of IBS, not for making a clinical diagnosis. Therefore, a population study that uses a standardized comprehensive structured diagnostic interview should be undertaken to confirm the influence of IBS on MS. Second, because this study was a cross-sectional study, we could not conclude that IBS increases the occurrence of MS or that MS leads to episodes of IBS in adult populations. Therefore, a prospective study or trial should be undertaken to confirm the existence of a relationship between IBS, and MS and elevated TG.

In the present study, IBS was significantly related to a higher prevalence of MS and elevated TG among an adult population. The differences in dietary patterns are not likely to explain our findings. The findings suggest that the treatment of IBS may be a potentially beneficial factor for the development and prevention of MS. Further study is required to clarify this causality.

Supporting Information

Table S1 Principal components analysis varimax-rotated 39 food groups factor loading scores (n = 1,096).

Table S2 Daily food and nutrient consumption of the participants according to the tertiles of dietary pattern factor score (n = 1,096).

Acknowledgments

We gratefully acknowledge all the men and women who participated in the study and Sendai Oroshisho Center for the possibility to perform the study.

References

1. Camilleri M (2001) Management of the irritable bowel syndrome. Gastroenterology 120: 652–668.
2. Chang L (2004) Review article: epidemiology and quality of life in functional gastrointestinal disorders. Aliment Pharmacol Ther 20 Suppl 7: 31–39.
3. Hayes P, Corish C, O'Mahony E, Quigley EM (2013) A dietary survey of patients with irritable bowel syndrome. J Hum Nutr Diet: (in press. doi: 10.1111/jhn.12114).
4. Feinle-Bisset C, Azpiroz F (2013) Dietary lipids and functional gastrointestinal disorders. Am J Gastroenterol 108: 737–747.
5. Simren M, Mansson A, Langkilde AM, Svedlund J, Abrahamsson H, et al. (2001) Food-related gastrointestinal symptoms in the irritable bowel syndrome. Digestion 63: 108–115.
6. Park HJ, Jarrett M, Heitkemper M (2010) Quality of life and sugar and fiber intake in women with irritable bowel syndrome. West J Nurs Res 32: 218–232.
7. Aller R, de Luis DA, Izaola O, La Calle F, del Olmo L, et al. (2004) [Dietary intake of a group of patients with irritable bowel syndrome; relation between dietary fiber and symptoms]. An Med Interna 21: 577–580.
8. Bohn L, Storsrud S, Tornblom H, Bengtsson U, Simren M (2013) Self-reported food-related gastrointestinal symptoms in IBS are common and associated with more severe symptoms and reduced quality of life. Am J Gastroenterol 108: 634–641.
9. Marciani L, Cox EF, Hoad CL, Pritchard S, Totman JJ, et al. (2010) Postprandial changes in small bowel water content in healthy subjects and patients with irritable bowel syndrome. Gastroenterology 138: 469–477, 477 e461.
10. Keller J, Layer P (2009) Intestinal and anorectal motility and functional disorders. Best Pract Res Clin Gastroenterol 23: 407–423.
11. Yao CK, Tan HL, van Langenberg DR, Barrett JS, Rose R, et al. (2013) Dietary sorbitol and mannitol: food content and distinct absorption patterns between healthy individuals and patients with irritable bowel syndrome. J Hum Nutr Diet.
12. Nikolopoulou A, Kadoglou NP (2012) Obesity and metabolic syndrome as related to cardiovascular disease. Expert Rev Cardiovasc Ther 10: 933–939.
13. Lozano R, Naghavi M, Foreman K, Lim S, Shibuya K, et al. (2012) Global and regional mortality from 235 causes of death for 20 age groups in 1990 and 2010: a systematic analysis for the Global Burden of Disease Study 2010. Lancet 380: 2095–2128.
14. Nestel P (2004) Nutritional aspects in the causation and management of the metabolic syndrome. Endocrinol Metab Clin North Am 33: 483–492, v.
15. Josse AR, Jenkins DJ, Kendall CW (2008) Nutritional determinants of the metabolic syndrome. J Nutrigenet Nutrigenomics 1: 109–117.
16. Leao LS, de Moraes MM, de Carvalho GX, Koifman RJ (2011) Nutritional interventions in metabolic syndrome: a systematic review. Arq Bras Cardiol 97: 260–265.
17. Willett WC, McCullough ML (2008) Dietary pattern analysis for the evaluation of dietary guidelines. Asia Pac J Clin Nutr 17 Suppl 1: 75–78.
18. Hu FB (2002) Dietary pattern analysis: a new direction in nutritional epidemiology. Curr Opin Lipidol 13: 3–9.
19. Niu K, Kobayashi Y, Guan L, Monma H, Guo H, et al. (2013) Low-fat dairy, but not whole-/high-fat dairy, consumption is related with higher serum adiponectin levels in apparently healthy adults. Eur J Nutr 52: 771–778.
20. Niu K, Kobayashi Y, Guan L, Momma H, Guo H, et al. (2013) Longitudinal changes in the relationship between serum adiponectin concentration and cardiovascular risk factors among apparently healthy middle-aged adults. Int J Cardiol: (http://dx.doi.org/10.1016/j.bbr.2011.1003.1031).
21. Fukudo S, Hongo M, Matsueda K, Drossman DA (2008) The Japanese Version of Rome III: the Functional Gastrointestinal Disorders, 3rd edn. Tokyo: KYOWA KIKAKU Ltd,.
22. Alberti KG, Eckel RH, Grundy SM, Zimmet PZ, Cleeman JI, et al. (2009) Harmonizing the metabolic syndrome: a joint interim statement of the International Diabetes Federation Task Force on Epidemiology and Prevention; National Heart, Lung, and Blood Institute; American Heart Association; World Heart Federation; International Atherosclerosis Society; and International Association for the Study of Obesity. Circulation 120: 1640–1645.
23. Science and Technology Agency: Standard Tables of Food Composition in Japan (Fifth revised edition). Printing Buteau, Ministry of Finance, Tokyo, 2000 (in Japanese).
24. Sakai K, Nakajima M, Watanabe S, Kobayashi T (1995) Available data on assessments of dietary fatty acid intake. J Lipid Nutr 4: 97–103 (in Japanese).
25. Sasaki S (2005) Serum Biomarker-based Validation of a Brief-type Self-administered Diet History Questionnaire for Japanese Subjects, The Study Group of Ministry of Health, Labor and Welfare of Japan, Tanaka H, chairman, "A research for assessment of nutrition and dietary habit in "Kenko Nippon 21". Tokyo: 10–42 (in Japanese).
26. Guo H, Niu K, Monma H, Kobayashi Y, Guan L, et al. (2012) Association of Japanese dietary pattern with serum adiponectin concentration in Japanese adult men. Nutr Metab Cardiovasc Dis 22: 277–284.
27. Kim J, Mueller C (1978) Factor Analysis: Statistical Methods and Practical Issues. Beverly Hills, Calif: Sage Publications;.
28. Craig CL, Marshall AL, Sjostrom M, Bauman AE, Booth ML, et al. (2003) International physical activity questionnaire: 12-country reliability and validity. Med Sci Sports Exerc 35: 1381–1395.
29. Ishikawa-Takata K, Tabata I (2007) Exercise and Physical Activity Reference for Health Promotion 2006 (EPAR2006). J Epidemiol 17: 177.
30. Fukuda K, Kobayashi S (1973) A study on a self-rating depression scale. Psychiatria et Neurologia Japonica 75: 673–679 (in Japanese).
31. Barrett J, Hurst MW, DiScala C, Rose RM (1978) Prevalence of depression over a 12-month period in a nonpatient population. Arch Gen Psychiatry 35: 741–744.
32. Morcos A, Dinan T, Quigley EM (2009) Irritable bowel syndrome: role of food in pathogenesis and management. J Dig Dis 10: 237–246.
33. Cena H, Fonte ML, Turconi G (2011) Relationship between smoking and metabolic syndrome. Nutr Rev 69: 745–753.
34. Alkerwi A, Boutsen M, Vaillant M, Barre J, Lair ML, et al. (2009) Alcohol consumption and the prevalence of metabolic syndrome: a meta-analysis of observational studies. Atherosclerosis 204: 624–635.
35. Yamaoka K, Tango T (2012) Effects of lifestyle modification on metabolic syndrome: a systematic review and meta-analysis. BMC Med 10: 138.
36. Li YQ, Zhao LQ, Liu XY, Wang HL, Wang XH, et al. (2013) Prevalence and distribution of metabolic syndrome in a southern Chinese population. Relation to exercise, smoking, and educational level. Saudi Med J 34: 929–936.
37. Pan A, Keum N, Okereke OI, Sun Q, Kivimaki M, et al. (2012) Bidirectional association between depression and metabolic syndrome: a systematic review and meta-analysis of epidemiological studies. Diabetes Care 35: 1171–1180.
38. Jung HJ, Park MI, Moon W, Park SJ, Kim HH, et al. (2011) Are Food Constituents Relevant to the Irritable Bowel Syndrome in Young Adults? - A Rome III Based Prevalence Study of the Korean Medical Students. J Neurogastroenterol Motil 17: 294–299.
39. Williams EA, Nai X, Corfe BM (2011) Dietary intakes in people with irritable bowel syndrome. BMC Gastroenterol 11: 9.
40. Saito YA, Locke GR 3rd, Weaver AL, Zinsmeister AR, Talley NJ (2005) Diet and functional gastrointestinal disorders: a population-based case-control study. Am J Gastroenterol 100: 2743–2748.
41. Camilleri M, Lasch K, Zhou W (2012) Irritable bowel syndrome: methods, mechanisms, and pathophysiology. The confluence of increased permeability, inflammation, and pain in irritable bowel syndrome. Am J Physiol Gastrointest Liver Physiol 303: G775–785.
42. Simren M, Abrahamsson H, Bjornsson ES (2007) Lipid-induced colonic hypersensitivity in the irritable bowel syndrome: the role of bowel habit, sex, and psychologic factors. Clin Gastroenterol Hepatol 5: 201–208.
43. Ghoshal UC, Shukla R, Ghoshal U, Gwee KA, Ng SC, et al. (2012) The gut microbiota and irritable bowel syndrome: friend or foe? Int J Inflam 2012: 151085.
44. Xiao S, Fei N, Pang X, Shen J, Wang L, et al. (2013) A gut microbiota-targeted dietary intervention for amelioration of chronic inflammation underlying metabolic syndrome. FEMS Microbiol Ecol: (in press. doi: 10.1111/1574-6941.12228).
45. D'Aversa F, Tortora A, Ianiro G, Ponziani FR, Annicchiarico BE, et al. (2013) Gut microbiota and metabolic syndrome. Intern Emerg Med 8 Suppl 1: S11–15.
46. Gulcan E, Taser F, Toker A, Korkmaz U, Alcelik A (2009) Increased frequency of prediabetes in patients with irritable bowel syndrome. Am J Med Sci 338: 116–119.
47. Shimazu T, Kuriyama S, Hozawa A, Ohmori K, Sato Y, et al. (2007) Dietary patterns and cardiovascular disease mortality in Japan: a prospective cohort study. Int J Epidemiol 36: 600–609.

Author Contributions

Conceived and designed the experiments: KN SF RN. Performed the experiments: YG KN HM YK MC AO. Analyzed the data: YG KN. Contributed reagents/materials/analysis tools: YG KN. Wrote the paper: YG KN. Critical revision of the manuscript for important intellectual content: YG KN SF RN.

14

Phenolic Acid Composition, Antiatherogenic and Anticancer Potential of Honeys Derived from Various Regions in Greece

Eliana Spilioti[1], Mari Jaakkola[2], Tiina Tolonen[2], Maija Lipponen[2], Vesa Virtanen[2], Ioanna Chinou[3], Eva Kassi[1], Sofia Karabournioti[1], Paraskevi Moutsatsou[1]*

1 Department of Biological Chemistry, Medical School, University of Athens, Athens, Greece, **2** CEMIS-Oulu, Kajaani University Consortium, University of Oulu, Sotkamo, Finland, **3** Laboratory of Pharmacognosy and Chemistry of Natural Products, Department of Pharmacy, University of Athens, Panepistimioupolis, Athens, Greece

Abstract

The phenolic acid profile of honey depends greatly on its botanical and geographical origin. In this study, we carried out a quantitative analysis of phenolic acids in the ethyl acetate extract of 12 honeys collected from various regions in Greece. Our findings indicate that protocatechuic acid, p-hydroxybenzoic acid, vanillic acid, caffeic acid and p-coumaric acid are the major phenolic acids of the honeys examined. Conifer tree honey (from pine and fir) contained significantly higher concentrations of protocatechuic and caffeic acid (mean: 6640 and 397 µg/kg honey respectively) than thyme and citrus honey (mean of protocatechuic and caffeic acid: 437.6 and 116 µg/kg honey respectively). p-Hydroxybenzoic acid was the dominant compound in thyme honeys (mean: 1252.5 µg/kg honey). We further examined the antioxidant potential (ORAC assay) of the extracts, their ability to influence viability of prostate cancer (PC-3) and breast cancer (MCF-7) cells as well as their lowering effect on TNF-α-induced adhesion molecule expression in endothelial cells (HAEC). ORAC values of Greek honeys ranged from 415 to 2129 µmol Trolox equivalent/kg honey and correlated significantly with their content in protocatechuic acid ($p < 0.001$), p-hydroxybenzoic acid ($p < 0.01$), vanillic acid ($p < 0.05$), caffeic acid ($p < 0.01$), p-coumaric acid ($p < 0.001$) and their total phenolic content ($p < 0.001$). Honey extracts reduced significantly the viability of PC-3 and MCF-7 cells as well as the expression of adhesion molecules in HAEC. Importantly, vanillic acid content correlated significantly with anticancer activity in PC-3 and MCF-7 cells ($p < 0.01$, $p < 0.05$ respectively). Protocatechuic acid, vanillic acid and total phenolic content correlated significantly with the inhibition of VCAM-1 expression ($p < 0.05$, $p < 0.05$ and $p < 0.01$ respectively). In conclusion, Greek honeys are rich in phenolic acids, in particular protocatechuic and p-hydroxybenzoic acid and exhibit significant antioxidant, anticancer and antiatherogenic activities which may be attributed, at least in part, to their phenolic acid content.

Editor: Giovanni Li Volti, University of Catania, Italy

Funding: Greek Secretariat of Research and Technology, Ministry of Development for financial support (Grant ESPA, SMEs 2009) in cooperation with the company "Attiki" Alex Pittas SA. The funders had no role in study design, data collection and analysis, decision to publish, or preparation of the manuscript.

Competing Interests: The authors have declared that no competing interests exist.

* E-mail: pmoutsatsou@med.uoa.gr

Introduction

Honey is a highly nutritious natural food product which has been used in various medicinal traditions throughout the world for its healing, antibacterial and antiinflammatory properties. Emerging evidence suggests that honey possesses chemopreventive, antiatherogenic and immunoregulatory properties as well as a great potential to serve as a natural food antioxidant [1–7].

Characterization of components in honey that might be responsible for its biological properties is of great interest. Honey contains about 200 substances including sugars, phenolic acids, flavonoids, amino acids, proteins, vitamins and enzymes [8]. Phenolic compounds are considered among the main constituents contributing to the antioxidant and other beneficial properties of honey [9–13]. Phenolic acid profile has been determined in various honeys and is considered as a useful tool for determination of the floral origin of honey. Phenolic acids like caffeic acid and p-coumaric acid in chestnut honey as well as protocatechuic acid in

honeydew honeys have been used as floral markers [14,15]. Phenolic acids are compounds with multiple biological activities, including anticancer, antiinflammatory, antioxidant and anti-atherogenic properties. Hydroxybenzoic acid derivatives like p-hydroxybenzoic, protocatechuic and vanillic acid as well as hydroxycinnamic acid forms like p-coumaric and caffeic acid, are components with important anticancer activity [16,17]. Interestingly, protocatechuic and caffeic acid have been also shown to exhibit a significant potential as antidiabetic and cardioprotective agents [18–20].

Greece is one of the main producing countries of honey within the EU. In our previous study, we determined the total phenolic content and phenolic acid profile (qualitative analysis) in three Greek honeys [21]. In this study, we carried out a quantitative analysis of phenolic acids in 12 honeys from different collection regions in Greece. Furthermore we evaluated a) their antioxidant potential as oxygen radical absorbance capacity (ORAC), b) their antiinflammatory activity in reducing the expression of adhesion molecules VCAM-1 and ICAM-1 in endothelial cells and c) their ability to influence cell viability of prostate cancer (PC-3) and breast cancer (MCF-7) cells. We also examined possible associations between the composition of honeys in phenolic acids and total phenolic content with their antioxidant, anticancer and antiatherogenic activity.

Materials and Methods

Honey Samples

Twelve honey samples harvested from different regions in Greece were obtained and used in this study. These include four samples from commercial honeys (H1–H4) and eight honey samples of different botanical origin (H5–H12) which were provided by certified Greek bee-keepers. Pollen analysis through microscopic examination (see Table S1), characterized the honey samples as follows: four thyme honeys with pollen grains of *Thymus capitatus* in the range of 35% to 62%, six conifer honeys (fir and pine), one honey comprised of a mixture of wildflowers, forest and thyme and one honey from citrus. The honey floral sources and regions of collection are given in Table 1.

Extraction

Each honey sample (3 kg) was diluted with water (1 L). The solution was extracted three times with butanol (3×1 L) and butanol was collected and evaporated to dryness. In order to eliminate the sugar content, it was further extracted with ethyl acetate (3×1 L). The ethyl acetate phase was collected and evaporated to dryness. Subsamples (0.5 g) of dried honey extracts were dissolved in 50% methanol in a volumetric flask (5 ml) and further diluted as needed prior to analysis of total phenolic content, phenolic acids, hydroxy methyl furfural, sugars and oxygen radical absorbance capacity as described below. Subsamples of dried ethyl acetate honey extracts were diluted in DMSO prior to cell culture analysis.

Chemicals and Reagents

All cell-culture materials, such as HBSS (Hanks balanced salt solution), trypsin-EDTA solution, Dulbecco's modified essential medium and FBS were obtained from Invitrogen Life Technol-ogies (Carlsbad, CA, USA). TNF-α (T0157), α-Tocotrienol (07205), 3-(4,5-dimethylthiazol-2-yl)-2,5-diphenyltetrazolium bromide (MTT; M5655) and the peroxidase substrate o-phenylen-diamine hydrochloride (FASTe OPD; P9187) were obtained from Sigma-Aldrich (St Louis, MO, USA). VCAM-1 antibody (BBA5) and ICAM-1 antibody (BBA3) were purchased from R&D Systems (Minneapolis, MN, USA). Sheep anti-mouse IgG secondary antibody (NA931) was purchased from Amersham (Little Chalfont, Bucks, UK). 2,2′-azobis [2-methylpropioamidine] dihydrochloride (AAPH, 97%), fluorescein (puriss.p.a.) and ($\pm$)-6-hydroxy-2,5,7,8-tetramethylchromane-2-carboxylic acid (Trolox, 97%) were obtained from Sigma-Aldrich Chemie GmbH (Steinheim, Germany). Gallic acid, phenolic acids (syringic, vanillic, sinapic, benzoic, chlorogenic, gallic, p-coumaric, cinnamic, caffeic, ferulic, hydroxybenzoic and protocatechuic), 5-(hydroxymethyl)-2-furaldehyde (hydroxymethylfurfural; HMF) and Folin- Ciocalteu's phenol reagent were purchased from Sigma-Aldrich (St Louis, MO, USA).

Quantitative Analysis of Total Phenolics

Total phenolic content was determined by Folin-Ciocalteau method [22], where gallic acid was used as a calibration standard in photometric measurement performed by microplate reader (Varioskan Flash, Thermo Scientific, Finland).

HPLC Analysis of Phenolic Acids and Hydroxymethylfurfural (HMF)

Ethyl acetate honey extracts in methanol solution were diluted fivefold with water, filtered (0.45 µm) and analyzed by HPLC (high-performance liquid chromatography) (Agilent 1100 Series HPLC-MSD, Agilent Technology) connected to a diode array detector (DAD) as described earlier [21]. Analyses were performed on a HyperClone ODS (C18) column (2.0 mm, 200 mm, 5 µm, Phenomenex) using the following gradient run (0.2 ml/min) with methanol (A) and 0.3% formic acid (B): 90–75% B from 0 to 5 min, 75–69% B from 5 to 20 min and 69–40% B from 20 to 40 min. The run was stopped at 65 min and post time was 25 min. Phenolic acids were quantified at 260 nm or 320 nm, and 5- (hydroxymethyl)-2-furaldehyde (HMF) at 280 nm. The authentic chemical compounds used for identification and quantification of phenolic acids were protocatechuic, benzoic vanillic, syringic,

Table 1. Floral source and regions of honey collection.

Sample Code	Floral Source	Localization
H1	Thyme Attiki (commercial)	Various Greek islands
H2	Wild flowers-forest-thyme Attiki (commercial)	Various Greek regions
H3	Forest Fino (commercial)	Various Greek regions
H4	Fir Attiki (commercial)	Various Greek regions
H5	Thyme 45%[a]	Chania, Crete
H6	Pine	Euboea
H7	Fir	Menalon
H8	Thyme 62%[a]	Iraklio, Crete
H9	Fir	Karpenisi
H10	Pine	Chalkidiki
H11	Thyme 55%[a]	Chania, Crete
H12	Citrus	Argos

[a]Indicating pollen grains of *Thymus capitatus*.

sinapic, trans-cinnamic, p-coumaric, caffeic, p-hydroxybenzoic, gallic, ferulic and chlorogenic acid.

ORAC Assay

Oxygen radical absorbance capacity (ORAC) of the ethyl acetate honey extracts was analyzed by modified method of Huang et al. [23] as described in our earlier publication [24]. Ethyl acetate extracts in the methanol solutions were diluted (1000–3000 fold) in a phosphate buffer (75 mM, pH 7.4). Blank (methanol) and Trolox standards (15–100 μM) were also prepared in a phosphate buffer. Area under oxidation curve of each sample dilution was calculated and the ORAC values of ethyl acetate honey extracts were calculated using linear calibration curve of Trolox standard.

Analysis of Sugars

Carbohydrates (glucose, fructose and sucrose) were analyzed by capillary electrophoresis (CE) modified from a method of Rovio et al. [25], with a P/ACE MDQ CE instrument (Beckman-Coulter, Fullerton, CA, USA) using diode array detection at 270 nm. Uncoated fused-silica capillaries of I.D. 25 μm and length 30/40 cm (effective/total length) were used. The ethyl acetate honey extracts were injected at a pressure of 0.5 psi for 10 s. The separation voltage was raised linearly within a 1 min ramp time from 0 to +16 kV. Calibration curves for the quantification of glucose, fructose and sucrose were prepared using standard solutions with concentrations between 10 and 430 mg/L.

Cell Culture

Human aortic endothelial cells (HAEC), MCF-7 and PC-3 were obtained as cryopreserved cells from European Collection of Cell Cultures (ECACC) and were maintained in T-75 culture flasks at 37°C in a humidified 95% air-5% CO$_2$ atmosphere. HAEC were grown in endothelial cell growth medium M200 (Cascade Biologics, Portland, OR) supplemented with 2% LSGS (low serum growth supplement, Cascade Biologics, Portland, OR) according to the manufacturer's recommended protocol. Cells between passages 4–8 were used in this study. MCF-7 and PC-3 cells were maintained in Dulbecco's minimal essential medium (DMEM) supplemented with 10% FBS, 100 U/ml penicillin, and 100 mg/ml streptomycin. Cells were cultured and were split according to standard procedures.

Cell Viability and Cytotoxicity

Cell viability was determined using the colorimetric MTT metabolic activity assay [26]. MTT is cleaved in the mitochondria of metabolically active cells to form a colored, water-insoluble formazan salt. Briefly, breast and prostate cancer cells were grown in 96 flat-bottomed well plates and incubated for 48 hours with various concentrations of extracts (20–500 μg). As a positive control, MCF-7 and PC-3 cells were cultured with ICI 182780 and Doxorubicin respectively. Cells in their growth medium were used as control samples (vehicle). After the incubation, medium was replaced with MTT (1 mg/ml) dissolved in serum-free, phenol red-free medium, and incubation continued at 37°C for an additional 4 h. The MTT-formazan product was solubilized thoroughly in isopropanol and the optical density was measured at a test wavelength of 550 nm and a reference wavelength of 690 nm. The absorbance was used as a measurement of cell viability, normalized to cells incubated in control medium, which were considered 100% viable. Cell viability assay was also carried out for normal endothelial cells (HAEC) under the same experimental conditions.

Cell ELISA

To measure the cell-surface expression of ICAM-1 and VCAM-1, cell ELISA was conducted. Confluent HAEC cultured in 96-well plates were pretreated with or without honey extracts for 18 h and then stimulated with 1 ng/ml TNF-α for 6 hours at 37°C. Subsequently, cells were fixed by 0.1% glutaraldehyde in PBS for 30 min at 4°C and plates were blocked at 37°C for 1 h with 5% skimmed milk powder in PBS. Monoclonal antibodies to ICAM-1 or VCAM-1 in 5% skimmed milk in PBS were then added to the wells and incubated at 4°C overnight. After washing away the excess unbound primary antibody, cells were further incubated with a horseradish peroxidase-conjugated sheep anti-mouse secondary antibody for 1.5 h at room temperature. The expression of cell adhesion molecules was quantified by the addition of the peroxidase substrate o-phenylenediamine hydrochloride. As a positive control we have used α-Tocotrienol (αT3, 25 μM), the most effective vitamin E analog in the reduction of cellular adhesion molecule expression and monocytic cell adherence [27]. The absorption of each well was measured at 450 nm using a microplate spectrophotometer.

Statistical Analysis

Data is represented as mean±S.D. Statistical analysis was performed using Student's t-test, two-tailed distribution, assuming two-sample unequal variance. Pearson's correlation was carried out to identify relationships between phenolic acids and biological properties. The minimum level of significance was set at $p<0.05$.

Results

Phenolic Acid Composition of Honey Extracts

For the HPLC analysis of phenolic acids a total of 12 standards including protocatechuic, benzoic, vanillic, syringic, sinapic, trans-cinnamic, p-coumaric, caffeic, p-hydroxybenzoic, gallic, ferulic and chlorogenic acid were used. Examination of the HPLC chromatogram of honey ethyl acetate extracts revealed that Greek honeys are rich in phenolic acids. The phenolic acid pattern of the extracts was confirmed to contain protocatechuic, p-hydroxybenzoic, vanillic, caffeic and p-coumaric acid while gallic, ferulic, sinapic, syringic, trans-cinnamic and chlorogenic acid were not detected (Figure 1). The main constituent was protocatechuic acid. Protocatechuic and caffeic acid levels were higher in pine (mean values: 4513 and 558.5 μg/kg honey respectively) and fir honey (mean values: 8058.3 and 289.3 μg/kg honey respectively) when compared to thyme (mean value of protocatechuic acid: 471.3 μg/kg honey, mean value of caffeic acid: 122 μg/kg honey) or citrus honey (protocatechuic acid content: 303 μg/kg honey, caffeic acid content: 92 μg/kg honey). p-Hydroxybenzoic acid was the dominant constituent in thyme honeys (mean: 1252.5 μg/kg honey). In all samples, the amount of vanillic acid content ranged from 71 to 376 μg/kg honey and p-coumaric acid content ranged from 135 to 701 μg/kg honey (Table 2).

Total Phenolic Content, Hydroxymethylfurfural and Sugar Content of Honey Extracts

Hydroxymethylfurfural (HMF) content ranged between 0.2–3 mg/kg in conifer tree honey and between 7.5–11.1 mg/kg in thyme and citrus honey (Table 2). Total phenolic content of ethyl acetate extracts ranged from 11 to 52 mg of gallic acid/kg honey (Table 3). The mean glucose and fructose content was 112 and 481 mg/kg honey respectively. Sucrose was not detected in the

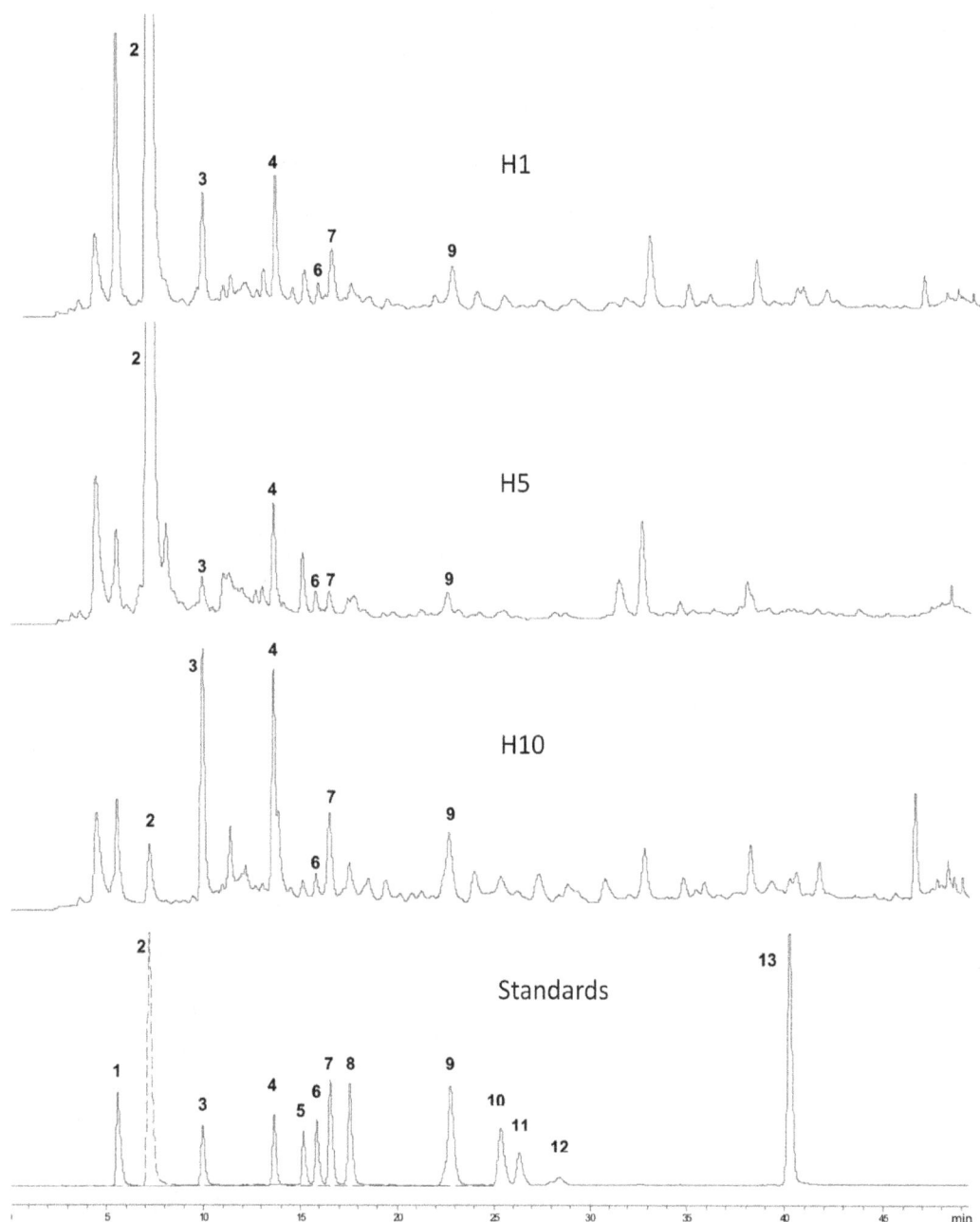

Figure 1. Representative chromatograms of honey extracts and standards. HPLC-DAD chromatograms (280 nm) obtained from honey samples (H1, H5, H10) and a standard mixture of phenolic acids and hydroxymethylfurfural (HMF). Peak identification: (1) gallic acid; (2) HMF; (3) protocatechuic acid; (4) p-hydroxybenzoic acid; (5) chlorogenic acid; (6) vanillic acid; (7) caffeic acid; (8) syringic acid; (9) p-coumaric acid; (10) ferrulic acid; (11) sinapic acid; (12) benzoic acid; (13) trans-cinnamic acid.

extracts. It is important to mention that although sugars are expressed as mg/kg honey, they do not represent the real amount of glucose and fructose in honey samples. The values concern the concentration of sugars in the ethyl acetate extracts and are important because the biological tests were performed with these extracts.

Antioxidant Activity of Honey Extracts

The antioxidant potential of the extracts was measured using the ORAC assay (Table 3). ORAC values, expressed as Trolox equivalent (TE)/kg honey ranged from 619 to 2129 μmol TE/kg

for pine and fir tree honeys and from 415 to 692 μmol TE/kg for thyme and citrus honeys indicating a higher antioxidant capacity for conifer tree honeys.

Effect of Honey Extracts on ICAM-1 and VCAM-1 Expression

TNF-α (1 ng/ml), increased the basal expression of ICAM-1 and VCAM-1 in HAEC. α-Tocotrienol decreased significantly the TNF-α-induced endothelial expression of both ICAM-1 and VCAM-1 as expected. Pretreatment of HAEC with different concentrations of honey extracts (20–500 μg/ml) for 18 h followed

Table 2. Concentration of phenolic acids and hydroxymethylfurfural detected in honey ethyl acetate extracts[a,b].

Floral source	PCA µg/kg honey	p-HBA µg/kg honey	VA µg/kg honey	CA µg/kg honey	p-COUA µg/kg honey	HMF mg/kg honey
Thyme Attiki (H1)	649±15	1070±25	225±6	134±3	146±4	7.5±0.2
Thyme 45% (H5)	346±25	1101±9	236±2	113±1	143±1	8.8±0.2
Thyme 55% (H11)	300±22	1724±79	245±11	121±5	252±11	11.1±0.5
Thyme 62% (H8)	590±21	1115±7	103±1	120±1	176±2	8.5±0.3
Fir (H9)	16777±780	1438±71	307±14	377±18	514±28	0.7±0.1
Fir (H7)	3258±24	1121±7	149±2	254±2	222±1	0.8±0.04
Fir Attiki (H4)	4140±131	1122±37	262±8	237±8	219±4	1.4±0.1
Pine (H10)	5967±57	4059±39	376±2	741±7	701±6	0.2±0.003
Pine (H6)	3058±111	1460±59	340±13	376±15	211±8	0.7±0.02
Forest Fino (H3)	2394±50	1503±41	237±7	445±12	288±11	3.0±0.1
Wild flowers-forest-thyme Attiki (H2)	1046±27	1124±25	201±5	255±5	193±5	6.1±0.2
Citrus (H12)	303±7	889±27	71±2	92±2	135±7	9.8±0.4

[a]PCA, protochatechuic acid, P-HBA, p-hydroxybenzoic acis, VA, vanillic acid, CA, caffeic acid, p-COUA, p-coumaric acid, HMF, hydroxymethylfurfural.
[b]All data expressed on a honey weight basis as means ± SD (n = 3 independent determinations).

by TNF-α stimulation for another 6 hours inhibited significantly ICAM-1 and VCAM-1 cell surface expression (Figure 2A and 2B respectively). Conifer tree honeys (fir honey Attiki, fir honey from Karpenisi and pine honey from Chalkidiki) caused a significant VCAM-1 inhibition (30%) at a low concentration (20 µg extract/ml) whereas VCAM-1 inhibition was higher (40%) at a concentration range 200–500 µg extract/ml. Other honeys (thyme honeys from Crete containing 45% and 62% of pollen grains and forest honey Fino) inhibited significantly VCAM-1 expression (30%) only at higher concentrations (200–500 µg extract/ml). Incubation with 20–200 µg/ml of honey extracts did not result in endothelial cell cytotoxicity. Incubation with 500 µg/ml of the extracts led to a 7–12% inhibition of endothelial cell viability (data not shown); however, this percentage inhibition is considerably lower than the magnitude of the effect of extracts on adhesion molecule expression at this concentration.

Effect of Honey Extracts on Viability of Cancer Cells

The inhibitory effect of honey extracts on the viability of PC-3 and MCF-7 cells is shown in Figure 3A and 3B respectively. Most honeys i.e. conifer tree honey as well as thyme honey, caused a significant PC-3 viability inhibition (30–60%) at a high concentration (500 µg extract/ml) whereas at a lower concentration (200 µg extract/ml) only four honeys (pine honey from Euboea, forest honey Fino, pine honey from Chalkidiki and thyme honey from Crete containing 55% of pollen grains) caused a significant cell viability inhibition (25–30%). Similarly, most honeys significantly reduced MCF-7 cell viability (20–50% inhibition) at a high concentration (500 µg extract/ml). Four honeys (pine honey from Euboea, forest honey Fino, wildflowers/forest/thyme honey Attiki and thyme honey from Crete containing 55% of pollen grains) caused a significant MCF-7 cell viability inhibition (15%, $p < 0.01$) at a lower concentration (200 µg extract/ml).

Table 3. Glucose and fructose content, total phenolic content and ORAC values of honey ethyl acetate extracts[a,b].

Floral source	Glucose mg/kg honey	Fructose mg/kg honey	TP (mg of GA/kg honey)	ORAC (µmol of TE/kg honey)
Thyme Attiki (H1)	90±9	308±57	17±0.3	421±45
Thyme 45% (H5)	97±5	279±27	18±0.2	415±29
Thyme 55% (H11)	117±2	311±5	24±0.9	692±93
Thyme 62% (H8)	124±2	325±15	18±0.6	469±69
Fir (H9)	109±4	319±16	52±0.2	2129±195
Fir (H7)	54±2	189±14	20±0.5	619±51
Fir Attiki (H4)	105±6	344±23	33±0.6	661±63
Pine (H10)	141±9	261±16	50±0.6	2068±314
Pine (H6)	75±1	144±3	11±0.2	712±107
Forest Fino (H3)	73±5	224±5	24±0.2	557±41
Wild flowers-forest-thyme Attiki (H2)	83±11	252±18	17±0.5	415±21
Citrus (H12)	112±6	481±18	14±0.4	455±46

[a]TP, total phenolic content, GA, gallic acid, TE, Trolox equivalent.
[b]All data expressed on a honey weight basis as means ± SD (n = 3 independent determinations).

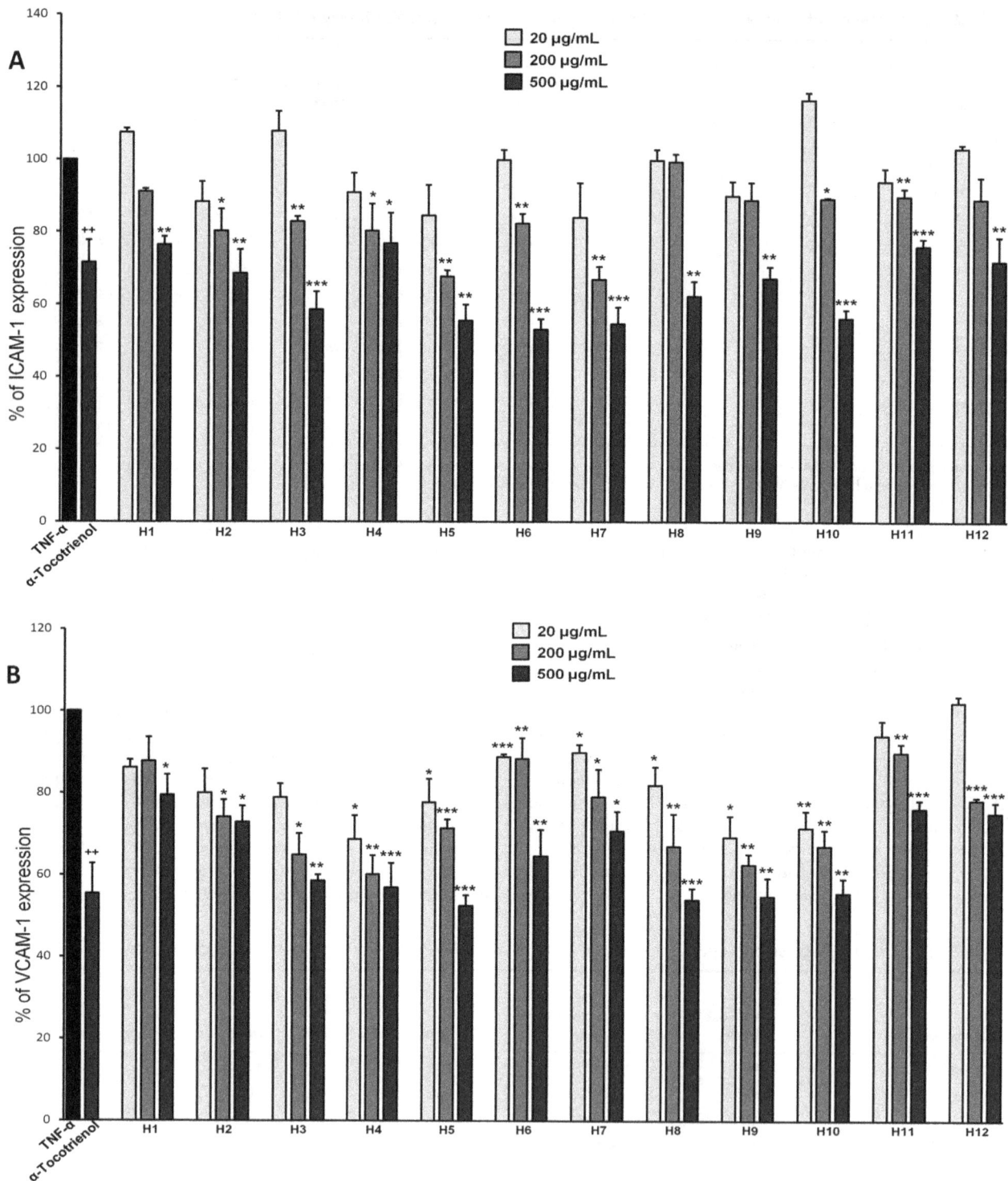

Figure 2. Greek honey extracts inhibit TNF-α-induced adhesion molecule expression. Greek honey extracts (H1–H12) inhibit TNF-α-induced ICAM-1 (A) and VCAM-1 (B) protein expression in HAEC. HAEC were incubated in the absence of TNF-α or compounds (control), with αT3 (25 μM), or with different concentrations (20–500 μg/ml) of honey extracts for 18 h, followed by stimulation with TNF-α (1 ng/mL) for up to 24 h. Adhesion molecules were measured by cell ELISA. Results are expressed as percent of control. A *p<0.05 value was considered statistically significant when compared to TNF-α-treated cells (**p<0.01, ***p<0.001). Values represent mean ± SD based on three independent experiments performed in triplicate.

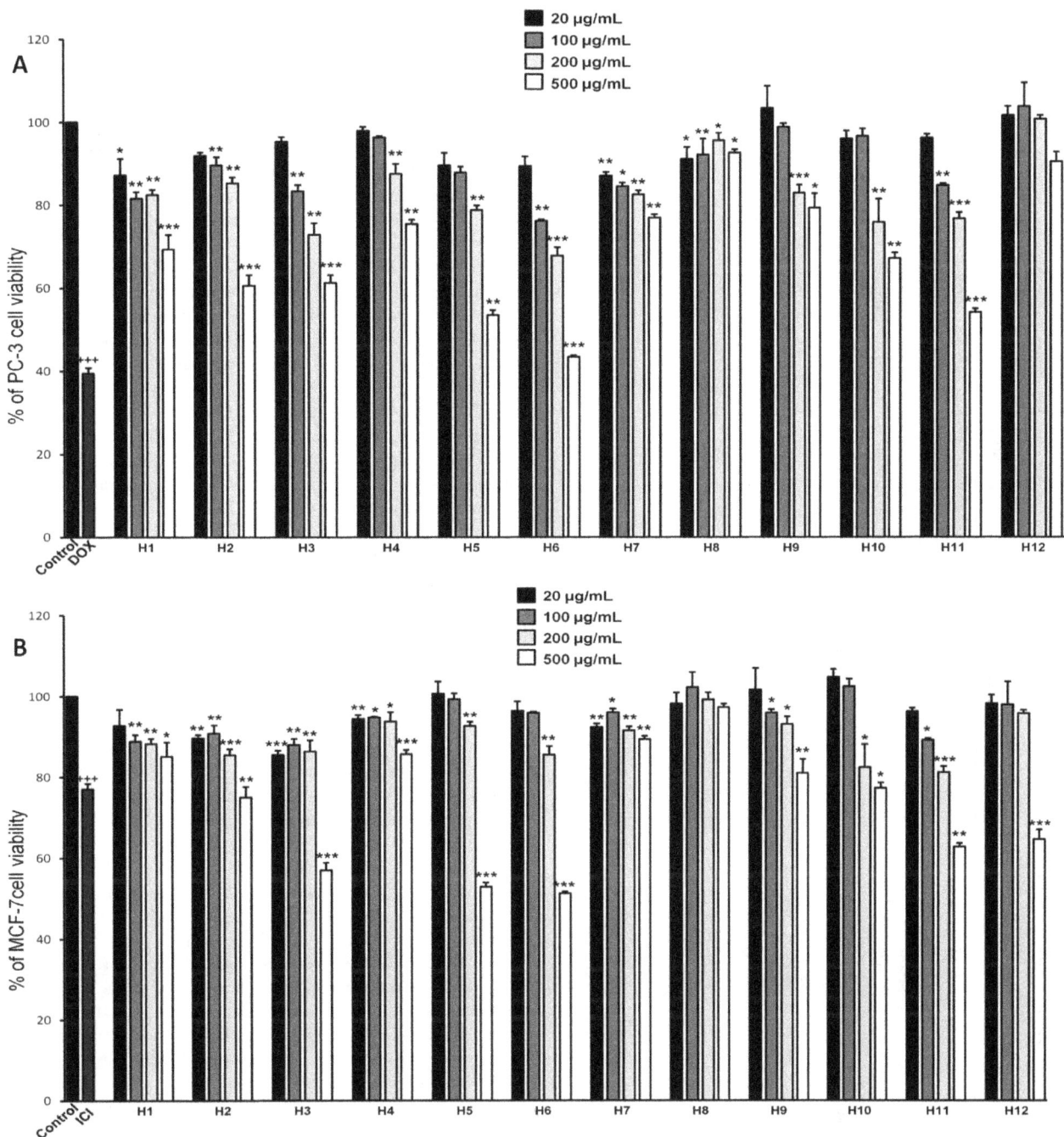

Figure 3. Greek honey extracts inhibit viability of prostate and breast cancer cells. Greek honey extracts (H1–H12) inhibit viability of PC-3 (A) and MCF-7 (B) cells. Cells were incubated in the absence of compounds (control) or with different concentrations (20–500 µg/ml) of honey extracts for 48 h. As a positive control, MCF-7 and PC-3 cells were cultured with ICI 182780 (0.1 µM) and doxorubicin (1 µM) respectively. After treatment, cells were subjected to the MTT assay. Results are expressed as percent of control. A *p<0.05 value was considered statistically significant when compared to TNF-α-treated cells (**p<0.01, ***p<0.001). Values represent mean ± SD based on three independent experiments performed in triplicate.

Association between Phenolic Acid Content and Biological Activities

As shown in Table 4, there were significant correlations between the antioxidant activity of honeys and protocatechuic acid content (r: 0.8497, p<0.001), p-hydroxybenzoic acid (r: 0.7079, p<0.01), vanillic acid (r: 0.6426, p<0.05), caffeic acid (r: 0.7315, p<0.01), p-coumaric acid (r: 0.9478, p<0.001) as well as with the total phenolic content of honeys (r: 0.926, p<0.001). There was also a significant correlation between VCAM-1 expression and total phenolic content (r: −0.6979, p<0.01) besides protocatechuic and vanillic acid content of honeys (r: −0.5749, p<0.05 and r: −0.5747, p<0.05 respectively). Vanillic acid content correlated significantly with MCF-7 and PC-3 cell viability inhibition (r: 0.6239, p<0.05, r: 0.7867, p<0.01 respectively).

Table 4. Correlation study between the results of biological tests and phenolic acid content[a,b,c].

	PCA	p-HBA	VA	CA	p-COUA	TP	ORAC
ORAC	0.8497 (***)	0.7079 (**)	0.6426 (*)	0.7315 (**)	0.9478 (***)	0.926 (***)	1
VCAM-1 expression	−0.5749 (*)	−0.3482	−0.5747 (*)	−0.5174	−0.5493	−0.6979 (**)	−0.5357
PC-3 inhibition	0.1331	0.4126	0.7867 (**)	0.5355	0.3255	0.1283	0.2323
MCF-7 inhibition	−0.0695	0.5620	0.6239 (*)	0.4888	0.3694	0.1322	0.2048

[a]PCA, protochatechuic acid, p-HBA, p-hydroxybenzoic acid, VA, vanillic acid, CA, caffeic acid, p-COUA, p-coumaric acid, TP, total phenolic content.
[b]Values represent Pearson's correlation coefficient (r).
[c]The minimum level of significance was set at $p < 0.05$ (*), $p < 0.01$ (**), $p < 0.001$ (***).

Discussion

Honey's beneficial health effects result from its active constituents including flavonoids and phenolic acids. Various honeys have been analyzed regarding their phenolic acid content which is rather variable and depends mainly on the floral and geographical origin of honey. In this study, we investigated the phenolic acid profile as well as the antioxidant, anticancer and antiinflammatory/antiatherogenic properties of twelve ethyl acetate extracts derived from Greek honeys, collected from different regions in Greece.

Different extraction methods or solvents yield different concentrations of phenolic compounds. In our study we have used ethyl acetate extracts because ethyl acetate results in higher recovery of phenolic acids than diethyl ether and ethyl acetate extracts seem to possess higher biological activity compared to other extracts [4,28].

HPLC analysis indicated that Greek honey ethyl acetate extracts contain high amounts of phenolic acids including protocatechuic acid, p-hydroxybenzoic acid, vanillic acid, caffeic acid and p-coumaric acid. The most abundant phenolic acid in this study was found to be protocatechuic acid. To our knowledge, protocatechuic acid is not detectable in most of the examined honeys derived from other countries, not even in "active Manuka honey", a well-known for its antibacterial properties honey from New Zealand which has been used as a gold standard for comparison with other honeys. Diethyl ether extracts from Turkish honeydew honeys only (pine and oak) have been shown to contain protocatechuic acid in the range of 1639 to 5986 μg/kg honey [15]. The presence of protocatechuic acid in all Greek honeys implicates its potential to be used as a characteristic indicator of the origin of honey. The concentrations of protocatechuic acid ranged from 3058 to 5967 μg/kg honey for pine honeys while in fir honeys the concentration ranged from 3258 to 16777 μg/kg honey. The content of protocatechuic acid in Greek thyme honey (300–649 μg/kg honey) and citrus honey (303 μg/kg honey) was considerably lower than in conifer honeys. This variation might be attributed to the botanical origin of honey. In agreement with our results, Haroun et al. (2012) reported the presence of protocatechuic acid in conifer tree honeys but not in honeys from other floral sources, supporting that protocatechuic acid may be used for the differentiation of conifer tree honey from floral honeys [15]. The mean vanillic, caffeic and p-coumaric acid levels (229, 272.1 and 266.67 μg/kg honey respectively) in Greek honey extracts were similar to those reported for ethyl acetate or ether extracts from honeys derived from other geographical regions [4,15,29]. However, the mean caffeic acid levels in conifer honeys were significantly higher than thyme and citrus honeys. p-Hydroxybenzoic acid was the second most abundant phenolic acid (after protocatechuic acid) in Greek conifer honey extracts (mean: 1840 μg/kg honey), but it was the dominant phenolic acid in Greek thyme honeys (mean: 1252.5 μg/kg honey). Quantitative analysis of phenolic acids in ethyl acetate or ether extracts from Turkish, Cuban and Malaysian honeys revealed that p-hydroxybenzoic acid was not among the phenolic acids detected [4,15,29]. Our data indicate that the phenolic acid composition of Greek honeys differs significantly than that of honeys originated from other countries, implicating that geography influences the honey phenolic acid profile.

Honey supplementation has been reported to be beneficial in diabetes mellitus [30]. Evidence suggests that protocatechuic and caffeic acid are potent antidiabetic compounds [19,20], which implicates that protocatechuic acid and caffeic acid may significantly contribute to the antidiabetic properties of honey. The current study reveals the unique composition of Greek conifer honey in protocatechuic acid, thus highlighting its utility as a potential antidiabetic agent.

In order to decide whether a physiologically achievable serving of honey (2–3 teaspoons) would result in biologically active protocatechuic acid levels we considered it important to take into account the following: 1) one teaspoon corresponds approximately to 20 g of honey, 2) our quantitative analysis data indicate that 50 g (2–3 teaspoons) of Greek conifer honey may contain 0.15–0.84 mg protocatechuic acid, 3) a bioavailability study in humans revealed that consumption of 350 ml wine containing 0.56 mg protocatechuic acid resulted in plasma concentration 0.2 μM, 4 hours after ingestion [31], 4) a recent study in mice supports that protocatechuic acid is a potent antidiabetic agent when plasma protocatechuic acid levels range from 0.06 to 0.13 μM. Importantly, protocatechuic acid administration to mice has been shown to increase its deposit in plasma and organs [19]. Taken together, we suggest that a daily intake of 2–3 teaspoons of conifer honey may result in plasma protocatechuic acid concentrations with potential antidiabetic activity. However, clinical studies are necessary to evaluate the utility of conifer honey in diabetes.

For the determination of the antioxidant activity of honey extracts we used the ORAC assay, which measures peroxyl radical scavenging capacity [23] and is widely used in food sector. The ORAC values of the extracts ranged between 415 and 2129 μmol TE/kg honey and were higher in conifer honeys (mean: 1237.8 μmol TE/kg honey) than thyme and citrus honeys (mean: 490.4 μmol TE/kg honey). The use of different honey fractions i.e. total honey, water soluble or organic solvent fractions have provided a wide range of antioxidant activity values. Comparison of our data with ORAC values from studies using similar fractions revealed that Greek honey possess high antioxidant potential; for example, ether extracted methanol fractions of native monofloral Cuban honeys varied from 430 to 1220 μmol TE/kg honey while in another study, the ORAC values for ether extracts of various monofloral commercial honeys from North America ranged from 110 to 900 μmol TE/kg honey [5,29].

Accumulating evidence indicates that honey is cardiopotective and vasoactive in vivo [32] while in vitro experiments demonstrate honey's protective activity in a cultured endothelial cell line subjected to oxidative stress [10]. However, the effect of honey on the inflammatory process of atherosclerosis remains unknown. In the present study, we have evaluated the antiinflammatory action of ethyl acetate honey extracts, by assessing their potential to inhibit TNF-α-induced activation of endothelial cells. For this purpose, cell ELISA, a widely used in vitro assay measuring the antiinflammatory effects of examined compounds or extracts in endothelial cells was conducted. As a positive control we have used α-Tocotrienol, a novel vitamin E with superior activity over tocopherol with regard to the attenuation of the expression of adhesion molecules. Incubation with Greek honey extracts caused a significant reducing effect on TNF-α-activated adhesion molecule expression. It is important to say that conifer tree honeys showed the highest inhibitory effect. These data indicate the potential of Greek honeys to inhibit the initial step of atherosclerotic process.

Honey has been previously shown to protect against various types of cancer, like bladder, colon and breast cancer [2,6,7,33,34]. In the present study, we demonstrate that Greek honey extracts caused cell death of breast and prostate cancer cell lines. Particularly, the ethyl acetate honey extracts inhibited significantly MCF-7 and PC-3 cell viability at concentrations ranging from 200 to 500 μg of extract/ml which correspond to 20–50% w/v of entire honey. Manuka honey has also been shown to inhibit significantly MCF-7 cell viability when an aqueous solution 2.5–5% w/v of entire honey was added to the cells [6]. Similarly, native Tualang and Indian honey when added directly to cells inhibited MCF-7 cell viability showing IC_{50} values of 2.4 and 4% respectively [7,34]. In another study, honey inhibited proliferation of PC-3 cells and IC_{50} concentration was 2.4% [35]. The anticancer activity shown by our ethyl acetate extracts is not comparable to the activity demonstrated by the entire honey because entire honey is a supersaturated solution of sugars and contains more active substances such as other phenolic compounds embedded in sugars and terpenes [1,10].

Given that phenolic acids are considered strong antioxidants and known to inhibit cancer related pathways, we further looked for features that could account for the differences in the biological activities between the various honey extracts. A significant correlation was found between vanillic acid content and inhibition of breast and prostate cancer cell viability. This suggests that vanillic acid may be the major component responsible for the anticancer activity observed. Vanillic acid has been previously shown to be an effective anticancer agent in vitro [36]. Each of the single phenolic acids present in the ethyl acetate extracts (protocatechuic acid, p-hydroxybenzoic acid, vanillic acid, caffeic acid and coumaric acid) as well as the total phenolic content, were highly correlated with the ORAC values. Thus, antioxidant property of the extracts can be attributed, at least in part, to the presence of these phenolic acids. The inhibitory effect of honey extracts on TNF-α-induced VCAM-1 was also significantly correlated with the total phenolic content as well as with the protocatechuic and vanillic acid content. This is consistent with the fact that antioxidant agents including protocatechuic aldehyde are inhibitors of TNF-α-inducible VCAM-1 expression in endothelial cells [37].

In conclusion, our findings suggest that Greek honeys are particularly rich in phenolic acids and exhibit significant antioxidant, anticancer and antiatherogenic activity. Greek conifer tree honeys are a rich source of protocatechuic and caffeic acid, implicating their beneficial use in patients with diabetes mellitus.

Supporting Information

Table S1 Plant species recorded in honey samples through microscopic examination. Pollen grains of *Thymus capitatus* were found in four honeys in the range of 35% to 62%. Six conifer honeys (fir and pine), one honey comprised of a mixture of wildflowers, forest and thyme and one honey from citrus were also characterized.

Author Contributions

Conceived and designed the experiments: PM. Performed the experiments: ES MJ TT ML IC SK. Analyzed the data: VV ES EK MJ PM. Contributed reagents/materials/analysis tools: PM VV EK IC SK. Wrote the paper: PM ES MJ.

References

1. Weston RJ, Mitchell KR, Allen KL (1999) Antibacterial phenolic components of Nea Zealand manuka honey. Food Chem 64: 295–301.

2. Swellam T, Miyanaga N, Onozawa M, Hattori K, Kawai K, et al. (2003) Antineoplastic activity of honey in an experimental bladder cancer implantation model: in vivo and in vitro studies. Int J Urol 10: 213–219.

3. Molan PC (2006) The evidence supporting the use of honey as a wound dressing. Int J Low Extrem Wounds 5: 40–54.

4. Kassim M, Achoui M, Mustafa MR, Mohd MA, Yusoff KM (2010) Ellagic acid, phenolic acids, and flavonoids in Malaysian honey extracts demonstrate in vitro anti-inflammatory activity. Nutr Res 30: 650–659.

5. Gheldof N, Wang XH, Engeseth NJ (2002) Identification and quantification of antioxidant components of honeys from various floral sources. J Agric Food Chem 50: 5870–5877.

6. Fernandez-Cabezudo MJ, El-Kharrag R, Torab F, Bashir G, George JA, et al. (2013) Intravenous administration of manuka honey inhibits tumor growth and improves host survival when used in combination with chemotherapy in a melanoma mouse model. PLoS One 8: e55993.

7. Fauzi AN, Norazmi MN, Yaacob NS (2011) Tualang honey induces apoptosis and disrupts the mitochondrial membrane potential of human breast and cervical cancer cell lines. Food Chem Toxicol 49: 871–878.

8. Wang J, Li QX (2011) Chemical composition, characterization, and differentiation of honey botanical and geographical origins. Adv Food Nutr Res 62: 89–137.

9. Stephens JM, Schlothauer RC, Morris BD, Yang D, Fearnley L, et al. (2010) Phenolic compounds and methylglyoxal in some New Zealand manuka and kanuka honeys. Food Chemi 120: 78–86.

10. Beretta G, Orioli M, Facino RM (2007) Antioxidant and radical scavenging activity of honey in endothelial cell cultures (EA.hy926). Planta Med 73: 1182–1189.

11. Gheldof N, Engeseth NJ (2002) Antioxidant capacity of honeys from various floral sources based on the determination of oxygen radical absorbance capacity and inhibition of in vitro lipoprotein oxidation in human serum samples. J Agric Food Chem 50: 3050–3055.

12. Al-Mamary M, Al-Meeri A, Al-Habori M (2002) Antioxidant activities and total phenolics of different types of honey. Nutr Res 22: 2041–2047.

13. Aljadi AM, Kamaruddin MY (2004) Evaluation of the phenolic contents and antioxidant capacities of two Malaysian floral honeys. Food Chem 85: 513–518.

14. Tomas-Barberan FA, Martos I, Ferreres F, Radovic BS, Anklam E (2001) HPLC flavonoid profiles as markers for the botanical origin of European unifloral honeys. J Sci Food Agric 81.

15. Haroun MI, Poyrazoglu ES, Konar N, Artik N (2012) Phenolic acids and flavonoids profiles of some Turkish honeydew and floral honeys. J Food Technol 10: 39–45.

16. Rocha LD, Monteiro MC, Anderson JT (2012) Anticancer properties of hydroxycinnamic acids-A Review. Cancer and clinical oncology 1: 1927–4866.

17. Tanaka T, Tanaka T, Tanaka M (2011) Potential Cancer Chemopreventive Activity of Protocatechuic Acid. J Exp Clin Med 3: 27–33.

18. Wang D, Wei X, Yan X, Jin T, Ling W (2010) Protocatechuic acid, a metabolite of anthocyanins, inhibits monocyte adhesion and reduces atherosclerosis in apolipoprotein E-deficient mice. J Agric Food Chem 58: 12722–12728.

19. Lin CY, Tsai SJ, Huang CS, Yin MC (2011) Antiglycative effects of protocatechuic acid in the kidneys of diabetic mice. J Agric Food Chem 59: 5117–5124.

20. Jung UJ, Lee MK, Park YB, Jeon SM, Choi MS (2006) Antihyperglycemic and antioxidant properties of caffeic acid in db/db mice. J Pharmacol Exp Ther 318: 476–483.

21. Tsiapara A, Jaakkola M, Chinou I, Graikou K, Tolonen T, et al. (2009) Bioactivity of Greek honey extracts on breast cancer (MCF-7), prostate cancer(PC-3) and endometrial cancer (Ishikawa) cells: Profile analysis of extracts. Food Chem 116: 702–708.

22. Magalhaes LM, Santos F, Segundo MA, Reis S, Lima JL (2010) Rapid microplate high-throughput methodology for assessment of Folin-Ciocalteu reducing capacity. Talanta 83: 441–447.

23. Huang D, Ou B, Prior RL (2005) The chemistry behind antioxidant capacity assays. J Agric Food Chem 53: 1841–1856.

24. Kallio T, Kallio J, Jaakkola M, Maki M, Kilpelainen P, et al. (2013) Urolithins Display both Antioxidant and Pro-oxidant Activities Depending on Assay System and Conditions. J Agric Food Chem 61: 10720–10729.

25. Rovio S, Yli-Kauhaluoma J, Siren H (2007) Determination of neutral carbohydrates by CZE with direct UV detection. Electrophoresis 28: 3129–3135.

26. Denizot F, Lang R (1986) Rapid colorimetric assay for cell growth and survival. Modifications to the tetrazolium dye procedure giving improved sensitivity and reliability. J Immunol Methods 89: 271–277.

27. Theriault A, Chao JT, Gapor A (2002) Tocotrienol is the most effective vitamin E for reducing endothelial expression of adhesion molecules and adhesion to monocytes. Atherosclerosis 160: 21–30.

28. Zaghloul AA, el-Shattawy HH, Kassem AA, Ibrahim EA, Reddy IK, et al. (2001) Honey, a prospective antibiotic: extraction, formulation, and stability. Pharmazie 56: 643–647.

29. Alvarez-Suarez JM, Gonzalez-Paramas AM, Santos-Buelga C, Battino M (2010) Antioxidant characterization of native monofloral Cuban honeys. J Agric Food Chem 58: 9817–9824.

30. Erejuwa OO, Sulaiman SA, Wahab MS (2012) Honey–a novel antidiabetic agent. Int J Biol Sci 8: 913–934.

31. Caccetta RA, Croft KD, Beilin LJ, Puddey IB (2000) Ingestion of red wine significantly increases plasma phenolic acid concentrations but does not acutely affect ex vivo lipoprotein oxidizability. Am J Clin Nutr 71: 67–74.

32. Rakha MK, Nabil ZI, Hussein AA (2008) Cardioactive and vasoactive effects of natural wild honey against cardiac malperformance induced by hyperadrenergic activity. J Med Food 11: 91–98.

33. Jaganathan SK, Mandal M (2009) Antiproliferative effects of honey and of its polyphenols: a review. J Biomed Biotechnol 2009: 830616.

34. Jaganathan SK, Mandal SM, Jana SK, Das S, Mandal M (2010) Studies on the phenolic profiling, anti-oxidant and cytotoxic activity of Indian honey: in vitro evaluation. Nat Prod Res 24: 1295–1306.

35. Samarghandian S, Afshari JT, Davoodi S (2011) Chrysin reduces proliferation and induces apoptosis in the human prostate cancer cell line pc-3. Clinics (Sao Paulo) 66: 1073–1079.

36. Kumar PPBS, Ammani K, Mahammad A, Gosala J (2013) Vanillic acid induces oxidative stress and apoptosis in non-small lung cancer cell line. International Journal of Recent Scientific Research 4: 1077–1083.

37. Zhou Z, Liu Y, Miao AD, Wang SQ (2005) Protocatechuic aldehyde suppresses TNF-alpha-induced ICAM-1 and VCAM-1 expression in human umbilical vein endothelial cells. Eur J Pharmacol 513: 1–8.

15

Effect of Cafeteria Diet History on Cue-, Pellet-Priming-, and Stress-Induced Reinstatement of Food Seeking in Female Rats

Yu-Wei Chen, Kimberly A. Fiscella, Samuel Z. Bacharach, Donna J. Calu*

Intramural Research Program, NIDA/NIH, Baltimore, Maryland, United States of America

Abstract

Background: Relapse to unhealthy eating habits is a major problem in human dietary treatment. The individuals most commonly seeking dietary treatment are overweight or obese women, yet the commonly used rat reinstatement model to study relapse to palatable food seeking during dieting primarily uses normal-weight male rats. To increase the clinical relevance of the relapse to palatable food seeking model, here we pre-expose female rats to a calorically-dense cafeteria diet in the home-cage to make them overweight prior to examining the effect of this diet history on cue-, pellet-priming- and footshock-induced reinstatement of food seeking.

Methods: Post-natal day 32 female Long-Evans rats had seven weeks of home-cage access to either chow only or daily or intermittent cafeteria diet alongside chow. Next, they were trained to self-administer normally preferred 45 mg food pellets accompanied by a tone-light cue. After extinction, all rats were tested for reinstatement induced by discrete cue, pellet-priming, and intermittent footshock under extinction conditions.

Results: Access to daily cafeteria diet and to a lesser degree access to intermittent cafeteria diet decreased food pellet self-administration compared to chow-only. Prior history of these cafeteria diets also reduced extinction responding, cue- and pellet-priming-induced reinstatement. In contrast, modest stress-induced reinstatement was only observed in rats with a history of daily cafeteria diet.

Conclusion: A history of cafeteria diet does not increase the propensity for cue- and pellet-priming-induced relapse in the rat reinstatement model but does appear to make rats more susceptible to footshock stress-induced reinstatement.

Editor: Judith Homberg, Radboud University, Netherlands

Funding: The work was supported by the Intramural Research Program of the National Institute on Drug Abuse. The funders had no role in study design, data collection and analysis, decision to publish, or preparation of the manuscript.

Competing Interests: The authors have declared that no competing interests exist.

* Email: donna.calu@nih.gov

Introduction

Excessive consumption of unhealthy palatable foods is a major public health problem contributing to obesity and obesity-related diseases [1,2]. While many people attempt to control their food intake through dieting, most relapse to maladaptive eating habits within a short time [3–6]. This relapse is often triggered by acute exposure to palatable food, food-associated cues, or stress [6–10]. In order to model this clinical situation, we and the others have adopted the rat reinstatement model [11], which has been used extensively to study relapse to drug seeking [12–14]. In this model, rats are trained to lever-press for palatable food. After extinction of the food-reinforced responding, the rats are tested for reinstatement induced by non-contingent exposure to palatable food, food-associated cues, or stressors [15,16].

The reinstatement procedure, which is relatively simple to implement, has proven to be very suitable for studying relapse to palatable food seeking during dieting because it reliably induces robust reinstatement effects across a variety of reinstating stimuli [16]. Nevertheless, this reinstatement model fails to capture certain aspects of the clinical problem. First, the target population seeking treatment for relapse to unhealthy eating after dieting is primarily comprised of overweight and obese individuals, and the proportion of women taking dietary supplements or seeking dietary treatment is more than twice that of men [17,18]. Yet most pre-clinical relapse studies to date use normal-weight male rats to study relapse to palatable food seeking [15,19]. Second, in the reinstatement procedure rats are typically given limited access to nutritionally balanced chow during the training and extinction/ testing phases in order to promote enhanced motivation in the operant chamber. While humans with unhealthy eating habits generally consume limited amounts of nutritionally balanced foods and greater amounts of calorically-dense palatable foods, during dieting, subjects not only restrict the amount of calorically-dense food they eat but also tend to consume nutritionally balanced foods. These aspects of the human situation are not accurately modeled using standard chow food deprivation throughout the reinstatement procedure.

To increase the clinical relevance of the relapse to palatable food seeking model, we gave young female rats daily or intermittent (Monday, Wednesday, Friday) access to a calorically-dense cafeteria diet in the home-cage prior to and throughout the training period in order to make the rats overweight. Exposure to highly palatable and energy-dense food not only leads to obesity in rodents, but also causes dysregulation of reward processing and stress responses [20–22], both of which implicated in relapse behavior [16,23]. Therefore, to more accurately model human dieting, we terminated access to cafeteria diets at the end of the training phase and examined the effect of diet history on discrete cue-, pellet priming-, and footshock-induced reinstatement of food seeking after extinction.

Materials and Methods

Subjects and apparatus

Female Long-Evans (total N = 24) rats (Charles River Laboratories, Wilmington, MA; 26 days old at time of arrival) were maintained on a reverse 12 h light/dark cycle (lights off at 8 AM) and were given free access to standard laboratory chow and water during the experiment. Rats were individually housed on postnatal day (PND) 30 and were weight matched across groups before being assigned to each diet condition (n = 8 per group). Diet manipulations began on PND 32. Following 50 days of diet exposure, rats began behavioral training. We performed the experiments in accordance to the "Guide for the care and use of laboratory animals" (8th edition, 2011, US National Research Council) and experimental protocols were approved by the Intramural Research Program (NIDA) Animal Care and Use Committee.

Behavioral experiments were conducted in standard self-administration chambers (Med Associates). Each chamber had two levers 9 cm above the floor, but only one lever (the "active," retractable lever) activated the pellet dispenser, which delivered 45 mg food pellets containing 12.7% fat, 66.7% carbohydrate, and 20.6% protein (catalog #1811155; Test Diet). This pellet type was chosen based on pellet preference tests in food-restricted female rats, using six pellet types (obtained from Test Diet and Bioserv) with different compositions of fat (0–35%) and carbohydrate (45–91% sugar pellets) and different flavors (no flavor, banana, chocolate, grape) [15]. Rats were housed in the animal facility and transferred to the self-administration chambers prior to the training sessions, and returned to the facility at the end of the 2-h sessions.

Diet manipulations

The chow only rats received free access to standard laboratory chow (Teklad global 18% protein rodent diet, Harlan Laboratories) (58% carbohydrate, 24% protein, 18% fat).

The daily cafeteria diet rats also had free access to standard laboratory chow in addition to a rotating schedule of highly palatable "cafeteria items." This diet consisted of two items per day—one savory item and one sweet item. Two weeks on this diet allowed for a full rotation of cafeteria diet items including potato chips, corn snacks, chocolate cake, peanut butter cookies, pepperoni, pretzels, chocolate sandwich cookies, chocolate-chip cookies, and breakfast pastries. Both the chow and the cafeteria items were pre-weighed and the savory and sweet items were placed in two separate ceramic ramekins in the rats' home-cages, while the chow was available in the home-cage food hopper. Twenty-four hours later any chow or cafeteria items remaining in

the cage were independently weighed to determine daily intake for each item. Daily intake of cafeteria items and chow was monitored Monday through Thursday. The amount of each item consumed per rat was converted to calories (kcal) using data provided by the manufacturers. The intermittent cafeteria diet rats received the same rotation of highly palatable cafeteria items that the daily cafeteria rats received in addition to free access to standard laboratory chow. However, these rats had 24 h of forced abstinence from the cafeteria items every Tuesday, Thursday, Saturday, and Sunday, during which the chow was available. Using the identical procedure described above, daily intake of cafeteria items (Monday, Wednesday, Friday) and chow (every day) was monitored and the caloric intake (kcal) from each item was calculated, except that the consumption was monitored Monday through Saturday. This was due to the intermittent schedule of this diet, in which cafeteria items had to be removed on Saturday morning following the 24 h access that began Friday morning. All rats were weighed daily during home-cage diet exposure and before each training, extinction, and reinstatement testing session.

Training, extinction and reinstatement of food-reinforced responding

The experimental conditions used during training, extinction and reinstatement phases were similar to those used in previous studies in male and female rats [19,24,25], with the exception of food restriction and duration of sessions.

Training and extinction

Behavioral training began with 2 days of magazine training which involved 2-h sessions during which pellets (Test Diet Purified Rodent Tablet (5TUL)) were delivered non-contingently, every 5 min into a food cup located to the right of the active lever. Pellet delivery was accompanied by a compound 5-s tone (2900 Hz)-light (7.5W white light) cue, both located above the active lever. Subsequently, rats were trained to self-administer the pellets on a fixed-ratio-1 (FR-1), 20-s timeout reinforcement schedule. At the start of each 2-h session, the red house light was turned on and the active lever was extended. Reinforced active lever presses resulted in delivery of one pellet, accompanied by the compound 5-s tone-light cue. Active lever presses during the 20-s timeout or presses on the inactive lever had no programmed consequences. Rats underwent 10 training sessions under these conditions. During training, rats were maintained on their respective diets.

Following ten 2-h food self-administration training sessions, all rats underwent no-cue extinction sessions in which active lever presses were no longer reinforced with a food pellet or the tone-light cue. During the extinction phase, in order to mimic human dieting, daily cafeteria and intermittent cafeteria rats were switched to free access chow diet, no longer receiving access to the cafeteria items in the home-cage. Chow rats experienced no change in the free access chow home-cage diet condition.

All rats were given 11 extinction sessions (no cue), and reached an extinction criterion (mean active lever pressing across the last three extinction sessions <20% of extinction day 1 responding) before the cue-induced reinstatement test began. Before the pellet-priming- and footshock-induced reinstatement tests started, all rats underwent another 7 and 2 sessions, respectively, of 2-h extinction, in which responses on the previously active lever led to tone-light cue presentations but no pellets (extinction with cue). On the day of the pellet-priming or footshock-induced reinstatement test, all rats underwent additional within-session extinction, and the reinstatement manipulation started after an extinction criterion was reached (<20 active lever responses within 1 h).

Cue-induced reinstatement

After all rats reached extinction criterion, they underwent one day of cue-induced reinstatement testing. Cue reinstatement testing consisted of a single 2-h session beginning with 1 non-contingent tone-light cue presentation immediately after the session started [26–28]. During the session, active lever presses resulted in tone-light cue but no pellet (20 s timeout). Active lever presses during the 20-s timeout or presses on the inactive lever had no programmed consequences.

Pellet-priming-induced reinstatement

After 7 sessions of 2-h daily extinction (with cue) session, the rats underwent one day of within-session extinction/pellet priming-induced reinstatement testing. The pellet priming test consisted of six 60 min mini-sessions separated by 5 min time out, in which the active lever was retracted and the house light was extinguished. The first 3 mini-sessions were conducted under extinction conditions (with cue) with the third hour serving as the 0 pellet baseline mini-session. At the start of the remaining 3 mini-sessions 1, 2, or 4 pellets were delivered non-contingently (each was separated by 20 s) into the food hopper, in ascending order. The within-session priming procedure is based on previous studies with cocaine priming [29,30].

Footshock-induced reinstatement

The rats underwent 2 additional 2-h daily sessions of extinction with cue between pellet priming- and footshock-induced reinstatement tests. For the test, the rats underwent one day of within-session extinction/footshock-induced reinstatement testing, which consisted of five to six mini-sessions. The testing started with three to four 60 min extinction mini-sessions separated by 5 min time out, in which the active lever was retracted and the house light was extinguished. These mini-sessions were conducted under extinction conditions (with cue), with the third or fourth hour serving as the 0 min duration shock baseline. A 10 min time out in which the active lever was retracted and the house light was extinguished followed this baseline session. After this timeout, rats received 5 min of intermittent footshock (0.18–0.30 mA based on individual shock reactivity; 0.5 s ON and an average of 40 s OFF period, 10–70 s intershock interval) with house light extinguished and lever retracted. Subsequently, the effect of 5-min footshock on reinstatement was assessed in a 60-min session under extinction conditions. After another 10 min time out, they received 10 min of intermittent footshock (0.18–0.30 mA based on individual shock reactivity). Reinstatement effect was again assessed in a 60-min session post-footshock. The within-session footshock-induced reinstatement procedure is based on a previous study with heroin [31] and the shock ON and OFF duration is based on previous studies on footshock-induced reinstatement of drug seeking [32,33].

Statistical analyses

Data were analyzed using the SPSS statistical software (IBM). The factors used in the statistical analyses are described in the Results section below. Results are presented as group means (±SEM). Significant main effects and interaction effects (p<0.05) from ANOVAs were followed by post-hoc Fisher PLSD tests when appropriate.

Results

Effect of home-cage diet history on body weight gain

Rats that had daily access to cafeteria items gained significantly more body weight and consumed more calories during the diet exposure, but this difference in body weight disappeared once the diet was no longer available. From PND 32 all rats had home-cage access to their designated diet (daily cafeteria, intermittent cafeteria, chow only) for 64 days before cafeteria diets were removed and extinction began. The cumulative body weight gain and caloric intake during these phases are shown in **Fig. 1**. For the body weight gain during the first phase, when the cafeteria items were available in home-cages (weeks 1 to 9), we analyzed the data using the mixed ANOVA, with the between-subjects factor of Diet Condition and within-subjects factor of Time (weeks 1 to 9). Statistical analysis on cumulative body weight gain showed significant main effects of Diet Condition and Time ($F_{2,21} = 6.8$ and $F_{8,168} = 839.0$, respectively, p<0.01), and a significant interaction of Diet Condition x Time ($F_{16,168} = 8.5$, p<0.01). Post-hoc analysis, as depicted in **Fig. 1A**, showed that the daily cafeteria group gained significantly more weight than the chow-only group from week 3 to the end of the diet exposure period. Similarly, analysis of caloric intake showed significant main effects of Diet Condition and Time ($F_{2,21} = 39.5$ and $F_{8,168} = 24.5$, respectively, p<0.01), and a significant interaction of Diet Condition x Time ($F_{16,168} = 4.8$, p<0.01). Post-hoc analysis (**Fig. 1B**) showed that the daily cafeteria group had significantly more caloric intake than chow-only groups from week 1.

Once the cafeteria items were removed (weeks 10–14), a similar analysis on cumulative body weight gain as stated above showed a significant main effect of Time ($F_{4,84} = 7.9$, p<0.01) and interaction of Diet Condition x Time ($F_{8,64} = 17.6$, p<0.01), but not Diet Condition (p = 0.29), suggesting that there was no longer a body weight difference between groups. Post-hoc analysis (**Fig. 1A grey shaded region**) showed that the daily cafeteria group weighed significantly more only during the first week following cessation of cafeteria diet, and there were no group differences during reinstatement testing. A similar analysis on caloric intake after the cessation of cafeteria diet as stated above showed significant main effects of Diet Condition and Time ($F_{2,21} = 9.0$ and $F_{4,84} = 6.3$, respectively, p<0.01) and interaction of Diet Condition x Time ($F_{8,84} = 4.7$, p<0.01). Post-hoc analysis (**Fig. 1B grey shaded region**) showed that the daily cafeteria group had significantly less caloric intake than the chow-only group for the first three weeks following cessation of cafeteria diet.

Effect of cafeteria diet history on pellet self-administration and extinction

Training. All rats demonstrated reliable food-reinforced responding, but the magnitude and time course of responding over days differed between diet conditions with significantly lower food self-administration in rats in the daily cafeteria diet condition (**Fig. 2**). We analyzed the data using the mixed ANOVA, with the between-subjects factor of Diet Condition and within-subjects factor of Session. Analyses of pellets earned showed a significant main effect of Diet Condition and Session ($F_{2,21} = 16.6$ and $F_{9,189} = 2.3$, respectively, p<0.05), but no significant interaction of Diet Condition x Session (p = 0.96). Post-hoc differences are depicted in **Fig. 2A**. Analyses on time-out responses on the active lever showed a significant main effect of Diet Condition and Session ($F_{2,21} = 9.1$ and $F_{9,189} = 5.0$, respectively, p<0.01), and a significant interaction of Diet Condition x Session ($F_{18,189} = 2.1$, p<0.01). A post-hoc analysis comparing the time-out responses between sessions 1 and 10 for each diet condition showed that the

A. Cumulative Body Weight Gain

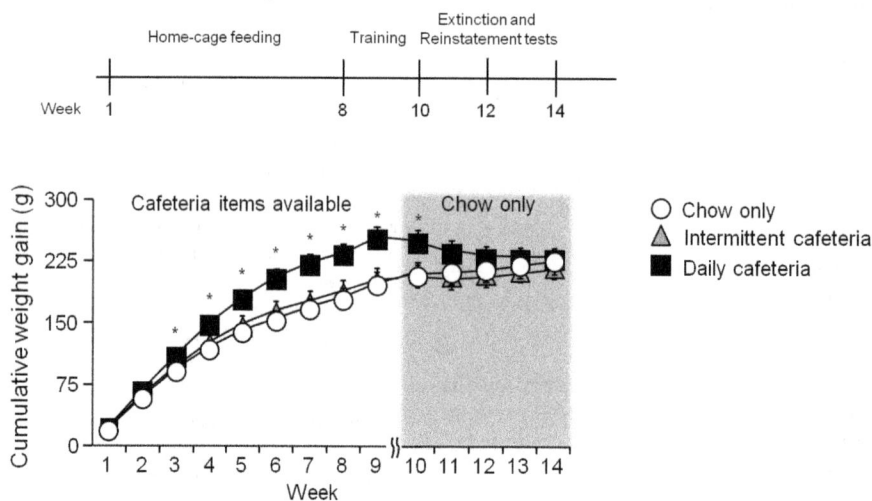

B. Energy Intake

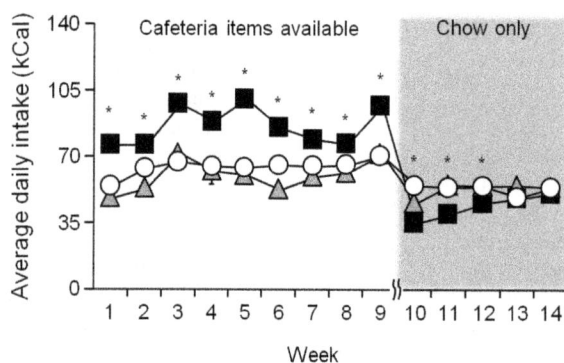

Figure 1. Cumulative body weight gain and average daily caloric intake in rats with different diet conditions. A. Weekly average cumulative body weight gain (mean±SEM) in rats with daily access to a cafeteria diet, intermittent access to a cafeteria diet or daily access to chow-only. Weeks 1–7 correspond to the home-cage access phase and Weeks 8–9 correspond to training phase in which access to home-cage diets remained unchanged. Extinction started on Week 10, from which time the cafeteria items were no longer available. All rats had *ad libitum* chow from Week 10 to the end of testing. Discrete-cue induced reinstatement test occurred in Week 12; pellet-priming- and intermittent footshock-induced reinstatement tests occurred in Week 14. **B.** Daily average caloric intake from diet (mean±SEM). *Different between chow-only and daily cafeteria group, p<0.05, n=8 per group.

chow-only group, but not the daily or intermittent cafeteria groups, demonstrated escalated time-out responding after 10 days of training ($t_7 = 2.4$, p<0.05) (**Fig. 2B, inset**). Additional analysis showed that while the intermittent group had access to cafeteria items before some training sessions, this did not affect their time-out responding (cafeteria diet available: 168.5±21.0 vs. cafeteria diet not available: 188.0±22.9). Inactive lever pressing data for the training phase are presented in table 1.

Extinction. All rats reduced active lever responding during the extinction phase, both in the absence and presence of the tone-light cue (prior to or after cue-induced reinstatement, respectively), but similar to food self-administration lever responding was significantly lower in the daily cafeteria group (**Fig. 2C**). The analysis using the mixed ANOVA, with the between-subjects factor of Diet Condition, and within-subjects factors of Lever (active, inactive) and Session, showed significant main effects of Diet Condition and Session ($F_{2,21} = 8.7$ and $F_{10,210} = 72.3$, p<

0.01) and interactions of Diet Condition x Lever, Diet Condition x Session, and Lever x Session ($F_{2,21} = 8.5$, $F_{20,210} = 9.0$, and $F_{10,210} = 62.2$, respectively, p<0.01), and Diet Condition x Lever x Session ($F_{20,210} = 6.8$, p<0.01). Post-hoc differences between groups were depicted in **Fig. 2C**, and consistent with the level of lever pressing observed during training, chow-only had greater lever responding compared to the daily cafeteria during the early extinction sessions. Inactive lever pressing data for the extinction phase are presented in table 1.

Effects of cafeteria diet history on cue-induced reinstatement

Discrete tone-light cues successfully reinstated active lever responding in all rats, but the magnitude of responding was lower in rats with a history of cafeteria diet (**Fig. 3A**). Data were analyzed using mixed-factor ANOVA, with the between-subjects

A. Pellets Earned

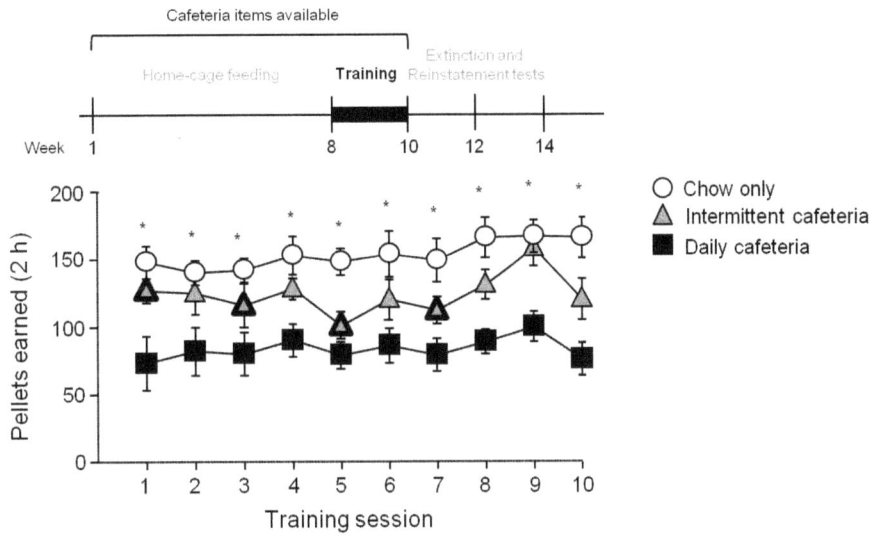

B. Time-out Responding

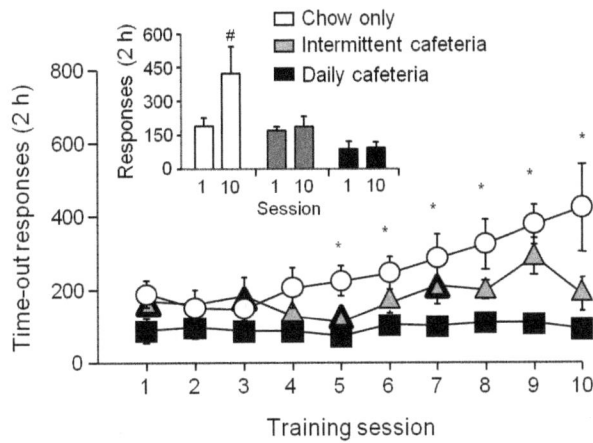

C. Extinction (no cue)

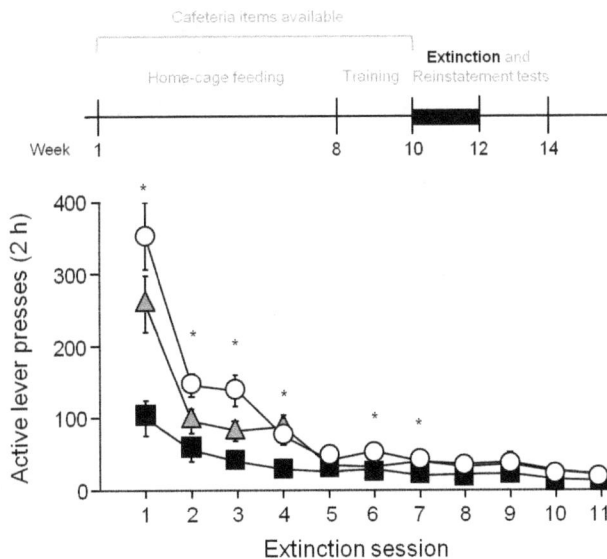

Figure 2. Effect of home-cage diet condition on food self-administration training. A. Reinforcers (pellets) earned (mean±SEM). **B.** Time-out responding during food self-administration training for different diet conditions. Inset: difference in time-out between training sessions 1 and 10 for each diet condition. [#] Different between sessions 1 and 10 within diet condition, p<0.05. **C.** Active lever responding (mean±SEM) under extinction conditions without cue (lever responding had no consequence).*Different between chow-only and daily cafeteria group, p<0.05, n = 8 per group. Data points with heavier outlines indicate days that intermittent group had access to cafeteria items prior to the session.

factor of Diet Condition, within-subjects factor of Lever (active, inactive) and Reinstatement Condition (last day extinction [no cue], cue). We found significant main effects of Lever and Reinstatement Condition ($F_{1,21} = 62.6$ and $F_{1,21} = 45.0$, respectively, p<0.01), but not Diet Condition (p = 0.15), and significant interactions of Diet Condition x Lever, Diet Condition x Reinstatement Condition, and Lever x Reinstatement Condition ($F_{2,21} = 4.9$, $F_{2,21} = 5.5$, $F_{1,21} = 35.4$, p<0.05). A post-hoc comparison showed an approaching significant difference between lever responding in the chow-only compared to the daily cafeteria group (p = 0.06). Overall, our results showed that while the cue-induced reinstatement effect is observed in all groups, this effect is most pronounced in the chow-only group and least pronounced in the daily cafeteria group.

Effects of cafeteria diet history on pellet priming-induced reinstatement

Similar to the cue-induced reinstatement test, pellet-priming-induced reinstatement was more pronounced in the chow-only group than in the rats with a history of cafeteria diet (**Fig. 3B**). Data were analyzed using mixed-factor ANOVA, with the between-subjects factor of Diet Condition, within-subjects factors of Lever (active, inactive) and Pellet Number (0, 1, 2, 4). The results showed significant main effects of Diet Condition, Lever, and Pellet Number ($F_{2,21} = 4.1$, $F_{1,21} = 79.1$, and $F_{3,63} = 16.5$, respectively, p<0.05), as well as significant interactions of Diet Condition x Lever, Lever x Pellet Number, and Diet Condition x Lever x Pellet Number ($F_{2,21} = 7.2$, $F_{3,63} = 18.6$, $F_{6,63} = 2.9$, p< 0.05). Lever responding across all pellet priming conditions in the chow-only group was significantly higher than either intermittent cafeteria or daily cafeteria groups (p<0.05). In order to further

understand the effect of Pellet Number on groups with cafeteria diet histories, we performed three separate repeated measures ANOVAs for chow-only, daily cafeteria and intermittent cafeteria groups, using the within-subjects factors of Lever (active, inactive) and Pellet Number (0, 1, 2, 4). Results from the analyses of daily and intermittent cafeteria diet groups showed significant main effects of Lever and Pellet Number (daily cafeteria: $F_{1,7} = 16.1$, $F_{3,21} = 3.3$, respectively, p<0.05; intermittent cafeteria: $F_{1,7} = 67.5$, $F_{3,21} = 5.9$, respectively, p<0.01), and interaction of Lever x Pellet Number ($F_{3,21} = 9.9$, p<0.01) in the intermittent cafeteria group only. Thus, similar to the effect of discrete cue, the reinstatement induced by pellet priming was most robust in chow-only and least robust in the daily cafeteria rats.

Effects of cafeteria diet history on intermittent footshock-induced reinstatement

Intermittent footshock induced modest reinstatement of lever pressing; this effect was significant in the daily cafeteria group, but not the intermittent cafeteria or chow-only groups (**Fig. 3C**). Data were analyzed using mixed-factor ANOVA, with the between-subjects factor of Diet Condition, within-subjects factor of Lever (active, inactive) and Shock Duration (0, 5, and 10 min). The results showed significant main effects of Lever and Shock Duration ($F_{1,21} = 39.0$, $F_{2,42} = 4.0$, p<0.05), but not Diet Condition (p = 0.93). In order to further understand the main effect of shock across groups, we performed two separate repeated measures ANOVAs, with Lever (active, inactive) and Shock Duration (0, 5, and 10 min) as within-subjects factors, focusing on rats that previously had access to cafeteria diets. The analysis of the daily cafeteria group only showed a Lever x Shock Duration interaction ($F_{2,14} = 3.7$, p = 0.05), and post-hoc tests on active lever

Table 1. Summary of inactive-lever presses during the training, extinction, and reinstatement testing.

		Chow only	Intermittent cafeteria	Daily cafeteria
Training	day 1	20.3±7.9	22.8±7.1	11.6±7.4
	day 10	10.5±4.8	3.4±1.1	2.5±1.1
Extinction (no cue)	day 1	19.6±8.7	6.3±2.9	4.1±2.3
	day 10	4.4±1.5	7.9±3.5	2.9±1.8
Cue-induced reinstatement	No cue	4.4±1.7	8.5±4.0	16.0±14.9
	Cue	9.1±3.5	6.6±3.4	2.4±2.0
Pellet priming-induced reinstatement	0 pellet	2.8±1.2	1.8±0.9	0.8±0.5
	1 pellet	2.9±1.7	2.5±1.7	1.6±1.2
	2 pellets	1.4±0.7	2.4±1.3	5.6±4.6
	4 pellets	4.6±2.3	1.3±0.6	9.8±7.0
Footshock-induced reinstatement	0 min	3.0±1.1	3.6±1.9	5.3±3.0
	5 min	7.5±2.7	4.3±1.6	8.8±6.0
	10 min	4.6±2.3	13.4±10.8	6.6±2.9

Data are mean±SEM.

A. Cue-induced Reinstatement

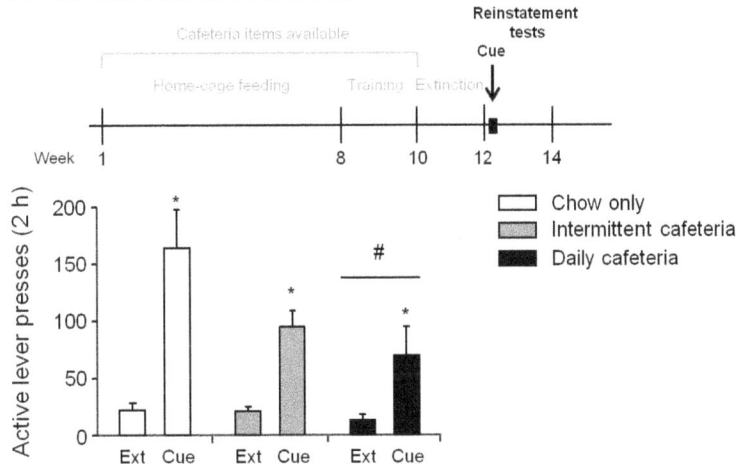

B. Pellet-priming-induced Reinstatement

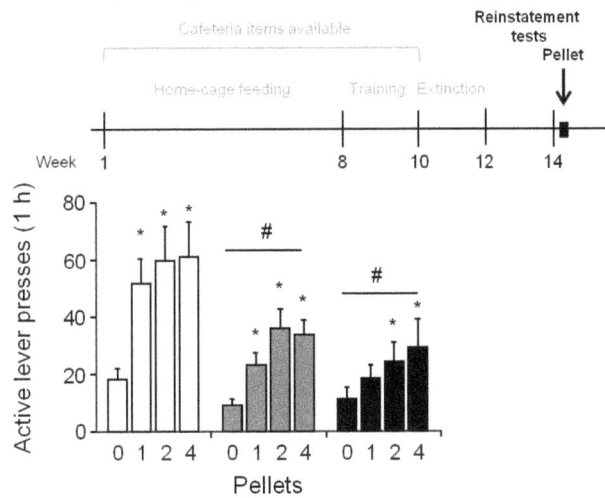

C. Footshock-induced Reinstatement

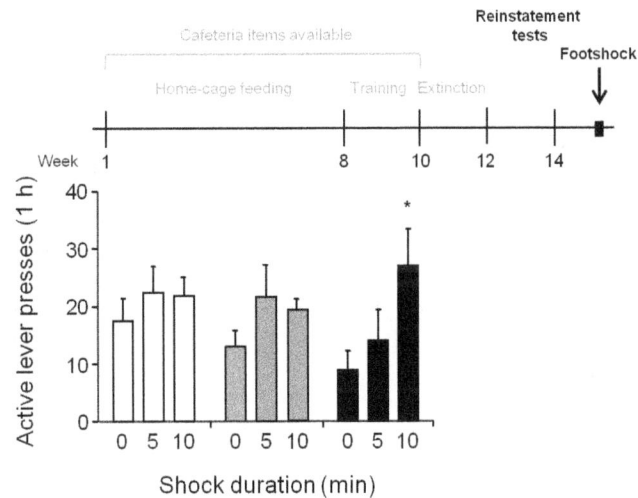

Figure 3. Effect of home-cage diet histories on reinstatement of food seeking. A. Effect of discrete cue presentation on reinstatement. Active lever responding (mean±SEM) under extinction (with cue) conditions following a single non-contingent cue presentation at the start of the session. *Different between the last day of extinction (without cue) and cue-test within subject, $p < 0.05$. # Different in overall lever pressing from chow-only group, $p = 0.06$. **B.** Effect of pellet priming on reinstatement. Active lever responding (mean±SEM) under extinction conditions (with cue) following 1, 2, and 4 non-contingent pellet deliveries at the beginning of an extinction session. *Different between 0 pellet baseline (last one hour

extinction session) and respective pellet test within subject, p<0.05. [#] Different in overall lever pressing from chow-only group, p<0.05. **C.** Effect of intermittent footshock on reinstatement. Active lever responding (mean±SEM) under extinction conditions (with cue) following 5 min and 10 min of intermittent footshock.

responding showed a significant difference between 0 min and 10 min shock (p<0.05). A similar analysis of the intermittent cafeteria group failed to show a significant interaction. Thus, in contrast to discrete cue or pellet-priming, the daily cafeteria diet group appears to be more sensitive to intermittent footshock-induced reinstatement than the intermittent cafeteria or chow-only groups.

Discussion

In the present study, we examined reinstatement of food seeking in a more clinically relevant rat model, which used unrestricted female rats that became overweight by prolonged exposure to a cafeteria diet. In agreement with previous reports [34,35], we found that daily exposure to cafeteria diet leads to increased caloric intake and significant weight gain. This diet manipulation, however, led to decreased concurrent self-administration of normally preferred food pellets. After cessation of diet exposure in the home-cage and removal of the preferred food pellets from the operant chambers during the extinction and reinstatement phases, rats with prior history of daily cafeteria diet, and to a lesser degree prior history of intermittent cafeteria diet, showed decreased extinction responding, and cue- and pellet-priming-induced reinstatement. In contrast, modest stress-induced reinstatement was only observed in rats with a history of daily cafeteria diet. Another finding in our study is that, independent of the history of home-cage diet, female rats with free access to standard chow not only demonstrated escalation in time-out responding during pellet self-administration, but also demonstrated robust cue- and pellet-priming-induced reinstatement. The increase in time-out responding is interesting in light of studies using drug or food reinforcers, which have suggested that lever pressing during the non-reinforced period reflects a compulsive reward seeking behavior that may predict the "addiction-prone" phenotype [36,37]. These data extend previous reinstatement studies in food-restricted male or female rats during the extinction and reinstatement phases [15,16] by examining these behaviors in unrestricted female rats.

Female rats with *ad libitum* access to cafeteria diet alongside standard chow show clinically relevant food intake patterns

The present home-cage food access procedures may provide a more clinically relevant picture of food intake prior to and during dieting than is typical of food reinstatement studies [15]. Initially, humans with unhealthy eating habits consume limited amounts of nutritionally balanced foods and greater amounts of calorically-dense palatable foods. Then during dieting human subjects not only restrict the amount of calorically-dense food they eat but also tend to consume nutritionally balanced foods. Traditional home-cage food restriction in reinstatement studies is an experimenter-administered manipulation that forces the rat to consume less nutritionally balanced food in order to accommodate for palatable food consumption. The current procedures (chow and cafeteria diet freely available) allowed the rats to choose how much nutritionally balanced and calorically-dense food to consume. Under these conditions, rats consumed less nutritionally balanced food and more palatable food (in the home-cage), a pattern similar to the unhealthy eating habits observed in humans prior to dieting.

Then during extinction and reinstatement, rats fed cafeteria diets were switched to a chow-only diet, a qualitative manipulation that more closely models human dieting than the maintenance of chow restriction commonly used in reinstatement studies.

The current procedures may prove valuable in revealing the involvement of neural mechanisms in reinstatement that are qualitatively different in overweight rats as compared to normal weight rats. Studies show that chronic access to a cafeteria diet leads to metabolic and immune dysfunctions commonly associated with human obesity [34,38]. Furthermore, unhealthy diet histories cause dysregulations in reward processing and stress systems [20,21], both of which are critical in controlling relapse behavior [16,23].

Female rats with cafeteria diet history show modest cue- and pellet-priming-induced reinstatement

We found that rats with prior daily and intermittent cafeteria diet history showed relatively modest discrete cue- and pellet-priming-induced reinstatement effects as compared to rats fed with chow only. This is perhaps not surprising, since both cafeteria groups demonstrated low levels of operant pellet self-administration during the training and extinction phases. These effects may be explained by the successive negative contrast effect, which describes a change in animals' behavior following a downshift in the qualitative or quantitative value of an expected reward [39] and is often observed in the context of instrumental settings when food pellets are used as reward [40,41]. Indeed, since rats compare the present reward with their previous experiences and respond according to its relative value [42], rats that previously had access to highly palatable cafeteria items may not perceive the operant pellet to be as rewarding as the rats fed chow-only.

If the value of outcome is lower for rats with cafeteria diet histories, this might explain the reduced motivation during self-administration and pellet-priming reinstatement testing. Previous work has shown that rats with access to sucrose during adolescence show a lower motivation to self-administer saccharin during adulthood [43]. Further, this less effective reinforcer would presumably support less incentive salience attribution to the pellet-associated cues [44], which may reduce the cue's effectiveness at reinstating behavior. The possible influence of such negative contrast effects in blunting motivation in the operant setting should be carefully considered in studies that seek to make rats overweight using palatable diets prior to assessing motivation for other natural and drug rewards.

Exposure to the daily cafeteria diet may also lead to a depressive-like, anhedonic state, which could contribute to the lack of responding during reinstatement tests observed in cafeteria rats. Exposure to energy dense diets elevates brain stimulation reward threshold in rats [35], and also increases depression-related behaviors in forced swim and sucrose preference tests in mice [45,46]. Withdrawal from palatable food, as experienced by the daily cafeteria rats during the extinction and testing phases of the present study, also leads to depressive-like behaviors in rats using similar diet manipulations [47]. Other studies have found that naltrexone and social defeat, both of which cause anhedonia [48,49], either suppress responding in reinstatement testing [50] or reduce lever responding during extinction [51]. Thus, an anhedonic state associated with withdrawal from cafeteria diets may contribute to the blunted reinstatement effects observed in the present study, another important consideration for future studies.

Female rats with prior cafeteria diet history show modest stress-induced reinstatement

Although rats that previously had daily access to cafeteria diet consistently showed less lever responding during training, extinction and cue- and pellet-priming-induced reinstatement, they were particularly sensitive to stress-induced reinstatement. Indeed, while chow-only rats did not respond to footshock-induced reinstatement of food seeking, a pattern similar to previous findings [52,53], the daily cafeteria group significantly increased lever pressing after 10 min of intermittent footshock. This heightened sensitivity to stress is consistent with studies in rodents showing that extended access to and withdrawal from palatable food causes adaptations in stress circuitry [22,54–57]. In the present study, rats previously fed cafeteria diet, which restrict intake of chow once palatable food is no longer available, show greater sensitivity to stress-induced reinstatement, a finding that parallels the clinical observation that restraint eaters have a greater tendency to overeat when encountering a stressful situation [58,59]. Future studies are needed to determine the underlying mechanisms and related physiological changes of this diet-induced hypersensitivity to stress. For example, while studies have shown that access to or removal of cafeteria diet did not affect baseline corticosterone levels [56], diet history can affect how stress circuitry responds to acute stressors [60].

The stress-effect in the cafeteria group should, however, be interpreted with caution. First, the amount of lever pressing observed in the cafeteria group after footshock, while significantly different than baseline responding, is not as robust as the cue and pellet priming reinstatement effects reported here and footshock stress effect reported previously with heroin and cocaine [31,32,61]. Second, all rats were repeatedly tested across multiple reinstating stimuli, and footshock-induced reinstatement testing was conducted last. However, repeated testing is unlikely to confound our interpretations, since all rats underwent additional extinction sessions before footshock testing, and the baseline lever-pressing for the cafeteria group remained consistent across all reinstatement testing. Similar repeated testing procedures have also been used in various reinstatement studies and reliable results are obtained [25,62–68].

Finally, footshock-induced reinstatement testing occurred 4–5 weeks after the cessation of cafeteria diets. At the time of footshock testing both body weight and caloric intake differences between cafeteria and chow-only groups had disappeared. When examining the impact of historical manipulations such as exposure to or withdrawal from cafeteria diets, it can be argued that it would be ideal to do so at an earlier time point than that used in the present study. However, similar withdrawal periods have been used in drug studies where footshock-induced reinstatement effects were found [31,32,61]. Further, the stress sensitivity observed here could be time-dependent, as is the incubation of drug or sucrose craving [69]. Future studies are needed to determine the time-course of this potentially enhanced sensitivity to stress in cafeteria rats.

Conclusions

Here, we demonstrate that it is possible to study reinstatement of food seeking in a clinically relevant model that makes rats overweight by giving them access to unhealthy foods commonly consumed by humans. Female rats serve as suitable subjects for reinstatement studies, as under the current procedures they demonstrated robust self-administration responding and reinstatement effects without the need for food restriction. Studies have shown that chronic caloric restriction can cause long-lasting changes in the stress- and feeding-related pathways [54], and the level of food restriction may contribute to the differential effect of pharmacological agents on reinstatement of food seeking [15]. A potential concern when using intact female rats is that fluctuations in ovarian hormones across the estrous cycle can influence reinstatement behavior [64,70,71]. In the present study we did not directly track estrous cycle phase, and thus we cannot rule out the possibility that cafeteria diet histories influence the estrous cycle, which might contribute to observed differences in reinstatement behavior between groups. Nevertheless, a previous study has shown that ovarian hormones and estrous cycle phase play relatively insignificant roles in reinstatement of palatable food seeking [72]. Together with a recent study showing that the behavior of female rodents is no more intrinsically variable than that of male rodents, as commonly assumed [73], our results provide a useful procedural validation for future studies that intend to study reinstatement under conditions that more accurately model clinical observations in individuals seeking treatment for relapse to unhealthy eating during dieting.

Acknowledgments

The work was supported by the Intramural Research Program of the National Institute on Drug Abuse. We thank Dr. Yavin Shaham for his insightful comments on the manuscript.

Author Contributions

Conceived and designed the experiments: YWC DJC. Performed the experiments: YWC KAF SZB. Analyzed the data: YWC KAF. Contributed to the writing of the manuscript: YWC KAF DJC.

References

1. Swinburn B, Sacks G, Ravussin E (2009) Increased food energy supply is more than sufficient to explain the US epidemic of obesity. Am J Clin Nutr 90: 1453–1456.

2. Swinburn BA, Sacks G, Lo SK, Westerterp KR, Rush EC, et al. (2009) Estimating the changes in energy flux that characterize the rise in obesity prevalence. Am J Clin Nutr 89: 1723–1728.

3. Kramer FM, Jeffery RW, Forster JL, Snell MK (1989) Long-term follow-up of behavioral treatment for obesity: patterns of weight regain among men and women. Int J Obes 13: 123–136.

4. Peterson CB, Mitchell JE (1999) Psychosocial and pharmacological treatment of eating disorders: a review of research findings. J Clin Psychol 55: 685–697.

5. Skender ML, Goodrick GK, Del Junco DJ, Reeves RS, Darnell L, et al. (1996) Comparison of 2-year weight loss trends in behavioral treatments of obesity: diet, exercise, and combination interventions. J Am Diet Assoc 96: 342–346.

6. McGuire MT, Wing RR, Klem ML, Lang W, Hill JO (1999) What predicts weight regain in a group of successful weight losers? J Consult Clin Psychol 67: 177–185.

7. Grilo CM, Shiffman S, Wing RR (1989) Relapse crises and coping among dieters. J Consult Clin Psychol 57: 488–495.

8. Kayman S, Bruvold W, Stern JS (1990) Maintenance and relapse after weight loss in women: behavioral aspects. Am J Clin Nutr 52: 800–807.

9. Polivy J, Herman CP (1999) Distress and eating: why do dieters overeat? Int J Eat Disord 26: 153–164.

10. Torres SJ, Nowson CA (2007) Relationship between stress, eating behavior, and obesity. Nutrition 23: 887–894.

11. Ghitza UE, Gray SM, Epstein DH, Rice KC, Shaham Y (2006) The anxiogenic drug yohimbine reinstates palatable food seeking in a rat relapse model: a role of CRF(1) receptors. Neuropsychopharmacology 31: 2188–2196.

12. See RE (2002) Neural substrates of conditioned-cued relapse to drug-seeking behavior. Pharmacol Biochem Behav 71: 517–529.

13. Self DW, Nestler EJ (1998) Relapse to drug-seeking: neural and molecular mechanisms. Drug Alcohol Depend 51: 49–60.

14. Shaham Y, Shalev U, Lu L, De Wit H, Stewart J (2003) The reinstatement model of drug relapse: history, methodology and major findings. Psychopharmacology 168: 3–20.

15. Calu DJ, Chen YW, Kawa AB, Nair SG, Shaham Y (2014) The use of the reinstatement model to study relapse to palatable food seeking during dieting. Neuropharmacology 76 Pt B: 395–406.

16. Nair SG, Adams-Deutsch T, Epstein DH, Shaham Y (2009) The neuropharmacology of relapse to food seeking: methodology, main findings, and comparison with relapse to drug seeking. Progress in neurobiology 89: 18–45.

17. Pillitteri JL, Shiffman S, Rohay JM, Harkins AM, Burton SL, et al. (2008) Use of dietary supplements for weight loss in the United States: results of a national survey. Obesity (Silver Spring) 16: 790–796.

18. Davy SR, Benes BA, Driskell JA (2006) Sex differences in dieting trends, eating habits, and nutrition beliefs of a group of midwestern college students. J Am Diet Assoc 106: 1673–1677.

19. Nair SG, Navarre BM, Cifani C, Pickens CL, Bossert JM, et al. (2011) Role of dorsal medial prefrontal cortex dopamine D1-family receptors in relapse to high-fat food seeking induced by the anxiogenic drug yohimbine. Neuropsychopharmacology 36: 497–510.

20. Kenny PJ (2011) Reward mechanisms in obesity: new insights and future directions. Neuron 69: 664–679.

21. Parylak SL, Koob GF, Zorrilla EP (2011) The dark side of food addiction. Physiol Behav 104: 149–156.

22. Cottone P, Sabino V, Roberto M, Bajo M, Pockros L, et al. (2009) CRF system recruitment mediates dark side of compulsive eating. Proc Natl Acad Sci U S A 106: 20016–20020.

23. Bossert JM, Marchant NJ, Calu DJ, Shaham Y (2013) The reinstatement model of drug relapse: recent neurobiological findings, emerging research topics, and translational research. Psychopharmacology 229: 453–476.

24. Nair SG, Adams-Deutsch T, Pickens CL, Smith DG, Shaham Y (2009) Effects of the MCH1 receptor antagonist SNAP 94847 on high-fat food-reinforced operant responding and reinstatement of food seeking in rats. Psychopharmacology 205: 129–140.

25. Pickens CL, Cifani C, Navarre BM, Eichenbaum H, Theberge FR, et al. (2012) Effect of fenfluramine on reinstatement of food seeking in female and male rats: implications for the predictive validity of the reinstatement model. Psychopharmacology 221: 341–353.

26. Grimm JW, See RE (2000) Dissociation of primary and secondary reward-relevant limbic nuclei in an animal model of relapse. Neuropsychopharmacology 22: 473–479.

27. See RE, Grimm JW, Kruzich PJ, Rustay N (1999) The importance of a compound stimulus in conditioned drug-seeking behavior following one week of extinction from self-administered cocaine in rats. Drug Alcohol Depend 57: 41–49.

28. Ghitza UE, Nair SG, Golden SA, Gray SM, Uejima JL, et al. (2007) Peptide YY3-36 decreases reinstatement of high-fat food seeking during dieting in a rat relapse model. J Neurosci 27: 11522–11532.

29. Deroche V, Le Moal M, Piazza PV (1999) Cocaine self-administration increases the incentive motivational properties of the drug in rats. Eur J Neurosci 11: 2731–2736.

30. Lu L, Grimm JW, Dempsey J, Shaham Y (2004) Cocaine seeking over extended withdrawal periods in rats: different time courses of responding induced by cocaine cues versus cocaine priming over the first 6 months. Psychopharmacology 176: 101–108.

31. Shalev U, Morales M, Hope B, Yap J, Shaham Y (2001) Time-dependent changes in extinction behavior and stress-induced reinstatement of drug seeking following withdrawal from heroin in rats. Psychopharmacology 156: 98–107.

32. Shaham Y, Stewart J (1995) Stress reinstates heroin-seeking in drug-free animals: an effect mimicking heroin, not withdrawal. Psychopharmacology 119: 334–341.

33. Shaham Y, Rajabi H, Stewart J (1996) Relapse to heroin-seeking in rats under opioid maintenance: the effects of stress, heroin priming, and withdrawal. J Neurosci 16: 1957–1963.

34. Sampey BP, Vanhoose AM, Winfield HM, Freemerman AJ, Muehlbauer MJ, et al. (2011) Cafeteria diet is a robust model of human metabolic syndrome with liver and adipose inflammation: comparison to high-fat diet. Obesity 19: 1109–1117.

35. Johnson PM, Kenny PJ (2010) Dopamine D2 receptors in addiction-like reward dysfunction and compulsive eating in obese rats. Nat Neurosci 13: 635–641.

36. Deroche-Gamonet V, Belin D, Piazza PV (2004) Evidence for addiction-like behavior in the rat. Science 305: 1014–1017.

37. Nair SG, Gray SM, Ghitza UE (2006) Role of food type in yohimbine- and pellet-priming-induced reinstatement of food seeking. Physiol Behav 88: 559–566.

38. Rolls BJ, Rowe EA, Turner RC (1980) Persistent obesity in rats following a period of consumption of a mixed, high energy diet. J Physiol 298: 415–427.

39. Crespi LP (1942) Quantitative variation of incentive and performance in the white rat. American Journal of Psychology 55: 467–517.

40. Mitchell EN, Marston HM, Nutt DJ, Robinson ES (2012) Evaluation of an operant successive negative contrast task as a method to study affective state in rodents. Behav Brain Res 234: 155–160.

41. Sastre A, Reilly S (2006) Excitotoxic lesions of the gustatory thalamus eliminate consummatory but not instrumental successive negative contrast in rats. Behav Brain Res 170: 34–40.

42. Flaherty CF (1996) Incentive relativity. Cambridge, UK Cambridge University Press.

43. Vendruscolo LF, Gueye AB, Darnaudery M, Ahmed SH, Cador M (2010) Sugar overconsumption during adolescence selectively alters motivation and reward function in adult rats. PLoS One 5: e9296.

44. Berridge KC, Robinson TE (1998) What is the role of dopamine in reward: hedonic impact, reward learning, or incentive salience? Brain Res Brain Res Rev 28: 309–369.

45. Sharma S, Fulton S (2013) Diet-induced obesity promotes depressive-like behaviour that is associated with neural adaptations in brain reward circuitry. Int J Obes (Lond) 37: 382–389.

46. Yamada N, Katsuura G, Ochi Y, Ebihara K, Kusakabe T, et al. (2011) Impaired CNS leptin action is implicated in depression associated with obesity. Endocrinology 152: 2634–2643.

47. Iemolo A, Valenza M, Tozier L, Knapp CM, Kornetsky C, et al. (2012) Withdrawal from chronic, intermittent access to a highly palatable food induces depressive-like behavior in compulsive eating rats. Behav Pharmacol 23: 593–602.

48. West TE, Wise RA (1988) Effects of naltrexone on nucleus accumbens, lateral hypothalamic and ventral tegmental self-stimulation rate-frequency functions. Brain Res 462: 126–133.

49. Donahue RJ, Muschamp JW, Russo SJ, Nestler EJ, Carlezon WA Jr (2014) Effects of Striatal DeltaFosB Overexpression and Ketamine on Social Defeat Stress-Induced Anhedonia in Mice. Biol Psychiatry.

50. Stewart J, Wise RA (1992) Reinstatement of heroin self-administration habits: morphine prompts and naltrexone discourages renewed responding after extinction. Psychopharmacology 108: 79–84.

51. Funk D, Harding S, Juzytsch W, Le AD (2005) Effects of unconditioned and conditioned social defeat on alcohol self-administration and reinstatement of alcohol seeking in rats. Psychopharmacology 183: 341–349.

52. Ahmed SH, Koob GF (1997) Cocaine- but not food-seeking behavior is reinstated by stress after extinction. Psychopharmacology 132: 289–295.

53. Buczek Y, Le AD, Wang A, Stewart J, Shaham Y (1999) Stress reinstates nicotine seeking but not sucrose solution seeking in rats. Psychopharmacology 144: 183–188.

54. Pankevich DE, Teegarden SL, Hedin AD, Jensen CL, Bale TL (2010) Caloric restriction experience reprograms stress and orexigenic pathways and promotes binge eating. J Neurosci 30: 16399–16407.

55. Teegarden SL, Bale TL (2007) Decreases in dietary preference produce increased emotionality and risk for dietary relapse. Biol Psychiatry 61: 1021–1029.

56. Martire SI, Maniam J, South T, Holmes N, Westbrook RF, et al. (2014) Extended exposure to a palatable cafeteria diet alters gene expression in brain regions implicated in reward, and withdrawal from this diet alters gene expression in brain regions associated with stress. Behav Brain Res 265: 132–141.

57. South T, Westbrook F, Morris MJ (2012) Neurological and stress related effects of shifting obese rats from a palatable diet to chow and lean rats from chow to a palatable diet. Physiol Behav 105: 1052–1057.

58. Greeno CG, Wing RR (1994) Stress-induced eating. Psychol Bull 115: 444–464.

59. Heatherton TF, Herman CP, Polivy J (1991) Effects of physical threat and ego threat on eating behavior. J Pers Soc Psychol 60: 138–143.

60. Tannenbaum BM, Brindley DN, Tannenbaum GS, Dallman MF, McArthur MD, et al. (1997) High-fat feeding alters both basal and stress-induced hypothalamic-pituitary-adrenal activity in the rat. Am J Physiol 273: E1168–1177.

61. Erb S, Shaham Y, Stewart J (1996) Stress reinstates cocaine-seeking behavior after prolonged extinction and a drug-free period. Psychopharmacology 128: 408–412.

62. De Vries TJ, Schoffelmeer AN, Binnekade R, Raaso H, Vanderschuren LJ (2002) Relapse to cocaine- and heroin-seeking behavior mediated by dopamine D2 receptors is time-dependent and associated with behavioral sensitization. Neuropsychopharmacology 26: 18–26.

63. de Wit H, Stewart J (1981) Reinstatement of cocaine-reinforced responding in the rat. Psychopharmacology 75: 134–143.

64. Fuchs RA, Evans KA, Mehta RH, Case JM, See RE (2005) Influence of sex and estrous cyclicity on conditioned cue-induced reinstatement of cocaine-seeking behavior in rats. Psychopharmacology 179: 662–672.

65. Self DW, Barnhart WJ, Lehman DA, Nestler EJ (1996) Opposite modulation of cocaine-seeking behavior by D1- and D2-like dopamine receptor agonists. Science 271: 1586–1589.

66. Shaham Y, Erb S, Leung S, Buczek Y, Stewart J (1998) CP-154,526, a selective, non-peptide antagonist of the corticotropin-releasing factor1 receptor attenuates stress-induced relapse to drug seeking in cocaine- and heroin-trained rats. Psychopharmacology 137: 184–190.

67. Spealman RD, Barrett-Larimore RL, Rowlett JK, Platt DM, Khroyan TV (1999) Pharmacological and environmental determinants of relapse to cocaine-seeking behavior. Pharmacol Biochem Behav 64: 327–336.

68. Weiss F, Martin-Fardon R, Ciccocioppo R, Kerr TM, Smith DL, et al. (2001) Enduring resistance to extinction of cocaine-seeking behavior induced by drug-related cues. Neuropsychopharmacology 25: 361–372.

69. Pickens CL, Airavaara M, Theberge F, Fanous S, Hope BT, et al. (2011) Neurobiology of the incubation of drug craving. Trends Neurosci 34: 411–420.

70. Anker JJ, Carroll ME (2011) Females are more vulnerable to drug abuse than males: evidence from preclinical studies and the role of ovarian hormones. Curr Top Behav Neurosci 8: 73–96.

71. Feltenstein MW, Henderson AR, See RE (2011) Enhancement of cue-induced reinstatement of cocaine-seeking in rats by yohimbine: sex differences and the role of the estrous cycle. Psychopharmacology 216: 53–62.

72. Cifani C, Koya E, Navarre BM, Calu DJ, Baumann MH, et al. (2012) Medial prefrontal cortex neuronal activation and synaptic alterations after stress-induced reinstatement of palatable food seeking: a study using c-fos-GFP transgenic female rats. J Neurosci 32: 8480–8490.

73. Prendergast BJ, Onishi KG, Zucker I (2014) Female mice liberated for inclusion in neuroscience and biomedical research. Neurosci Biobehav Rev 40C: 1–5.

16

DHA Serum Levels Were Significantly Higher in Celiac Disease Patients Compared to Healthy Controls and Were Unrelated to Depression

Nathalie J. M. van Hees[1]*, Erik J. Giltay[2], Johanna M. Geleijnse[3], Nadine Janssen[1], Willem van der Does[1,2,4]

1 Institute of Psychology, Leiden University, Leiden, The Netherlands, **2** Department of Psychiatry, Leiden University Medical Center, Leiden, The Netherlands, **3** Division of Human Nutrition, Wageningen University, Wageningen, The Netherlands, **4** Leiden Institute of Brain and Cognition, Leiden, The Netherlands

Abstract

Objectives: Celiac disease (CD), a genetically predisposed intolerance for gluten, is associated with an increased risk of major depressive disorder (MDD). We investigated whether dietary intake and serum levels of the essential n-3 polyunsaturated fatty acids (PUFA) eicosapentaenoic acid (EPA) and docosahexanoic acid (DHA) found in fatty fish play a role in this association.

Methods: Cross-sectional study in 71 adult CD patients and 31 healthy volunteers, matched on age, gender and level of education, who were not using n-3 PUFA supplements. Dietary intake, as assessed using a 203-item food frequency questionnaire, and serum levels of EPA and DHA were compared in analyses of covariance, adjusting for potential confounders. Serum PUFA were determined using gas chromatography.

Results: Mean serum DHA was significantly higher in CD patients (1.72 mass%) than controls (1.28 mass%) after multivariable adjustment (mean diff. 0.45 mass%; 95% CI: 0.22–0.68; $p = 0.001$). The mean intake of EPA plus DHA did not differ between CD patients and controls after multivariable adjustment (0.15 and 0.22 g/d, respectively; $p = 0.10$). There were no significant differences in intake or serum levels of EPA and DHA between any of the CD patient groups (never depressed, current MDD, minor/partially remitted MDD, remitted MDD) and controls.

Conclusions: Patients on a long term gluten-free diet had similar intakes of EPA plus DHA compared to controls. Contrary to expectations, DHA serum levels were significantly higher in CD patients compared to healthy controls and were unrelated to MDD status.

Editor: Kenji Hashimoto, Chiba University Center for Forensic Mental Health, Japan

Funding: This project was funded by a grant from the Netherlands Science Foundation (N.W.O-MaGW Vici grant #453-06-005 to WVDD) (http://www.nwo.nl/en), and grants from the Leiden University Fund (LUF) and Gratama Foundation (http://www.luf.nl/subsidies/information-in-english). The funders had no role in study design, data collection and analysis, decision to publish, or preparation of the manuscript.

Competing Interests: The authors have declared that no competing interests exist.

* E-mail: heesnjmvan@fsw.leidenuniv.nl

Introduction

Celiac disease (CD) is a genetically predisposed intolerance for gluten that affects approximately 1 in 160 people [1]. CD is caused by an inappropriate enhanced immune response of the T-lymphocytes of the small intestines to gluten peptides. This results in intestinal malabsorption, atrophy of the intestinal villi and chronic inflammation of the jejunal mucosa of the small intestine. There is currently no cure for CD, but a gluten-free diet improves the histopathology as well as symptoms like weight loss, steatorrhea, diarrhea, abdominal distension, and pain [2]. Besides these intestinal problems, CD is associated with an almost doubled prevalence of major depressive disorder (MDD) [3–9]. Its prevalence rate remains high when a gluten-free diet is initiated [10,11], and may even increase after initiation of the gluten-free diet [12–15].

Although the burden of having a chronic disease might be sufficient to cause MDD in some patients nutrient deficiencies due to malabsorption and the mandatory restrictive diet may also contribute. Treated CD patients often obtain restoration of the function and structure of their atrophied intestinal villi which should correct their malabsorption problems [16], but the strict gluten-free diet may induce nutrient deficiencies in itself. The gluten-free diet has been found to be low in micronutrients and fatty acids like iron, calcium, B vitamins, alpha-linolenic acid and arachidonic acid [17–19], and CD patients may avoid high fat meals (including fatty fish) that induces steatorrhea and other intestinal problems. Eicosapentaenoic acid (EPA, 20:5n-3), docosahexaenoic acid (DHA, 22:6n-3) and alpha-linolenic acid (ALA, 18:3n-3) are essential long-chain n-3 polyunsaturated fatty acids (PUFA) that are important components of the human diet. EPA and DHA are found in fatty fish, while ALA is found in green vegetables, nuts (e.g. walnuts), and vegetable oils (e.g. canola and soybean oils). There is only a minor pathway of biosynthesis from the precursor ALA to EPA and DHA with an approximately 10–15% efficiency [20,21], and vegetarians and persons who do not eat fish may depend on this metabolic pathway for their n-3 PUFA. EPA and DHA concentrations in plasma phospholipid have been found to largely reflect dietary intakes of these fatty acids. DHA comprises about 30% of the fatty tissue in the central nervous system [22] and is a precursor to the signaling eicosanoid molecules prostaglandins and leukotrines involved in the regulation of inflammation and microvascular control. EPA and DHA are considered to have anti-inflammatory effects in the human body [23].

There is evidence that an increased dietary intake of DHA and EPA, and possibly ALA, may lower the risk of MDD [24–26]. Also, circulating levels of n-3 PUFA (or their ratio to n-6 unsaturated fatty acids) have been inversely associated with MDD [27,28] and depressive symptoms [29]. Randomized trials with n-3 PUFA supplementation studies have shown mixed results [30–34].

CD is associated with a higher prevalence of MDD [3–9]. Several studies have found that the daily intake of total fat is significantly higher in CD patients. Furthermore, CD patients' total energy intake is significantly lower than that of healthy controls [35–37] but no previous study has analyzed the intake of n-3 PUFA in CD patients or its relationship with MDD. Several paediatric studies suggest that the lipid profile is different in CD patients than in healthy controls, but most did not focus on n-3 [38–40]. A small paediatric study found no significant difference in total serum n-3 fatty acids among 7 patients with active CD, 6 patients in remission and 11 controls, however arachidonic acid to DHA ratio in patients in remission was significantly higher than in controls [41]. In adults, DHA and EPA serum levels were significantly lower than in controls at time of diagnosis and after one year of gluten-free diet treatment [42].

In summary, n-3 fatty acid intake and blood levels seem to be associated with MDD. Some studies suggest an association between CD and circulating n-3 fatty acid levels, but studies are small and mainly done in children. N-3 fatty acid intake has not previously been measured in CD patients. Our aim was to investigate whether dietary intake and serum levels of the essential n-3 polyunsaturated fatty acids (PUFA) eicosapentaenoic acid (EPA) and docosahexanoic acid (DHA), play a role in the association between CD and depression. We hypothesized that the gluten-free diet may cause low EPA and DHA intake and serum levels, resulting in an increased risk of MDD in patients with CD. Therefore, we compared intake and serum levels of EPA and DHA among groups of CD patients with and without MDD, and compared dietary intake of EPA and DHA in CD patients with healthy controls matched on age, gender and level of education.

Methods

Participants

A cohort of CD patients was recruited from the 2,265 participants (age 18–93 y) of a previous survey study [9] performed among adult members of the Dutch Celiac Association (NCV), as shown in a flow chart (Figure 1).

216 CD patients in the regions Leiden and Amsterdam were contacted and screened for assignment to the never depressed, remitted depressed and currently depressed CD patient study conditions. Participants with self-reported depressive symptoms were oversampled in order to obtain equal group sizes. Healthy, never-depressed controls were recruited from the 1,295 partici-

pants of another study that aimed to gather reference data from the general population ('Normquest' study) [43]. Eligible healthy participants in the Leiden region (N = 615) were pre-screened for mood disorders and were matched for age, gender and level of education. Between October 2010 and April 2011, 85 CD patients and 42 controls took part. Written confirmation of CD diagnosis was requested from the treating specialists and was obtained for all but 8 (11.2%) participants. Participants were excluded if they were younger than 18, had ulcerative colitis, Crohn's disease, current chemotherapy or conditions which would make the testing session unreliable or impossible such as severe psychosis, mental retardation, blindness or deafness. CD participants were excluded if they were on the gluten-free diet less than 2 years or had self-reported low adherence to the gluten-free diet. Healthy controls where excluded if they had celiac disease, shared a household or had a 1st degree family relation with a CD patient or had any mood disorder diagnosis on MINI-plus interview [43]. For the current analyses the following subjects were excluded (Figure 1): those with missing data on any main variables (n = 0), bipolar disorder (n = 3), current alcohol abuse (n = 3), current drug abuse (n = 1), not fasting on morning of testing (n = 1), healthy control with lifetime diagnosis on repeated MINI-plus interview of any mood disorder (n = 6), use of fatty acid (i.e., n-3 PUFA) food supplements (n = 11).

Procedure

In accordance with the declaration of Helsinki, this study was reviewed and approved by the Medical Ethics Committee of the Leiden University Medical Centre and all participants provided written informed consent before the start of data collection. Participants had the capacity to consent, as assessed during screening, and there was no surrogate consent procedure. The interview and the blood collection was performed at the Leiden University Medical Center or at a participating general practitioners office in Amsterdam. Six participants were tested at home due to their advanced age or disability. Participants were sent the study information and instructions, the food frequency questionnaire, the Celiac Disease Adherence Test, a Lifestyle and Health questionnaire, and an informed consent form two weeks before the day of testing. Participants were fasting and refrained from smoking in the hour prior to blood sampling. The testing day started with the physical examination and blood collection, after which participants consumed a light breakfast.

Instruments

Psychiatric diagnoses. The Dutch version [44] of the complete Mini International Neuropsychiatric Interview Plus 5.0.0-R (MINI-Plus) was administered [45]. The MINI is a structured clinical diagnostic interview of current and lifetime Axis-I disorders according to the criteria of the Diagnostic and Statistical Manual of Mental Disorders – Fourth Edition (DSM-IV) [46]. A minimal modification was made to the criteria for scoring 'MDD partially in remission', using the criteria suggested by Rush et al. [47]. Participants who had never experienced a mood disorder were placed in the 'never' group. Participants with dysthymia were placed in the MDD groups (n = 1 in the current MDD, and n = 2 remitted MDD group). Participants who were currently suffering from an episode of MDD or dysthymia were placed in the 'current' group. Participants who recently had an episode of MDD or dysthymia but who now had subclinical symptoms were placed in the 'partially remitted' group. Participants who had suffered from MDD, and currently experienced an absence of both sad mood and reduced interest and no more than three of the remaining seven symptoms of MDD for three or more

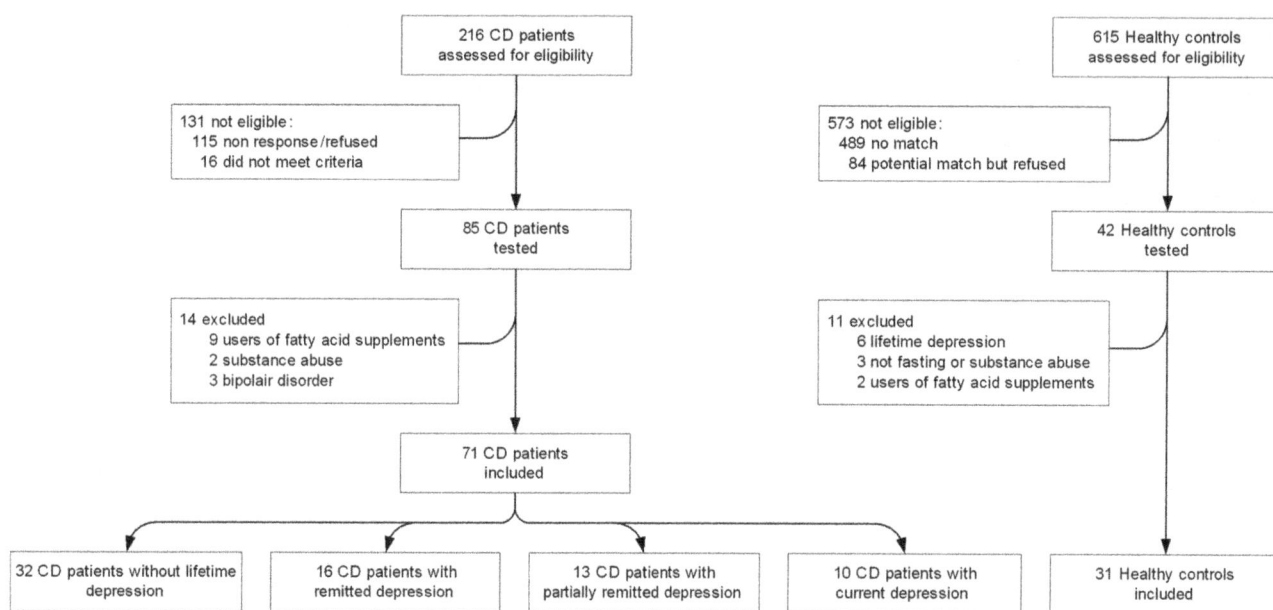

Figure 1. Flow chart of participants in the study.

weeks were placed in the 'remitted' group. All participants were tested and interviewed by two interviewers (from five interviewers in total). At the end of each session both raters tried to reach agreement on all diagnoses. When agreement could not be reached, an intervision meeting was scheduled with the full research team during which consensus was reached.

Food frequency questionnaire. We used an updated version of a validated semi-quantitative food frequency questionnaire that was previously used in epidemiological studies in the Netherlands (45,47,48) The questionnaire covers the 1-month intake of 203 food items and beverages. Information on use of food supplements is also obtained. The questionnaire was sent to the participants and filled in at home. A data check for completeness was performed during the visit to the study center. The food frequency questionnaire was not designed to take the gluten-free diet into account, and therefore additional questions on the ingredients of the gluten-free food products were included. Also, we asked participants to provide the packaging and labels of the gluten-free products that they had used in the past month. Nutrient intake was calculated using the Dutch Food Composition Table ('Nederlands Voedingsstoffenbestand'; NEVO, 2006), which was extended for gluten free products by a dietician [48].

Other variables. Body weight (kg) and height (cm) were measured and body mass index (kg/m^2) was computed. Blood pressure was measured twice after a 5-minute rest, once lying down before breakfast and once sitting up after having breakfast. Physical activity was assessed using the Physical Activity Scale for the Elderly (PASE) [49], to estimate 'metabolic equivalents of task' (MET) minutes. Smoking behavior, alcohol consumption, and self-reported medication use was assessed using questionnaires. Furthermore the nature and method of CD diagnosis were assessed, as well as current and lifetime medical disorders.

Blood sampling and other measures. Fasting venous blood samples were obtained on ice, centrifuged and serum was kept at $-80°C$ within 3 hours after collection. The fatty acids (omega-3, omega-6 and omega-6:omega-3 ratio) from total lipids and high-sensitivity C-reactive protein (hsCRP) were assessed. hsCRP concentrations (mg/L) were measured using nephelome-

try. Fasting serum fatty acids were determined as percentage of total fatty acids by a slightly modified gas chromatographic procedure as described by Lepage and Roy [50]. To 100 μL of serum 15 μL of a 1.0 mg/ml solution of heptadecanoic acid in chloroform/methanol (1:1; v/v) was added as an internal standard and subsequently 2.0 ml of methanol/benzene (4:1; v/v). Then 200 μL of acetylchloride was added slowly and derivatization was performed for 1 h at 100°C. After adding 5 ml of a 6% (w/v) potassium carbonate solution in water and cooling of the mixture it was centrifuged and as much as possible of the benzene upper layer was isolated. This was dried under a gentle nitrogen flow and the residue was taken up in 50 μL of hexane. Finally 1 μL of this sample was injected using split-injection (1:20) on a Trace/Focus gaschromatograph (Interscience, Breda, The Netherlands) using a 30 m capillary BPX-70 column (SGE, Ringwood, Australia) and fatty acids were quantified by calculating the peak area ratios of the fatty acids and the internal standard.

Assessment of adherence to gluten-free diet and CD diagnosis. The current level of adherence to the gluten-free diet was assessed with a single self-report question [9] as well as with the Celiac Disease Adherence Test (CDAT) [51]. In addition a series of questions was asked to assess diet history (duration, age at onset, and any diet interruptions).

Statistical Analysis

Group differences were analyzed using chi-squared (χ^2) tests for categorical variables and analysis of variance (ANOVA) for continuous variables. Analysis of covariance (ANCOVA) was used to adjust for age, gender, level of education, BMI, smoking, alcohol use and statin use in model 1. To assess the potential mediation by differences in dietary intake, we additionally adjusted for daily intake of EPA and DHA when analyzing serum levels of n-3 PUFA in model 2. Odds ratios were calculated using logistic regression analysis assessing the risk of MDD according to EPA and DHA intake and serum EPA and DHA levels. Statistical significance was inferred at a two-sided $p<0.05$. Analyses were done with SPSS software (Version 19.0. Armonk, NY: IBM Corp).

Results

Participant Characteristics

No participant had missing data on the main study variables. Patients with CD were on average 54 years old (range 20–86 years) and 76% was female (Table 1). Healthy controls were on average 51 years old (range 22–66 years), 65% was female. Both groups had an above average education level. The CD group (n = 71) comprised 46 participants (65%) who had one or more (up to 4) current psychiatric diagnoses, mainly anxiety disorders. CD patients were engaged in physical activity 60 MET hours per month less than healthy controls ($p = 0.03$). On average, CD patients were maintaining a gluten-free diet for an uninterrupted period of 15.1 years (SD = 11.5), ranging between 2.6 and 52 years. Length of current gluten-free diet did not differ significantly among the 4 depression groups of CD patients. Self-reported diet adherence in our sample could be categorized as 'very strict' in 74% of participants, 'strict' in 24%, and only 1% for 'moderately well' to 'poor'. Diet adherence according to Celiac Disease Adherence Test results was categorized as 'excellent' or 'very good' in 63% of participants, 'very good' to 'fair' in 26% and 'fair' to 'poor' in 11%.

Dietary Intake of Fat and Fatty Acids

Nutrient value tables specifically for the gluten-free diet were used, which did not affect the estimations of EPA plus DHA intake. Table 2 shows a no significant difference in overall intake of fat and fatty acids nor were there significant differences after controlling for covariates. The intake of EPA plus DHA was not significantly different in CD patients compared to controls (mean 0.17 and 0.21 g/d, respectively; $F(1,100) = 1.06$; $p = 0.31$) nor after controlling for covariates (mean 0.15 and 0.22 g/d, respectively; mean diff. 0.073 g/d; 95% CI: −0.015–0.161; $p = 0.10$). The MDD groups did not differ from controls on this variable either. The intakes of total energy, total fat, unsaturated fatty acids, ALA and of EPA plus DHA did not differ significantly between CD patients and controls, nor between the CD depression groups and controls. After controlling for confounders, energy intake seemed to differ among groups, but when comparing the CD patient group as a whole to controls this difference was not significant ($p = 0.67$).

Serum Levels of Fatty Acids

The n-6: n-3 ratio was approximately 17:1 in both groups with no significant difference between CD patients and controls after controlling for confounders. Intake of EPA plus DHA showed a small but significant Pearson correlation to serum EPA as well as to serum DHA concentrations in our total sample (n = 102; r = 0.18, $p = 0.08$ and r = 0.27, $p = 0.007$, respectively). This indicates that fatty fish intake was indeed associated with serum DHA, more so than with serum EPA concentrations, but that it could only explain a small part of its variance. None of the tested serum PUFA levels differed significantly between CD patients (n = 71) and healthy controls (n = 31), except for DHA (table 3). DHA serum level in CD patients was significantly different from controls (mean 1.72 and 1.28 mass%, respectively; $F(1,100) = 15,47$; $p = 0.001$). After controlling for confounders the mean level of DHA in CD patients was on average 1.72 mass%

Table 1. Socio-demographic and medical characteristics in celiac disease patients and matched controls.

	Controls (n = 31)	Celiac disease (n = 71)	P-value*
Age (years)	51.1±13.3	53.9±18.7	0.45
Gender			
- Male	11 (35%)	17 (24%)	0.23
- Female	20 (65%)	54 (76%)	
Level of education			
- Low	7 (22.6%)	18 (25.4%)	0.52
- Intermediate	6 (19.4%)	20 (28.2%)	
- High	18 (58.1%)	33 (46.5%)	
Body mass index (kg/m²)	24.9±3.7	24.6±4.0	0.72
Blood pressure			
- Systolic (mmHg)	120.5±20.3	125.7±21.6	0.25
- Diastolic (mmHg)	72.6±10.5	69.2±10.8	0.14
Statin use	6 (19.4%)	7 (9.9%)	0.19
Current smoker	7 (22.6%)	7 (9.9%)	0.09
Alcohol intake			
- No	10 (32.3%)	31 (43.7%)	0.28
- 1–2 glasses/d	14 (45.2%)	32 (45.1%)	
- ≥2 glasses/d	7 (22.6%)	8 (11.3%)	
Number comorbid diseases	1.0 (0.0–2.0)	2.0 (1.0–3.0)	0.02
hsCRP (mg/L)	0.94 (0.50–2.31)	0.95 (0.31–2.13)	0.11
Physical activity (MET hours/week)	44.5±35.1	30.6±25.1	0.03

Data are presented as n (%), mean (± SD) or median (Q₁–Q₃), when appropriate. ALA denotes alpha-linolenic acid; DHA, docosahexaenoic acid; EPA, Eicosapentaenoic acid; hsCRP, High-sensitivity C-reactive protein; MET, metabolic equivalents of task.
*P-values by chi-squared test for categorical variables and by ANOVA for continuous variables.

Table 2. Daily dietary intakes in 31 controls and 71 patients with celiac disease with and without depression.

Dietary intake	Controls (n = 31)	Patients with celiac disease				P-value*
		Never MDD (n = 32)	Remitted MDD (n = 16)	Partially remitted MDD (n = 13)	Current MDD (n = 10)	
Total energy (kcal/d)	1979±98	2025±102	1798±153	1872±149	1815±165	0.64
Total fat (g/d)	72.9±4.7	73.2±5.1	66.5±6.2	76.5±7.2	70.7±7.9	0.89
Unsaturated fatty acids (g/d)	14.0±1.8	16.3±2.0	12.5±1.3	15.5±2.1	14.3±2.0	0.61
ALA (g/d)	1.03±0.08	1.23±0.16	1.09±0.12	1.08±0.11	1.05±0.15	0.78
EPA plus DHA (g/d)	0.17±0.03	0.20±0.03	0.15±0.03	0.32±0.08	0.20±0.08	0.18

Data are presented as mean (± SE), when appropriate. ALA denotes alpha-linolenic acid; DHA, docosahexaenoic acid; EPA, eicosapentaenoic acid; MDD, major depressive disorder.
*P-values by ANOVA for continuous variables; adjusted for gender, age, education, BMI, smoking, alcohol use, and statin use.

versus 1.28 mass% for controls (mean diff. 0.45 mass%; 95% CI: 0.22–0.68 $p<0.001$). This difference remained significant after controlling for dietary intake of EPA and DHA as an additional covariate (mean diff. 0.39 mass%; 95% CI: 0.17–0.62 $p<0.001$), where intake of EPA and DHA contributed significantly to the regression model and explained 8% of the variance and group membership explained 12% of the variance.

In post-hoc tests, we compared the mean serum levels of DHA between the 4 CD groups using ANCOVA adjusting for covariates. We did not find a significant difference in serum levels among CD depression categories (figure 2). When comparing CD patients with current and partially remitted MDD (cases) versus CD patients with no or remitted MDD (controls), continuous EPA levels were not associated with a higher risk of MDD with an odds ratio of 0.90 (95% CI: 0.44–1.85; $p=0.77$) nor were DHA levels with an odds ratio of 1.33 (95% CI: 0.58–3.08; $p=0.50$).

To take the influence of possible current inflammation into account we performed a sensitivity analysis where we additionally adjusted our multivariate model for log-transformed hsCRP levels, which did not alter the results for serum EPA and DHA levels (data not shown). In another sensitivity analysis, we excluded the 6 CD participants for whom the potential confirmed CD diagnosis could not be retrieved. Again this did not alter our results for serum EPA and DHA levels (data not shown). We investigated the relationship between gluten-free diet characteristics and DHA serum levels and DHA intake. Adherence to the gluten-free diet as measured by the Celiac Disease Adherence Test did not predict DHA serum levels ($p=0.27$), nor DHA intake ($p=0.72$). Length of gluten-free diet did not predict DHA serum levels ($p=0.77$) or DHA intake ($p=0.70$).

Discussion

Our study showed that treated CD patients had a higher serum DHA level than healthy controls. This does not seem to reflect an increased intake of EPA and DHA by CD patients on a gluten-free

Table 3. Serum fatty acid content in 31 controls and 71 patients with celiac disease with and without depression.

	Controls (n = 31)	Patients with celiac disease				P-value*
		Never MDD (n = 32)	Remitted MDD (n = 16)	Partially remitted MDD (n = 13)	Current MDD (n = 10)	
Total fatty acids	10.01±2.06	9.47±2.14	9.86±1.52	9.86±1.76	10.27±2.56	.62
C12:0 (Lauric A)	0.15±0.01	0.13±0.01	0.15±0.02	0.10±0.01	0.12±0.02	.16
C14:0 (Myrisitic A)	1.26±0.06	1.19±0.06	1.22±0.11	0.97±0.08	1.21±0.16	.45
C16:0 (Palmitic A)	25.06±0.34	24.34±0.25	24.54±0.60	24.20±0.38	24.24±0.77	.48
C16:1. n-7 (Palmitoleic A)	2.41±0.12	2.47±0.16	2.17±0.14	1.99±0.16	2.13±0.31	.25
C18:0 (Stearic A)	6.67±0.11	6.76±0.12	6.71±0.21	6.51±0.23	6.88±0.29	.78
C18:1. n-9 (Oleic A)	20.95±0.50	21.30±0.44	20.63±0.73	20.73±0.70	22.00±0.89	.71
C18:2. n-6 (LA)	28.65±0.79	27.71±0.80	28.79±1.11	29.99±1.04	28.35±1.99	.66
C18:3. n-3 (ALA)	0.56±0.03	0.61±0.05	0.48±0.03	0.57±0.06	0.53±0.06	.35
C20:4. n-6 (AA)	5.89±0.21	6.36±0.21	5.91±0.34	5.69±0.35	5.67±0.42	.34
C20:5. n-3 (EPA)	0.72±0.05	0.92±0.12	0.94±0.18	0.94±0.25	0.79±0.17	.65
C22:5. n-3 (DPA)	0.35±0.02	0.40±0.02	0.36±0.03	0.34±0.04	0.36±0.03	.22
C22:6. n-3 (DHA)	1.28±0.06	1.65±0.09[†]	1.78±0.16[†]	1.87±0.20[†]	1.69±0.18[†]	.003**

Data are (adjusted) mean ± standard errors (SE), total fatty acids in mmol/L, individual fatty acids as a % of total fatty acids. MDD denotes major depressive disorder.
[†]significantly different in post-hoc tests versus the controls.
*P-values by ANOVA for continuous variables; adjusted for gender, age, education, BMI, smoking, alcohol use, and statin use.
**additionally adjusted for daily intake of EPA and DHA.

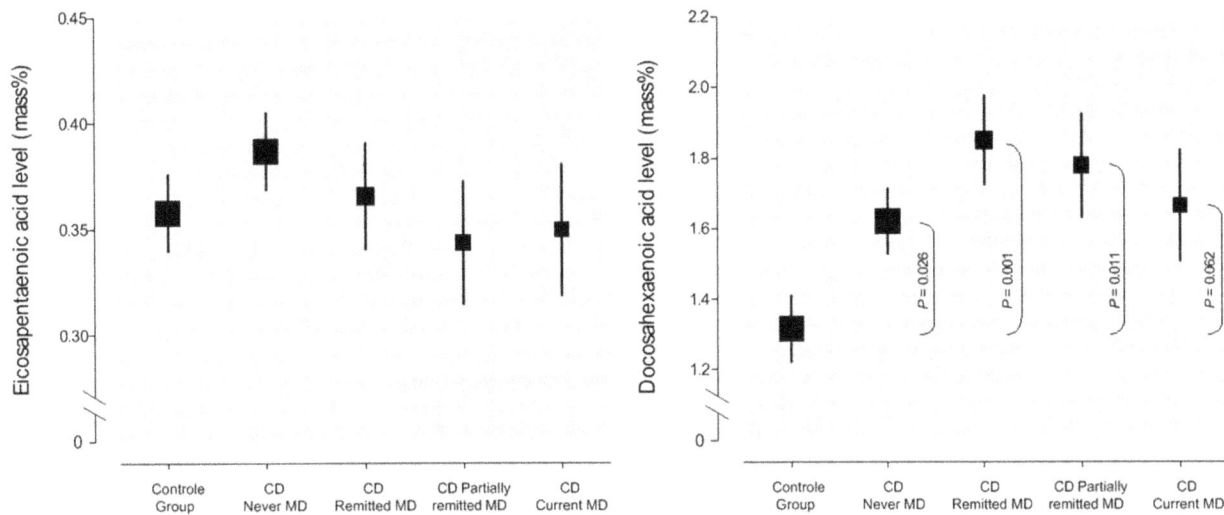

Figure 2. Mean standard scores (with error bars representing standard errors) for plasma levels of doxosahexaenoic acid (DHA) and eicosapentaenoic acid (EPA) in % of total fatty acids. The size of each square is proportional to the number of participants. Mean scores are adjusted for gender, age, education, BMI, smoking, alcohol use, statin use, and daily intake of EPA and DHA. *P*-values by analysis of covariance.

diet, as fish fatty acid intake did not differ significantly among these groups. We found no association between dietary intake or serum levels of EPA or DHA and MDD status within the group of CD patients, and therefore the differences in DHA levels could not explain the differences in occurrence of MDD.

There is evidence that an increased dietary intake of DHA and EPA, and possibly ALA, may lower the risk of MDD [24–26]. Also, circulating levels of n-3 PUFA (or their ratio to n-6 unsaturated fatty acids) have been inversely associated with MDD [27,28] and depressive symptoms [29]. Randomized trials with n-3 PUFA supplementation studies have shown mixed results [30–34]. Some reviews have found lower EPA and DHA plasma levels in depressed patients [30] and a small but significant benefit of EPA and DHA supplementation in MDD patients [31]. Other reviews found no effect of EPA and DHA supplementation on MDD [32] and no beneficial effect of EPA and DHA supplementation on mood in women with perinatal depressive symptoms [33] or subjects not suffering from current MDD [34].

We found increased serum DHA in our CD patients, but no difference in serum EPA, compared to controls which is inconsistent with the literature. Although the literature on n-3 fatty acid metabolism in patients with CD, or chronic diseases in general, is very limited, data suggest that severe malabsorption and chronic gastrointestinal disorder is associated with essential fatty acid deficiencies; in particular linoleic acid and DHA [52,53]. Some studies even propose permanent changes in fatty acid metabolism [52–55]. Studies on n-3 fatty acid plasma levels and fatty acid profiles in CD patients have shown mixed results, in particular between studies done in paediatric [38,39,41] and adult samples [42,56]. All studies however show unfavorable differences in CD patients' fatty acid profiles when comparing them to healthy groups. For example, a study assessing recently diagnosed adult CD patients found that patients' DHA, EPA and arachidonic acid serum levels increased after a one-year strict gluten-free diet but stayed significantly lower than those of controls. Serum arachidonic acid and DHA levels improved most. As the authors propose in their discussion, essential fatty acid concentration may continue to increase after following a gluten-free diet for a longer period of time. [42]. Studies into CD or the gluten-free diet are difficult to compare however since differences in stages of disease activity of

CD, different length of the gluten-free diet and level of adherence to the gluten-free diet need to be taken into account. For example, paediatric patients with active CD had significant signs of essential fatty acid deficiency, but when these patients were in remission and on a gluten-free diet for one year or longer, their DHA levels were not significantly lower than those of controls [41]. In contrast, in our study in CD patients on a long term gluten-free diet (mean 15 years) serum DHA levels were significantly higher than healthy controls. As some authors have previously suggested permanent changes in fatty acid metabolism in chronically ill samples [52–55], we speculate that such a change may have occurred in our sample of CD patients in remission. Possibly through sustained activation of counterbalancing (e.g. anti-inflammatory) processes that help to restore homeostasis, which might have involved the formation of DHA [57–60]. After antigen exposure is eliminated, chronic inflammation might slowly be reduced by anti-inflammatory mechanisms including the increased production of DHA. We hypothesize that the prolonged activation of this process might have resulted in a permanent up-regulation of DHA formation.

An alternative explanation is a change in PUFA intake due to the exclusive nature of the gluten-free diet. Our questionnaire data did however not reveal a significantly different dietary intake of PUFA or total fat between patients and controls. It is therefore less likely that our finding of an elevated DHA serum level is attributable to differences in DHA intake as a result of the gluten-free diet. We also found that using nutrient tables designed for the gluten-free diet did not really alter the outcome of the FFQ on variables involving fatty acids. This leads us to conclude that the gluten-free diet does not really have an impact on main dietary sources of fatty acids. Contrary to our findings most previous studies found an increased intake of total fat in treated CD patients [61–63]. But one study found equal fat intake when comparing female participants only [64]. The lack of difference in energy intake between treated CD patients and healthy controls we found in our study is in line with earlier findings [36,61,62,64]. Some other studies however found a significantly lower intake of energy [65] or higher intake of energy [63,64]. Our findings point to a normalization of fat and energy intake in CD patients who have been living with the gluten-free diet for a long time. Whether this finding is generalizable to samples from other populations remains to be seen.

EPA and DHA supplementation in chronic inflammatory diseases such as rheumatoid arthritis, Crohn's disease and multiple sclerosis may have beneficial effects on disease activity [66], but it is unclear whether this also applies to CD. Because serum EPA and DHA in our CD patients with depressive symptoms were similar to levels in non-depressed CD patients, we consider a beneficial effect of supplemental n-3 PUFA intake less likely. However, well-designed randomized trials in CD patients with MDD are warranted to definitely refute this hypothesis.

There are some limitations that need to be addressed. First, our sample was relatively small. Second, as we did not include patients with high dietary gluten exposure, our findings cannot be extrapolated to CD patients not on a gluten-free diet or with poor gluten-free diet adherence. However, previous research did not show a relationship between level of diet adherence and psychopathology [9,67] nor is a relation likely between very small dietary transgressions and medical symptoms [68]. Third, both interviewers and participants were aware of the purpose of the study giving room to bias in the assessment of MDD. This possible bias was addressed to some extent by having every diagnostic interview observed by a second rater and discussed in intervision.

Clinical Implications and Future Research

DHA serum levels were significantly higher in CD patients on a long term strict gluten-free diet and presumed in remission than in healthy controls, which may reflect alterations in fatty acid metabolism in response to the prolonged period of intestinal inflammation. Within the group of CD patients, we found no association between dietary or serum EPA plus DHA and depression status. Therefore, our findings do not support the hypothesis that supplementation of n-3 PUFA in CD patients after the first years of gluten-free diet is warranted to reduce the risk of MDD. Nevertheless, our findings warrant confirmation by other studies, preferably randomized controlled trials.

Acknowledgments

We gratefully acknowledge Els Siebelink, Susanne Tielemans and Eveline Waterham for their invaluable help in calculating the FFQ gluten-free nutrient tables, as well as Dr. W. Onkenhout for the analysis of the fatty acids. We also acknowledge the contribution of all students who were involved in this project.

Author Contributions

Conceived and designed the experiments: NVH. Performed the experiments: NVH NJ. Analyzed the data: NVH EG NJ. Contributed reagents/materials/analysis tools: JG. Wrote the paper: NVH EG JG NJ WVDD.

References

1. Biagi F, Biagi C, Klersy D, Balduzzi G, Corazza (2010) Are we not over-estimating the prevalence of coeliac disease in the general population? Annals of medicine (Helsinki) 42: 557–561.
2. Rubio-Tapia A, Rahim MW, See JA, Lahr BD, Wu T-T, et al. (2010) Mucosal Recovery and Mortality in Adults With Celiac Disease After Treatment With a Gluten-Free Diet. Am J Gastroenterol 105: 1412–1420.
3. Addolorato G, Leggio L, D'Angelo C, Mirijello A, Ferrulli A, et al. (2008) Affective and psychiatric disorders in celiac disease. Dig Dis 26: 140–148.
4. Addolorato G, Stefanini GF, Capristo E, Caputo F, Gasbarrini A, et al. (1996) Anxiety and depression in adult untreated celiac subjects and in patients affected by inflammatory bowel disease: A personality "trait" or a reactive illness? Hepato-Gastroenterology 43: 1513–1517.
5. Carta MG, Hardoy MC, Boi MF, Mariotti S, Carpiniello B, et al. (2002) Association between panic disorder, major depressive disorder and celiac disease: a possible role of thyroid autoimmunity. J Psychosom Res 53: 789–793.
6. Ciacci C, Iavarone A, Mazzacca G, De Rosa A (1998) Depressive symptoms in adult coeliac disease. Scand J Gastroenterol 33: 247–250.
7. Hernanz A, Polanco I (1991) Plasma precursor amino acids of central nervous system monoamines in children with coeliac disease. Gut 32: 1478–1481.
8. Rostom A, Murray JA, Kagnoff MF (2006) American Gastroenterological Association (AGA) Institute Technical Review on the Diagnosis and Management of Celiac Disease. Gastroenterology 131: 1981–2002.
9. van Hees NJM, Van der Does W, Giltay EJ (2013) Coeliac disease, diet adherence and depressive symptoms. J Psychosom Res 74: 155–160.
10. Addolorato G, Mirijello A, D'Angelo C, Leggio L, Ferrulli A, et al. (2008) State and trait anxiety and depression in patients affected by gastrointestinal diseases: psychometric evaluation of 1641 patients referred to an internal medicine outpatient setting. International journal of clinical practice 62: 1063–1069.
11. Karwautz A, Wagner G, Berger G, Sinnreich U, Grylli V, et al. (2008) Eating pathology in adolescents with celiac disease. Psychosomatics 49: 399–406.
12. Hallert C, Gotthard R, Jansson G, Norrby K, Walan A (1983) Similar prevalence of coeliac disease in children and middle-aged adults in a district of Sweden. Gut 24: 389–391.
13. Corvaglia L, Catamo R, Pepe G, Lazzari R, Corvaglia E (1999) Depression in adult untreated celiac subjects: diagnosis by the pediatrician. Am J Gastroenterol 94: 839–843.
14. Addolorato G, Capristo E, Ghittoni G, Valeri C, Masciana R, et al. (2001) Anxiety but not depression decreases in coeliac patients after one-year gluten-free diet: A longitudinal study. Scandinavian journal of gastroenterology 36: 502–506.
15. Pynnonen P, Isometsa E, Verkasalo M, Savilahti E, Aalberg V (2002) Untreated celiac disease and development of mental disorders in children and adolescents. Psychosomatics 43: 331–334.
16. Abenavoli L, Proietti I, Leggio L, Ferrulli A, Vonghia L, et al. (2006) Cutaneous manifestations in celiac disease. World Journal of Gastroenterology 12: 843–852.
17. Thompson T, Dennis M, Higgins LA, Lee AR, Sharrett MK (2005) Gluten-free diet survey: are Americans with coeliac disease consuming recommended amounts of fibre, iron, calcium and grain foods? Journal of Human Nutrition and Dietetics 18: 163–169.
18. Alvarez-Jubete L, Arendt EK, Gallagher E (2009) Nutritive value and chemical composition of pseudocereals as gluten-free ingredients. International Journal of Food Sciences and Nutrition 60: 240–257.
19. Thompson T (2000) Folate, iron, and dietary fiber contents of the gluten-free diet. J Am Diet Assoc 100: 1389–1396.
20. Emken EA, Rohwedder WK, Adlof RO, Rakoff H, Gulley RM (1987) Metabolism in humans of cis-12, trans-15-octadecadienoic acid relative to palmitic, stearic, oleic and linoleic acids. Lipids 22: 495–504.
21. Emken EA, Adlof RO, Gulley RM (1994) Dietary linoleic acid influences desaturation and acylation of deuterium-labeled linoleic and linolenic acids in young adult males. Biochimica et biophysica acta 1213: 277–288.
22. Innis SM (2000) The role of dietary n-6 and n-3 fatty acids in the developing brain. Developmental neuroscience 22: 474–480.
23. Holub BJ (2002) Clinical nutrition: 4. Omega-3 fatty acids in cardiovascular care. CMAJ: Canadian Medical Association journal = journal de l'Association medicale canadienne 166: 608–615.
24. DeMar JC Jr, Ma K, Bell JM, Igarashi M, Greenstein D, et al. (2006) One generation of n-3 polyunsaturated fatty acid deprivation increases depression and aggression test scores in rats. J Lipid Res 47: 172–180.
25. Freeman MP (2000) Omega-3 fatty acids in psychiatry: a review. Annals of clinical psychiatry: official journal of the American Academy of Clinical Psychiatrists 12: 159–165.
26. Lesperance F, Frasure-Smith N, St-Andre E, Turecki G, Lesperance P, et al. (2011) The efficacy of omega-3 supplementation for major depression: a randomized controlled trial. J Clin Psychiatry 72: 1054–1062.
27. Maes M, Smith R, Christophe A, Cosyns P, Desnyder R, et al. (1996) Fatty acid composition in major depression: decreased omega 3 fractions in cholesteryl esters and increased C20: 4 omega 6/C20: 5 omega 3 ratio in cholesteryl esters and phospholipids. Journal of affective disorders 38: 35–46.
28. Peet M, Horrobin DF (2002) A dose-ranging study of the effects of ethyl-eicosapentaenoate in patients with ongoing depression despite apparently adequate treatment with standard drugs. Arch Gen Psychiatry 59: 913–919.
29. Tiemeier H, van Tuijl HR, Hofman A, Kiliaan AJ, Breteler MM (2003) Plasma fatty acid composition and depression are associated in the elderly: the Rotterdam Study. Am J Clin Nutr 78: 40–46.
30. Lin PY, Huang SY, Su KP (2010) A meta-analytic review of polyunsaturated fatty acid compositions in patients with depression. Biological psychiatry 68: 140–147.
31. Appleton KM, Rogers PJ, Ness AR (2010) Updated systematic review and meta-analysis of the effects of n-3 long-chain polyunsaturated fatty acids on depressed mood. Am J Clin Nutr 91: 757–770.
32. Bloch MH, Hannestad J (2012) Omega-3 fatty acids for the treatment of depression: systematic review and meta-analysis. Molecular psychiatry 17: 1272–1282.

33. Jans LA, Giltay EJ, Van der Does AJ (2010) The efficacy of n-3 fatty acids DHA and EPA (fish oil) for perinatal depression. The British journal of nutrition 104: 1577–1585.

34. Giltay EJ, Geleijnse JM, Kromhout D (2011) Effects of n-3 fatty acids on depressive symptoms and dispositional optimism after myocardial infarction. Am J Clin Nutr 94: 1442–1450.

35. Ferrara P, Cicala M, Tiberi E, Spadaccio C, Marcella L, et al. (2009) High fat consumption in children with celiac disease. Acta Gastroenterol Belg 72: 296–300.

36. Hallert C, Grant C, Grehn S, Granno C, Hulten S, et al. (2002) Evidence of poor vitamin status in coeliac patients on a gluten-free diet for 10 years. Aliment Pharmacol Ther 16: 1333–1339.

37. Mariani P, Viti MG, Montouri M, La Vecchia A, Cipolletta E, et al. (1998) The Gluten-Free Diet: A Nutritional Risk Factor for Adolescents with Celiac Disease? Journal of Pediatric Gastroenterology and Nutrition 27: 519–523.

38. Rey J, Frezal J, Polonovski J, Lamy M (1965) [Modifications of plasma lipids in disorders of intestinal absorption in children]. Rev Fr Etud Clin Biol 10: 488–494.

39. Jaskiewicz J, Szafran H, Kruszewska M, Brylska U, Krol M (1987) The effect of gluten-free diet supplemented with Humana-MCT on the level of lipid fractions in blood serum of infants with coeliac disease. Acta Physiol Pol 38: 22–30.

40. Rosenthal E, Hoffman R, Aviram M, Benderly A, Erde P, et al. (1990) Serum lipoprotein profile in children with celiac disease. J Pediatr Gastroenterol Nutr 11: 58–62.

41. Steel DM, Ryd W, Ascher H, Strandvik B (2006) Abnormal fatty acid pattern in intestinal mucosa of children with celiac disease is not reflected in serum phospholipids. J Pediatr Gastroenterol Nutr 43: 318–323.

42. Solakivi T, Kaukinen K, Kunnas T, Lehtimaki T, Maki M, et al. (2009) Serum fatty acid profile in celiac disease patients before and after a gluten-free diet. Scand J Gastroenterol 44: 826–830.

43. Schulte-van Maaren YW, Carlier IV, Giltay EJ, van Noorden MS, de Waal MW, et al. (2013) Reference values for mental health assessment instruments: objectives and methods of the Leiden Routine Outcome Monitoring Study. Journal of evaluation in clinical practice 19: 342–350.

44. van Vliet IM (2007) The MINI-International Neuropsychiatric Interview. A brief structured diagnostic psychiatric interview for DSM-IV en ICD-10 psychiatric disorders. Tijdschrift voor psychiatrie 49: 393–397.

45. Sheehan DV (1998) The Mini-International Neuropsychiatric Interview (M.I.N.I.): the development and validation of a structured diagnostic psychiatric interview for DSM-IV and ICD-10. The Journal of clinical psychiatry 59 Suppl 20: 22–33; quiz 34–57.

46. American Psychiatric Association APATFoDSMIV (2000) Diagnostic and statistical manual of mental disorders: DSM-IV-TR. Washington, DC: American Psychiatric Association.

47. Rush AJ, Kraemer HC, Sackeim HA, Fava M, Trivedi MH, et al. (2006) Report by the ACNP Task Force on response and remission in major depressive disorder. Neuropsychopharmacology: official publication of the American College of Neuropsychopharmacology 31: 1841–1853.

48. Stichting Nederlands Voedingsstoffenbestand (NEVO foundation). Dutch Food Composition Table 2006. The Netherlands Nutrition Centre, The Hague, The Netherlands.

49. Washburn RA, Smith KW, Jette AM, Janney CA (1993) The Physical Activity Scale for the Elderly (PASE): development and evaluation. J Clin Epidemiol 46: 153–162.

50. Lepage G, Roy CC (1986) Direct transesterification of all classes of lipids in a one-step reaction. Journal of lipid research 27: 114–120.

51. Leffler DA, Dennis M, Edwards George JB, Jamma S, Magge S, et al. (2009) A simple validated gluten-free diet adherence survey for adults with celiac disease. Clinical gastroenterology and hepatology: the official clinical practice journal of the American Gastroenterological Association 7: 530–536, 536 e531–532.

52. Siguel EN, Lerman RH (1996) Prevalence of essential fatty acid deficiency in patients with chronic gastrointestinal disorders. Metabolism: clinical and experimental 45: 12–23.

53. Chambrier C, Garcia I, Bannier E, Gerard-Boncompain M, Bouletreau P (2002) Specific changes in n-6 fatty acid metabolism in patients with chronic intestinal failure. Clinical nutrition 21: 67–72.

54. Färkkilä MA, Tilvis RS, Miettinen TA (1987) Plasma fatty acid composition in patients with ileal dysfunction. Scandinavian journal of gastroenterology 22: 411–419.

55. Solakivi T, Kaukinen K, Kunnas T, Lehtimaki T, Maki M, et al. (2011) Serum fatty acid profile in subjects with irritable bowel syndrome. Scandinavian journal of gastroenterology 46: 299–303.

56. Jakobsdottir G, Jakobsdottir J, Bjerregaard H, Skovbjerg M (2013) Fasting serum concentration of short-chain fatty acids in subjects with microscopic colitis and celiac disease: no difference compared with controls, but between genders. Scandinavian journal of gastroenterology 48: 696–701.

57. Forsberg G, Hernell O, Melgar S, Israelsson A, Hammarstrom S, et al. (2002) Paradoxical coexpression of proinflammatory and down-regulatory cytokines in intestinal T cells in childhood celiac disease. Gastroenterology 123: 667–678.

58. Forsberg G, Hernell O, Hammarstrom S, Hammarstrom ML (2007) Concomitant increase of IL-10 and pro-inflammatory cytokines in intraepithelial lymphocyte subsets in celiac disease. Int Immunol 19: 993–1001.

59. Fornari MC, Pedreira S, Niveloni S, Gonzalez D, Diez RA, et al. (1998) Pre- and post-treatment serum levels of cytokines IL-1beta, IL-6, and IL-1 receptor antagonist in celiac disease. Are they related to the associated osteopenia? Am J Gastroenterol 93: 413–418.

60. Lahat Shapiro, Karban Gerstein, Kinarty, et al. (1999) Cytokine Profile in Coeliac Disease. Scandinavian Journal of Immunology 49: 441–447.

61. Capristo E, Mingrone G, Addolorato G, Greco AV, Corazza GR, et al. (1997) Differences in metabolic variables between adult coeliac patients at diagnosis and patients on a gluten-free diet. Scandinavian journal of gastroenterology 32: 1222–1229.

62. Capristo E, Addolorato G, Mingrone G, De Gaetano A, Greco AV, et al. (2000) Changes in body composition, substrate oxidation, and resting metabolic rate in adult celiac disease patients after a 1-y gluten-free diet treatment. Am J Clin Nutr 72: 76–81.

63. Kemppainen T, Uusitupa M, Janatuinen E, Jarvinen R, Julkunen R, et al. (1995) Intakes of nutrients and nutritional-status in celiac patients. Scandinavian journal of gastroenterology 30: 575–579.

64. Wild D, Robins GG, Burley VJ, Howdle PD (2010) Evidence of high sugar intake, and low fibre and mineral intake, in the gluten-free diet. Alimentary Pharmacology & Therapeutics 32: 573–581.

65. Bardella MT (2000) Body composition and dietary intakes in adult celiac disease patients consuming a strict gluten-free diet. The American journal of clinical nutrition 72: 937–939.

66. Simopoulos AP (2002) Omega-3 fatty acids in inflammation and autoimmune diseases. J Am Coll Nutr 21: 495–505.

67. Leffler DA, Edwards-George J, Dennis M, Schuppan D, Cook F, et al. (2008) Factors that influence adherence to a gluten-free diet in adults with celiac disease. Digestive Diseases and Sciences 53: 1573–1581.

68. Biagi F, Andrealli A, Bianchi PI, Marchese A, Klersy C, et al. (2009) A gluten-free diet score to evaluate dietary compliance in patients with coeliac disease. The British journal of nutrition 102: 882–887.

17

Population Growth of the Cladoceran, *Daphnia magna*: A Quantitative Analysis of the Effects of Different Algal Food

Jong-Yun Choi[1], Seong-Ki Kim[1], Kwang-Hyeon Chang[2], Myoung-Chul Kim[3], Geung-Hwan La[4], Gea-Jae Joo[1], Kwang-Seuk Jeong[1,5]*

1 Department of Biological Sciences, Pusan National University, Busan, Republic of Korea, 2 Departments of Environmental Science and Engineering, Kyung-Hee University, Yongin, Republic of Korea, 3 Institutes of Environmental Ecology, Chemtopia Co. Ltd., Seoul, Republic of Korea, 4 Department of Environmental Education, Suncheon National University, Suncheon, Republic of Korea, 5 Institute of Environmental Science & Technology, Pusan National University, Busan, Republic of Korea

Abstract

In this study, we examined the effects of two phytoplankton species, *Chlorella vulgaris* and *Stephanodiscus hantzschii*, on growth of the zooplankton *Daphnia magna*. Our experimental approach utilized stable isotopes to determine the contribution of food algae to offspring characteristics and to the size of adult *D. magna* individuals. When equal amounts of food algae were provided (in terms of carbon content), the size of individuals, adult zooplankton, and their offspring increased significantly following the provision of *S. hantzschii*, but not after the provision of *C. vulgaris* or of a combination of the two species. Offspring size was unaffected when *C. vulgaris* or a mixture of the two algal species was delivered, whereas providing only *S. hantzschii* increased the production of larger-sized offspring. Stable isotope analysis revealed significant assimilation of diatom-derived materials that was important for the growth of *D. magna* populations. Our results confirm the applicability of stable isotope approaches for clarifying the contribution of different food algae and elucidate the importance of food quality for growth of *D. magna* individuals and populations. Furthermore, we expect that stable isotope analysis will help to further precisely examine the contribution of prey to predators or grazers in controlled experiments.

Editor: Syuhei Ban, University of Shiga Prefecture, Japan

Funding: This research was fully supported by Basic Science Research Program through the National Research Foundation of Korea (NRF) funded by the Ministry of Education (grant number: NRF-2012-R1A6A3A04040793; http://www.nrf.re.kr). The funders had no role in study design, data collection and analysis, decision to publish, or preparation of the manuscript.

Competing Interests: Dr. Myoung-Chul Kim one of the authors of the manuscript is an employee of Chemtopia Co. Ltd. and has acknowledged that Chemtopia Co. Ltd. has no ownership or rights to the data for the aforementioned paper. His company provided no inputs or support for the research or paper writing. In the process of writing the manuscript, he received no financial support for the research or grants from the company Chemtopia Co. Ltd.

* E-mail: kknd.ecoinfo@gmail.com

Introduction

Cladocerans in freshwater ecosystems are among the most important biological entities that contribute to the complexity of food web structure and function [1]. They are typically primary consumers that utilize phytoplankton as their food source. Many species of cladocerans are filter-feeders that obtain food from the water by filtration [2], [3], and are sometimes used as control agents against phytoplankton proliferation in a method known as biomanipulation [4], [5]. Some cladocerans consume organic particles or rotifers [6], [7]; however, most species rely on energy obtained from phytoplankton for population growth.

The nutrient content and biomass of prey phytoplankton are important factors in cladoceran growth [8], [9]. Therefore, the quality and quantity of phytoplankton consumed [10], [11] are crucial factors controlling the growth of cladoceran populations. Previous studies have investigated the size and morphology of algal species (e.g., [12], [13]), and Ahlgren et al. [8] provided comprehensive data on phytoplankton nutritional status. These studies suggest that the quality of prey phytoplankton affects

cladoceran population growth. Urabe and Waki [14] provided evidence that changes in biochemical composition of the diet clearly affected the growth of herbivorous species such as *Daphnia*. However, comparisons of algal growth and composition with cladoceran growth are required to quantify the direct contribution of algal intake to cladocerans.

Even if highly nutritious algae are available, algae do not contribute to growth of individuals or populations unless high assimilation rates are maintained, and the majority of this energy resource will be confined to the gut contents and ultimately ejected. Determining the quantitative contribution of prey to consumers is challenging, and few studies addressed this topic [15], [2], [16] prior to the emergence of stable isotope analysis. Phillips and Koch [17] recommended the isotope mixing model, which enables determination of the contribution of the most abundant algal species to growth of an individual consumer. Although the stable isotope signature does not accurately represent assimilation rate, results obtained using this approach can provide information on the quantitative contribution of prey to the consumer, which can be recognized as the assimilation rate. Stable isotope analysis can be used to explain the contribution of algal species to the population growth of cladocerans.

In this study, we experimentally investigated the relationship between a cladoceran species and its prey phytoplankton from the perspective of the algal contribution to offspring characteristics and size of adult individual zooplankton. Two phytoplankton species, *Chlorella vulgaris* and *Stephanodiscus hantzschii*, were used as food algae; the zooplankton studied was the cladoceran *Daphnia magna*. Offspring characteristics and adult individual size were measured. *Daphnia magna* is one of the most popular herbivorous cladocerans for use in culture experiments and *C. vulgaris* is frequently used in *Daphnia* growth experiments. *Stephanodiscus hantzschii* is an important phytoplankton species, particularly in Far-East Asian countries, where the species proliferates in the winter [18], [19], [20]. Therefore, these species were conducive to understanding the contribution of two algal species to a common grazer. Stable isotope analysis was conducted to quantitatively determine the contribution of food algae to *D. magna*.

Materials and Methods

Plankton Subculture

D. magna obtained from the National Institute of Environmental Research (NIER) of South Korea were grown in Elendt M4 medium [21] in a growth chamber (Eyela FLI-2000, Japan) at 20°C, with a 12L:12D light-dark cycle. Subcultures of the green alga *C. vulgaris* (strain number UMACC 001) and the diatom *S. hantzschii* (strain number CPCC 267) were maintained in a growth chamber (Eyela FLI-301N, Japan) at 10°C, with a 12L:12D light-dark cycle. *S. hantzschii* tolerates a wide range of temperatures, but favors relatively low temperatures [22], [23]. In contrast to *S. hantzschii*, the optimal temperature for *C. vulgaris* growth is >20°C. Excessive population growth often occurs at optimal temperatures, which may affect the constancy of food algae provision (see experimental protocol). Therefore, we maintained the *C. vulgaris* subculture at 10°C for the experiment.

We chose approximately 100 *D. magna* offspring born within 24 h (most ~0.8 mm long) with similar life-history traits (e.g., birth time and size) from the stock culture, and allowed them to reach the offspring-production stage (hereafter referred as SC, sampled culture). Generally, daphniid females that are larger at birth grow more rapidly and are larger at maturity than those that are smaller at birth [24]. To obtain a similar-sized cohort, we first sorted and eliminated extraordinarily larger or smaller maternal *Daphnia* from the SC (29 individuals were removed). We then randomly selected 30 offspring from the remaining *D. magna* individuals in the SC by adapting a scaled loupe (unit, mm). The 30 adults were between 3 and 4 mm and contained eggs in their brood chamber. We transferred the 30 *D. manga* to a new beaker filled with fresh Elendt M4 medium, and provided sufficient food (*C. vulgaris*) until offspring were produced. In determining the quantity of food algae to provide, we considered the supply level that would be appropriate for zooplankton population growth. Previous research [25] indicated that algal carbon content of approximately 2.5 mg Carbon L^{-1} (units shown as mg C L^{-1} hereafter) in a given volume of zooplankton medium would be sufficient for zooplankton survival and population growth. The first reproduction event occurred at 4 to 5 days after selection. However, the number of neonates from the first reproduction was small, and we used the second clutch from the selected *D. magna* for the experiment. In summary, the initially selected *D. magna* adults were used to produce the offspring employed in the main experimental procedure. The offspring from the second clutch

were collected after birth (within 6 h) and were used for the main experiment.

Experimental Protocol

The overall experimental design is shown in Figure 1. We transferred 150 of the collected offspring into three experimental groups as follows: (1) *C. vulgaris* only (CHL); (2) *S. hantzschii* only (STE); and (3) mixed algae (MIX). For each group, we prepared 10 replicates in 500-mL sterilized beakers filled with 500 mL Elendt M4 medium, and five acclimated *D. magna* individuals were placed in each beaker.

Food algae were supplied at a quantity sufficient to maintain 2.5 mg C L^{-1} in each beaker throughout the experiment. This quantity can be determined from the relationship between algal density and carbon content [25] when algal cell size is known. Thus, before the experiment, we obtained the size information of *C. vulgaris* and *S. hantzschii* by measuring their diameter 50 times, and calculated average size of the algal species (Table 1). Based on the size information, we determined daily food algal supply amount, that the number of cells equal to 2.5 mg C L^{-1} was 8652 cells for *C. vulgaris*, and 8802 cells for *S. hantzschii*. The total daily injection volume was determined accordingly. For example, if daily density of *S. hantzschii* was 687 cells mL^{-1}, we injected ca 12.81 mL of *S. hantzschii* stock (687 cells mL^{-1} * 12.81 mL = 8800 cells). The density of food algae changed during culture maintenance, so the calculated injection volume was determined daily prior to administering the food supply. We injected food algae between 3 and 4 PM.

Unlike supplying a single algal species, for one treatment (MIX), a mixture of two food algae was used. In the MIX group, careful determination of the food algal supply was performed. To maintain the daily dosage of food algae at 2.5 mg C L^{-1} for the two different algal species, each species was supplied at 1.25 mg C L^{-1}. We determined the required quantity of each algal species by daily enumeration of food algal density.

The experiment was conducted using a plant growth chamber (Eyela FLI-2000, Japan). The aforementioned maintenance conditions for *D. magna* were also applied to the experiment (20°C; photon flux density = 30 μmol·m^{-2}·sec^{-1}; 12L:12D light-dark cycle). The experiment was terminated when adult *D. magna* produced offspring from the second clutch. The shortest duration in which the second clutch was obtained was 8 d, which is generally accepted as an appropriate turnover time for the assimilation of carbon and nitrogen from *D. magna* food [26]. Each day during the experiment, we transferred *D. magna* to fresh Elendt M4 medium before providing food algae, to maintain the supply of algae at 2.5 mg C L^{-1}. After termination of the experiment, we randomly collected two adults from each beaker of the three experimental groups (total $n = 20$ for each experimental group), and measured their body length and counted food algal cells in their guts. We took microscope images (×200 magnification, Axioskop 40, Carl Zeiss Microscopy, Germany) of the adult *D. magna* and used an image processing program (AxioVision Rel 4.8, Carl Zeiss Microscopy, Germany) to measure body length following the manufacturer's protocol for calibrating image length to actual length.

Gut contents were examined to determine the pattern of food algal consumption, particularly in the MIX group. For this investigation, we eviscerated the guts of *D. magna* ($n = 20$) from the MIX group and counted algal cell numbers of each species in the guts of each individual. Food algal cells tended to be broken as they progressed through the gut, which could complicate enumeration of algal cells. From an empirical approach, just after consumption, algal cells resided in the upper part of the gut

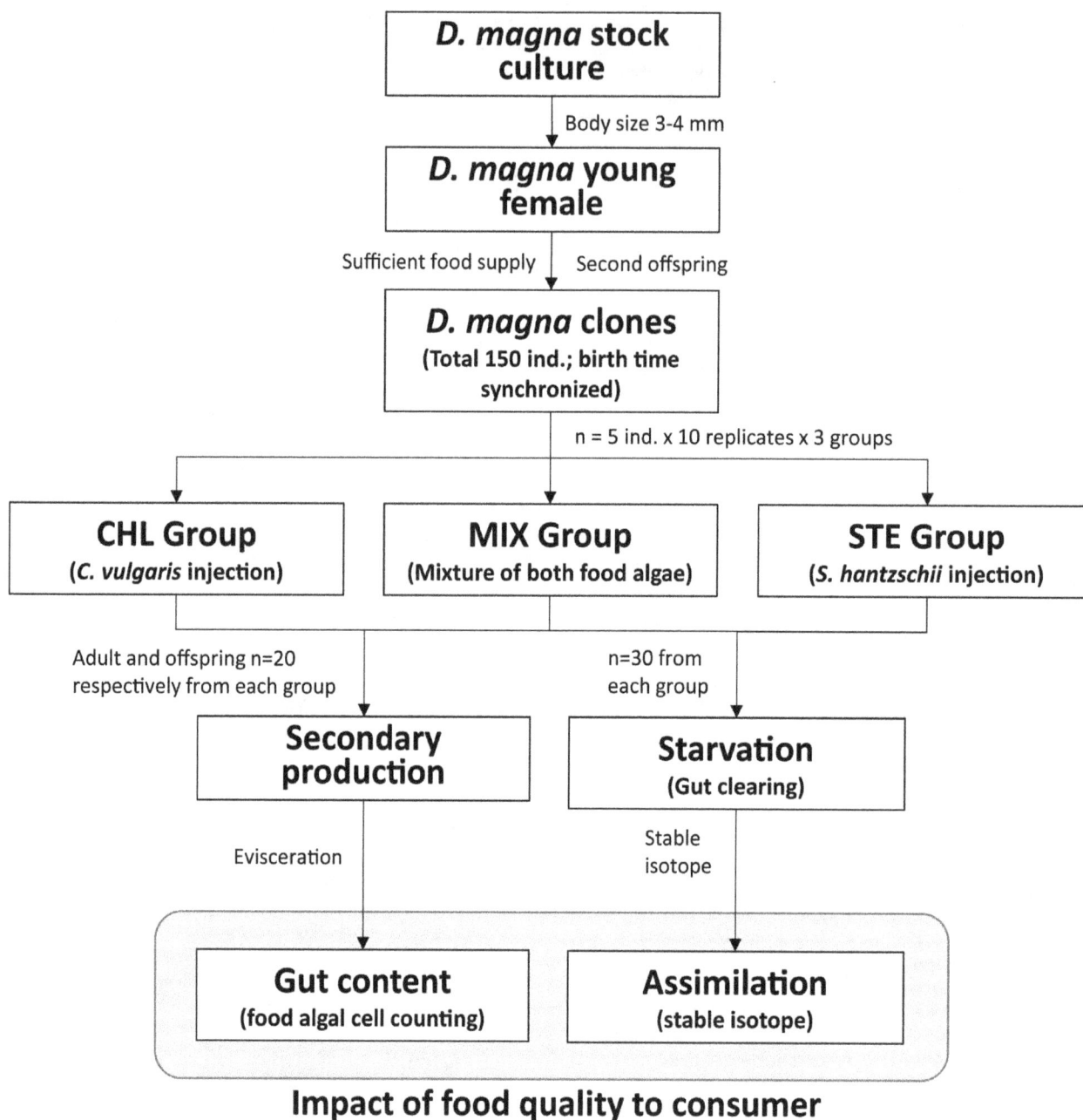

Figure 1. Schematic flowchart of the study. The shaded box indicates analytical processes that used the prepared samples in earlier stages of the experiment.

Table 1. Size and density of two food algal species used in the experiment.

Prey species	Size (μm)	Density (cells mL^{-1})
C. vulgaris	3.84±0.6	4333.3±629.1
S. hantzschii	4.03±0.4	687.3±142.3

Density of the two algal species was the average daily dose provided to *Daphnia magna* in the growth experiment.

(approximately 1 mm from the mouth) and were relatively fresh and unbroken. Therefore, we divided the total gut length (approximately 3 mm) into three sections (fore-, mid-, and rear-gut; each approximately 1 mm), and counted algal cells in the foreguts to minimize enumeration error due to broken cells. Typically, filter feeders such as daphniids are not selective feeders; therefore, the ratio of the two consumed algal species in the foregut would be maintained during passage through the gut.

To determine offspring size, two randomly sampled offspring from each beaker (total $n = 20$ in each experimental group) were measured. The size measurement of offspring was based on application of the image-processing program, as performed for the size analysis of *D. magna* adults. To determine the total number of offspring per adult, we counted the number of offspring in each beaker and divided that number by five (i.e., the number of adults in each beaker).

The density and size measurements of zooplankton individuals and algal cells in the gut samples were carried out using a microscope (Axioskop 40, Carl Zeiss Microscopy, Germany) under ×200, and ×400 magnification, respectively.

Stable Isotope Analysis

The remaining 30 adult *D. magna* individuals in each experimental group and the two food algae were used for stable isotope analysis. The *D. magna* samples contained phytoplankton in their guts; therefore, we transferred those individuals into fresh Elendt M4 medium for more than 24 h without provision of additional food algae. This allowed these individuals to eject their gut contents; they were then included in the stable isotope analysis. The 30 *D. magna* individuals were divided into six groups ($n = 5$ per group); three of these groups ($n = 15$ individuals) were used for detection of the carbon signature and three groups were used for detection of the nitrogen signature.

Carbon and nitrogen measurements of *D. magna* were conducted separately. It is necessary to extract tissue lipids for accurate interpretation of trophodynamics using carbon stable isotope data. The carbon isotope signature depends on protein content in tissue; the presence of lipids can affect the reliability of the isotope analysis. Lipid content varies in accordance with tissue type and is ^{13}C-depleted relative to proteins. Therefore, tissue samples that contain lipid may produce an unstable carbon isotope signature. In contrast, lipid extraction affects δ^{15}N. Therefore, we divided the samples into separate groups comprising carbon- and nitrogen-signature samples. Lipids were removed only from the carbon-signature samples. Comparison between the two samples was accomplished by δ^{13}C and δ^{15}N analyses [27]. The carbon-signature samples were placed in a solution of methanol-chloroform-triple-distilled water (2:1:0.8 v/v/v) for 24 h.

For stable isotope analysis of food algae, we prepared 5 mL of algal suspension from each species and analyzed in triplicate. The algal samples were treated with 1 mol·L^{-1} hydrochloric acid (HCl) to remove inorganic carbon. The samples were then rinsed with ultrapure water to remove the HCl.

The prepared samples (two algal species and *D. magna*) were freeze-dried and then ground with a mortar and pestle. The powdered samples were maintained at−70°C until analysis. When all samples were collected, carbon and nitrogen isotope ratios were determined using continuous-flow isotope mass spectrometry. Dried samples (approximately 0.5 mg of animal samples and 1.0 mg of algae) were combusted in an elemental analyzer (EuroVector), and the resultant gases (CO_2 and N_2) were introduced into an isotope ratio mass spectrometer (CF-IRMS, model-ISOPRIME 100, Micromass Isoprime) in a continuous flow, using helium as the carrier gas. Data were expressed as the relative per-mil (‰) difference between sample and conventional standards of Pee Dee belemnite (PDB) carbonate for carbon and air N_2 for nitrogen, according to the following equation:

$$\delta X(‰) = \left[\left(\frac{R_{\text{sample}}}{R_{\text{standard}}} \right) - 1 \right] \times 1000$$

where X is ^{13}C or ^{15}N, and R is the ^{13}C:^{12}C or ^{15}N:^{14}N ratio. A secondary standard of known relation to the international standard was used as a reference material. The standard deviations of δ^{13}C and δ^{15}N for 20 replicate analyses of the 'Peptone (δ^{13}C = −15.8‰ and δ^{15}N = 7.0‰, Merck)' standard were ±0.1 and ±0.2 (‰), respectively.

To determine which of the two food sources (*C. vulgaris*, and *S. hantzschii*) was assimilated more readily by *D. magna*, we calculated two-source isotope mixing models. The carbon isotope values of *C. vulgaris* and *S. hantzschii* differed significantly. The model is defined as:

$$\delta^{15}C_M = f_X \left(\delta^{13}C_X + \Delta^{13}N \right) + f_Y \left(\delta^{13}C_Y + \Delta^{13}C \right); 1 = f_X + f_Y$$

where X, Y, and M represent the two food sources and the mixture, respectively; f represents the proportion of N from each food source in the consumer's diet; and Δ^{15}C is the assumed trophic fractionation (i.e., the change in δ^{15}C over one trophic step from prey to predator) [28], [17]. Trophic fractionation was assumed to be constant, and either 3.4‰ or 2.4‰ [29].

Statistical Analysis

For statistical assessment of the experimental groups, we applied one-way nested ANOVA (two-tailed, $\alpha = 0.05$) to compare the size of adults and offspring. Although we prepared 10 replicates (beakers) for each experimental group, pseudo-replication had to be carefully considered [30]. Therefore, we set the different food algal treatments as the primary factors and the 10 beakers as nested subgroups for every treatment.

Comparison of numbers of *D. magna* offspring was performed using one-way ANOVA. Student's *t*-tests (two-tailed, $\alpha = 0.05$) were used to compare cell numbers of algal species in the guts of the MIX group. Tukey's post-hoc tests were employed to identify groups with different average values. All statistical tests were performed using the package SPSS Statistics ver. 20.

Results

D. magna Response to Different Food Algae Resources

A clear difference in size of adult *D. magna* was observed between the experimental groups (Figure 2 and Table 2). Adult *D. magna* that consumed *S. hantzschii* were significantly larger than those that fed on *C. vulgaris* (mean ± standard deviation; STE: 3.18±0.05 mm; CHL: 2.78±0.07 mm). Adult size in the MIX group was intermediate (2.99±0.07 mm; Figure 2A). Although subgroups (i.e., beakers) showed statistical differences, post-hoc tests revealed a significant difference in average values of the three groups (three comparisons, $P<0.001$). The size of *D. magna* offspring also differed significantly between groups (Figure 2B and Table 2). As for adults, offspring from the STE group (1.14±0.05 mm) were significantly larger than offspring from the other groups (CHL: 1.08±0.06 mm; MIX: 1.08±0.07 mm). The sizes of individuals in the CHL and MIX groups were similar. Post-hoc tests revealed significant differences between CHL and STE, and between MIX and STE ($P<0.001$), but not between CHL and MIX ($P>0.05$).

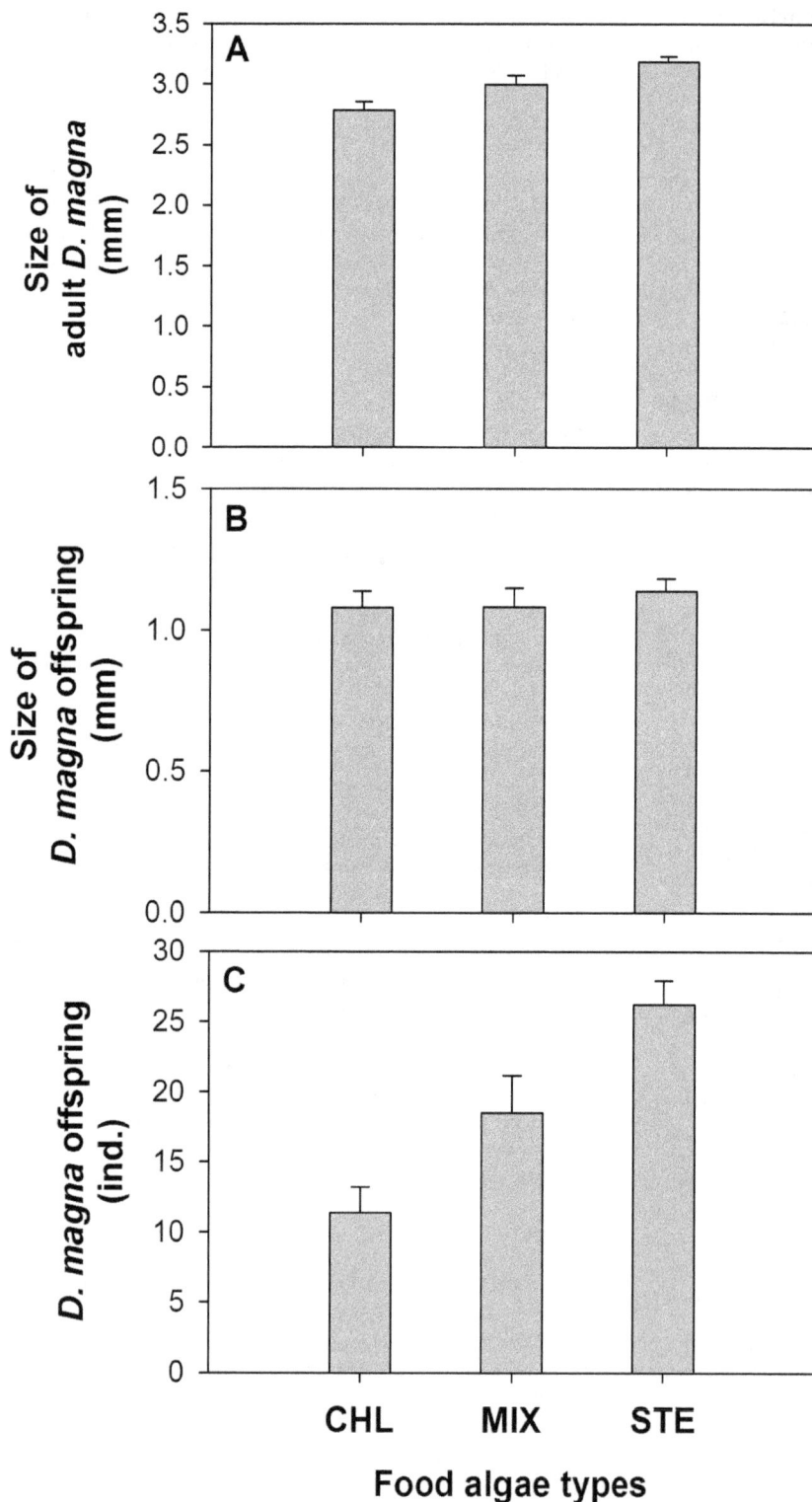

Figure 2. Size and number of individual *Daphnia magna* according to the prey species provided. A) size of adults; B) size of offspring; C) number of individual offspring. CHL, *C. vulgaris*; STE, *S. hantzschii*; MIX, a mixture of both food algal species; *n* = 20, respectively.

Difference in food algae also resulted in variations in offspring number (Figure 2C). Adult *D. magna* that consumed *S. hantzschii* produced more offspring than did those of the other groups (STE: 26.2±1.8 ind. per adult; CHL: 11.4±1.8 ind. per adult; MIX: 18.5±2.67 ind. per adult). The three groups differed significantly from one another (one-way ANOVA; F = 120.76, P<0.001, d.f. = 2), which was supported by the results of post-hoc tests (all three cases, P<0.001). Gut content analysis showed that *D. magna*

Table 2. Two-way nested ANOVA results for the effects of main groups (i.e. food algae *Chlorella vulgaris, Stephanodiscus hantzschii,* and Mixture) and subgroups (i.e. beakers) on adults and offspring of *D. magna* size.

	Factors	d.f.	F	p
Size of adult *D. magna*	Food algae	2	91.85	P<0.001
	Beaker	18	2.95	P<0.01
Size of *D. magna* offspring	Food algae	2	5.67	P<0.05
	Beaker	18	0.91	P>0.05

adults in the MIX group consumed both *C. vulgaris* and *S. hantzschii* during the experiment (*C. vulgaris*, 1,966.0±235.3 cells per gut; *S. hantzschii*, 2,130.0±460.4 cells per gut; t=−1.457, *P*>0.05; n=20).

Stable Isotope Analysis of Food Algae Assimilation

Stable isotope analysis revealed that *D. magna* depended more on *S. hantzschii* than on *C. vulgaris* when they fed on a mixture of these algae (Figure 3). The $\delta^{13}C$ and $\delta^{15}N$ ratio indicated the contribution of prey phytoplankton to *D. magna*. *D. magna* adults in the two groups fed only one species (CHL and STE) depended on either *C. vulgaris* or *S. hantzschii*, respectively. However, *D. magna* in the MIX group relied more on *S. hantzschii*. In addition, when the contribution rates of the two food algal species in the MIX group were calculated from the isotope analyses, the *S. hantzschii* contribution rate (92%) was higher than that of *C. vulgaris* (8%) from the two-source mixing model. Therefore, diatom species (*S. hantzschii*) contributed most to the growth of *D. magna* individuals and populations.

Discussion

Contribution of Different Algal Species to *D. magna* Growth

Of the two algal species studied, the diatom *S. hantzschii* appears to be the more suitable food item for *D. magna*; this was true for both population growth and for individual growth. Despite similar consumption rates of the two algae (see Table 1), the size of *D. magna* individuals increased much more when they utilized *S. hantzschii*. There are several explanations for why this may occur. Diatoms are commonly regarded as good sources of lipids and serve as a food source for zooplankton [31], [8]. They are known to contain large amounts of eicosapentaenoic acid [32], a fatty acid required in the diet of many animals that may not be able to synthesize the compound [33], [34]. However, green algae are known to contain relatively lower nutrient contents compared to diatoms.

However, a greater availability of food algae (quantity-wise) does not always guarantee increased growth of grazer populations. A second possibility involves the digestive capacity of *D. magna*.

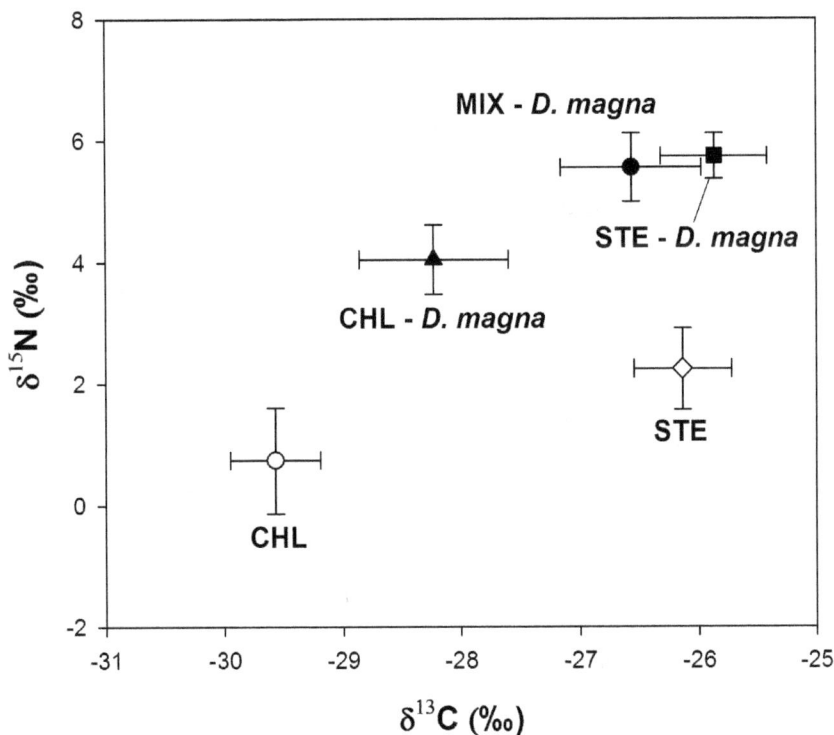

Figure 3. Results of the stable isotope analysis. CHL, *C. vulgaris*; STE, *S. hantzschii*; MIX, a mixture of both prey species.

The digestion rate of *D. magna* differs depending on the phytoplankton species consumed. Van Donk et al. [35] suggested that the cell wall morphology of green algae might reduce their digestibility by *Daphnia*. Our stable isotope results suggest that the greater dependence of *D. magna* on *S. hantzschii* is attributable to more effective absorption of nutrients from the diatom species. Therefore, it may be assumed that a balance between quality and absorbability of food algae is important for individual and population growth in zooplankton. Further research should investigate the characteristics of this balance.

Offspring Characteristics

An interesting finding of this experiment was the changing pattern of offspring size and number according to the species of algae consumed. Both abundance and size of offspring in the STE group were greater than in the other groups. Although the size and number of offspring depend on clutch size and reproductive capability of adults [36], we suggest that *D. magna* adults may respond flexibly to the quality of the energy sources they capture, resulting in changes in the size and number of their offspring in accordance with different algal species consumed. That *D. magna* actively responds to food algal quality was not definitively shown, but the response was clear. When algal resources are of low quality (less nutritious algae, such as *C. vulgaris* in the present study), female *Daphnia* may limit offspring size to ensure survival. Enhanced nutritional conditions (inclusion of *S. hantzschii* in the MIX group) resulted in a slight increase in offspring number. We maintained the supported amount of carbon in all of treatments at 2.5 mg C L^{-1}, but the algal species that comprised this carbon differed. Although availability of carbon allowed survival and reproduction of *D. magna*, offspring characteristics were further improved when the proportion of *S. hantzschii* was increased. The provision of more nutritious food algae caused this pattern to emerge. We thought that a semi-restricted diet would result in moderate changes in size and fecundity (i.e., the MIX group) and that a nutritious diet fed to a growing individuals would increase the size and fecundity of that individual, thereby also increasing the population (i.e., the STE group). In previous studies, offspring size was affected by the quantity of food algae and the presence of predators [37], [38]. Although we did not consider the presence of predators, the quality of food algae plays a key role in the population growth of *D. magna*, at least when food algae are sufficiently abundant.

The quality of food algae may be very important to filter-feeding zooplankton. Recent studies have found that other zooplankton groups (mainly copepods) are not affected by the quality of algae resources during population growth [39]. In one study, egg production was not significantly related to lipid content

in six phytoplankton species; the authors suggested that slow transit time in the gut (i.e., increased opportunity to absorb nutrients) could explain this result. In contrast to copepods, *Daphnia* typically shows relatively fast gut-passage time, which does not allow optimal absorption of nutrients [40]. Therefore, the quality of food algae, as well as its absorbability, is crucial for *Daphnia* population growth, and food algae that are fully assimilated will result in maintenance of or increases in zooplankton population levels.

Appropriate Food Selection using Stable Isotope Analysis

Based on the results of this study, it is possible to quantify the energy channeled from primary producers to primary consumers, expanding on basic understandings of connectivity. The traditional method of investigating food web structure involves visual inspection of gut contents [41], [42], but recently, DNA barcoding has increased the resolution of prey identification to the species level [43]. From a functional perspective, prey consumption is related to the growth of grazers and predators [44], [45]. Despite apparent evidence, these methods are limited in their ability to quantify the contribution of prey to grazers. Assimilation indicates how grazers utilize prey for growth, and the results of the present study further elucidate microbial food web structure. Consequently, the consumption rate (including qualitative and quantitative aspects) and the contribution rate of food items should be examined simultaneously to further precisely elucidate food web functions. In addition, as more information on multi-species prey and grazer relationships becomes available, understanding of ecological integrity will be improved.

Conclusion

The growth of *D. magna* individuals and offspring was significantly improved by the consumption of *S. hantzschii* but not *C. vulgaris*. Stable isotope analysis revealed substantial assimilation of diatom-derived materials in *D. magna*, indicating that diatoms are important to the population growth of this species. These results confirm the applicability of stable isotope approaches for clarifying the contribution of different food algae and for elucidating the importance of food quality for *D. magna* population growth.

Author Contributions

Conceived and designed the experiments: JC KJ MK GL. Performed the experiments: JC SK. Analyzed the data: JC KJ SK. Contributed reagents/materials/analysis tools: KJ GJ KC. Wrote the paper: JC SK KC MK KJ GL GJ.

References

1. Wetzel RG (1983) Limnology. – Saunders College Publishing. Philadelphia.
2. Gliwicz WR, Lampert W (1990) Food thresholds in *Daphnia* species in the absence and presence of blue-green filaments. Ecology 71: 691–702.
3. Hart RC, Jarvis AC (1993) In situ determinations of bacterial selectivity and filtration rates by five cladoceran zooplankters in a hypertrophic subtropical reservoir. J Plankton Res 15: 295–315.
4. Lampert W, Fleckner W, Rai H, Taylor BE (1986) Phytoplankton control by grazing zooplankton: a study on the clear water phase. Limnol Oceanogr 31: 478–490.
5. Dawidowicz P (1990) Effectiveness of phytoplankton control by large-bodied and small-bodied zooplankton. Hydrobiologia 200/201: 43–47.
6. Branstrator DK, Lehman JT (1991) Invertebrate predation in Lake Michigan: Regulation of *Bosmina longirostris* by *Leptodora kindtii*. Limnol Oceanogr 36: 483–495.
7. Herzig A, Koste W (1989) The development of *Hexarthra* spp. in a shallow alkaline lake. Hydrobiologia 186/187: 129–136.
8. Ahlgren G, Lundstedt L, Brett M, Forsberg C (1990) Lipid composition and food quality of some freshwater phytoplankton for cladoceran zooplankters. J Plankton Res 12: 809–818.
9. Müller-Navarra DC, Lampert W (1996) Seasonal patterns of food limitation in *Daphnia galeata*: separating food quantity and food quality effects. J Plankton Res 18: 1137–1157.
10. Ahlgren G, Gustafsson IB, Boberg M (1992) Fatty acid content and chemical composition of freshwater microlgae. J Phycol 28: 37–50.
11. Müller-Navarra DC (1995) Biochemical vs. mineral limitation in *Daphnia*. Limnol Oceanogr 40: 1209–1214.
12. Burns CW (1968) The relationship between body size of filter-feeding cladocera and the maximum size of particle ingested. Limnol Oceanogr 13: 675–678.
13. Geller W, Müller H (1981) The filtration apparatus of Cladocera: Filter mesh-sizes and their implications on food selectivity. Oecologia 49: 316–321.
14. Urabe J, Waki N (2009) Mitigation of adverse effects of rising CO_2 on a planktonic herbivore by mixed algal diets. Glob Change Biol 15: 523–531.
15. Bohrer RN, Lampert W (1988) Simultaneous measurement of the effect of food concentration on assimilation and respiration in *Daphnia magna* Straus. Funct Ecol 2: 463–471.
16. Yu RQ, Wang WX (2002) Trace metal assimilation and release budget in *Daphnia magna*. Limnol Oceanogr 47: 495–504.

17. Phillips DL, Koch PL (2002) Incorporating concentration dependence in stable isotope mixing models. Oecologia 130: 114–125.
18. Jeong KS, Kim DK, Joo GJ (2007) Delayed influence of dam storage and discharge on the determination of seasonal proliferations of *Microcystis aeruginosa* and *Stephanodiscus hantzschii* in a regulated river system of the lower Nakdong River (South Korea). Water Res 41: 1269–1279.
19. Kim DK, Jeong KS, Whigham PA, Joo GJ (2007) Winter diatom blooms in a regulated river in South Korea: explanations based on evolutionary computation. Freshwater Biol 52: 2021–2041.
20. Jung S, Kang YH, Katano T, Kim BH, Cho SY, et al. (2010) Testing addition of Pseudomonas fluorescens HYK0210-SK09 to mitigate blooms of the diatom *Stephanodiscus hantzschii* in small- and large-scale mesocosms. J Appl Phycol 22: 409–419.
21. Elendt BP (1990) Selenium deficiency in Crustacea: An ultrastructural approach to antennal damage in *Daphnia magna* Straus. Protoplasma 154: 25–33.
22. Kilham P, Kilham SS, Hecky RE (1986) Hypothesized resource relationships among African planktonic diatoms. Limnol Oceanogr 31: 1169–1181.
23. van Donk E, Kilham SS (1990) Temperature effects on silicon-and phosphorous-limited growth and competitive interactions among three diatoms1. J Phycol 26: 40–50.
24. Lampert W (1993) Phenotypic plasticity of the size at first reproduction in *Daphnia*: the importance of maternal size. Ecology 74: 1455–1466.
25. Strathman RR (1967) Estimating the organic carbon content of phytoplankton from cell volume or plasma volume. Limnol Oceanogr 12: 4 11–418.
26. O'Reilly CM, Hecky RE, Cohen AS, Plisnier PD (2002) Interpreting stable isotopes in food webs: recognizing the role of time averaging at different trophic levels. Limnol Oceanogr 47: 306–309.
27. Post DM, Layman CA, Arrington DA, Takimoto G, Quattrochi J, et al. (2007). Getting to the fat of the matter: models, methods and assumptions for dealing with lipids in stable isotope analyses. Oecologia 152: 179–189.
28. Phillips DL, Gregg JW (2001) Uncertainty in source partitioning using stable isotopes. Oecologia 127: 171–179.
29. Minagawa M, Wada E (1984) Stepwise enrichment of ^{15}N along food chains: further evidence and the relation between δ^{15}N and animal age. Geochem Cosmochim Acta 48: 1135–1140.
30. Hurlbert SH (1984) Pseudoreplication and the design of ecological field experiments. Ecol Monogr 54: 187–211.
31. Bourdier GG, Amblard CA (1989) Lipids in *Acanthodiaptomus denticornis* during starvation and fed on three different algae. J Plankton Res 11: 1201–1212.
32. Renaud SM, Parry DL, Thinh LV (1994) Microalgae for use in tropical aquaculture I: Gross chemical composition and fatty acid composition of twelve species of microalgae from the Northern territory, Australia. J Appl Phycol 6: 337–345.
33. Sicko-Goad L, Andresen NA (1991) Effect of growth and light/dark cycles on diatom lipid content and composition. J Phycol 27: 710–718.
34. Dunstan GA, Volkman JK, Barrett SM, Garland CD (1993) Changes in the lipid composition and maximisation of the polyunsaturated fatty acid content of three microalgae grown in mass culture. J Appl Phycol 5: 71–83.
35. Van Donk E, Lürling M, Hessen DO, Lokhorst GM (1997) Altered cell wall morphology in nutrient-deficient phytoplankton and its impact on grazers. Limnol Oceanogr 42: 357–364.
36. Scheiner SM, Berrigan D (1998) The genetics of phenotypic plasticity. VIII. The cost of plasticity in *Daphnia pulex*. Evolution 52: 368–378.
37. Spitze K (1991) Chaoborus Predation and life-History evolution in *Daphnia pulex*: temporal pattern of population diversity, fitness, and mean life history. Evolution 45: 82–92.
38. Reede T (1995) Life history shifts in response to different levels of fish kairomones in *Daphnia*. J Plankton Res 17: 1661–1667.
39. Dutz J, Koski M, Jónasdóttir SH (2008) Copepod reproduction is unaffected by diatom aldehydes or lipid composition. Limnol Oceanogr 53: 225–235.
40. Murtaugh PA (1985) The influence of food concentration and feeding rate on the gut residence time of *Daphnia*. J Plankton Res 7: 415–420.
41. Kimball DC, Helm WT (1971) A method of estimating fish stomach capacity. T Am Fish Soc 100: 572–575.
42. Drenner RW, Threlkeld ST, McCracken MD (1986) Experimental analysis of the direct and indirect effects of an omnivorous filter-feeding clupeid on plankton community structure. Can J Fish Aquat Sci 43: 1935–1945.
43. Jo H, Gim JA, Jeong KS, Kim HS, Joo GJ (2014) Application of DNA barcoding for identification of freshwater carnivorous fish diets: Is number of prey items dependent on size class for *Micropterus salmoides*?. Ecology and Evolution. 4: 219–229.
44. Chant DA (1961) The effect of prey density on prey consumption and oviposition in adults of *Thphlodromus* (T.) *Occidentalis Nesbitt* (Acrina: Phytoseiidae) in the laboratory. Can J of zoolog 39: 311–315.
45. Toft S, Wise DH (1999) Growth, development, and survival of a generalist predator fed single- and mixed-species of different quality. Oecologia 119: 191–197.

18

Comparison of *spa* Types, SCC*mec* Types and Antimicrobial Resistance Profiles of MRSA Isolated from Turkeys at Farm, Slaughter and from Retail Meat Indicates Transmission along the Production Chain

Birgit Vossenkuhl*, Jörgen Brandt¤a, Alexandra Fetsch, Annemarie Käsbohrer, Britta Kraushaar, Katja Alt¤b, Bernd-Alois Tenhagen

Federal Institute for Risk Assessment, Berlin, Germany

Abstract

The prevalence of MRSA in the turkey meat production chain in Germany was estimated within the national monitoring for zoonotic agents in 2010. In total 22/112 (19.6%) dust samples from turkey farms, 235/359 (65.5%) swabs from turkey carcasses after slaughter and 147/460 (32.0%) turkey meat samples at retail were tested positive for MRSA. The specific distributions of *spa* types, SCC*mec* types and antimicrobial resistance profiles of MRSA isolated from these three different origins were compared using chi square statistics and the proportional similarity index (Czekanowski index). No significant differences between *spa* types, SCC*mec* types and antimicrobial resistance profiles of MRSA from different steps of the German turkey meat production chain were observed using Chi-Square test statistics. The Czekanowski index which can obtain values between 0 (no similarity) and 1 (perfect agreement) was consistently high (0.79–0.86) for the distribution of *spa* types and SCC*mec* types between the different processing stages indicating high degrees of similarity. The comparison of antimicrobial resistance profiles between the different process steps revealed the lowest Czekanowski index values (0.42–0.56). However, the Czekanowski index values were substantially higher than the index when isolates from the turkey meat production chain were compared to isolates from wild boar meat (0.13–0.19), an example of a separated population of MRSA used as control group. This result indicates that the proposed statistical method is valid to detect existing differences in the distribution of the tested characteristics of MRSA. The degree of similarity in the distribution of *spa* types, SCC*mec* types and antimicrobial resistance profiles between MRSA isolates from different process stages of turkey meat production may reflect MRSA transmission along the chain.

Editor: Herminia de Lencastre, Rockefeller University, United States of America

Funding: This work was carried out within the Project MedVet-Staph funded by the German Bundesministerium für Bildung und Forschung, Grant Nr. 01KI1014C. The funders had no role in study design, data collection and analysis, decision to publish, or preparation of the manuscript.

Competing Interests: The authors have declared that no competing interests exist.

* E-mail: 43@bfr.bund.de

¤a Current address: Humboldt University, Unter den Linden 6, Berlin, Germany
¤b Current address: Robert Koch Institute, DGZ-Ring 1, Berlin, Germany

Introduction

Staphylococcus (S.) aureus is a common cause of food poisoning due to the production of various enterotoxins. *S. aureus* is a frequent colonizer of the skin and mucous membranes and therefore, personnel and food-producing animals are the main sources of *S. aureus* in food [1]. The control of *S. aureus* is routinely considered in the food producing industry if standard food safety management systems are operated. In recent years, methicillin-resistant *Staphylococcus aureus* (MRSA), previously known as a multidrug resistant pathogen causing severe healthcare associated and community acquired infections, [2] has been observed worldwide in livestock husbandry as well as in food of different animal origins raising concerns about a possible farm to fork transmission.

First reported from pigs in the Netherlands [3] and France [4] a distinct MRSA lineage, Clonal Complex (CC) 398, has emerged in food producing animals in Europe especially in herds of pigs [5–8],

veal calves [9] broiler flocks [10,11] and turkeys [12]. Therefore, the term "livestock-associated MRSA" (LA-MRSA) was introduced considering livestock to form a new and separate reservoir for MRSA [13]. In Asian countries, however, sequence type ST9, a separate genetic linage, is predominating among MRSA isolates from livestock animals [14,15]. Different DNA sequencing methods are used for typing MRSA strains. In order to define MRSA clones, Multilocus sequence typing (MLST), a method of classifying MRSA strains by the allelic profile of seven housekeeping genes, is used in conjunction with PCR analysis of the staphylococcal chromosomal cassette *mec* (SCC*mec*), a mobile genetic element that contains the *mec A* gene encoding for resistance to methicillin [16]. 11 different SCC*mec* types have been described, so far. The class of *mec* gene complex and the type of ccr gene complex carrying a set of recombinase genes responsible for integration and excision of the cassette characterize the different types of SCCmec elements [17]. Whereas SCCmec I-X harbor *mecA* SCCmec XI carries a divergent *mecA* homologue (*mecA*$_{LGA251}$) [18]. *Spa* typing differentiates MRSA strains by the number of tandem repeats and the sequence variation in region X of the protein A gene (spa) and can be used for reliable and discriminatory typing of MRSA [19]. As particular MLST have shown to be associated with specific repeats and repeat successions it is, with few exceptions, possible to infer an MLST type from the spa type. (http://www.spaserver.ridom.de). The frequent use of antimicrobials in animal production is suspected to facilitate the emergence and spread of MRSA due to antimicrobial selection pressure [20–22]. High stocking density in intensive food animal production holdings and intensive animal trading promote the rapid spread of MRSA between livestock populations [23,24]. LA-MRSA strains have also been detected in raw meat at retail including beef, veal, pork and poultry [25–33] indicating potential transmission along the chain due to cross contamination during slaughter and processing. However, the extent of this transmission is so far poorly understood.

In Germany, the national monitoring for zoonotic agents aims at characterizing the prevalence of potential zoonotic pathogens at different stages of various food chains. The monitoring is part of the official control of foodstuffs and fulfills the requirements of EU Directive 2003/99/EC [34]. In 2010, the turkey meat production chain was addressed in this monitoring scheme.

The objective of the present study was to use data from the national monitoring of zoonotic agents in the food chain to obtain a comprehensive insight into the presence and transmission of MRSA in the German turkey meat production chain. A new approach is proposed for analyzing a cross sectional MRSA data set from different stages of the food chain in order to draw conclusions on potential farm to fork transmission. For this purpose, the prevalence of MRSA and the distribution of *spa* types, SCC*mec* types and antimicrobial resistance profiles among MRSA isolated from different steps of the turkey meat production chain were compared. It is proposed that the degree of similarity in the distribution of *spa* types, SCC*mec* types and antimicrobial resistance profiles between the samples from the three process steps may be interpreted as reflecting MRSA transmission along the chain.

Materials and Methods

1. Study Design

Sampling was conducted in 2010 by the competent authorities of the federal states according to a pre-defined protocol in the framework of the national monitoring for zoonotic agents. All participating competent authorities are listed in table S1. Dust samples from 112 German turkey flocks were collected in order to quantify the presence of MRSA in primary production and to assess the introduction of MRSA into the slaughterhouses. Samples at slaughterhouses (n = 359) were analyzed to estimate the transfer to carcasses during slaughter and to determine the transmission of MRSA from carcasses to fresh turkey meat during further processing. Finally, 460 turkey meat portions were sampled to evaluate the MRSA exposure of consumers via contaminated turkey meat.

Turkey pens were sampled by pooling 5 dust swab samples, collected from different sections representing an area of 500 cm^2, each. At the slaughterhouse, at least 30 g neck skin was sampled from turkey carcasses after slaughter and chilling, but prior to further processing. Samples of 25 g of fresh turkey meat (with or without skin) were collected at retail. In order to ensure a high level of representativity, the distribution of the samples in primary production and at slaughter across Germany was proportional to the number of turkey flocks and the slaughter capacity of the respective federal state. Meat samples at retail were distributed according to the human population size of the executive federal state. A more detailed description of the principles of the national monitoring for zoonotic agents has been published before [35].

2. MRSA Isolation

MRSA were isolated by the regional laboratories according to the recommended method of the National Reference Laboratory (NRL) for staphylococci including S. aureus at the Federal Institute for Risk Assessment (BfR). The dust samples were pooled per turkey house in 100 ml Mueller Hinton broth supplemented with 6.5% NaCl for pre-enrichment. Neck skin samples (at least 30 g), fresh meat (25 g) and meat preparations (25 g) were pre-enriched in 225 ml Mueller Hinton broth supplemented with 6.5% NaCl. After incubation for 16–20 h at 37°C, 1 ml pre-enrichment broth was transferred into 9 ml of tryptic soy broth supplemented with 50 mg/l aztreonam and 3.5 mg/l cefoxitin. After incubation of this selective-enrichment broth for a further 16–20 h at 37°C one loopful was plated onto sheep blood agar and chromogenic MRSA screening agar respectively, and incubated for 24–48 h at 37°C. Presumptive MRSA isolates were sent to the National Reference Laboratory (NRL) for staphylococci including S. aureus at the Federal Institute for Risk Assessment (BfR) for MRSA confirmation and characterization. The number of MRSA isolates included in further analyses is not exactly congruent to the amount of positive samples obtained within the national monitoring for zoonotic agents because first, the NRL did not always receive the corresponding isolate from the competent authorities of the federal states or second, isolates which did not exactly correspond to the monitoring sampling plan in terms of completeness of data reporting to the national level but were obtained from the correct matrix were excluded from prevalence estimations but included in further typing and strain comparisons.

Twenty one MRSA isolates from wild boar meat within the national monitoring for zoonotic agents of 2011 were used in the analyses as a control group (data not shown in detail). The control group was selected to ensure wide differences with the population under study concerning the distribution of MRSA strains in order to evaluate if the used analytical approach is appropriate to differentiate between the matrices.

3. Molecular Typing

Presumptive MRSA isolates were confirmed by an in-house multiplex PCR simultaneously targeting the 23S rDNA specific for *Staphylococcus* species [36], the nuclease gene *nuc* which is specific for *S. aureus*, and the resistance gene *mecA* [37]. Template DNA was extracted using the "RTP Bacteria DNA Mini Kit" (Invitek, Berlin, Germany). All MRSA isolates were further characterized using *spa* typing [38] and SCC*mec*-typing [39]. The method applied for typing of the *SCC*mec differentiates SCC*mec* types I to V and their subtypes. However, isolates of the CC398 characterized as type III by the method have been shown to rather be a variant of type V [40]. The software Ridom Staphytype (Ridom GmbH, Würzburg, Germany) was used to assign *spa* types. *Spa* types which have not been identified and assigned to a clonal complex (CC) by the NRL before were additionally subjected to multilocus sequence typing (MLST) [41].

4. Antimicrobial Susceptibility Testing

All isolates were tested for the susceptibility to antimicrobials using broth microdilution in accordance with Clinical and Laboratory Standards Institute guidelines [42]. Commercial microtitre plates were used (TREK Diagnostic Systems, Magellan Biosciences, West Sussex, England). Minimum inhibitory concentrations (MIC) were evaluated according to epidemiological cut-off values (ECOFFs) published for MRSA and *S. aureus* by the European committee for antimicrobial susceptibility testing (www.eucast.org). MIC values above the ECOFFs indicated microbiological resistance. MIC lower or equal to the ECOFFs characterised susceptible strains. *S. aureus* strain ATCC 25923 was used for quality assurance Resistance testing included gentamicin, kanamycin, streptomycin, chloramphenicol, ciprofloxacin, tetracycline, clindamycin, erythromycin, mupirocin, linezolid, vancomycin, quinupristin/dalfopristin, penicillin, fusidic acid, cefoxitin, trimethoprim, sulfamethoxazole, rifampicin and tiamulin.

5. Statistical Analysis

The chi square test of homogeneity was used to analyze differences in the distribution of *spa* types and antibiotic resistance profiles between MRSA strains from the turkey flocks, carcasses at slaughter and meat. Isolates were grouped according to their *spa* types and antibiotic resistance profiles to assure appropriate numbers of isolates in all categories. All *spa* types were aggregated in accordance to their frequency of occurrence. The phenotypic antimicrobial resistance profiles were grouped by hierarchical cluster analysis using Ward's minimum variance and squared Euclidean distance. The MIC values for each isolate were categorized into resistant or susceptible according to the ECOFFs to generate a binary data set. The final amount of clusters was determined using the Pseudo-F [43] and Pseudo-T [44] statistics. Both tests indicate possible breakpoints for splitting the data into the appropriate amount of clusters. The distribution of SCC*mec* types in the different matrices were compared using Fisher's exact test as 33.3% of the cells of the contingency table had an expected value below 5. P-values of <0.05 were considered statistically significant. Chi square test, Fisher's exact test and cluster analysis were calculated using the statistical software package SPSS 18.0 (SPSS Inc. Munich, Germany). Pseudo-F and Pseudo-T statistics were performed using SAS/STAT software 9.2 (SAS Institute Inc., Cary, NC, USA).

The degree of similarity between the frequency distributions of *spa* types, SCC*mec* types and resistance profiles of MRSA among the sample sets from the turkey primary production, carcasses at slaughterhouse and turkey meat at retail was estimated using the Czekanowski index or proportional similarity index (PSI) [45]. It is calculated by:

$$PS = 1 - 0.5 \sum_i |p_i - q_i| = \sum_i \min(p_i, q_i)$$

where p_i and q_i represent the proportion of strains out of all strains among the data sets P and Q which agree in the realization i of the variable of interest. The values for PS range from 1 for identical frequency distributions of the variable of interest to zero for no similarities between the data sets. Since the size of the samples is rather small, a realization of the PSI index may deviate largely from its true value. Thus, the PSI was bootstrapped obtaining a probability density distribution from which we derived the 95% confidence interval for the PSI. The statistic open source software R (available at: http://www.R-project.org) was used to calculate the approximate confidence interval of the Czekanowski index using the non-parametric boostrap BC_α method utilizing 2000 iterations [46].

Results

Twenty two (19.6%) of 112 dust samples from the turkey primary production, 235 (65.5%) of 359 turkey carcasses after slaughter and 147 (32.0%) of 460 turkey meat samples at retail were tested positive for MRSA [47]. A set of 32 isolates from dust samples, 248 isolates from turkey carcasses and 241 isolates from turkey meat was used for further laboratory analyses (Table 1).

A total of 16 different *spa* types were identified. The number of different *spa* types increased during processing from 5 different types in dust samples over 8 in carcasses to 15 different types in meat samples. The proportion of strains assigned to CC398 ranged between 85.9 and 90.6%. Among CC398, t011 (43.8–46.9%) and t034 (32.0–43.8%) were the predominating *spa*-types on every process step. *Spa* types t1430 (4.0–6.3%) and t002 (3.1–9.1%) were dominating within the group of non CC398 strains.

Most of the strains carried SCC*mec*-type V (58.1–71.9%) followed by type IVa (19–27.0%). Type III (0–1.2%) was identified sporadically (Table 1). However, there is evidence in former literature that CC398 strains which were identified as SCC*mec* type III by the typing scheme of Zhang et al. [39] are rather assigned to a separate variant of SCC*mec* type V [40,48,49]. In 5.7–17.6% of the strains the SCC*mec* type could not be identified by the method used.

Susceptibility to 19 different antimicrobial agents was determined (Figure 1). Throughout the turkey production chain, the vast majority of isolates was resistant to tetracycline (98.8%–100%). High resistance rates were obtained to clindamycin (79.4–93.8%), erythromycin (73.8–87.5%), trimethoprim (65.7–78.1%), quinupristin/dalfopristin (62.2–66.1%) and tiamulin (52.3–65.6%). Resistances to mupirocin, linezolid, sulfamethoxazole and rifampicin were observed sporadically in individual isolates from all steps of the process chain. All isolates were susceptible to vancomycin. Resistance to tiamulin (62.2 versus 8.2%), gentamicin (25.2 versus 6.6%) and trimethoprim (72.0 versus 36.1%) was considerably more frequent among CC398 than among non-CC398 strains. Resistance to ciprofloxacin was common among non-CC398 strains (98.4 versus 26.1% in CC398 strains) (Figure 1).

All 521 MRSA strains were included in further similarity estimations. In accordance to the frequency of their occurrence all *spa* types were aggregated in 4 different categories for further statistical analysis. The most prevalent *spa* types t011 and t034 built their own group whereas rare *spa* types of CC398 and all non CC398 strains were summarized in separate groups. The chi square distribution of the *spa* type groups did not significantly differ between primary production, carcasses at slaughter and meat at retail (p = 0.06). Likewise, no significant difference was identified in the distribution of SCC*mec* types between the origins using fisher's exact test (p = 0.095). A total of 101 different resistance profiles were identified among the MRSA isolates including resistance to 2 to 12 different antimicrobial substances. The hierarchical cluster algorithm of Wards minimum variance combined with squared Euclidean distance separated the antimicrobial resistance profiles into homogenous clusters. Identical resistance phenotypes did not appear in more than one cluster. Based on the Pseudo-F and Pseudo-T statistics the 3 cluster solution containing 33, 44 and 24 different phenotypic resistance profiles, respectively, was identified to best describe the binary data set. Detailed characteristics of the cluster composition, concerning antimicrobial resistance and the distribution of groups of *spa* types and SCC*mec* types, is summarized in table S2. The antimicrobial resistance clusters did not significantly differ in their chi square distribution between the MRSA samples from the three origins (p = 0.295).

Table 1. MRSA prevalence and distribution of *spa* types, SCC*mec* types and antimicrobial resistance clusters at different steps of the German turkey meat production chain in 2010.

Process step		Primary production		Slaughter		Meat		total
Samples (n)		112		359		460		931
MRSA positive samples (n)		22		235		147		404
MRSA prevalence (%)		19,6		65,5		32		
No. of isolates included in further statistics[a]		32		248		241		521
Genetic Typing								
CC398	*spa* types	n	%	n	%	n	%	n
	t011	14	43.8	113	45.6	113	46.9	240
	t034	14	43.8	105	42.3	77	32.0	196
	t108	-		-		1	0.4	1
	t571	1	3.1	-		1	0.4	2
	t899	-		1	0.4	5	2.1	6
	t1255	-		-		3	1.2	3
	t1344	-		-		3	1.2	3
	t1580	-		-		1	0.4	1
	t2510	-		-		1	0.4	1
	t2576	-		1	0.4	-		1
	t2970	-		3	1.2	1	0.4	4
	t4652	-		1	0.4	1	0.4	2
	total	29	90.6	224	90.3	207	85.9	460
non CC398								
assigned MLST types	*spa* types							
ST5	t002	1	3.1	14	5.6	22	9.1	37
ST5	t010	-		-		1	0.4	1
ST45	t015	-		-		1	0.4	1
ST9	t1430	2	6.3	10	4.0	10	4.1	22
	total	3	9.4	24	9.7	34	14.1	61
	total	32	100	248	100	241	100	521
SCC*mec*Types								
	n.t.[b]	2	6.3	24	9.7	33	13.7	59
	mec III	-		1	0.4	3	1.2	4
	mec IVa	7	21.9	47	19.0	65	27.0	119
	mec V	23	71.9	176	71.0	140	58.1	339
	total	32	100	248	100	241	100	521
Resistance profiles[c]								

Table 1. Cont.

Process step		Primary production		Slaughter		Meat		total
	Cluster A	17	53.1	121	48.8	97	40.2	235
	Cluster B	10	31.3	82	33.1	88	36.5	180
	Cluster C	5	15.6	45	18.1	56	23.2	106
total		32	100	248	100	241	100	521

[a]MRSA isolates which did not exactly correspond to the monitoring sampling plan in terms of completeness of data reporting to the national level were excluded from prevalence estimations but included in further typing and strain comparisons.
[b]Not typable.
[c]Resistance cluster were calculated using Ward's minimum variance with squared Euclidean distance.

The distribution of *spa* types, SCC*mec* types and antimicrobial resistance profiles within the sample collections from the three process steps and the control group were compared pair wise using the Czekanowski index (Table 2). High index values were obtained for the distribution of *spa* types (PSI 0.79–0.86) among MRSA from the turkey meat chain. The comparison of the distribution of antimicrobial resistance profiles resulted in the lowest index values (PSI 0.42–0.56). The distribution of *spa* types and antimicrobial resistance profiles showed remarkably higher similarity between the different production steps of the turkey meat chain as to samples from the control group (PSI 0.55–0.56 and 0.13–0.19 resp.). High similarity in the distributions of SCC*mec* types was calculated between all process steps of the turkey meat production chain (PSI 0.85–0.91). However, a strong association was also received with SCC*mec* types of the control group (PSI 0.83–0.85).

Discussion

In the present study, a new approach is proposed for analyzing a cross sectional set of MRSA isolates originating from three consecutive stages of the turkey meat production chain in order to draw conclusions on a potential farm to fork transmission. In the course of the German national monitoring for zoonotic agents in 2010 MRSA was isolated at all stages of the turkey meat production chain with prevalences ranging from 19.6% to 65.5%. To our knowledge, this is the first representative national MRSA prevalence study in the turkey production chain. In a regional prevalence study among fattening turkeys in southern Germany in 2009, a considerably higher prevalence of 90% MRSA positive flocks was observed using the same sampling procedure [12]. The difference might be explained by the regional restriction of sampling and the small sample size in that study. The proportion of positive meat samples is in line with results from the Netherlands [25]. Outside of Europe, low MRSA contamination rates of 3.85% [32] and 1.7% [50] were reported among US turkey meat.

The high MRSA prevalence in turkey carcasses after slaughter in comparison to the flock prevalence is in contrast to the situation in pigs [12,51] and indicates that the turkey slaughter process may play an important role in the transmission of MRSA. Turkeys are slaughtered highly automated at a speed of line up to 3,600 turkey hens and up to 2,700 turkey toms per hour which leads to a permanent introduction of MRSA into the poultry processing plants [52]. During the process, MRSA on animal surfaces can get transmitted via direct contact or indirect via surface processing machinery, scalding water or the hands of staff. Scalding takes place at a constant water temperature between 50 and 65°C for 60 to 210 sec [52]. Although the surface of the carcasses is exposed to a heat treatment during scalding, the temperature and duration of the process might be insufficient to substantially reduce superficial MRSA counts. The selective growth of *S. aureus* after the elimination of less heat resistant microbial flora in the scalding water has been discussed [53]. As bacterial counts increase in the tanks throughout the slaughter day scalding can contribute to cross contamination [54]. After scalding, the birds go through the plucking machines consisting of revolving drums with rubber beaters or discs with plucking fingers. The birds are flailed and scraped for 30–90 sec while being sprayed with warm or cold water [52]. Plucking equipment is difficult to clean and a persisting microbiological flora can get established [55]. Cross contamination during slaughter and meat processing might lead to an extensive distribution of *spa* types between different animals and slaughter flocks. In addition, the increase in manual handling during processing facilitates the entry of human MRSA strains into

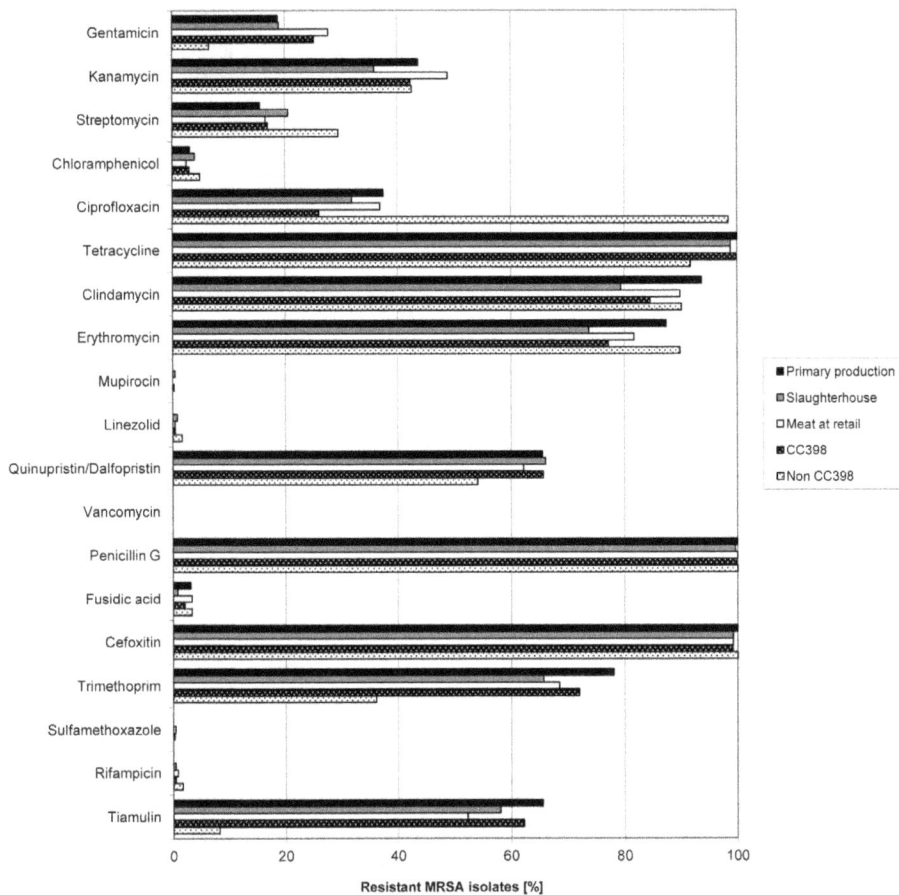

Figure 1. Antimicrobial resistance of MRSA in the German turkey meat production chain. Distribution of antimicrobial resistance of MRSA strains separated into CC398 and non CC 398 strains as well as different steps of the turkey meat production chain isolated from dust samples at turkey primary production (n = 32), carcasses at slaughter (n = 248) and meat at retail (n = 241). The MRSA strains were isolated in the course of the national monitoring for zoonotic agents in Germany in 2010.

the production units. This can explain the increase in the variability of *spa* types along the chain and is in line with the increase in the proportion of non CC398 strains in meat samples compared to dust or carcasses. *Spa* types t002 and t1430 were also present in primary production and therefore probably have been transmitted along the food chain. In contrast, *spa* types t010, t015 were first observed in meat samples.

The majority of MRSA from the German turkey production chain was assigned to the livestock associated CC398 with the predominant *spa* types t011 and t034. This is in line with results from other livestock like veal calves [9], dairy cattle [56,57] and pigs [6] as well as in food [25]. In the present study, 37 of the 521 MRSA strains (7.1%) were identified as t002. This *spa* type t002 is assigned to CC5. In Germany, CC5 is one of the epidemic MRSA strains among humans [58]. Finding t002 in turkey flocks and in turkey meat is in line with other studies from central Europe [12,25,59]. So far, it is not known, whether this strain originates from the "human" strain and is introduced into the food chain on different levels or whether it got established in the turkey population and is transmitted along the chain. Detailed molecular-epidemiological investigations are needed to compare strains both from human and farm to fork origin. In the present study, 4.2% of the MRSA isolates were characterized as *spa* type t1430, a MRSA strain which was also frequently isolated from chicken

meat [25] and broilers at slaughter [60] in the Netherlands. However, it was has also been detected in turkey flocks at farm level [12]. The strain is assigned to ST9, a lineage genetically unrelated to ST398. ST9 is the predominating sequence type among MRSA from pigs in Asian countries [14,15,61–64]. Outside of Europe, MRSA contamination was reported among US turkey meat [32,50]. In both surveys, all isolates belonged to USA 300 (ST8), the most common community associated MRSA strain in the USA, suggesting human contamination during processing.

The frequent use of antimicrobials at farm is discussed as a risk factor for the wide dissemination of MRSA in livestock production chains [65]. In recent studies antimicrobials were identified to be used in more than 90% of the investigated turkey flocks and animals received on average 33 daily doses of antimicrobials during raising and fattening [66]. With a share of 21% β-lactams were most often used followed by polypeptides (15.2%), macrolides (13.4%), tetracyclines and aminoglycosides (12.4% both). Fluoroquinolones were used in 6.5% of the investigated flocks. The common application of antimicrobials via drinking water bears the risk of under dosing of individual animals and contamination of the barn environment with antimicrobials which also facilitates the selection of resistance [67].

Table 2. Similarity matrix of *spa* types, SCC*mec* types and resistance profiles of MRSA isolated from the German turkey meat production chain in the course of the national monitoring for zoonotic agents in 2010 (95% confidence intervals).

		Primary production	Slaughterhouse	Meat at retail	Control Group Wild boar meat
		av. PSI[a] (CI 95%)[b]	av. PSI[a] (CI 95%)[b]	av. PSI[a] (CI 95%)[b]	av. PSI[a] (CI 95%)[b]
Primary production	*spa* types	1			
	SCC*mec* types	1			
	resistance profiles	1			
Slaughterhouse	*spa* types	0.86 (0.72, 0.95)	1		
	SCC*mec* types	0.91 (0.79, 0.98)	1		
	resistance profiles	0.43 (0.30, 0.53)	1		
Meat at retail	*spa* types	0.79 (0.64, 0.90)	0.86 (0.79, 0.92)	1	
	SCC*mec* types	0.85 (0.70, 0.96)	0.87 (0.79, 0.95)	1	
	resistance profiles	0.42 (0.33, 0.51)	0.56 (0.49, 0.62)	1	
Control Group	*spa* types	0.55 (0.33, 0.71)	0.56 (0.38, 0.74)	0.56 (0.38, 0.74)	1
Wild boar meat	SCC*mec* types	0.84 (0.62, 0.98)	0.83 (0.64, 0.96)	0.85 (0.70, 0.95)	1
	resistance profiles	0.13 (0.03, 0.27)	0.19 (0.06, 0.34)	0.14 (0.04, 0.23)	1

[a]PSI: Czekanowski index or proportional similarity index.
[b]CI 95%: 95% confidence interval.

Cluster analysis was used to better describe the multidimensional data set of antibiotic resistance profiles grouping all MRSA strains within 3 different clusters. As the ordinal MIC values generated by two-fold dilutions in substance concentration are difficult to describe by cluster analysis a binary interpretation of the data set was used. Ward's minimum variance with squared Euclidian distance was proven to be the best method to produce well separated cluster in binary antimicrobial resistance data sets [68,69]. No resistance phenotype simultaneously appeared in several clusters. The distribution of *spa* types, SCC*mec* types and the three clusters of antimicrobial resistance types did not significantly differ in the MRSA samples from the three origins. The chi square value was approaching significance with respect to the *spa*-types, which was presumably due to the slightly higher proportion of other CC398 and non CC398. However, considering all three features it cannot be rejected on the basis of the included data that the MRSA isolates from different steps of the turkey meat production chain originate from the same population of strains. This result might rather indicate farm to fork transmission of MRSA of the same pool of strains than development of separate MRSA populations at each step of the chain. The calculation of the Czekanowski index for *spa* type and SCC*mec* type data results in consistently high similarity values between the matrices whereas the comparison of antimicrobial resistance phenotypes observed medium index values. Higher values of similarity were obtained between the adjacent process steps primary production/slaughter and slaughter/meat than between samples from primary production and meat. This result was expected as an increase in the variability of the MRSA isolates might be conceivable at each process stage due to external introduction of new strains via human or environmental contamination or due to spontaneous mutations in the strains.

The lower values of similarity between the distribution of *spa* types and antimicrobial resistance profiles of samples from the turkey meat production chain and the control group indicate that the proposed statistical method is valid to detect existing differences in the distribution of these characteristics of MRSA.

Concerning SCC*mec* types, high index values were also observed in comparison to the control group which might be explained by the insufficient discriminatory power of SCC*mec* typing. In addition, MRSA isolates with not typeable SCC*mec* cassettes were considered as equal that might lead to an overestimation of similarity.

It can be concluded that MRSA is present at every step of the turkey meat production chain in Germany. Using the Czekanowski index it is possible to quantify the similarity of the distribution of *spa* types, SCC*mec* types and antimicrobial resistance phenotypes between MRSA data sets from different stages of turkey meat production chain. Combined with chi square statistics, the high level of similarity suggests MRSA transmission along the chain.

Supporting Information

Table S1 List of competent authorities of the German federal states which were responsible for collecting the samples.

Table S2 Distribution of resistance against 19 different antimicrobials, grouped *spa* types and SCC*mec* types within the binary phenotypic resistance clusters of 521 MRSA isolates sampled at different steps of the German turkey meat production chain in 2010.

Author Contributions

Conceived and designed the experiments: AK BAT KA. Performed the experiments: BK. Analyzed the data: BV. Contributed reagents/materials/analysis tools: JB. Wrote the paper: BV. Designed the software used for bootstrap method: JB. Administration and organisation of the project: AF.

References

1. Hennekinne JA, De Buyser ML, Dragacci S (2012) *Staphylococcus aureus* and its food poisoning toxins: Characterization and outbreak investigation. FEMS Microbiology Reviews 36: 815–836.

2. Köck R, Becker K, Cookson B, van Gemert-Pijnen JE, Harbarth S, et al. (2010) Methicillin-resistant *Staphylococcus aureus* (MRSA): burden of disease and control challenges in Europe. Euro surveillance: bulletin européen sur les maladies transmissibles = European communicable disease bulletin 15: 19688.

3. Voss A, Loeffen F, Bakker J, Klaassen C, Wulf M (2005) Methicillin-resistant *Staphylococcus aureus* in Pig Farming. Emerging Infectious Diseases 11: 1965–1966.

4. Armand-Lefevre L, Ruimy R, Andremont A (2005) Clonal comparison of Staphylococcus from healthy pig farmers, human controls, and pigs. Emerging Infectious Diseases 11: 711–714.

5. Crombé F, Willems G, Dispas M, Hallin M, Denis O, et al. (2012) Prevalence and antimicrobial susceptibility of methicillin-resistant *Staphylococcus aureus* among pigs in Belgium. Microbial Drug Resistance 18: 125–131.

6. EFSA (2009) Analysis of the baseline survey on the prevalence of methicillin-resistant *Staphylococcus aureus* (MRSA) in holdings with breeding pigs, in the EU, 2008, Part A: MRSA prevalence estimates; on request from the European Commission. The EFSA Jounal 11: 1376.

7. Khanna T, Friendship R, Dewey C, Weese JS (2008) Methicillin resistant *Staphylococcus aureus* colonization in pigs and pig farmers. Vet Microbiol 128: 298–303.

8. Smith TC, Male MJ, Harper AL, Kroeger JS, Tinkler GP, et al. (2009) Methicillin-resistant *Staphylococcus aureus* (MRSA) strain ST398 is present in midwestern U.S. swine and swine workers. PloS one 4.

9. Graveland H, Wagenaar JA, Heesterbeek H, Mevius D, van Duijkeren E, et al. (2010) Methicillin resistant *Staphylococcus aureus* ST398 in veal calf farming: human MRSA carriage related with animal antimicrobial usage and farm hygiene. PloS one 5.

10. Nemati M, Hermans K, Lipinska U, Denis O, Deplano A, et al. (2008) Antimicrobial resistance of old and recent *Staphylococcus aureus* isolates from poultry: First detection of livestock-associated methicillin-resistant strain ST398. Antimicrobial Agents and Chemotherapy 52: 3817–3819.

11. Persoons D, Van Hoorebeke S, Hermans K, Butaye P, De Kruif A, et al. (2009) Methicillin-resistant *Staphylococcus aureus* in poultry. Emerging Infectious Diseases 15: 452–453.

12. Richter A, Sting R, Popp C, Rau J, Tenhagen BA, et al. (2012) Prevalence of types of methicillin-resistant *Staphylococcus aureus* in turkey flocks and personnel attending the animals. Epidemiology and Infection 140: 2223–2232.

13. Reischl U, Frick J, Hoermansdorfer S, Melzl H, Bollwein M, et al. (2009) Single-nucleotide polymorphism in the SCC*mec*-orfX junction distinguishes between livestock-associated MRSA CC398 and human epidemic MRSA strains. Euro surveillance: bulletin european sur les maladies transmissibles = European communicable disease bulletin 14.

14. Anukool U, O'Neill CE, Butr-Indr B, Hawkey PM, Gaze WH, et al. (2011) Meticillin-resistant *Staphylococcus aureus* in pigs from Thailand. International Journal of Antimicrobial Agents 38: 86–87.

15. Wagenaar JA, Yue H, Pritchard J, Broekhuizen-Stins M, Huijsdens X, et al. (2009) Unexpected sequence types in livestock associated methicillin-resistant *Staphylococcus aureus* (MRSA): MRSA ST9 and a single locus variant of ST9 in pig farming in China. Veterinary Microbiology 139: 405–409.

16. Cookson BD, Robinson DA, Monk AB, Murchan S, Deplano A, et al. (2007) Evaluation of molecular typing methods in characterizing a European collection of epidemic methicillin-resistant *Staphylococcus aureus* strains: the HARMONY collection. J Clin Microbiol 45: 1830–1837.

17. (IWG-SCC) IWGotCoSCCE (2009) Classification of staphylococcal cassette chromosome mec (SCCmec): guidelines for reporting novel SCCmec elements. Antimicrobial Agents and Chemotherapy 53: 4961–4967.

18. Garcia-Alvarez L, Holden MT, Lindsay H, Webb CR, Brown DF, et al. (2011) Meticillin-resistant *Staphylococcus aureus* with a novel mecA homologue in human and bovine populations in the UK and Denmark: a descriptive study. Lancet Infect Dis 11: 595–603.

19. Frenay HM, Bunschoten AE, Schouls LM, van Leeuwen WJ, Vandenbroucke-Grauls CM, et al. (1996) Molecular typing of methicillin-resistant *Staphylococcus aureus* on the basis of protein A gene polymorphism. Eur J Clin Microbiol Infect Dis 15: 60–64.

20. de Neeling AJ, van den Broek MJ, Spalburg EC, van Santen-Verheuvel MG, Dam-Deisz WD, et al. (2007) High prevalence of methicillin resistant *Staphylococcus aureus* in pigs. Vet Microbiol 120: 366–372.

21. Pires SM, Evers EG, van PW, Ayers T, Scallan E, et al. (2009) Attributing the human disease burden of foodborne infections to specific sources. Foodborne PathogDis 6: 417–424.

22. WHO (2001) World Health Organisation Global Principles for the Containment of Antimicrobial Resistance in Animals intended for Food. http://wwwwhoint/drugresistance/WHO_Global_Strategyhtm/en/.

23. Alt K, Fetsch A, Schroeter A, Guerra B, Hammerl J, et al. (2011) Factors associated with the occurrence of MRSA CC398 in herds of fattening pigs in Germany. BMC Veterinary Research 7: 69.

24. Broens EM, Graat EAM, Van der Wolf PJ, Van de Giessen AW, de Jong MCM (2011) Prevalence and risk factor analysis of livestock associated MRSA-positive pig herds in The Netherlands. Preventive Veterinary Medicine 102: 41–49.

25. de Boer E, Zwartkruis-Nahuis JTM, Wit B, Huijsdens XW, de Neeling AJ, et al. (2009) Prevalence of methicillin-resistant *Staphylococcus aureus* in meat. International Journal of Food Microbiology 134: 52–56.

26. Hanson BM, Dressler AE, Harper AL, Scheibel RP, Wardyn SE, et al. (2011) Prevalence of *Staphylococcus aureus* and methicillin-resistant *Staphylococcus aureus* (MRSA) on retail meat in Iowa. Journal of Infection and Public Health 4: 169–174.

27. Lim SK, Nam HM, Park HJ, Lee HS, Choi MJ, et al. (2010) Prevalence and characterization of methicillin-resistant *Staphylococcus aureus* in raw meat in Korea. Journal of Microbiology and Biotechnology 20: 775–778.

28. Lozano C, López M, Gómez-Sanz E, Ruiz-Larrea F, Torres C, et al. (2009) Detection of methicillin-resistant *Staphylococcus aureus* ST398 in food samples of animal origin in Spain. Journal of Antimicrobial Chemotherapy 64: 1325–1326.

29. O'Donoghue M, Chan M, Ho J, Moodley A, Boost M. Prevalence of Mehicillin-Resistant Staphylococcus aureus in Meat from Hong Kong Shops and Markets. 27.

30. Pu S, Han F, Ge B (2009) Isolation and characterization of methicillin-resistant *Staphylococcus aureus* strains from louisiana retail meats. Applied and Environmental Microbiology 75: 265–267.

31. Van Loo IHM, Diederen BMW, Savelkoul PHM, Woudenberg JHC, Roosendaal R, et al. (2007) Methicillin-resistant *Staphylococcus aureus* in meat products, the Netherlands. Emerging Infectious Diseases 13: 1753–1755.

32. Waters AE, Contente-Cuomo T, Buchhagen J, Liu CM, Watson L, et al. (2011) Multidrug-resistant *staphylococcus aureus* in US meat and poultry. Clinical Infectious Diseases 52: 1227–1230.

33. Weese JS, Reid-Smith R, Rousseau J, Avery B (2010) Methicillin-resistant *Staphylococcus aureus* (MRSA) contamination of retail pork. Canadian Veterinary Journal-Revue Veterinaire Canadienne 51: 749–752.

34. EC (2003) Directive 2003/99/EC of the European Parliament and of the Council of 17 November 2003 on the monitoring of zoonoses and zoonotic agents. Official Journal of the European Union, L325/31. http://eur-lex.europa.eu/LexUriServ/LexUriServ.do?uri = OJ:L:2003:325:0031:0040:EN:PDF.

35. Käsbohrer A, Wegeler C, Tenhagen BA (2009) EU-weite und nationale Monitoringprogramme zu Zoonoseerregern in Deutschland. Journal für Verbraucherschutz und Lebensmittelsicherheit. EU-weite und nationale Monitoringprogramme zu Zoonoseerregern in Deutschland. 41–45.

36. Straub JA, Hertel C, Hammes WP (1999) A 23S rDNA-targeted polymerase chain reaction-based system for detection of *Staphylococcus aureus* in meat starter cultures and dairy products. Journal of Food Protection 62: 1150–1156.

37. Poulsen AB, Skov R, Pallesen LV (2003) Detection of methicillin resistance in coagulase-negative staphylococci and in staphylococci directly from simulated blood cultures using the EVIGENE MRSA Detection Kit. Journal of Antimicrobial Chemotherapy 51: 419–421.

38. Shopsin B, Gomez M, Montgomery SO, Smith DH, Waddington M, et al. (1999) Evaluation of protein A gene polymorphic region DNA sequencing for typing of *Staphylococcus aureus* strains. Journal of Clinical Microbiology 37: 3556–3563.

39. Zhang K, McClure JA, Elsayed S, Louie T, Conly JM (2005) Novel multiplex PCR assay for characterization and concomitant subtyping of staphylococcal cassette chromosome *mec* types I to V in methicillin-resistant *Staphylococcus aureus*. Journal of Clinical Microbiology 43: 5026–5033.

40. Jansen MD, Box ATA, Fluit AC (2009) SCCmec typing in methicillin-resistant *Staphylococcus aureus* strains of animal origin. Emerging Infectious Diseases 15: 136.

41. Enright MC, Day NPJ, Davies CE, Peacock SJ, Spratt BG (2000) Multilocus sequence typing for characterization of methicillin-resistant and methicillin-susceptible clones of *Staphylococcus aureus*. Journal of Clinical Microbiology 38: 1008–1015.

42. CLSI (2006) Performance Standards for Antimicrobial Disk Susceptibility Tests; Approved Standard.

43. Calinski T, Harabasz J (1974) A dendrite method for cluster analysis. Communications in Statistics: Taylor & Francis. 1–27.

44. Duda RO, Hart PE (1973) Pattern Classification and Scene Analysis. New York: Wiley.

45. Rosef O, Kapperud G, Lauwers S, Gondrosen B (1985) Serotyping of *Campylobacter jejuni*, *Campylobacter coli*, and *Campylobacter laridis* from domestic and wild animals. Applied and Environmental Microbiology 49: 1507–1510.

46. Efron B, Tibshirani R (1986) Bootstrap methods for standard errors, confidence invervals and other measures of statistical accuracy. Statistical Science 1: 54–75.

47. Bundesamt für Verbraucherschutz und Lebensmittelsicherheit (2012) Berichte zur Lebensmittelsicherheit 2010. BVL Reporte http://www.bvl.bund.de/SharedDocs/Downloads/01_Lebensmittel/04_Zoonosen_Monitoring/Zoonosen_Monitoring_Bericht_2010.pdf?__blob = publicationFile&v = 6.

48. Kreausukon K, Fetsch A, Kraushaar B, Alt K, Müller K, et al. (2012) Prevalence, antimicrobial resistance, and molecular characterization of methicillin-resistant *Staphylococcus aureus* from bulk tank milk of dairy herds. Journal of Dairy Science 95: 4382–4388.

49. Argudin MA, Rodicio MR, Guerra B (2010) The emerging methicillin-resistant *Staphylococcus aureus* ST398 clone can easily be typed using the Cfr9I SmaI-neoschizomer. Lett Appl Microbiol 50: 127–130.

50. Bhargava K, Wang X, Donabedian S, Zervos M, da Rocha L, et al. (2011) Methicillin-resistant *staphylococcus aureus* in retail meat, Detroit, Michigan, USA. Emerging Infectious Diseases 17: 1135–1137.

51. Lassok B, Tenhagen BA (2013) From pig to pork: Methicillin-resistant *staphylococcus aureus* in the pork production chain. Journal of Food Protection 76: 1095–1108.

52. Löhren U (2012) Overview on current practices of poultry slaughtering and poultry meat inspection Supporting Publications 2012: EN-298. http://wwwefsaeuropaeu/en/supporting/doc/298epdf.

53. Hentschel S, Kusch D, Sinell HJ (1979) *Staphylococcus aureus* in Poultry-biochemical characteristics, antibiotic resistance and phage pattern. Zentralblatt für Bakteriologie, Parasitenkunde, Infektionskrankheiten und Hygiene. Erste Abteilung Originale. Reihe B: Hygiene, Betriebshygiene, präventive Medizin 07, 168 (5–6): 546–561.

54. Großklaus D, Lessing G (1972) Hygieneprobleme beim Schlachtgeflügel. Fleischwirtschaft 52: 1011–1013.

55. Berrang ME, Buhr RJ, Cason JA, Dickens JA (2001) Broiler carcass contamination with Campylobacter from feces during defeathering. Journal of Food Protection 64: 2063–2066.

56. Spohr M, Rau J, Friedrich A, Klittich G, Fetsch A, et al. (2011) Methicillin-resistant *Staphylococcus aureus* (MRSA) in three dairy herds in southwest Germany. Zoonoses and Public Health 58: 252–261.

57. Vanderhaeghen W, Hermans K, Haesebrouck F, Butaye P (2010) Methicillin-resistant *Staphylococcus aureus* (MRSA) in food production animals. Epidemiol Infect 138: 606–625.

58. Robert Koch-Institut (2011) Auftreten und Verbreitung von MRSA in Deutschland 2010. Epidemiologisches Bulletin 26: 233–244.

59. Feßler AT, Kadlec K, Hassel M, Hauschild T, Eidam C, et al. (2011) Characterization of methicillin-resistant *Staphylococcus aureus* isolates from food and food products of poultry origin in Germany. Applied and Environmental Microbiology 77: 7151–7157.

60. Mulders MN, Haenen APJ, Geenen PL, Vesseur PC, Poldervaart ES, et al. (2010) Prevalence of livestock-associated MRSA in broiler flocks and risk factors for slaughterhouse personnel in the Netherlands. Epidemiology and Infection 138: 743–755.

61. Cui S, Li J, Hu C, Jin S, Li F, et al. (2009) Isolation and characterization of methicillin-resistant *Staphylococcus aureus* from swine and workers in China. Journal of Antimicrobial Chemotherapy 64: 680–683.

62. Larsen J, Imanishi M, Hinjoy S, Tharavichitkul P, Duangsong K, et al. (2012) Methicillin-resistant *Staphylococcus aureus* ST9 in pigs in Thailand. PloS one 7.

63. Neela V, Zafrul AM, Mariana NS, Van Belkum A, Liew YK, et al. (2009) Prevalence of ST9 methicillin-resistant *Staphylococcus aureus* among pigs and pig handlers in Malaysia. Journal of Clinical Microbiology 47: 4138–4140.

64. Tsai HY, Liao CH, Cheng A, Liu CY, Huang YT, et al. (2012) Isolation of meticillin-resistant *Staphylococcus aureus* sequence type 9 in pigs in Taiwan. International Journal of Antimicrobial Agents 39: 449–451.

65. Schwarz S, Chaslus-Dancla E (2001) Use of antimicrobials in veterinary medicine and mechanisms of resistance. Veterinary Research 32: 201–225.

66. State Office for Consumer Protection and Food Safety in Lower Saxony G (2011) Bericht über den Antibiotikaeinsatz in der landwitschaftlichen Nutztierhaltung in Niedersachsen November 2011. http://wwwmlniedersachsende/portal/livephp?navigation_id=27751&article_id=102202&_psmand=7.

67. Richter A, Hafez HM, Böttner A, Gangl A, Hartmann K, et al. (2009) Applications of antibiotics in poultry. Tierarztliche Praxis Ausgabe G: Grosstiere – Nutztiere Verabreichung von Antibiotika in Geflügelbeständen. 321–329.

68. Berge ACB, Atwill ER, Sischo WM (2003) Assessing antibiotic resistance in fecal *Escherichia coli* in young calves using cluster analysis techniques. Preventive Veterinary Medicine 61: 91–102.

69. Milligan GW (1981) A monte carlo study of thirty internal criterion measures for cluster analysis. Psychometrika 46: 187–199.

19

Protracted Effects of Juvenile Stressor Exposure Are Mitigated by Access to Palatable Food

Jennifer Christine MacKay[1,5], Jonathan Stewart James[1,5], Christian Cayer[1,5], Pamela Kent[1,5], Hymie Anisman[4], Zul Merali[1,2,3,5]*

1 School of Psychology, University of Ottawa, Ottawa, Ontario, Canada, 2 Department of Psychiatry, University of Ottawa, Ottawa, Ontario, Canada, 3 Department of Cellular and Molecular Medicine, University of Ottawa, Ottawa, Ontario, Canada, 4 Institute of Neuroscience, Carleton University, Ottawa, Ontario, Canada, 5 University of Ottawa Institute of Mental Health Research, Ottawa, Ontario, Canada

Abstract

Stressor experiences during the juvenile period may increase vulnerability to anxiety and depressive-like symptoms in adulthood. Stressors may also promote palatable feeding, possibly reflecting a form of self-medication. The current study investigated the short- and long-term consequences of a stressor applied during the juvenile period on anxiety- and depressive-like behavior measured by the elevated plus maze (EPM), social interaction and forced swim test (FST). Furthermore, the effects of stress on caloric intake, preference for a palatable food and indices of metabolic syndrome and obesity were assessed. Male Wistar rats exposed to 3 consecutive days of variable stressors on postnatal days (PD) 27–29, displayed elevated anxiety-like behaviors as adults, which could be attenuated by consumption of a palatable high-fat diet. However, consumption of a palatable food in response to a stressor appeared to contribute to increased adiposity.

Editor: Alessandro Bartolomucci, University of Minnesota, United States of America

Funding: This project was funded by an operating grant (Application No. 275228) from the Canadian Institute for Health Research (CIHR: http://www.cihr-irsc.gc.ca) awarded to Zul Merali, and by a Canadian Health Research Institutes doctoral Canada Graduate Scholarship awarded to J. Christine MacKay. The funders had no role in study design, data collection and analysis, decision to publish, or preparation of the manuscript.

Competing Interests: The authors have declared that no competing interests exist.

* E-mail: Zul.Merali@uottawa.ca

Introduction

Adolescent and childhood obesity has become a worldwide epidemic to the extent that globally, approximately 200 million school aged children can be classified as either overweight or obese [1]. Further, childhood and adolescent obesity is a strong predictor of adult obesity [2–4], and have been associated with adverse long-term health outcomes such as, hypercholesterolemia, insulin resistance [5], hypertension [6], type-2 diabetes [7], nonalcoholic fatty liver disease [8] and various cancers [9,10].

Recent increases in the prevalence and incidence of obesity have been attributed to the interplay of a variety of different factors, such as genetics, the family environment, levels of physical activity, advertising, and sedentary behaviors [11]. Of particular interest is the role of stress in the development and maintenance of obesity, as the level of daily stressors individuals have been experiencing continues to increase [12–14]. In this regard, a stressor-induced preference for palatable foods (especially those with a high fat and/or sugar content) has been documented in both animals and humans [15–18]. This preference was proposed to serve as a form of self-medication to protect against the adverse effects of stress [19]. Access to a palatable food has been shown to protect against behavioral disturbances elicited by inescapable foot shock [20], reduce sympathetic responses to a stressor [21–30], and diminish hypothalamic-pituitary adrenal (HPA) axis activity [31–34]. Consumption of palatable foods may also limit some of the long-term negative effects of an early life stressor (maternal separation) [35].

Although self-medication with a palatable food can have some beneficial effects on psychological functioning, it may be a counterproductive long-term stress coping strategy. The excess calories gained from consumption of a high fat and/or sugar diet leads to an increase in adipose tissue [36], which secretes several hormones and signaling factors [37–39] involved in the regulation of food intake [37,40]. Increases of adipose tissue and the resulting endocrinological consequences have been implicated in atherosclerosis [36], elevated plasma levels of inflammatory cytokines [39], and elevated concentrations of free fatty acids, which can contribute to reduced muscle glucose uptake [41] and the development of insulin resistance, which has been directly linked to type 2 diabetes, cardiovascular disease, and cancer [42,43].

The relationship between stress, eating behavior and obesity during adolescence may be of particular significance given the increased prevalence of obesity in this population. Indeed, in both humans and rodents, the juvenile phase represents a critical period in development during which there is substantial cerebral development and reorganization as well as altered HPA axis function [44,45]. It has been suggested that this major biological transition period renders adolescents more sensitive to the effects of stressors and the subsequent development of stressor-related psychopathologies [46]. Animal studies have shown unique effects of stressors on HPA activity during adolescence as pre-pubertal rats exhibited higher or prolonged adrenocorticotrophin releasing hormone and corticosterone release compared to adults [44]. However, these effects were sex dependent and also varied with the stressor employed [44]. Adolescence is further characterized as a unique stage in brain development, as regions related to emotional and learning processes, such as the prefrontal cortex, hippocampus, and the amygdala, all undergo substantial remodeling during this phase and have also been proposed to be exquisitely sensitive to the effects of stress [47]. Importantly, these regions have also been implicated in the homeostatic mechanisms underlying energy balance and feeding behaviors [18,47].

The present study was conducted to further characterize the short- and long-term consequences of stressor exposure during the juvenile period (PD 27–29) on behavioral indices of depression and anxiety using a modified version of the juvenile stress protocol developed previously [48]. A second objective was to determine the effects of juvenile stressors on feeding behavior and preference for a palatable food. In this regard, we aimed to elucidate whether the consumption of a palatable food during adolescence could attenuate the long-term behavioral consequences of stressor exposure. A final objective of this study was to investigate the long term consequences of stress-induced palatable feeding on indices of metabolic syndrome and obesity.

Materials and Methods

1. Animals

Ninety-six 21 day old male Wistar rats were obtained from Charles River (Quebec, Canada). Upon arrival, rats were randomly assigned to four conditions: (1) Chow + Control; (2) Chow + Stress; (3) Palatable + Control; and (4) Palatable + Stress. Rats were double housed until PD 30 or 60 in standard plastic cages with bedding at a room temperature of $22\pm1°C$ on a 12 hour light-dark cycle (lights on at 0700hr and off at 1900hr). On PD 30 or 60 all animals were singly housed in order to prepare for social interaction testing. Animals remained singly housed for the remainder of testing. Weight gain was measured on PDs 25, 30, 40, 50, 60, 70. All procedures met the guidelines established by the Canadian Council on Animal Care and were approved by the Animal Care Committee of the University of Ottawa Institute of Mental Health Research.

2. Experiments

Forty rats were used in Experiment 1 (juvenile testing, PD-30-37) and fifty-six rats were used in Experiment 2 (adult testing, PD 60-67). Animals in both experiments underwent all behavioral tests. Weight gain, caloric intake, comfort preference, glucose tolerance, plasma corticosterone and adiposity were only collected in Experiment 2. An overview of each experiment is presented in Figure 1.

3. Diet

All rats were given free access to standard laboratory chow (3.4 kcal/g, 4.5% fat, 18.1% protein, 57.3% carbohydrate, Charles River Rodent Diet 5075, Agribrand Purina Canada, Woodstock, Ontario, Canada) and rats assigned to the palatable food condition were given limited daily access to a 45%Kcal Fat diet (4.7 Kcal/g, 23.2% fat, 17.3% protein, 47.6% carbohydrates, TD.08811, Harlan Laboratories, Madison, Wisconsin, USA). Palatable food was given daily at 8:00 and removed at 10:00 and intake was measured. Every three days, chow consumption was measured by subtracting the weight of the chow in the cage from its weight 24-hours earlier. Total caloric intake was calculated by multiplying the energy content (Kcal/g) of individual diets by the amount consumed and summing them. Preference for palatable food was calculated by taking the ratio of calories of palatable food consumed over total calories consumed. Food

intake data was collected from PD 21 to PD 60. As rats were housed in pairs, equal food intake by cage mates was assumed.

4. Juvenile Stress Paradigm

The juvenile stress procedure used was a modification of the 3 day procedure described by Jacobson-Pick and Ritcher-Levin [48]. This procedure comprised 3 consecutive days of exposure to a different stressor per day throughout PD 27–29.

PD 27 - Forced swimming. Rats were individually placed in a circular water basin (diameter 48 cm; height 42 cm; water depth 29 cm and a temperature of $22\pm2°C$) for 10 min, during which they swam or floated continuously.

PD 28 – Elevated platform. Rats were individually placed on a small elevated platform (12 cm×12 cm; 70 cm elevation from water level) for three separate 30 min sessions. Platforms stood within a basin filled with water for the animals' protection should they fall off. During the intersession interval of 60 min rats were returned to their home cage. Animals which fell during the testing session were immediately returned to the platform for the remainder of the test session.

PD 29 – Restraint. Rats were placed for 30 min in a plastic restraining bag that prevented side-to-side movement and limited forward-backward movement. The plastic bag had a hole at the end closest to the rat's nose to allow for ventilation.

5. Behavioral Testing

Behavioral testing was conducted under low illumination (30–40 lux) between 10:00–13:00 daily following a 1 hr period during which rats habituated to the testing room. Behavior was monitored via a video camera mounted above the arena. Rats were tested in all behavioral paradigms. Testing was conducted between PD 30–37 or 60–67. An overview of the stress procedure and behavioral testing schedule is presented in Figure 1.

5.1 Open Field. On PD 30 or 60, prior to elevated plus maze (EPM) testing, rats were placed in the center of the arena and its behavior monitored for 5 min. The arena consisted of a square Plexiglas arena measuring 60×60 cm with 30 cm high walls. Using lines, the floor was divided up into 16 squares (5×5 cm). The total number of squares crossed was recorded as an index for general locomotor activity [49].

5.2 Elevated Plus Maze. On PD 30 or 60, rats were placed on the center platform of the EPM. The EPM consisted of two open arms (50×10 cm) and two perpendicularly situated arms enclosed by 40 cm high walls, elevated approximately 66 cm above the floor. Immediately after OF testing, rats were placed onto the open central platform of the EPM (facing a closed arm). Behavior scored during the 5 min test included time spent on the open and closed arms and risk assessment behavior (unprotected head dips; head protruding over the edge of an open arm). Time in the open arms and unprotected head dips are validated measures of reduced anxiety-like behavior [50].

5.3 Social Interaction. The social interaction test was conducted over a total duration of three days. On PD 30 or 60 rats were individually housed 2 hours after the completion of EPM testing in order to increase the level of social interaction [51]. On PD 33 or 63, rats were matched to a partner from the same diet x stress condition (but from a different cage) on the basis of body weight, such that members of a pair did not differ by >10 g. On habituation day 1 (PD 33 or 63), rats along with their test day partner were placed in the arena (60×60 cm; 30 cm high walls) for five minutes. On habituation day 2 (PD 34 or 64), rats were individually placed in the arena for a period of 3 min. On test day (PD 35 or 65), pairs of rats were placed in the test chamber and the behavior of both rats was observed for 7 min. Total time spent in

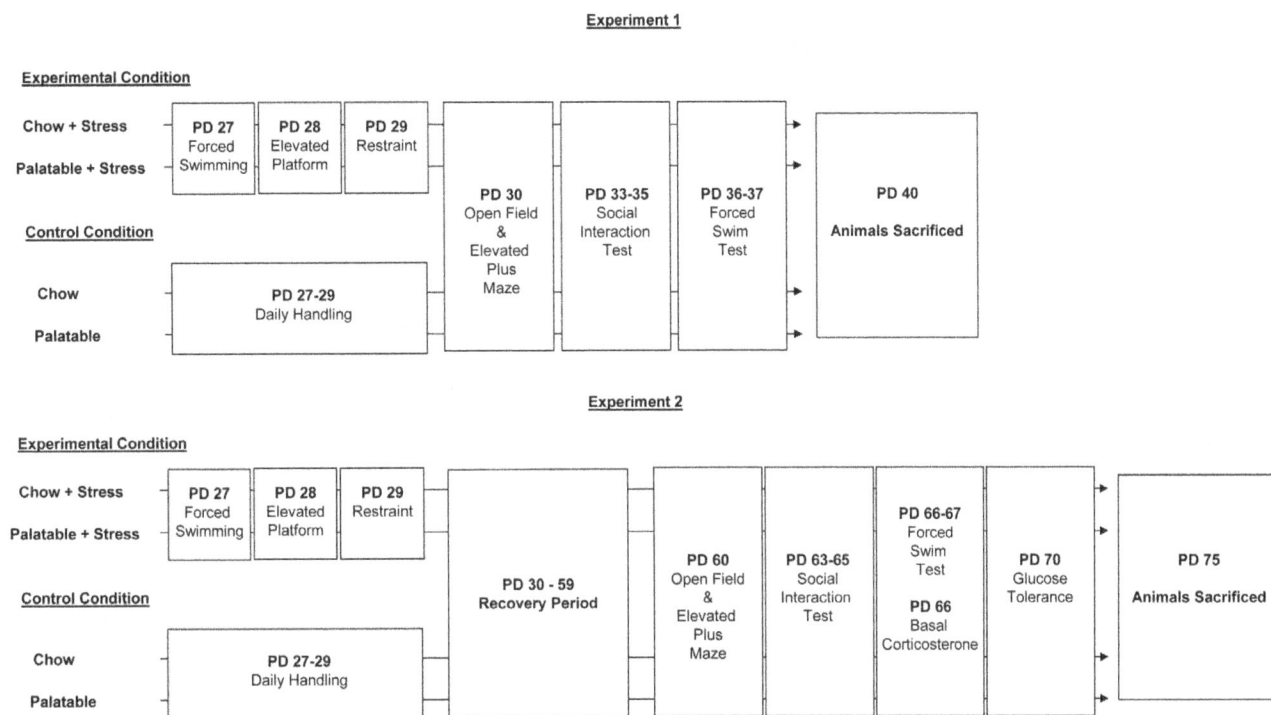

Figure 1. Summary of study design. Animals assigned to stress condition were exposed to a different stressor per day from PD 27–29 while control animals received daily handling. All animals were tested in the indicated behavioral paradigms as described in each experiment.

social interaction (including sniffing, climbing over each other, following, allogrooming, and play fighting) was recorded. Decreases in social interaction are reflective of an anxiogenic profile [51].

5.4 Forced Swim Test. The forced swim test is a widely used behavioral despair paradigm used to evaluate the effectiveness of antidepressant drugs [52–54]. The forced swim arena consisted of a clear Plexiglas cylinder (20×45 cm, water height 30 cm, temperature 25°C; Stoelting Co., Wood Dale, Illinois). A habituation session (PD 36 or 66: 15 min) was performed twenty-four hours prior to the test session. On test day (PD 37 or 67) the animal was returned to the same cylinder for 5 minutes. The time spent immobile was recorded during the test session.

6. Basal Corticosterone Levels and Assay

On PD 66, prior to the FST training session, rats from Experiment 2 were moved in their home cages to the testing room and allowed to rest for a one hour. Blood samples were collected from rats individually using tail venipuncture. The time elapsed from retrieving the rat from their home cage to the depositing of the blood sample on the filter paper was approximately 1–2 min per rat. Blood droplets were deposited onto 903 ProteinSaver filter paper (GE Healthcare Bio-Sciences Corp, MA, USA), allowed to dry at room temperature then stored at −20°C.

Collected samples were analyzed using a radioimmunoassay (RIA). Two days prior to the RIA procedure, blood was eluted from the filter paper by placing one 3 mm punch (per time point) of filter paper in a 12×75 culture tube containing 200 μL Dulbecco's Phosphate Buffered Saline (sigma, item D-5773) w/ 0.1% gelatine, covered with parafilm in a fridge at 4°C. On the day of the RIA procedure, culture tubes containing the samples were placed on an orbital shaker for 1 hour at room temp. CORT levels were the determined from the eluted blood sample using

commercial RIA kits as per the manufacturer's instructions (MP Biomedicals, CA). The inter- and intra-assay variability was 7.3% and 7.4%, respectively.

7. Glucose Tolerance

On PD 70, following 12 hr of fasting, rats in Experiment 2 were given an intraperitoneal injection of a 0.75 g/mL dextrose solution (dose: 1.75 g/kg). Blood glucose was measured by applying a drop of blood (via tail venipuncture) onto a test strip, then taking a reading with a blood glucose meter (Accu-Chek Aviva Nano, Roche Diagnostics, Mannheim, Germany). Levels were assessed immediately before the injection, and 15, 30, 60 and 120 min post-injection.

8. Adiposity

On PD 75, carcasses of animals in Experiment 2 were collected following sacrifice and then stored at −20°C. The carcasses were later thawed and fat pads hand dissected and weighed. The fat pads collected included: mesenteric, retroperitoneal, subcutaneous inguinal white fat, and inter-scapular brown fat.

9. Statistical analysis

Data obtained from the open field, EPM, social interaction, FST, corticosterone and adiposity tests were analyzed by 2 (Diet)×2 (Stress) analysis of variance (ANOVA) for each measure. Data for total caloric intake, palatable food preference, weight, and glucose tolerance were analyzed using three-way (Diet × Stress × Time) repeated-measures ANOVAs where Diet and Stress were the between group variables and Time the (repeated) within group variable. Subsequent follow-up comparisons were conducted using t tests with a Bonferroni correction to maintain the alpha level at 0.05. For some variables (weight, caloric intake, palatable food preference, behavioural tests, glucose intolerance and adiposity) a priori predictions that access to the palatable diet would alter the impact of the stressor were made. Follow-up comparisons for interactions for the aforementioned predictions were conducted irrespective of whether the F value for an interaction reached significance [55]. Data points ±3 standard deviations from calculated means were considered as outliers and not included in statistical analysis [56]. Some rats were removed from the statistical analysis of behavioral indices as a result of missing data (e.g. rat fell off the EPM), thus the N and df associated with these measures vary across outcomes.

Results

1. Effect of stress and diet on weight and food intake

Juvenile stressor exposure and diet had a significant effect on body weight. Repeated measures ANOVA revealed a significant Stress ($F_{5,255} = 7.82$, p<.001) and Diet effect ($F_{5,255} = 9.89$, p< .001) (Figure 2A and 2B). Follow-up comparisons were completed based on an a priori hypothesis that a significant interaction ($F_{5,255} = 1.47$, p = .201) would be present. Simple effects analysis revealed that previously stressed chow fed rats weighed significantly less than controls at PDs 40 (p = .021), 60 (p = .007) and 70 (p = .002). Among previously stressed rats with access to the palatable diet, a significant decrease in weight relative to the palatable control group was observed on PDs 30 (p = .040), 40 (p = .045), and 50 (p = .033). Among previously stressed rats, those with access to palatable food weighed significantly more than chow fed rats on PD 70 (p = .011).

Repeated measures ANOVA revealed a significant effect of Diet ($F_{9,207} = 2.83$, p = .026) on total caloric intake (Figure 2C). In general, rats with access to the palatable diet consumed more calories. Follow-up comparisons were completed based on an a priori hypothesis that a significant interaction ($F_{9,207} = 1.23$, p = .280) would be present. Simple effects analysis showed that previously stressed rats with access to chow only displayed significant reductions in total caloric intake on PD 32 (p = .003) and PD 47 (p = .024). A significant difference between the two stress groups was observed on PD 41 (p = .023) and PD 48 (p = .018). No significant effect of Stress on preference for palatable food ($F_{9,99} = 1.895$, p = .061; Figure 2D) was observed.

2. Effect of stress and diet on anxiety- and depressive-like behavior

Locomotor behavior across groups was comparable in both Experiment 1 and 2. No significant effects of Stress or Diet were found regarding number of squares crossed in the open field at both the juvenile and adult time point. Means (±SEM) for the juvenile time point were as follows: Chow + Control 123.10±14.26; Chow + Stress 118.00±10.25; Palatable + Control 140.10±11.36 and Palatable + Stress 128.10±3.85. Means (±SEM) for the adult time point were as follows: Chow + Control 104.19±6.41; Chow + Stress 98.57±7.81; Palatable + Control 103.85±6.16 and Palatable + Stress 93.86±4.02.

In the EPM (Figure 3, A–D), a significant main effect for Diet on the amount of time spent on the open arms ($F_{1,35} = 4.717$, p = .037) was observed in Experiment 1. In general, juvenile rats with access to the palatable diet spent more time on the open arms of the maze. No significant differences in time spent in the closed arm, and number of risk assessments was observed. No significant effects of Stress or Diet were found regarding entries into the closed arm suggesting that locomotor activity across groups was comparable.

In Experiment 2, a significant main effects for Stress was observed in the EPM (Figure 3, E–H) for time spent in the open

arms ($F_{1,50} = 12.271$, p = .001), and time spent in the closed arms ($F_{1,50} = 13.55$, p = .001) suggesting that the juvenile stressor induced long-term anxiety-like behaviors. Follow-up simple effects comparisons on time spent on the open and closed arms were completed based on an a priori hypothesis that significant interactions ($F_{1,50} = 2.52$, p = .118 for closed; $F_{1,50} = .911$, p = .345 for open) would be present. Interestingly, previously stressed rats with access to chow spent significantly less time in the open arm (p = .003), and spent more time in the closed arms of the maze (p<.001) compared to their controls. Among the two stress groups, chow fed rats spent significantly more time in the closed arms relative to those with access to the palatable diet (p = .039). No significant differences were observed in terms of the number of risk assessments made and number of entries into the closed arm entries, again suggesting that locomotor activity was comparable across groups.

In the social interaction test, no significant differences in total time spent engaging in active social behaviors was observed in juvenile rats (Figure 4, A–B). In regard to individual social behaviors, a significant Diet effect was observed in time spent play fighting ($F_{1,35} = 5.250$, p = .028) as rats with access to the palatable diet engaged in more play fighting.

In Experiment 2, juvenile stress induced a long-term increase in anxiety-like behavior measured in the social interaction test (Figure 4, C–D). A significant interaction between Stress and Diet was found with regard to total time spent engaged in social behaviors ($F_{1,51} = 9.085$, p = .004), following ($F_{1,51} = 6.23$, p = .016), moving over/under one another ($F_{1,51} = 4.93$, p = .031) and time spent sniffing one another ($F_{1,51} = 6.41$, p = .014). The follow-up simple effects analyses confirmed that previously stressed rats with access to chow spent significantly less time engaged in active social interaction compared to non-stressed controls (p = .004). This decreased activity was also observed in individual social behaviors, including following (p = .015), and moving over/under one another (p = .013) and time spent sniffing on another (p = .037). Stressed rats with access to chow also spent significantly less time total time engaged in social behaviors (p = .029), as well as following one another (p = .003) compared to previously stressed rats with access to palatable food.

For both the juvenile and adult time points, no significant effects of Stress or Diet were found regarding time spent immobile in the FST. Mean (±SEM) time spent immobile for the juvenile time point was as follows: Chow + Control 5.208±1.20; Chow + Stress 3.805±1.20; Palatable + Control 4.060±1.27 and Palatable + Stress 3.298±1.20. Mean (±SEM) time spent immobile for the adult time point was as follows: Chow + Control 21.29±3.96; Chow + Stress 20.89±3.96; Palatable + Control 27.52±4.11 and Palatable + Stress 19.47±4.11.

3. Effect of stress and diet on endocrine and metabolic indices

A significant Diet effect was observed in basal corticosterone levels measured on PD 66, $F_{1,50} = 10.56$, p = .002 (Figure 5A), wherein rats with access to the palatable diet displayed significantly reduced basal corticosterone.

Repeated measures ANOVA revealed a significant Time × Stress interaction for blood glucose levels in response to a glucose challenge, ($F_{4,200} = 2.621$, p = .036; Figure 5B). Although previously stressed rats with access to the palatable diet tended to show increased blood glucose relative to their controls follow-up simple effects analysis did not reach significance.

In general all rats with access to the palatable diet displayed an increase in adiposity (Figure 5C). A significant Diet effect was observed in the weight of retroperitoneal ($F_{1,52} = 10.93$ p = .002)

Body Weight

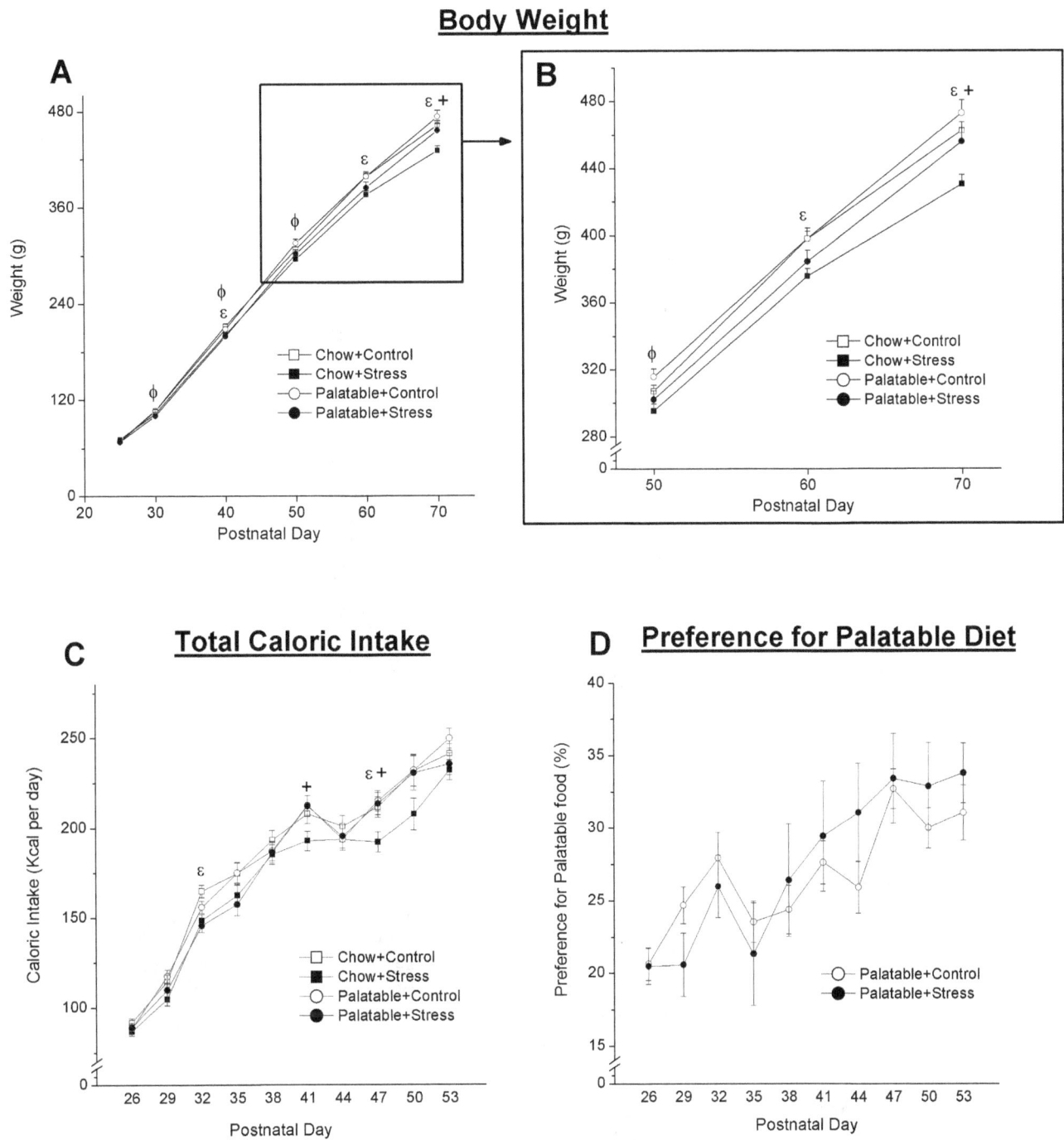

Figure 2. Effect of stress and diet on weight and food intake. Initially, both stress groups gained weight (A and B) at the slower rate relative to their controls; however, this reduction in the rate of weight gain persisted past PD50 in the Chow+Stress group only. A significant difference between stress groups was observed on PD 70. Overall, rats with access to the palatable diet consumed more calories relative to chow-fed rats (C). No significant difference in preference for the palatable diet was observed (D). Lines represent mean ± SEM. ε Significant difference between Chow and Chow+Stress. φ Significant difference between Palatable and Palatable+Stress. + Significant difference between the two stress groups receiving opposite diets.

and subcutaneous inguinal ($F_{1,52} = 10.93$ p = .002) fat pads. Two-way ANOVA of mesenteric fat pad weight revealed a significant Diet × Stress interaction ($F_{1,52} = 4.75$ p = .034). Follow-up simple effects analysis revealed that previously stressed rats with access to chow displayed significant decreases in mesenteric fat relative to previously stressed rats with access to the palatable diet (p<.001),

as well as the chow-fed control group (p = .006). No significant differences in inter-scapular brown fat were observed ($F_{1,52} = 1.69$ p>.1)

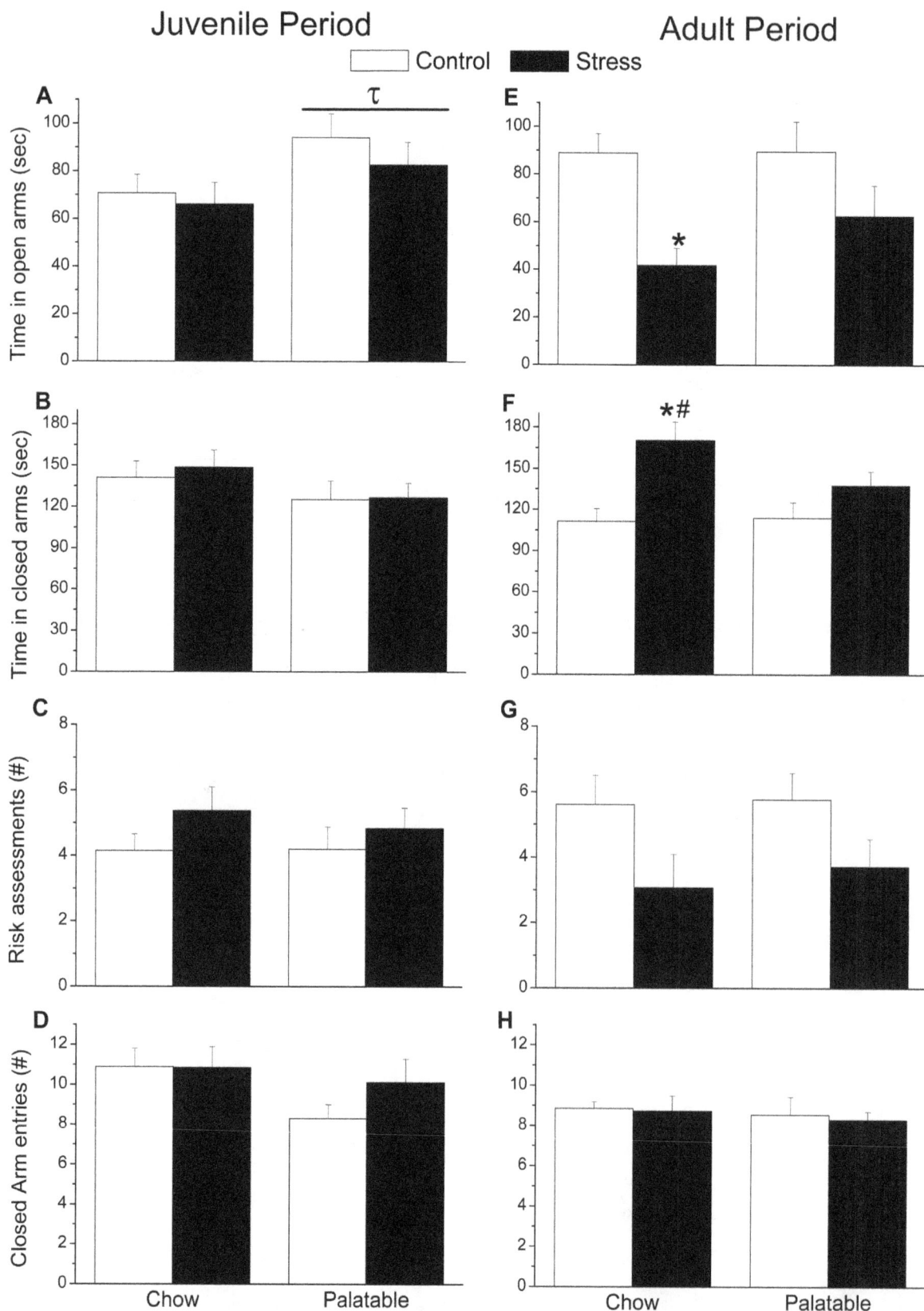

Figure 3. Anxiety-like behavior in the elevated plus maze. A significant effect of the palatable diet was observed in the juvenile time point. In general, rats with access to the palatable diet spent more time in the open arms of the maze. In adulthood, previously stressed chow fed rats (E) spent less time in the open arms, and (F) more time in the closed arms compared to chow fed controls. A significant difference between stress groups was observed in terms of time spent in the close arm with chow-fed rats spending more time relative to palatable food fed counterparts. No significant difference across groups was observed regarding number of (G) risk assessments and closed arms entries (H), suggesting no significant

differences in locomotor activity. Bars represent mean ± SEM. τ Significant diet effect. * Significantly different from controls. # Significantly different from condition matched rats receiving opposite diet.

Discussion

The present study investigated the short- and long-term consequences of unpredictable physical stress applied during the juvenile period on subsequent behavioral markers of depression and anxiety and indices of metabolic syndrome and obesity. No short-term effects of juvenile stress on behavioral markers of anxiety-like behaviors were observed in the open field, EPM and social interaction test. In contrast, juvenile stress resulted in decreased exploration of high-risk areas of the EPM in adulthood, which is consistent with results reported by Jacobson-Pick and Ritcher-Levin [48]. Reduced social exploration was also observed in the present study, as rats previously exposed to juvenile stress exhibited less social behavior compared to controls. Decreased social behavior in adulthood following juvenile stress has been

previously reported in both mice [57] and rats [58]. Exposure to juvenile stress did not appear to impact behavior in the forced swim test at both time points; however, it is possible that previous exposure to the swim stress in the juvenile stress protocol hay have altered behavior in the FST. This particular stress paradigm may also be more in keeping with a model of anxiety as opposed to depressive symptoms. Together, these findings indicate that juvenile stress can have lasting effects on behavior in adulthood [46,48,57–62].

Consistent with the view that eating palatable foods may serve as a means of coping with distress in humans [63,64], in the present study palatable food mitigated the behavioral effects of the juvenile stress. In adulthood, chow-fed rats exposed to the stressor displayed significantly higher levels of anxiety-like behavior compared to rats with access to palatable food. No significant

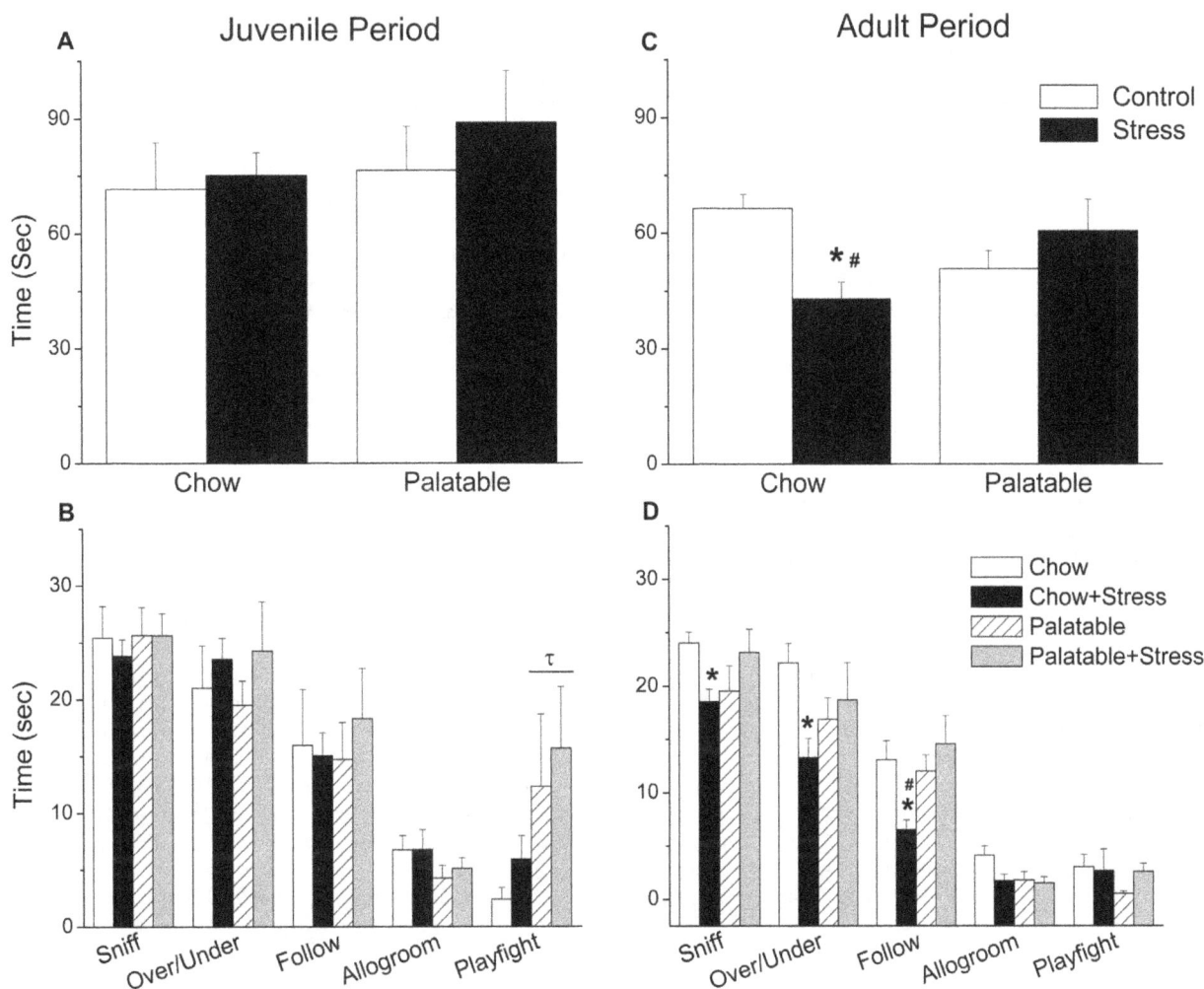

Figure 4. Social behavior in the social interaction test. No significant differences in total time engaged in social behaviors were observed in the juvenile phase (A); however, in general rats with access to the palatable diet engaged in more play fighting (B). In adulthood, previously stressed chow fed rats spent less total time engaged in social behaviors compared to chow-fed controls and previously stressed rats with access to the palatable diet (C). In regards to specific behaviors, chow fed rats exposed to the juvenile stress spent less time sniffing, following and moving over/under one another (D) relative to their controls. The difference between stress groups was significant for following behavior as well (D). Bars represent mean ± SEM. τ Significant diet effect. * Significantly different from controls. # Significantly different from condition matched rats receiving opposite diet.

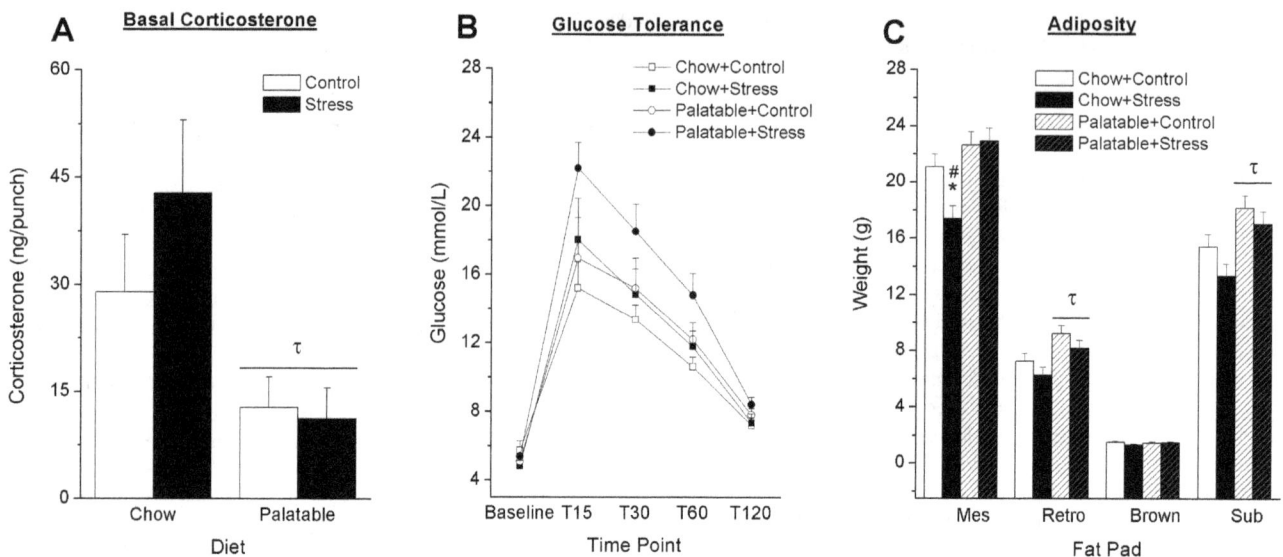

Figure 5. Effect of stress and diet on endocrine and metabolic indices. Rats with access to palatable food displayed lower plasma levels of CORT compared to chow fed rats on PD 66 (A). Blood glucose levels over time following an I.P. injection of a 0.75 g/mL dextrose solution did not differ significantly across groups (B). Rats with access to the palatable diet displayed an increased in retroperitoneal and inguinal subcutaneous fat (C). A decreased weight of the mesenteric fat pad was observed in the Chow+Stress condition relative to their controls and the Palatable+Stress group. Bars represent mean ± SEM. τ Significant diet effect. * Significantly different from controls. # Significantly different from condition matched rats receiving opposite diet. Mes = Mesenteric fat pad. Retro = retroperitoneal fat pad. Brown = Inter-scapular brown fat. Sub = Subcutaneous inguinal white fat pad.

differences occurred in closed arm entries across groups suggesting that the observed results were independent of locomotor activity. Similar results were obtained in the social interaction test with previously stressed chow-fed rats displaying reduced social behavior relative to those with access to the palatable diet. Indeed, access to a palatable diet has been shown to mitigate the long term effects of maternal separation in the sucrose preference test and EPM [35], as well as the effects of short-term restraint stress in adulthood on anxiety-like behavior in the EPM [25]. Similar beneficial effects have been described among adult rats with access to a sucrose solution following a restraint stress [23]. Taken together with these earlier reports, our results provide evidence for the notion of palatable food consumption acts as a buffer against the negative behavioral effects of stress. Our results also provide preliminary evidence that such a diet may have proactive effects that are still evident in adulthood.

While no effects of the stressor were observed when animals were tested as juveniles, a diet effect in the EPM and social interaction test was observed, suggesting that the palatable diet may have had an anxiolytic effect. The lack of a stress effect among chow-fed rats contrasts with previous findings, as stressed juvenile rats were observed to display increased activity and exploration in both the open field and elevated plus maze which the authors characterized as "non-classic anxious behavior" [48; p. 274]. One possible explanation for the observed differences between studies is the modification made to the stressor paradigm. Reducing the duration of restraint stress from 2 hrs in the original protocol to 30 min may have diminished the acute effects of the stressor, resulting in a corresponding change in the behavior in the open field and EPM. Testing windows were also slightly different (24 hours post stress in the current study vs. 1–6 hrs post stress).

Stressors ordinarily increase HPA activity [65,66] and palatable foods consumption has been associated with lower resting and stress-evoked cortisol levels in humans [67,68]. In rodents, previous studies have reported normalized HPA activity in

adrenalectomized animals [69–71], reduced HPA axis response and hypothalamic corticotrophin releasing factor (CRF) mRNA expression following acute restraint stress [34] and diminished HPA axis response to repeated restraint stress [25] when animals are given access to a palatable diet. Moreover, a blunted corticosterone response, as well as reductions in hypothalamic CRF mRNA expression has been reported in response to restraint stress [23]. HPA axis dampening is proposed to be a result of the hedonic properties of the sucrose rather than the increase caloric value, as reductions of plasma corticosterone in response to restraint are not observed when rats were given the same amount and schedule of sucrose solution via intragastric gavage [23]. Consistent with previously reported results, we observed a significant decrease in basal plasma corticosterone concentrations among rats with access to palatable food.

Access to the palatable diet resulted in an overall increase in total caloric intake regardless of stressor exposure. Of greater interest, however, is the proportion of calories obtained from the palatable food to that of the standard chow. As stressor-induced preference for more pleasurable or palatable foods has been documented in both animals and humans [18,19,25], we hypothesized that following the 3-day stressor, rats would show a preference for the palatable food. However, despite the mitigating effects of the palatable food on behavioral and neuroendocrine indices of anxiety, no such preference was observed. The lack of preference may be a result of the limited access to the palatable diet, as well as its availability during the light phase. Furthermore, as palatable food was provided in the morning prior to each stressor, it is possible that the timing of access to the palatable food might have influenced our results, as previous research has shown an increase in consumption when the palatable food is provided after the stressor [34]. In addition, palatable preference in the present study was determined on the basis of intake per cage, rather than individual intake as rats were doubly housed. Although this was done to reduce stress stemming from isolation [69–71], the result of this decision is a reduction in statistical power, as well as confounding of individual eating patterns. Indeed, previous animal studies documenting a stress-induced increase in palatable feeding provided rats with *ad libitum* access to the favored diet [18,19,25].

As expected, juvenile stressor exposure resulted in a decreased rate of weight gain, which is a well-established consequence of stressor exposure in rats [72]. While initially both stress groups displayed a reduction in weight gain, rats with access to the palatable diet displayed a rate of weight gain comparable to their control group from PD 60 onwards, suggesting that access to the palatable food had a mitigating effect on this outcome. This pattern of recovered weight gain has been previously reported following restraint stress [25].

Consistent with findings concerning the negative health consequences of high fat/sugar diets, rats with access to palatable food displayed increased adiposity compared to chow fed rats, despite the absence of any differences in body weight. In addition, previously stressed rats with access to the palatable diet also displayed a significant increase in the mesenteric fat pad compared to the previously stressed chow fed rats. These findings are in line with the perspective that stress alters how fat is distributed throughout the body. It seems that when poor diet is coupled with chronic stress, reorganization of energy stores from peripheral storage to central storage, primarily as abdominal fat, is facilitated, accompanied by elevated levels of glucocorticoids and insulin [19]. High levels of abdominal fat have been associated with hypertension, cardiovascular disease, metabolic syndromes, type 2 diabetes, and morbidity and mortality [73,74]. However, selectively increased adiposity in the abdominal region is not reported in all studies that examined the effects of palatable food in the rat. Studies investigating the short-term effects of restraint stress [25,34] and maternal separation [35] demonstrated general increases in fat pad weights among rats with access to palatable foods relative to chow fed rats, but not a selective increases in

abdominal fat. Our results may be more germane to the long-term effects of access to a palatable diet. In the present study, rats were sacrificed 6 weeks following termination of the stressor and thus the observed differences in fat distribution between studies might be attributable to the prolonged access to the palatable diet.

High fat diets have been proposed as one of the factors which may lead to reduced insulin sensitivity, followed by insulin resistance and ultimately the development of type 2 diabetes [3]. This has a corresponding effect on the ability of the body to process glucose. Reduced ability to tolerate a glucose load was not observed in the present study; however, the development of glucose intolerance following access to a high fat diet has been previously reported [75–78]. Exposure to a stressor may also contribute to the development of glucose intolerance. Dysregulation of the HPA axis as a result of stressor exposure, in conjunction with chronically elevated insulin levels, contribute to the development of insulin resistance, abdominal obesity, as well as metabolic syndrome [79].

Conclusion

The present study provides additional evidence that stressor exposure during juvenility can have long lasting effects on anxiety-like behaviors in adulthood, and that access to palatable food may have mitigating effects on the anxiety and corticosterone effects of juvenile stress. Not only do these results provide further support for the notion that palatable foods may be protective against the negative effects of stress, but also that these effects last into adulthood. However, the use of palatable foods as coping-strategy has long term negative effects on adiposity.

Author Contributions

Conceived and designed the experiments: JCM PK ZM. Performed the experiments: JCM JSJ CC. Analyzed the data: JCM PK CC JSJ. Wrote the paper: JCM JSJ PK HA.

References

1. International Obesity Task Force (2010) Obesity: The Global Epidemic. http://www.iaso.org/iotf/obesity/obesitytheglobalepidemic/. Acessed 2013 November 1.
2. Singh AS, Mulder C, Twisk JW, van Mechelen W, Chinapaw MJ (2008) Tracking of childhood overweight into adulthood: a systematic review of the literature. Obes Rev 9: 474–488.
3. Spruijt-Metz D (2011) Etiology, treatment, and prevention of obesity in childhood and adolescence: A decade in review. Journal of Reserach on Adolescence 21: 129–152.
4. Whitaker RC, Wright JA, Pepe MS, Seidel KD, Dietz WH (1997) Predicting obesity in young adulthood from childhood and parental obesity. N Engl J Med 337: 869–873.
5. Shaibi GQ, Goran MI (2008) Examining metabolic syndrome definitions in overweight Hispanic youth: a focus on insulin resistance. J Pediatr 152: 171–176.
6. Freedman DS, Dietz WH, Srinivasan SR, Berenson GS (1999) The relation of overweight to cardiovascular risk factors among children and adolescents: the Bogalusa Heart Study. Pediatrics 103: 1175–1182.
7. Pinhas-Hamiel O, Zeitler P (1996) Insulin resistance, obesity, and related disorders among black adolescents. J Pediatr 129: 319–320.
8. Cruz ML, Shaibi GQ, Weigensberg MJ, Spruijt-Metz D, Ball GD, et al. (2005) Pediatric obesity and insulin resistance: chronic disease risk and implications for treatment and prevention beyond body weight modification. Annu Rev Nutr 25: 435–468.
9. Calle EE, Thun MJ (2004) Obesity and cancer. Oncogene 23: 6365–6378.
10. Calle EE, Kaaks R (2004) Overweight, obesity and cancer: epidemiological evidence and proposed mechanisms. Nat Rev Cancer 4: 579–591.
11. Hills AP, Andersen LB, Byrne NM (2011) Physical activity and obesity in children. Br J Sports Med 45: 866–870.
12. Coccurello R, D'Amato FR, Moles A (2009) Chronic social stress, hedonism and vulnerability to obesity: lessons from rodents. Neurosci Biobehav Rev 33: 537–550.
13. Hill JO, Peters JC (1998) Environmental contributions to the obesity epidemic. Science 280: 1371–1374.
14. Torres SJ, Nowson CA (2007) Relationship between stress, eating behavior, and obesity. Nutrition 23: 887–894.
15. Zellner DA, Loaiza S, Gonzalez Z, Pita J, Morales J, et al. (2006) Food selection changes under stress. Physiol Behav 87: 789–793.
16. O'Connor DB, Jones F, Conner M, McMillan B, Ferguson E (2008) Effects of daily hassles and eating style on eating behavior. Health Psychol 27: S20–S31.
17. Gibson EL (2006) Emotional influences on food choice: sensory, physiological and psychological pathways. Physiol Behav 89: 53–61.
18. Dallman MF (2010) Stress-induced obesity and the emotional nervous system. Trends Endocrinol Metab 21: 159–165.
19. Dallman MF, Pecoraro NC, la Fleur SE (2005) Chronic stress and comfort foods: self-medication and abdominal obesity. Brain Behav Immun 19: 275–280.
20. Dess NK (1992) Divergent responses to saccharin vs. sucrose availability after stress in rats. Physiol Behav 52: 115–125.
21. Young JB (2000) Effects of neonatal handling on sympathoadrenal activity and body composition in adult male rats. Am J Physiol Regul Integr Comp Physiol 279: R1745–R1752.
22. Ulrich-Lai YM, Ostrander MM, Thomas IM, Packard BA, Furay AR, et al. (2007) Daily limited access to sweetened drink attenuates hypothalamic-pituitary-adrenocortical axis stress responses. Endocrinology 148: 1823–1834.
23. Ulrich-Lai YM, Christiansen AM, Ostrander MM, Jones AA, Jones KR, et al. (2010) Pleasurable behaviors reduce stress via brain reward pathways. Proc Natl Acad Sci U S A 107: 20529–20534.
24. Ulrich-Lai YM, Ostrander MM, Herman JP (2011) HPA axis dampening by limited sucrose intake: reward frequency vs. caloric consumption. Physiol Behav 103: 104–110.
25. Pecoraro N, Reyes F, Gomez F, Bhargava A, Dallman MF (2004) Chronic stress promotes palatable feeding, which reduces signs of stress: feedforward and feedback effects of chronic stress. Endocrinology 145: 3754–3762.
26. la Fleur SE, Houshyar H, Roy M, Dallman MF (2005) Choice of lard, but not total lard calories, damps adrenocorticotropin responses to restraint. Endocrinology 146: 2193–2199.
27. Kant GJ, Bauman RA (1993) Effects of chronic stress and time of day on preference for sucrose. Physiol Behav 54: 499–502.
28. Fachin A, Silva RK, Noschang CG, Pettenuzzo L, Bertinetti L, et al. (2008) Stress effects on rats chronically receiving a highly palatable diet are sex-specific. Appetite 51: 592–598.

29. Christiansen AM, Dekloet AD, Ulrich-Lai YM, Herman JP (2011) "Snacking" causes long term attenuation of HPA axis stress responses and enhancement of brain FosB/deltaFosB expression in rats. Physiol Behav 103: 111–116.

30. Buwalda B, Blom WA, Koolhaas JM, van Dijk G (2001) Behavioral and physiological responses to stress are affected by high-fat feeding in male rats. Physiol Behav 73: 371–377.

31. Zeeni N, Daher C, Fromentin G, Tome D, Darcel N, et al. (2012) A cafeteria diet modifies the response to chronic variable stress in rats. Stress 16: 211–219.

32. Strack AM, Akana SF, Horsley CJ, Dallman MF (1997) A hypercaloric load induces thermogenesis but inhibits stress responses in the SNS and HPA system. Am J Physiol 272: R840–R848.

33. Levin BE (1996) Reduced paraventricular nucleus norepinephrine responsiveness in obesity-prone rats. Am J Physiol 270: R456–R461.

34. Foster MT, Warne JP, Ginsberg AB, Horneman HF, Pecoraro NC, et al. (2009) Palatable foods, stress, and energy stores sculpt corticotropin-releasing factor, adrenocorticotropin, and corticosterone concentrations after restraint. Endocrinology 150: 2325–2333.

35. Maniam J, Morris MJ (2010) Palatable cafeteria diet ameliorates anxiety and depression-like symptoms following an adverse early environment. Psychoneuroendocrinology 35: 717–728.

36. de Ferranti S, Mozaffarian D (2008) The perfect storm: obesity, adipocyte dysfunction, and metabolic consequences. Clin Chem 54: 945–955.

37. Woods SC, Seeley RJ (2000) Adiposity signals and the control of energy homeostasis. Nutrition 16: 894–902.

38. Kriketos AD, Greenfield JR, Peake PW, Furler SM, Denyer GS, et al. (2004) Inflammation, insulin resistance, and adiposity: a study of first-degree relatives of type 2 diabetic subjects. Diabetes Care 27: 2033–2040.

39. Bastard JP, Maachi M, Lagathu C, Kim MJ, Caron M, et al. (2006) Recent advances in the relationship between obesity, inflammation, and insulin resistance. Eur Cytokine Netw 17: 4–12.

40. Korner J, Leibel RL (2003) To eat or not to eat - how the gut talks to the brain. N Engl J Med 349: 926–928.

41. Randle PJ (1998) Regulatory interactions between lipids and carbohydrates: the glucose fatty acid cycle after 35 years. Diabetes Metab Rev 14: 263–283.

42. Jee SH, Kim HJ, Lee J (2005) Obesity, insulin resistance and cancer risk. Yonsei Med J 46: 449–455.

43. DeFronzo RA, Ferrannini E (1991) Insulin resistance. A multifaceted syndrome responsible for NIDDM, obesity, hypertension, dyslipidemia, and atherosclerotic cardiovascular disease. Diabetes Care 14: 173–194.

44. McCormick CM, Mathews IZ (2010) Adolescent development, hypothalamic-pituitary-adrenal function, and programming of adult learning and memory. Prog Neuropsychopharmacol Biol Psychiatry 34: 756–765.

45. Bingham B, Gray M, Sun T, Viau V (2011) Postnatal blockade of androgen receptors or aromatase impair the expression of stress hypothalamic-pituitary-adrenal axis habituation in adult male rats. Psychoneuroendocrinology 36: 249–257.

46. Avital A, Richter-Levin G (2005) Exposure to juvenile stress exacerbates the behavioural consequences of exposure to stress in the adult rat. Int J Neuropsychopharmacol 8: 163–173.

47. Spear LP (2000) The adolescent brain and age-related behavioral manifestations. Neurosci Biobehav Rev 24: 417–463.

48. Jacobson-Pick S, Richter-Levin G (2010) Differential impact of juvenile stress and corticosterone in juvenility and in adulthood, in male and female rats. Behav Brain Res 214: 268–276.

49. Correa M, Arizzi MN, Betz A, Mingote S, Salamone JD (2003) Open field locomotor effects in rats after intraventricular injections of ethanol and the ethnaol metabolites acetaldehyde and acetate. Br Res Bull 62: 197–202.

50. Carobrez AP, Bertoglio LJ (2005) Ethological and temporal analyses of anxiety-like behavior: the elevated plus-maze model 20 years on. Neurosci Biobehav Rev 29: 1193–1205.

51. File SE, Seth P (2003) A review of 25 years of the social interaction test. Eur J Pharmacol 463: 35–53.

52. Schiller GD, Pucilowski O, Wienicke C, Overstreet DH (1992) Immobility-reducing effects of antidepressants in a genetic animal model of depression. Brain Res Bull 28: 821–823.

53. Porsolt RD, Bertin A, Jalfre M (1977) Behavioral despair in mice: a primary screening test for antidepressants. Arch Int Pharmacodyn Ther 229: 327–336.

54. Porsolt RD, Anton G, Blavet N, Jalfre M (1978) Behavioural despair in rats: a new model sensitive to antidepressant treatments. Eur J Pharmacol 47: 379–391.

55. Winer BJ (1962) Statistical Principles in Experimental Design. New York: McGraw-Hill.

56. Taylor JR (1997) An Introduction to Error Analysis. Sausolito, California: University Science Books.

57. Jacobson-Pick S, Audet MC, Nathoo N, Anisman H (2011) Stressor experiences during the juvenile period increase stressor responsivity in adulthood: transmission of stressor experiences. Behav Brain Res 216: 365–374.

58. Toth E, Avital A, Leshem M, Richter-Levin G, Braun K (2008) Neonatal and juvenile stress induces changes in adult social behavior without affecting cognitive function. Behav Brain Res 190: 135–139.

59. Avital A, Ram E, Maayan R, Weizman A, Richter-Levin G (2006) Effects of early-life stress on behavior and neurosteroid levels in the rat hypothalamus and entorhinal cortex. Brain Res Bull 68: 419–424.

60. Jacobson-Pick S, Elkobi A, Vander S, Rosenblum K, Richter-Levin G (2008) Juvenile stress-induced alteration of maturation of the GABAA receptor alpha subunit in the rat. Int J Neuropsychopharmacol 11: 891–903.

61. Taylor SE, Klein LC, Lewis BP, Gruenewald TL, Gurung RA, et al. (2000) Biobehavioral responses to stress in females: tend-and-befriend, not fight-or-flight. Psychol Rev 107: 411–429.

62. Toledo-Rodriguez M, Sandi C (2007) Stress before puberty exerts a sex- and age-related impact on auditory and contextual fear conditioning in the rat. Neural Plast 2007: 71203.

63. Dube L, LeBel JL, Lu J (2005) Affect asymmetry and comfort food consumption. Physiol Behav 86: 559–567.

64. Macht M (2008) How emotions affect eating: a five-way model. Appetite 50: 1–11.

65. Greaves-Lord K, Ferdinand RF, Oldehinkel AJ, Sondeijker FE, Ormel J, et al. (2007) Higher cortisol awakening response in young adolescents with persistent anxiety problems. Acta Psychiatr Scand 116: 137–144.

66. Kallen VL, Tulen JH, Utens EM, Treffers PD, De Jong FH, et al. (2008) Associations between HPA axis functioning and level of anxiety in children and adolescents with an anxiety disorder. Depress Anxiety 25: 131–141.

67. Deuster PA, Singh A, Hofmann A, Moses FM, Chrousos GC (1992) Hormonal responses to ingesting water or a carbohydrate beverage during a 2 h run. Med Sci Sports Exerc 24: 72–79.

68. Markus R, Panhuysen G, Tuiten A, Koppeschaar H (2000) Effects of food on cortisol and mood in vulnerable subjects under controllable and uncontrollable stress. Physiol Behav 70: 333–342.

69. Einon DF, Morgan MJ (1977) A critical period for social isolation in the rat. Dev Psychobiol 10: 123–132.

70. Fone KCF, Dixon DM (1991) Acute and chronic effects of intrathecal galanin on behavioural and biochemical markers of spinal motor function in adult rats. Brain Res 544: 118–125.

71. Weiss IC, Pryce CR, Jongen-Relo AL, Nanz-Bahr NI, Feldon J (2004) Effect of social isolation on stress-related behavioural and neuroendocrine state in the rat. Behav Brain Res 152: 279–295.

72. Marti O, Gavalda A, Jolin T, Armario A (1996) Acute stress attenuates but does not abolish circadian rhythmicity of serum thyrotrophin and growth hormone in the rat. Eur J Endocrinol 135: 703–708.

73. Stunkard AJ, Faith MS, Allison KC (2003) Depression and obesity. Biol Psychiatry 54: 330–337.

74. Friedman JM (2003) A war on obesity, not the obese. Science 299: 856–858.

75. Garg N, Thakur S, Alex MC, Adamo ML (2011) High fat diet induced insulin resistance and glucose intolerance are gender-specific in IGF-1R heterozygous mice. Biochem Biophys Res Commun 412: 476–480.

76. Cerf ME (2007) High fat diet modulation of glucose sensing in the beta-cell. Med Sci Monit 13: RA12–RA17.

77. Akyol A, McMullen S, Langley-Evans SC (2012) Glucose intolerance associated with early-life exposure to maternal cafeteria feeding is dependent upon post-weaning diet. Br J Nutr 107: 964–978.

78. Akerfeldt MC, Laybutt DR (2011) Inhibition of Id1 Augments Insulin Secretion and Protects Against High-Fat Diet-Induced Glucose Intolerance. Diabetes 60: 2506–2514.

79. Pervanidou P, Chrousos GP (2011) Stress and obesity/metabolic syndrome in childhood and adolescence. Int J Pediatr Obes 6 Suppl 1: 21–28.

20

Awareness of Climate Change and the Dietary Choices of Young Adults in Finland: A Population-Based Cross-Sectional Study

Essi A. E. Korkala[1,2], **Timo T. Hugg**[1,2,3], **Jouni J. K. Jaakkola**[1,2,3]*

1 Center for Environmental and Respiratory Health Research, University of Oulu, Oulu, Finland, **2** Medical Research Center Oulu, Oulu University Hospital and University of Oulu, Oulu, Finland, **3** Public Health, Institute of Health Sciences, University of Oulu, Oulu, Finland

Abstract

Climate change is a major public health threat that is exacerbated by food production. Food items differ substantially in the amount of greenhouse gases their production generates and therefore individuals, if willing, can mitigate climate change through dietary choices. We conducted a population-based cross-sectional study to assess if the understanding of climate change, concern over climate change or socio-economic characteristics are reflected in the frequencies of climate-friendly food choices. The study population comprised 1623 young adults in Finland who returned a self-administered questionnaire (response rate 64.0%). We constructed a Climate-Friendly Diet Score (CFDS) ranging theoretically from −14 to 14 based on the consumption of 14 food items. A higher CFDS indicated a climate-friendlier diet. Multivariate linear regression analyses on the determinants of CFDS revealed that medium concern raised CFDS on average by 0.51 points (95% confidence interval (CI) 0.03, 0.98) and high concern by 1.30 points (95% CI 0.80, 1.80) compared to low concern. Understanding had no effect on CFDS on its own. Female gender raised CFDS by 1.92 (95% CI 1.59, 2.25). Unemployment decreased CFDS by 0.92 (95% CI −1.68, −0.15). Separate analyses of genders revealed that high concern over climate change brought about a greater increase in CFDS in females than in males. Good understanding of climate change was weakly connected to climate-friendly diet among females only. Our results indicate that increasing awareness of climate change could lead to increased consumption of climate-friendly food, reduction in GHG emissions, and thus climate change mitigation.

Editor: Sisira Siribaddana, Faculty of Medicine & Allied Sciences, Rajarata Univeresity of Sri Lanka, Sri Lanka

Funding: This work was supported by Academy of Finland (grant no. 129419 of SALVE research program and grants no. 138691 and no. 266314). The funders had no role in study design, data collection and analysis, decision to publish, or preparation of the manuscript.

Competing Interests: The authors have declared that no competing interests exist.

* E-mail: jouni.jaakkola@oulu.fi

Introduction

Climate change has been characterized as the biggest global health threat of the 21st century [1]. The probable adverse health effects of climate change include more daily deaths due to temperature extremes, increased allergic disorders due to longer pollen season and increased risk of infectious disease due to flooding [2]. By mitigating climate change we can promote public health in the future.

From the public health perspective it is beneficial to promote especially those climate change mitigation actions that are good for health directly. Climate-friendly food consumption is an example of such behavior: through certain dietary choices one can mitigate climate change and promote his or her own health at the same time [3,4]. This is because adjusting into a more climate-friendly diet decreases the risk of many diseases such as coronary heart disease and cancer [3–5]. For these reasons encouraging climate-friendly eating is reasonable from the public health perspective.

Climate change and food production are closely connected. The production of food is a major contributor to anthropogenic greenhouse gas (GHG) emissions which are the most important cause of climate change [6]. Worldwide the agriculture sector is responsible for 22% of total GHG emissions [7] and together with food processing it causes approximately one-third of total GHG emissions [8].

Food products differ substantially in the amount of GHGs their production generates. The specific food items associated with high GHG emissions include beef, sheep, pork, cheese, rice and butter [9–11]. On the other hand food items like fresh vegetables, potatoes and margarine are associated with low GHG emissions [9–12]. In Finland, the typical diet of an adult (aged 25 to 74) is high in both climate-friendly and non-climate-friendly food items [13]. For example, majority of Finnish adults consume potatoes rather than rice as a side dish, which is a climate-friendly choice. On the other hand, many non-climate friendly food items such as red meat and cheese are also consumed by the majority of adults in Finland. Reducing the consumption of foods that are associated with high GHG emissions is a feasible and practical way to mitigate climate change [7,14,15]. Therefore predictors of climate-friendly food choices are of interest.

There is some research on the Finnish people's perceptions about climate change. According to the Eurobarometer 2008 [16], 78% of Finnish adults perceive climate change as a very serious problem. 73% of Finnish people agree that climate change is among the two biggest global problems at the moment. Therefore it seems that the public in Finland is aware of climate change and perceives climate change as a serious threat. Finnish people recognize the link between climate change and different consumption behaviors quite well [17]. For example, 93% of Finns agree that reducing car use would have quite a big effect or a big effect on climate change. More than 80% agree that residential heating, travelling and electricity consumption have quite a big effect or a big effect on climate change. In addition, 42% of Finnish adults agree that favoring a plant-based diet has quite a big effect or a big effect on climate change [17].

There is only a limited number of previous literature on the predictors of climate-friendly eating [18,19]. The perceived seriousness of climate change consequences seems to be a predictor of climate-friendly food choices among social science university students [18]. Pro-environmental self-identity has been found to predict climate-friendly shopping and eating [19]. It is not known if the determinants of climate-friendly eating differ between the genders or different socio-economic groups. In addition, previous studies base their assessment of climate-friendliness of a diet on only a few measures rather than diet-wide assessment of the intake frequencies of different food items [18,19]. Therefore there is a need for information on the determinants of climate-friendly eating across different socio-economic groups and genders with assessment of actual food intake frequencies. To add to the knowledge on the predictors of climate-friendly food choices, we conducted a study among young adults in Finland. Our primary aim was to assess if understanding of and concern over climate change are reflected in the frequencies of climate-friendly food choices. In addition, we studied the role of gender and socio-economic factors as determinants of climate-friendly food choices.

We hypothesized that people with high concern over climate change make climate-friendlier dietary choices than people who are not concerned over climate change. This hypothesis was based on the previous finding that high concern over climate change predicts mitigation actions [20,21]. We also hypothesized that good understanding of climate change is connected to climate-friendlier dietary choices since heightened knowledge about environmental problems is associated with pro-environmental behaviors [22]. However, we expected the effect of understanding to be smaller than that of concern because people's environmental behavior is not always in accordance with their knowledge [19,23]. Our hypothesis concerning the socio-demographics was that female gender and high educational level are connected to climate-friendly eating since these two factors have been found to be predictors of climate change mitigation action [21,24,25].

Methods

Study Population

This was a population-based cross-sectional study. The study population was the Espoo cohort established in 1991 when the cohort members were living in the city of Espoo in Southern Finland. The cohort consists of 2568 members born between January, 1984 and March, 1990. For this 20-year follow-up the contact information of the cohort members was acquired from the Population Register Centre (The Population Register Centre operates under the Ministry of Finance and contains basic identification information about all Finnish citizens) [26]. A self-administered, multiple choice questionnaire was sent to 2534 cohort members whose address was available between March 2010 and June 2011. The information gathering consisted of several posting rounds as well as phone contacts. 1623 completed questionnaires were received (response rate 64.0%). The respondents were a representative sample of the original baseline study

population as reported in another study on the 20-year follow up [27]. The questionnaire contained several sections and was partly based on questions used in the previous follow-ups and research projects [28,29]. The study was approved by the Ethics Committee of the Oulu University Hospital District.

Assessment of Awareness of Climate Change

Awareness of climate change was evaluated by assessing understanding of climate change and concern over climate change. Assessment of understanding of climate change was based on the question: *What do you think is meant by climate change?* Respondents were to choose their preferred definition of climate change from the following five definitions: *Global warming of the climate caused by 1 an increase in sunspot activity and sun's radiant energy, 2 a change in the axial tilt of the Earth, 3 an increase in population growth, energy consumption and exploitation of nature, 4 an increase in the greenhouse gas concentration of the atmosphere derived from human actions and 5 a natural fluctuation of climate periods on Earth.* We judged alternatives 3 and 4 to represent good understanding and alternatives 1, 2 and 5 poor understanding of climate change. This judgment was based on the causes of climate change as reported by the Intergovernmental Panel on Climate Change [6].

The degree of concern over climate change was assessed on the basis of the answer to the question: *If the climate is in some way changing, how a serious threat to the humankind do you think it is?* The five alternatives were: *1 A very great threat, 2 Quite a great threat, 3 Not a special threat, 4 Not a threat at all and 5 I do not know.* Alternative 1 indicated high concern, alternative 2 medium concern and the rest low concern over climate change.

Assessment of Dietary Choices

The food consumption during the past 12 months was assessed by asking the intake frequency of food items on a 5-point scale (*less than once a month, 1–3 times a month, 1–3 times a week, almost daily, at least once a day*). The intake frequency of organic food was asked on a different scale (*not at all, less frequently than once a month, 1–3 times a month, 1–3 times a week, daily or almost daily*). For the analysis we selected the food items especially climate-friendly and the ones non-climate-friendly. The climate-friendliness of a food item was defined by the GHG emissions created by the production of the food item from farm to table (as measured in CO_2 equivalents per 1 kg of food produced). The emissions of different food items were acquired from the literature [9,10,29–32] and compared to make the distinction between climate-friendly and non-climate friendly food items. We used emission data from European studies, mainly from Sweden, where the conditions are comparable with those in Finland. In the case of French fries information on energy consumption during production and preparation was used as an indicator of climate-friendliness [12] because information on GHG emissions was not available for this food item. Specific information on the GHGs emitted by the production of soy products was not available, so a general value for meat substitutes (tofu, tempeh, lupin and vegaburgers) was judged to apply to soy products [30]. In our study, climate-friendly food items included fresh vegetables/salad/root vegetables [10,31], soy products (such as tofu) [30], potatoes (cooked or mashed) [10], fresh fruits [10], margarines [11], vegetable oils [10] and organic food [32,33]. Non-climate-friendly food items included pork/beef/lamb [10,31], poultry [10], low fat cheese [10], other cheese [10], rice [10], butter [11] and French fries [12]. The individual intake frequencies of the food items were compared with the median intake frequency of the study population and were classified as high (>median), average and low (<median).

To assess the overall climate-friendliness of the respondents' diets, a novel measure, the climate-friendly diet score (CFDS) was generated. CFDS was constructed to be a comparative measure which uses the typical diet of a Finnish adult as the baseline. CFDS was calculated for each respondent based on how often they consume the food products in consideration. One point was given for high frequency intake of the climate-friendly food items whereas one minus point was given for low intake of these items. One minus point was given for high frequency intake of non-climate-friendly food items and one plus point for low frequency intake of these items. No points were given if the intake of the food item in question was average as this was judged to indicate no dietary adjustment into any direction. CFDS was calculated as a sum of the points given. Thus the theoretical range of the CFDS was from −14 to 14 and a higher score indicated a climate-friendlier diet.

Statistical Methods

The relations of interest were 1) the level of understanding of climate change and the consumption of climate-friendly food and 2) the level of concern about climate change and the consumption of climate-friendly food and 3) the socio-demographic factors and the consumption of climate-friendly food.

First, the intake frequencies of selected items were compared according to understanding (poor vs. good) and concern (low, medium and high). The role of chance in the differences between the frequency distributions was assessed applying Chi square-test and corresponding trend test. Second, the average CFDS's were compared between categories of understanding and concern of climate change. Finally, the determinants of CFDS were modeled with multivariate linear regression analysis using the following variables: understanding of climate change, concern over climate change, gender, education, occupation, marital status, parental status and average annual income. The presence of interaction between understanding and concern was assessed by fitting corresponding product terms (understanding*high concern) in addition to actual variables for independent effects. Weak positive interaction was observed but it turned out not to be statistically significant. The statistical software used for all analyses was SAS 9.3.

Results

Characteristics of the Study Population

Out of the whole study population (n = 1623) 89.3% had good understanding and 10.7% poor understanding of climate change. 35.6% had high concern over climate change, 47.5% medium concern and 16.9% low concern.

Women had on average better understanding and higher concern over climate change compared to men (Table 1). Age within this narrow range did not seem to have a clear effect on understanding nor concern. Higher vocational or academic degree holders were highly concerned about climate change whereas comprehensive school and vocational school degree holders were underrepresented in the highly concerned. Understanding of climate change was fairly good in all educational groups except vocational school degree holders, who were overrepresented among the people who understand climate change poorly. The unemployed and people working in the factory, mining or construction trade tended to have low concern over climate change. Marital status or the presence of children did not seem to affect understanding of or concern over climate change. People in the highest income category were less concerned about climate change.

The Determinants of Climate-friendly Food Consumption

Respondents with good understanding of climate change reported to eat fresh vegetables/salad/root vegetables, fresh fruits, soy products, vegetable oils, organic food and rice more frequently and French fries and pork/beef/lamb less frequently than respondents with poor understanding (Table S1). Respondents highly concerned about climate change reported to eat vegetables/salad/root vegetables, fresh fruits, soy products, vegetable oils, organic food and low fat cheese more frequently and pork/beef/lamb and French fries less frequently (Table S2).

The average CFDS of the whole study population was 0.56 (SD 3.09). CFDS's of the respondents ranged from −11 to 11. The people with good understanding of climate change had higher average CFDS (0.67, SD 3.11) than those with poor understanding (−0.46, SD 2.70). The average CFDS increased gradually as the concern over climate change increased: from −0.62 (low concern) to 0.37 (medium concern) to 1.37 (high concern) (Table 2).

Multivariate linear regression analyses on the determinants of CFDS revealed that medium concern raised CFDS on average by 0.51 points (95% CI 0.03, 0.98) and high concern by 1.30 points (95% CI 0.80, 1.80) (Table 2) compared to low concern. This result is in accordance with the hypothesis that people concerned with climate change make climate-friendlier dietary choices. Unlike we hypothesized, understanding of climate change did not affect CFDS on its own. Female gender raised CFDS by 1.92 (95% CI 1.59, 2.25) when compared to males. This result was in line with the hypothesis. Unemployment decreased CFDS by 0.92 (95% CI −1.68, −0.15) when compared to studying and income in the medium range (€8,401–16,800/year) by 0.42 (95% CI −0.80, −0.05) when compared to the lowest income category. Education did not have an effect on CFDS, unlike we hypothesized.

When females and males were analyzed separately (Table 2), it could be seen that high concern over climate change brought about a greater increase in CFDS in females (1.52, 95% CI 0.72, 2.32) than in males (1.16, 95% CI 0.48, 1.83). The effect of medium concern weakened in the separate analyses of the genders. Among females, good understanding of climate change weakly increased CFDS (0.79, 95% CI −0.11, 1.70) compared to poor understanding. Unemployment decreased CFDS by 1.82 (95% CI −3.01, −0.64) and income in category €8,401–16,800/year by 0.74 (95% CI −1.26, −0.21) among females.

Discussion

Main Findings

Respondents highly concerned about climate change made climate-friendlier dietary choices than people who were only slightly or not at all concerned. The high concern over climate change had a greater effect on dietary choices among females than males. The level of understanding of climate change was only weakly connected to climate-friendly dietary choices in females but not in males. Unemployed females had less climate-friendly diets than females in other occupational groups.

Validity of Results

The assessment of understanding of climate change was based on a question about the presumed causes of the phenomenon (see *Assessment of awareness of climate change*). Answer alternatives 3 and 4 were taken to indicate good understanding of climate change. As stated in alternative 4, anthropogenic GHG emissions are very likely the cause of the increase in global average temperatures [6]. Alternative 3 also indicates good understanding since it lists in a general way the human activities that cause these GHG emissions.

Table 1. The Espoo cohort study, 20-year follow-up 2010–2011: the understanding of and concern over climate change in different socio-demographic groups.

Characteristic	Understanding[a]			Concern[b]				Total n (%)
	Poor n (%)	Good n (%)	P value[c]	Low n (%)	Medium n (%)	High n (%)	P value[c]	
Gender			<.0001				<.0001	
Female	50 (29.6)	794 (56.2)		90 (33.3)	401 (52.7)	364 (64.0)		869 (53.5)
Male	119 (70.4)	620 (43.8)		180 (66.7)	359 (47.2)	205 (36.0)		754 (46.5)
Age group (yr)			0.7067				0.1587	
20–23	96 (56.8)	824 (58.3)		165 (61.1)	453 (59.6)	314 (55.2)		948 (58.4)
24–27	73 (43.2)	589 (41.7)		105 (38.9)	307 (40.9)	255 (44.8)		675 (41.6)
Highest qualification			0.1762				0.0104	
Comprehensive school degree	12 (7.1)	91 (6.5)		22 (8.2)	50 (6.6)	33 (5.8)		105 (6.5)
Upper secondary school degree	76 (45.0)	678 (48.1)		122 (45.2)	368 (48.5)	272 (48.0)		772 (47.6)
Vocational school degree	36 (21.3)	202 (14.3)		55 (20.4)	117 (15.4)	68 (12.0)		245 (15.1)
Upper secondary and vocational school degree	8 (4.7)	72 (5.1)		18 (6.7)	36 (4.7)	26 (4.6)		81 (5.0)
Higher vocational or academic degree	37 (21.9)	367 (26.0)		53 (19.6)	188 (24.8)	168 (29.6)		415 (25.7)
Missing information								5
Occupation			0.1611				<.0001	
Studying	84 (50.9)	768 (55.4)		125 (47.7)	407 (54.6)	325 (58.0)		861 (54.7)
Factory/mining/construction	17 (10.3)	107 (7.7)		41 (15.7)	53 (7.1)	30 (5.4)		124 (7.9)
Office/service	46 (27.9)	398 (28.7)		68 (26.0)	224 (30.1)	159 (28.4)		454 (28.8)
Stay-at-home mother/father	3 (1.82)	23 (1.66)		3 (1.2)	16 (2.2)	11 (2.0)		30 (1.9)
Unemployed	14 (8.5)	63 (4.55)		23 (8.8)	33 (4.4)	21 (3.8)		77 (4.9)
Other	1 (0.61)	27 (1.95)		2 (0.8)	12 (1.6)	14 (2.5)		28 (1.8)
Missing information								49
Marital status			0.1720				0.7435	
Single	116 (69.1)	866 (61.3)		173 (64.3)	460 (60.3)	365 (62.6)		1003 (62.0)
Married/civil partnership	9 (5.4)	70 (5.9)		14 (5.2)	41 (5.4)	26 (4.6)		82 (5.1)
Cohabitation	42 (25.0)	472 (33.4)		80 (29.7)	256 (33.7)	186 (32.7)		528 (32.6)
Divorced/separated	1 (0.60)	5 (0.35)		2 (0.7)	3 (0.39)	1 (0.2)		6 (0.4)
Missing information								4
Children			0.7194				0.3218	
No	161 (95.3)	1334 (94.6)		252 (93.3)	714 (94.1)	542 (95.6)		1518 (94.5)
Yes	8 (4.7)	76 (5.4)		18 (6.7)	45 (5.9)	25 (4.4)		88 (5.5)
Missing information								17
Income (€/yr)			0.7763				0.0297	
≤8400	62 (37.4)	484 (35.5)		88 (34.1)	267 (36.4)	198 (35.8)		556 (35.9)

Table 1. Cont.

Characteristic	Understanding[a]			Concern[b]				Total n (%)
	Poor n (%)	Good n (%)	P value[c]	Low n (%)	Medium n (%)	High n (%)	P value[c]	
8401–16800	59 (35.5)	474 (34.8)		92 (35.7)	231 (31.5)	216 (39.1)		541 (34.9)
≥16801	45 (27.1)	405 (29.7)		78 (30.2)	235 (32.1)	139 (25.1)		454 (29.3)
Missing information								72

[a]Missing information n = 41; [b]missing information n = 24; [c]the P value for X² test.

Alternatives 1 and 2 indicate poor understanding since it is very unlikely that climate change is caused by known natural external causes alone [6]. Alternative 5 also indicates weaker understanding because there is increased confidence that natural internal variability cannot be the cause of the observed changes in the climate [6].

The assessment of the climate-friendliness of the food items was based on information about the GHG emission generated during their life cycles. In the case of French fries information on energy consumption was used [12] because information on GHG emissions was not available. However, the energy consumption during the production and processing of a food item is not a very accurate measure of the product's climate-friendliness. This is because some GHGs emitted during the production of a food item may be non-energy related [9]. Organic food was categorized as climate-friendly but it is unclear whether organic food is always better for the climate than conventionally produced food. Organic farming increases carbon sequestration in soil [32] and generates less nitrous oxide emissions than conventional farming [8] but tends to have higher CO_2 emissions on per-unit output scale [34]. We classified fresh vegetables as climate-friendly food, even though they are not very climate-friendly if they are grown in heated greenhouses [31]. However, the majority of Finnish vegetables are grown in open fields (66.4% in 2012) [35], which causes very little GHG emissions [30]. Thus classifying fresh vegetables as climate-friendly is minimal erroneous.

One of the strengths of our study is that we were able to assess the consumption frequencies of a wide variety of food items. However, CFDS could be further improved to include information on consumption of local food products and avoiding food waste, since these food-related behaviors are relevant in climate change mitigation [12,36].

We compared the food consumption frequencies of our study population to the consumption frequencies reported in a national study of Finnish adults (aged 25–74 years) [13]. Our study population and the general Finnish adult population had the same median intake frequencies of rice, low fat cheese, other cheese, boiled or mashed potatoes, French fries, fresh vegetables/salad, poultry and red meat. This indicates that the dietary choices of our study population did not seem to substantially diverge from those of the general Finnish adult population.

We did not have information on the actual amounts (in e.g. grams) of the foods consumed by the study subjects. Only the intake frequencies could be taken into account when calculating the CFDS. Therefore the CFDS is not an absolute measure of the GHGs caused by a person's diet. Rather, CFDS helps to evaluate if a person has a tendency to consume climate-friendly food items. Therefore CFDS gives an approximation of the climate-friendliness of the dietary choices on the whole. Some food items such as cheese had a slightly pronounced influence on CFDS because we gave points on the intake frequency of low fat cheese and other types of cheese separately. On the other hand, we handled the red meats (pork/beef/lamb) causing huge GHG emissions as a single food choice. Therefore the choice of consuming several types of red meat might have influenced the CFDS too little.

The cohort members had to answer the dietary questions on the basis of their food consumption during the past 12 months. This is a potential source of information bias since it might have been hard for the cohort members to remember precisely their food consumption patterns for the past year. However, the possible error most probably is not systematic and therefore this it is not likely that the CFDSs are biased to any specific direction.

There are plenty of factors affecting dietary choices and we were able to take many of them into account. Age, sex, marital status,

Table 2. The Espoo cohort study, 20-year follow-up 2010–2011: linear regression analysis of CFDS against understanding of climate change, concern about climate change and socio-demographic variables.

Determinant	The whole study population (n = 1364)				Females (n = 726)			Males (n = 638)		
	Mean CFDS (SD)	Beta	95% CI	p value	Beta	95% CI	p value	Beta	95% CI	p value
Understanding										
Poor (ref)	−0.46 (2.70)									
Good	0.67 (3.11)	0.13	−0.40, 0.67	0.6264	0.79	−0.11, 1.70	0.0858	−0.18	−0.84, 0.49	0.6052
Concern										
Low (ref)	−0.62 (2.76)									
Medium	0.37 (3.00)	0.51	0.03, 0.98	0.0357	0.71	−0.08, 1.50	0.0766	0.42	−0.18, 1.03	0.1672
High	1.37 (3.15)	1.30	0.80, 1.80	<.0001	1.52	0.72, 2.32	0.0002	1.16	0.48, 1.83	0.0008
Gender										
Male (ref)	−0.54 (2.89)									
Female	1.50 (2.95)	1.92	1.59, 2.25	<.0001						
Highest qualification										
Comprehensive school (ref)	0.08 (2.98)									
Upper secondary school	0.71 (3.09)	0.50	−0.20, 1.20	0.1611	0.30	−0.68, 1.27	0.5478	0.63	−0.40, 1.66	0.2282
Vocational school	−0.21 (3.02)	0.02	−0.73, 0.78	0.9525	−0.12	−1.17, 0.94	0.8435	0.20	−0.91, 1.31	0.7239
Upper secondary and vocational school	0.42 (2.82)	0.11	−0.82, 1.04	0.8153	−0.35	−1.59, 0.90	0.5874	0.85	−0.59, 2.28	0.2472
Higher vocational or academic	0.86 (3.14)	0.33	−0.39, 1.06	0.3651	0.18	−0.81, 1.16	0.7264	0.33	−0.78, 1.43	0.5617
Occupation										
Studying (ref)	0.79 (3.19)									
Factory/mining/construction	−0.56 (2.42)	−0.13	−0.80, 0.54	0.7015	−1.25	−2.88, 0.39	0.1347	0.08	−0.71, 0.88	0.8400
Office/service	0.53 (3.01)	−0.30	−0.73, 0.12	0.1599	−0.53	−1.10, 0.03	0.0654	−0.06	−0.71, 0.60	0.8653
Stay-at-home mother/father	0.89 (2.49)	−0.18	−1.61, 1.26	0.8063	−0.19	−1.79, 1.42	0.8188	[a]		
Unemployed	−0.54 (2.72)	−0.92	−1.68, −0.15	0.0186	−1.82	−3.01, −0.64	0.0026	−0.31	−1.32, 0.70	0.5447
Other	1.12 (3.14)	−0.18	−1.31, 0.95	0.7549	−0.77	−2.12, 0.58	0.2612	0.97	−1.17, 3.12	0.3717
Marital status										
Single (ref)	0.57 (3.09)									
Married/civil partnership	0.90 (2.57)	0.46	−0.31, 1.23	0.2448	0.42	−0.59, 1.42	0.4151	0.44	−0.80, 1.69	0.4864
Cohabitation	0.51 (3.18)	−0.20	−0.55, 0.15	0.2520	−0.15	−0.62, 0.32	0.5306	−0.27	−0.79, 0.25	0.3121
Divorced/separated	−1.00 (2.61)	−1.40	−3.75, 0.94	0.2409	−1.12	−3.77, 1.52	0.4043	−1.16	−6.80, 4.47	0.6854
Children										
No (ref)	0.58 (3.08)									
Yes	0.04 (3.04)	−0.35	−1.22, 0.51	0.4258	−0.59	−1.75, 0.57	0.3176	−0.07	−1.44, 1.29	0.9143
Income (€/yr)										
≤8400 (ref)	0.78 (3.18)									

Table 2. Cont.

Determinant	The whole study population (n=1364)				Females (n=726)			Males (n=638)		
	Mean CFDS (SD)	Beta	95% CI	p value	Beta	95% CI	p value	Beta	95% CI	p value
8401–16800	0.53 (2.99)	−0.42	−0.80, −0.05	0.0272	−0.74	−1.26, −0.21	0.0054	0.00	−0.55, 0.54	0.9884
≥16801	0.30 (3.04)	−0.20	−0.66, 0.27	0.4059	−0.31	−0.97, 0.34	0.3486	−0.03	−0.70, 0.64	0.9265

ªNo data.

parental status, occupation, education, and income level affect food behavior [37,38]. In the present study the study subjects had a narrow age range (20 to 27 years) and the other above mentioned factors were included in the regression analyses. However, there are several psychological factors that may influence climate-friendly food choices that were not taken into account in our study. For example a recent study by Dowd and Burke (2013) indicates that positive moral attitude and ethical concern predict the intention to purchase sustainably sourced foods [39]. These kinds of factors may also affect the intention to consume and the actual consumption of climate-friendly food. In addition, because climate-friendly eating may delay or avert death of chronic diseases [4] health reasons rather than environmental reasons might be behind the climate-friendly food choices. Indicator of this is the fact that the highly concerned in our study population ate more frequently low-fat cheese, which is healthier than normal cheese but equally harmful to the climate. Personal preferences such as familiarity and sensory appeal may also play a role in food choice [40]. However, we believe that CFDS is not prone to error from such sources because CFDS is mostly based on the consumption of food categories (e.g. fresh fruits) rather than specific food products. Therefore there is room for personal preferences inside many of the food categories incorporated to CFDS.

There is a possibility of selection bias because all the cohort members did not answer the questionnaire (the response rate being 64.0%). The theme of the questionnaire was climate change, environment and health. Thus people especially interested in environmental issues might have been more eager to answer the questionnaire. This seems not to be the case with this study, because 35.6% of our sample was highly concerned about climate change whereas Eurobarometer 2008 found 78% of Finnish people to be highly concerned (both measured on a 3-point scale) [16]. This remarkable difference between the percentages might be due to different age groups studied or different study times.

It can be regarded as a weakness of our study that we did not ask the respondents if they are aware of the link between food consumption and climate change. However, according to a national study on the climate change perceptions of Finnish people, 42% of Finnish adults agree that favoring a plant-based diet has quite a big effect or a big effect on climate change [17]. Hence we argue that it is not a remarkable weakness of the present study that we assume Finnish people to recognize the link between dietary choices and climate change.

Synthesis with Previous Knowledge

There are only a few previous studies on the predictors of climate-friendly dietary choices. Mäkiniemi and Vainio (2013) found that the perceived seriousness of climate change consequences predicted climate-friendly food choices in their study population, that consisted of university students in the social and behavioral sciences of whom 80% were female [18]. Mäkiniemi's and Vainio's construct "Probable Seriousness of Consequences" is quite similar to the variable "concern" in our study: both of the measures aim to assess how a great threat climate change is perceived to be by the respondent. Because our study population was more representative of the general population (in terms of occupations, educational backgrounds and gender), our results add to the previous knowledge. In our study, too, the concern was connected to climate-friendlier diet. But we also found that this effect is stronger among females and that there are some special socio-demographic groups, such as unemployed females, that do not tend to make climate-friendly food choices.

It has previously been found that intention to change food consumption in order to mitigate climate change increases with worry about climate change consequences [41]. We did not study intentions but assessed the actual food intake frequencies. Therefore our study adds to the previous knowledge: the high concern about climate change might actually concretize the intentions to make dietary adjustments.

Other climate change mitigation behaviors and the factors affecting those behaviors have been studied more widely. In accordance with those studies our results indicate that high concern over climate change [20,21] and female gender [24,25] are strong predictors of climate change mitigation action.

It has been argued that when it comes to mitigating climate change people do not usually act in accordance with what they know or care about (the knowledge-action gap or the value-action gap) [19,23]. Our results are partially in line with this argument. We discovered that understanding of climate change was only weakly connected to the dietary choices. But then again high concern over climate change was clearly connected to climate-friendlier food consumption. This result is understandable since large majority of the respondents (89.3%) had a good understanding over climate change: the topic is widely discussed in the media and schools in Finland. Information about climate change is hence easily accessible without hard personal effort. Concern, on the other hand, requires more active personal reflection. Therefore it is understandable that concern over climate change has a more remarkable effect on the food choices.

A diet that is climate-friendly is likely to be healthier. Many studies support the fact that decreasing red meat consumption would reduce GHG emissions and the risk for several chronic diseases simultaneously. A study by Scarborough et al. found that reduction in meat and dairy consumption replaced by vegetables, fruit and cereals averts deaths from coronary heart disease, stroke and cancer as well as reduces GHG emissions [4]. Aston et al. (2012) calculated that a reduction in red meat and processed meat intake would remarkably decrease GHG emissions and the incidence of coronary heart disease, diabetes mellitus and colorectal cancer [3]. Friel et al. (2009) conclude that decreased livestock production would reduce GHG emissions and decrease deaths and disability caused by ischemic heart disease [5]. Thus adjusting into climate-friendlier diet can have positive effects on public health. This aspect should be more clearly emphasized when promoting climate-friendly lifestyles.

Conclusions

In this study among young Finnish adults, concern about climate change was connected to climate-friendly food choices among both genders. The level of understanding of climate change was only weakly connected to climate-friendly dietary choices among females but not among males. Our results indicate that increasing awareness of climate change could lead to increased use of climate-friendly food items, reduction in GHG emissions, and thus climate change mitigation.

Acknowledgments

We would like to thank Mr Jussi Tuomola alias Juba for permission to use his characters Viivi and Wagner to promote the data collection. We would also like to thank Dr Marika Kaakinen for her valuable tips on creating a dietary score.

Author Contributions

Conceived and designed the experiments: EK TH JJ. Performed the experiments: EK TH JJ. Analyzed the data: EK. Contributed reagents/materials/analysis tools: EK TH JJ. Wrote the paper: EK TH JJ.

References

1. Costello A, Abbas M, Allen A, Ball S, Bell S, et al. (2009) Managing the health effects of climate change. Lancet and University College London institute for global health commission. Lancet 373: 1693–1733.

2. McMichael AJ, Woodruff RE, Hales S (2006) Climate change and human health: Present and future risks (Review). Lancet 367: 859–869.

3. Aston LM, Smith JN, Powles JW (2012) Impact of a reduced red and processed meat dietary pattern on disease risks and greenhouse gas emissions in the UK: A modelling study. BMJ Open 2: e001072. doi:10.1136/bmjopen-2012-001072.

4. Scarborough P, Allender S, Clarke D, Wickramasinghe K, Rayner M (2012) Modelling the health impact of environmentally sustainable dietary scenarios in the UK. Eur J Clin Nutr 66: 710–715.

5. Friel S, Dangour AD, Garnett T, Lock K, Chalabi Z, et al. (2009) Public health benefits of strategies to reduce greenhouse-gas emissions: Food and agriculture. Lancet 374: 2016–2025.

6. IPCC (2007) Climate change 2007: Working group I: The physical science basis. In: Solomon S, Qin D, Manning M, Chen Z, Marquis M, et al, editors. Contribution of Working Group I to the Fourth Assessment Report of the Intergovernmental Panel on Climate Change. Cambridge: Cambridge University Press.

7. McMichael AJ, Powles JW, Butler CD, Uauy R (2007) Food, livestock production, energy, climate change, and health. Lancet 370: 1253–1263.

8. Scialabba NE, Müller-Lindenlauf M (2010) Organic agriculture and climate change. Renew Agric Food Syst 25: 158–169.

9. Carlsson-Kanyama A (1998) Climate change and dietary choices - how can emissions of greenhouse gases from food consumption be reduced? Food Policy 23: 277–293.

10. Carlsson-Kanyama A, González AD (2009) Potential contributions of food consumption patterns to climate change. Am J Clin Nutr 89: 1704S–1709S.

11. Nilsson K, Flysjö A, Davis J, Sim S, Unger N, et al. (2010) Comparative life cycle assessment of margarine and butter consumed in the UK, Germany and France. Int J Life Cycle Assess 15: 916–926.

12. Carlsson-Kanyama A, Ekström MP, Shanahan H (2003) Food and life cycle energy inputs: Consequences of diet and ways to increase efficiency. Ecol Econ 44: 293–307.

13. Peltonen M, Harald K, Männistö S, Saarikoski L, Peltomäki P, et al. (2008) Kansallinen FINRISKI 2007 -terveystutkimus. Tutkimuksen toteutus ja tulokset. Kansanterveyslaitoksen julkaisuja B34/2008. [The national FINRISK 2007 health study. The methods and results. Publications of the National Public Health Institute B34/2008].

14. Popp A, Lotze-Campen H, Bodirsky B (2010) Food consumption, diet shifts and associated non-CO2 greenhouse gases from agricultural production. Global Environ Change 20: 451–462.

15. Stehfest E, Bouwman L, Van Vuuren DP, Den Elzen MGJ, Eickhout B, et al. (2009) Climate benefits of changing diet. Clim Change 95: 83–102.

16. European Commission (2008) Europeans' attitudes towards climate change. Special Eurobarometer 300.

17. Tervonen K (2009) Ilmastotalkoot-tutkimus, helmikuu 2009. [The Ilmastotalkoot climate research, February 2009]. Available: http://www.slideshare.net/DemosHelsinki/ilmastotalkoot-tutkimuspresentaatio. Retrieved 2nd December 2013.

18. Mäkiniemi J, Vainio A (2013) Moral intensity and climate-friendly food choices. Appetite 66: 54–61.

19. Whitmarsh L, O'Neill S (2010) Green identity, green living? The role of pro-environmental self-identity in determining consistency across diverse pro-environmental behaviours. J Environ Psychol 30: 305–314.

20. Aitken C, Chapman R, McClure J (2011) Climate change, powerlessness and the commons dilemma: Assessing New Zealanders' preparedness to act. Global Environ Change 21: 752–760.

21. Semenza JC, Hall DE, Wilson DJ, Bontempo BD, Sailor DJ, et al. (2008) Public perception of climate change voluntary mitigation and barriers to behavior change. Am J Prev Med 35: 479–487.

22. Fielding KS, Head BW (2012) Determinants of young Australians' environmental actions: The role of responsibility attributions, locus of control, knowledge and attitudes. Environ Educ Res 18: 171–186.
23. Whitmarsh L, Seyfang G, O'Neill S (2011) Public engagement with carbon and climate change: To what extent is the public 'carbon capable'? Global Environ Change 21: 56–65.
24. Semenza JC, Ploubidis GB, George LA (2011) Climate change and climate variability: Personal motivation for adaptation and mitigation. Environ Health 10: 46.
25. Agho K, Stevens G, Taylor M, Barr M, Raphael B (2010) Population risk perceptions of global warming in Australia. Environ Res 110: 756–763.
26. Population Register Centre. http://www.vrk.fi/.
27. Hyrkäs H, Jaakkola MS, Ikäheimo TM, Hugg TT, Jaakkola JJK (2014) Asthma and allergic rhinitis increase respiratory symptoms in cold weather among young adults. Respir Med 108(1): 63–70.
28. Jaakkola JJ, Jaakkola N, Ruotsalainen R (1993) Home dampness and molds as determinants of respiratory symptoms and asthma in pre-school children. J Expo Anal Environ Epidemiol 3 Suppl 1: 129–142.
29. Jaakkola JJK, Hwang B, Jaakkola N (2005) Home dampness and molds, parental atopy, and asthma in childhood: A six-year population-based cohort study. Environ Health Perspect 113: 357–361.
30. Nijdam D, Rood T, Westhoek H (2012) The price of protein: Review of land use and carbon footprints from life cycle assessments of animal food products and their substitutes. Food Policy 37: 760–770.
31. González AD, Frostell B, Carlsson-Kanyama A (2011) Protein efficiency per unit energy and per unit greenhouse gas emissions: Potential contribution of diet choices to climate change mitigation. Food Policy 36: 562–570.
32. Goh KM (2011) Greater mitigation of climate change by organic than conventional agriculture: A review. Biol Agric Hortic 27: 205–230.
33. Hansen B, Alrøe HF, Kristensen ES (2001) Approaches to assess the environmental impact of organic farming with particular regard to Denmark. Agric Ecosyst Environ 83: 11–26.
34. Stolze M, Piorr A, Häring A, Dabbert S (2000) The Environmental Impacts of Organic Farming in Europe. Organic Farming in Europe: Economics and Policy. Available: https://www.uni-hohenheim.de/i410a/ofeurope/organicfarmingineurope-vol6.pdf. Retrieved 4th May 2012.
35. TIKE the Information Centre of the Ministry of Agriculture and Forestry. Available: http://185.20.137.77/taxonomy/term/50?q = taxonomy/term/50. Retrieved 3rd December 2013.
36. Grizzetti B, Pretato U, Lassaletta L, Billen G, Garnier J (2013) The contribution of food waste to global and European nitrogen pollution. Environ Sci Policy 33: 186–195.
37. Darmon N, Drewnowski A (2008) Does social class predict diet quality? Am J Clin Nutr 87: 1107–1117.
38. Roos E, Lahelma E, Virtanen M, Prättälä R, Pietinen P (1998) Gender, socioeconomic status and family status as determinants of food behaviour. Soc Sci Med 46: 1519–1529.
39. Dowd K, Burke KJ (2013) The influence of ethical values and food choice motivations on intentions to purchase sustainably sourced foods. Appetite 69: 137–144.
40. Steptoe A, Pollard TM, Wardle J (1995) Development of a measure of the motives underlying the selection of food: The food choice questionnaire. Appetite 25: 267–284.
41. Sundblad E, Biel A, Gärling T (2014) Intention to change activities that reduce carbon dioxide emissions related to worry about global climate change consequences. Rev Eur Psychol Appl In Press.

21

Dietary Protein Intake and Coronary Heart Disease in a Large Community Based Cohort: Results from the Atherosclerosis Risk in Communities (ARIC) Study

Bernhard Haring[1]*, Noelle Gronroos[2], Jennifer A. Nettleton[3], Moritz C. Wyler von Ballmoos[4], Elizabeth Selvin[5], Alvaro Alonso[2]

1 Department of Internal Medicine I, Comprehensive Heart Failure Center, University of Würzburg, Würzburg, Bavaria, Germany, 2 Division of Epidemiology and Community Health, University of Minnesota, Minneapolis, Minnesota, United States of America, 3 Division of Epidemiology, Human Genetics and Environmental Sciences, School of Public Health, University of Texas Health Science Center at Houston, Houston, Texas, United States of America, 4 Department of Surgery & Division of Cardiothoracic Surgery, Froedtert Memorial Hospital & Medical College of Wisconsin, Milwaukee, Wisconsin, United States of America, 5 Department of Epidemiology and the Welch Center for Prevention, Epidemiology and Clinical Research, Johns Hopkins Bloomberg School of Public Health, Baltimore, Maryland, United States of America

Abstract

Background: Prospective data examining the relationship between dietary protein intake and incident coronary heart disease (CHD) are inconclusive. Most evidence is derived from homogenous populations such as health professionals. Large community-based analyses in more diverse samples are lacking.

Methods: We studied the association of protein type and major dietary protein sources and risk for incident CHD in 12,066 middle-aged adults (aged 45–64 at baseline, 1987–1989) from four U.S. communities enrolled in the Atherosclerosis Risk in Communities (ARIC) Study who were free of diabetes mellitus and cardiovascular disease at baseline. Dietary protein intake was assessed at baseline and after 6 years of follow-up by food frequency questionnaire. Our primary outcome was adjudicated coronary heart disease events or deaths with following up through December 31, 2010. Cox proportional hazard models with multivariable adjustment were used for statistical analyses.

Results: During a median follow-up of 22 years, there were 1,147 CHD events. In multivariable analyses total, animal and vegetable protein were not associated with an increased risk for CHD before or after adjustment. In food group analyses of major dietary protein sources, protein intake from red and processed meat, dairy products, fish, nuts, eggs, and legumes were not significantly associated with CHD risk. The hazard ratios [with 95% confidence intervals] for risk of CHD across quintiles of protein from poultry were 1.00 [ref], 0.83 [0.70–0.99], 0.93 [0.75–1.15], 0.88 [0.73–1.06], 0.79 [0.64–0.98], P for trend = 0.16). Replacement analyses evaluating the association of substituting one source of dietary protein for another or of decreasing protein intake at the expense of carbohydrates or total fats did not show any statistically significant association with CHD risk.

Conclusion: Based on a large community cohort we found no overall relationship between protein type and major dietary protein sources and risk for CHD.

Editor: Antony Bayer, Cardiff University, United Kingdom

Funding: The Atherosclerosis Risk in Communities Study is carried out as a collaborative study supported by National Heart, Lung, and Blood Institute contracts (HHSN268201100005C, HHSN268201100006C, HHSN268201100007C, HHSN268201100008C, HHSN268201100009C, HHSN268201100010C, HHSN268201100011C, and HHSN268201100012C). There are no relationships with industry to declare. Open Access publication was funded by the German Research Foundation (DFG) and the University of Würzburg in the funding program Open Access Publishing. The funders had no role in study design, data collection and analysis, decision to publish, or preparation of the manuscript.

Competing Interests: The authors have declared that no competing interests exist.

* Email: Haring_B@ukw.de

Introduction

The relationship of dietary protein distinguished by animal versus vegetable origin with risk of coronary heart disease (CHD) has shown conflicting results [1,2,3,4,5,6]. This is surprising since the type of protein has been shown to influence cardiovascular risk factors such as hypertension [7,8,9,10,11,12]. Various observational studies [7,8,10,13] and feeding trials [9,11,12] have associated dietary protein of vegetable type inversely with blood pressure. To elucidate this apparent paradox, Bernstein et al. have focused on the effect of various food groups as major sources of dietary protein rather than on protein type in the Nurses' Health Study [2]. Greater consumption of red meat or processed meat products was associated with a higher risk of CHD, while higher intakes of poultry, fish and nuts were associated with lower risk [2]. These findings are again in contrast with other results that showed no association of red meat with CHD [14] or that showed beneficial effects of animal protein on vascular health [15,16]. The discordance in findings may be explained by several factors including research design and study populations. Large randomized controlled feeding trials on this topic are sparse and current evidence is mostly derived from observational studies which used nurses or health professionals as study populations [1,2,4,5]. Community based analyses are mostly missing [6,17]. Thus, conclusions regarding the relation of various sources of protein intake with cardiovascular health are difficult to draw. Analyses conducted in large general communities are warranted as these may provide greater exposure variability with more generalizable results.

In this study, we aimed to investigate the associations between total, animal, and plant-based dietary protein, as well as individual protein-rich food groups, and the risk for CHD in a large, community-based cohort of middle-aged adults. We hypothesized that intake of animal protein and proteins from processed meats would be associated with a higher risk of CHD and vegetable proteins and corresponding food groups with a lower CHD risk.

Methods

Study Population

The Atherosclerosis Risk in Communities Study (ARIC) is a community-based prospective cohort study of 15,792 middle-aged adults (aged 45–64 years at baseline) from four U.S. communities (Washington County, Md; Forsyth County, NC; Jackson, Miss; and suburbs of Minneapolis, Minn.) [18]. The first examination (visit 1) of participants occurred during 1987–1989, with three follow-up visits taking place each approximately every 3 years; response rates were 93%, 86%, and 81% at visits 2 (1990 to 1992), 3 (1993 to 1995), and 4 (1996 to 1998), respectively. A fifth exam (visit 5) took place in 2011–2013 among surviving participants. At all visits, participants received an extensive examination, including collection of medical, social, and demographic data [18]. For this analysis, only white and black adults were included; blacks from the Minneapolis and Washington County field centers were excluded due to small numbers. Individuals with self-reported diabetes, fasting blood glucose ≥126 mg/dL, non-fasting blood glucose ≥200 mg/dL or use of diabetes medication, a history of myocardial infarction, stroke, heart failure, coronary bypass surgery, angioplasty or with missing data on covariates of interest were excluded. Our final sample size included 12,066 persons.

The ARIC study was approved by the Institutional Review Boards (IRB) of all participating institutions, including the IRBs of the University of Minnesota, Johns Hopkins University, University of North Carolina, University of Mississippi Medical Center, and Wake Forest University. Written informed consent at each clinical site was obtained from all participants.

Assessment of protein intake

Protein intake was assessed using an interviewer-administered, 66-item food frequency questionnaire (FFQ) adapted from the 61-item FFQ developed by Willett et al. [19]. The FFQ was administered to all subjects at visit 1 at baseline (1987–1989) and at visit 3 (1993–1995). The usual frequency of food consumption was reported in 9 categories, from never or less than once a month to >6 times per day. The major contributors to protein intake included: unprocessed red meat, processed red meat, poultry, high-fat dairy, low-fat dairy, fish & seafood, eggs, nuts, and legumes. Average daily intake of nutrients was calculated by multiplying the frequency of consumption of each food item by its nutrient content and adding up the nutrient intake for all of the items. Vegetable protein intake was defined as the difference of total and animal protein intake. The residual method was used to adjust for total energy intake [20]. For assessing dietary behaviour, participants were divided into quintiles of cumulative average intake of various protein sources. Cumulative updating of the FFQ (i.e. visit 1 FFQ for follow-up between visit 1 and visit 3 and the average of visits 1 and 3 FFQ afterwards for those who attended both examinations, or visit 1 FFQ for those who did not attend visit 3) was used to reduce within-person variation and best represent long-term dietary behavior [2]. Participants with incomplete dietary information or with extreme calorie intake (<600 kcal or >4200 kcal per day for men, <500 kcal or >3600 kcal per day for women) were excluded from further analysis. We stopped updating a participant's cumulative average intake when the participant of our study was diagnosed with an intermediate variable on the causal pathway between diet and CHD such as hypercholesterolemia, hypertension, stroke and diabetes. This was done to avoid exposure misclassification due to short-term changes in dietary patterns.

Assessment of coronary heart disease

The primary end point for this study was CHD occurring after the completion of the first FFQ (between 1987 and 1989). CHD was defined as a definite or probable myocardial infarction or a death from coronary heart disease. CHD events were identified and adjudicated using information from study visits, yearly telephone follow-up calls, review of hospital discharge lists and medical charts, death certificates, next-of-kin interviews, and physician-completed questionnaires [18,21]. Follow-up for CHD was available until December 31, 2010.

Covariates

Height, weight, and waist circumference were measured following a standardized protocol [18,21]. ARIC participants underwent fasting venipuncture at each examination [18]. Diabetes was defined as current use of glucose-lowering medications, fasting blood glucose ≥126 mg/dL, non-fasting blood glucose ≥200 mg/dL or self-reported history of diabetes. Hypertension was defined as the average of the last two of three blood-pressure readings at the first visit (using 140 mmHg or higher for systolic and 90 mmHg or higher for diastolic as cut-off points). Current smoking, ethanol intake, education, intake of antihypertensive or lipid lowering medication were derived from standardized questionnaires [18]. Sports-related physical activity and leisure related physical activity were assessed with the use of Baecke's questionnaire and scoring systems [22].

Statistical Analysis

To assess the association of CHD and average cumulative intake of protein by quintiles, we calculated incidence rates (IR) of CHD events per 1000 person-years as the number of diagnosed cases of CHD occurring during the entire follow-up period divided by person-years of follow-up. Person-years of follow up were defined as time from the baseline examination to the date of the first coronary event, death, lost to follow-up, or December 31, 2010, whichever occurred earlier. Thereafter, corresponding rate ratios were calculated by dividing the rate among participants in each specific intake quintile by the rate among participants in the lowest quintile of intake (reference). Cox proportional hazards regression models were used to account for potential confounding. An initial model adjusted for age, race, sex, ARIC study center,

Table 1. Unadjusted baseline characteristics according to quintiles of total protein intake, ARIC 1987–1989.

	Q 1 (low)	Q 2	Q 3	Q 4	Q5 (high)	p-trend[a]
N	2412	2414	2413	2414	2413	
Protein intake, g/day (SD)	49.3 (10.2)	62.9 (3.9)	70.2(3.9)	77.8 (4.0)	93.5 (12.6)	<0.0001
Protein intake, % of total energy	12.4 (1.7)	15.7 (1.0)	17.8 (1.5)	19.8 (2.0)	22.8 (3.4)	<0.0001
Age, years (SD)	53.4 (5.7)	54.0 (5.7)	54.0 (5.8)	53.8 (5.7)	53.8 (5.7)	0.11
Women, %	55.9	55.8	55.8	55.8	55.8	0.99
Black, %	24.8	24.5	21.9	21.6	22.4	0.005
High school graduate, %	72.1	77.7	80.7	82.6	84.0	<0.0001
Current smoker, %	32.3	28.8	24.7	22.5	21.6	<0.0001
Hypertension, %	29.9	27.5	29.1	28.0	28.8	0.57
Body Mass Index, kg/m^2 (SD)	26.6 (5.1)	26.8 (4.9)	27.0 (5.0)	27.4 (5.0)	27.7 (5.1)	<0.0001
Waist-to-hip ratio (SD)	0.9 (0.1)	0.9 (0.1)	0.9 (0.1)	0.9 (0.1)	0.9 (0.1)	0.19
Baecke Sport Activity Score (SD)	2.4 (0.8)	2.4 (0.8)	2.5 (0.8)	2.5 (0.8)	2.6 (0.8)	<0.0001
Baecke Leisure Index (SD)	2.3 (0.6)	2.3 (0.6)	2.4 (0.6)	2.4 (0.6)	2.5 (0.6)	<0.0001
Systolic blood pressure, mmHg (SD)	120.7 (18.7)	119.9 (17.5)	119.9 (17.6)	119.2 (17.6)	119.4 (18.2)	0.004
Serum HDL, mmol/L (SD)	1.4 (0.4)	1.4 (0.4)	1.4 (0.4)	1.4 (0.5)	1.4 (0.4)	0.27
Serum cholesterol, mmol/L (SD)	5.5 (1.0)	5.5 (1.1)	5.6 (1.1)	5.5 (1.1)	5.5 (1.1)	0.23
Use of antihypertensive medication, %	19.4	18.3	20.4	18.4	19.6	0.89
Use of lipid lowering medication, %	1.3	1.7	2.2	3.2	2.7	<0.0001
Carbohydrate intake, g/day (SD)	231.2 (57.6)	208.0 (31.4)	197.8 (31.3)	188.5 (32.3)	172.4 (38.4)	<0.0001
Carbohydrate intake, % of total energy (SD)	56.2 (9.5)	51.3 (7.9)	48.3 (7.7)	45.8 (7.5)	42.8 (7.8)	<0.0001
Fiber intake, g/day (SD)	15.5 (7.4)	16.7 (5.6)	17.3 (5.6)	17.6 (5.9)	18.3 (7.0)	<0.0001
Magnesium intake, g/day (SD)	217.0 (66.1)	240.7 (52.5)	253.5 (53.9)	267.0 (57.9)	288.4 (65.6)	<0.0001
Alcohol intake, g/week (SD)	68.7 (148.7)	43.9 (89.0)	40.7 (78.9)	36.9 (70.4)	32.7 (65.4)	<0.0001
Total energy intake, kcal/day (SD)	1818.1 (693.9)	1488.7 (555.6)	1489.7 (538.1)	1565.9 (537.1)	1802.8 (606.7)	0.23
Total fat Intake (g/d) (Median ±SD)	53.9 (16.0)	59.9 (11.7)	61.4 (12.0)	62.6 (12.8)	63.4 (15.4)	<0.0001
Total fat intake, % of total energy (SD)	30.0 (7.2)	32.5 (6.2)	33.4 (6.3)	34.2 (6.2)	34.5 (6.5)	<0.0001

Values are % for categorical variables and mean (SD) for continuous variables.
[a]p-values from general linear models for continuous variables and Mantel-Haenszel 1-degree of freedom chi-square statistic.

and total energy intake (minimally adjusted model). A second model additionally adjusted for smoking (current, former, never), pack years of smoking, education (less than high school, high school, more than high school), systolic blood pressure (mmHg), use of antihypertensive medication, HDLc (mmol/l), total cholesterol (mmol/l), use of lipid lowering medication, body mass index (kg/m^2), waist-to-hip ratio, alcohol intake (g/week), Baecke's physical activity score, leisure-related physical activity, carbohydrate intake (quintiles), fiber intake (quintiles), and magnesium intake (quintiles) (fully adjusted model). Median protein intake of each quintile (g/d) modeled as a continuous variable was used to test for linear trend.

We further conducted food substitution analyses based on the fully adjusted model. Hazard ratios of CHD associated with increasing 1 serving/day in the consumption of protein sources at the expense of decreasing 1 serving/day in a different protein source were calculated. Similarly, we conducted nutrient substitution analyses by examining the risk for CHD when increasing 10% energy from carbohydrates or fat while decreasing 10% of energy from protein. Tests of the proportional hazards assumption were evaluated. All p-values were 2-tailed. Data were analyzed with SAS 9.3 (SAS Corp, Cary, NC).

Results

Baseline characteristics of the study participants according to quintiles of total protein intake at baseline are shown in Table 1. Compared with participants with low protein consumption, individuals with high protein consumption were less likely to be current smoker, to drink less alcohol per week, and more likely to conduct physical activity and to have graduated from high school. Furthermore, participants with high protein intake had higher BMI levels, higher intakes of fiber, magnesium and fat whereas decreased intake of carbohydrates.

During a median follow-up of 22 years, there were 1,147 CAD events among the 12,066 participants at baseline. In age, sex, race, study center and total energy adjusted analyses (minimally adjusted model) animal protein intake was not associated with an increased risk for CHD (Table 2). These results did not change significantly after full adjustment. In the minimally adjusted model, total and vegetable protein were associated with a significantly lower risk of CHD (Table 2, Model 1). This relationship was considerably attenuated and became non-significant after full adjustment (Table 2, Model 2).

In food-group analyses of major dietary protein sources using our minimally adjusted model, higher intake of red or processed meat was significantly associated with increased risk for CHD,

Table 2. Association of total, animal and vegetal protein intake with coronary heart disease incidence, ARIC 1987–2010.

	Q 1	Q 2	Q 3	Q 4	Q 5	p-trend
Total Protein Intake						
Events, n	241	230	231	230	215	
Person-time	46149	46720	46725	46991	47102	
Incidence, per 1000 py	5.2	4.9	4.9	4.9	4.6	
HR (95%CI)*	1 (ref)	0.84 (0.70, 1.01)	0.86 (0.72, 1.03)	0.89 (0.74, 1.06)	0.79 (0.66, 0.95)	0.04
HR (95%CI)**	1 (ref)	0.91 (0.75, 1.11)	0.93 (0.76, 1.14)	1.00 (0.80, 1.24)	0.84 (0.66, 1.07)	0.34
Animal Protein Intake						
Events, n	236	240	212	238	221	
Person-time	46175	46781	46915	47067	46750	
Incidence, per 1000 py	5.1	5.1	4.5	5.1	4.7	
HR (95%CI)*	1 (ref)	1.05 (0.88, 1.26)	0.97 (0.80, 1.17)	1.01 (0.84, 1.21)	0.95 (0.79, 1.15)	0.56
HR (95%CI)**	1 (ref)	1.16 (0.96, 1.40)	1.01 (0.82, 1.24)	1.11 (0.90, 1.37)	1.00 (0.79, 1.26)	0.94
Vegetable Protein Intake						
Events, n	247	228	253	215	204	
Person-time	45991	46902	46518	47070	47207	
Incidence, per 1000 py	5.4	4.9	5.4	4.6	4.3	
HR (95%CI)*	1 (ref)	0.92 (0.77, 1.10)	0.91 (0.76, 1.09)	0.82 (0.68, 0.98)	0.71 (0.59, 0.85)	0.0001
HR (95%CI)**	1 (ref)	1.08 (0.89, 1.31)	1.15 (0.94, 1.40)	1.04 (0.84, 1.29)	0.87 (0.68, 1.10)	0.17

*adjusted for age, sex, race, study, center, and total energy intake.
** adjusted for age, sex, race, study center, total energy intake, smoking, education, systolic blood pressure, use of antihypertensive medication, HDLc, total cholesterol, use of lipid lowering medication, body mass index, waist-to-hip ratio, alcohol intake, sports-related physical activity, leisure-related physical activity, carbohydrate intake, fiber intake, and magnesium intake.

whereas low-fat dairy, poultry and nuts consumption were significantly associated with decreased risk for CHD (Table 3, Model 1). After adjustment for potential confounders, only higher poultry intake remained associated with a lower risk of CHD (Table 3, Model 2).

Last, we conducted replacement analysis evaluating the association of substituting one source of dietary protein for another and of decreasing protein intake at the expense of carbohydrates or total fats (Table 4). Overall, these results did not show any statistically significant association with CHD risk, although they suggested that decreasing red meat and increasing intake of any other protein source was associated with non-significant lower risk of CHD (10–20% reduction per 1 serving/day change).

Discussion

In this prospective community based study with 22 years of follow-up, neither total, animal or vegetable protein intake was associated with risk of CHD. In food group analyses of major dietary protein sources, we found no significant trend between various sizes of intake of meat products, poultry, dairy, eggs, nuts, fish, or legumes and risk for CHD. Our results are contrary to our initial hypothesis as we expected food groups based on animal protein such as red or processed meat products to be significantly associated with an increased risk for CHD.

Thus far, the largest cohort analyses to examine an association between protein intake and coronary heart disease were undertaken using data from the Nurses' Health Study with 14, 16 and 26 years of follow up [2,5,23]. Interestingly, similar to our results it was found that after 14 to 16 years of follow-up, neither animal protein nor vegetable protein were associated with CHD [5,24].

Results from the Health Professional Follow-up Study also indicate no association between dietary protein and risk of coronary heart disease after 18 years of follow-up [4]. In a later analysis of the Nurses' Health Study spanning 26 years of follow-up higher intakes of red meat, red meat excluding processed meat, and high-fat dairy were indeed found to be significantly associated with an elevated risk of CHD while higher intakes of poultry, fish, and nuts were significantly associated with lower risk [2]. Other prospective studies using California Seventh Day Adventists or the NIH-AARP Diet and Health cohort as study base also report a positive association between (red) meat consumption and CHD risk [6,25]. Nonetheless, generalizability of the existing data is limited as the respective cohorts are characterized by well-educated, ethnically homogenous study populations. A recent meta-analysis summarizing 9 studies on red and processed meat consumption and risk for CHD found processed meats (RR 1.42, 95%CI 1.07, 1.89), but not red meats (RR 1.00, 95% CI 0.81,1.23) to increase incident coronary events [14]. The effects of other dietary protein sources or type of protein were not addressed in this analysis. Moreover all included studies were observational and among included studies only one was based in a general community setting in the UK [17]. Interventional studies such as the Bold Study suggest that dietary protein, also of animal origin, can exert positive effects on biomarkers of CHD [15,16]. Lean beef in an optimal lean diet has been shown to exhibit beneficial effects on systolic blood pressure and vascular elasticity [15,16].

In spite of the lack of strong epidemiologic evidence for an association between animal derived protein sources (in particular meat products) and risk for CHD, several arguments mainly based on contents of sodium and saturated fat have been previously made to potentially explain a harmful effect of animal derived protein products on the risk of CHD. Processed meats are known

Table 3. Association of major dietary protein sources with coronary heart disease, ARIC 1987–2010.

	Q 1	Q 2	Q 3	Q 4	Q 5	p-trend
Processed Meat						
Median svg/day	0	0.1	0.4	0.5	1.1	
HR (95%CI)*	1 (ref)	1.04 (0.86, 1.25)	1.22 (1.01, 1.49)	1.28 (1.06, 1.553)	1.40 (1.15, 1.71)	0.003
HR (95%CI)**	1 (ref)	0.95 (0.78. 1.15)	1.02 (0.84, 1.24)	1.04 (0.85, 1.265)	1.04 (0.85, 1.29)	0.49
Red Meat						
Median svg/day	0.1	0.3	0.5	0.6	1.1	
HR (95%CI)*	1 (ref)	1.00 (0.83, 1.21)	1.11 (0.93, 1.33)	1.18 (0.98, 1.43)	1.30 (1.06, 1.59)	0.004
HR (95%CI)**	1 (ref)	0.92 (0.76, 1.12)	1.02 (0.85, 1.24)	1.10 (0.90, 1.35)	1.13 (0.89, 1.44)	0.13
Red Meat & Processed Meat						
Median svg/day	0.2	0.5	0.8	1.2	1.9	
HR (95%CI)*	1 (ref)	1.15 (0.95, 1.40)	1.32 (1.09, 1.60)	1.51 (1.24, 1.84)	1.53 (1.23, 1.90)	<0.0001
HR (95%CI)**	1 (ref)	1.00 (0.82, 1.23)	1.084 (0.88, 1.33)	1.18 (0.95, 1.46)	1.15 (0.89, 1.48)	0.21
Poultry						
Median svg/day	0.1	0.1	0.3	0.4	0.8	
HR (95%CI)*	1 (ref)	0.80 (0.67, 0.95)	0.83 (0.67, 1.03)	0.75 (0.63, 0.90)	0.67 (0.55, 0.82)	0.0007
HR (95%CI)**	1 (ref)	0.83 (0.70, 0.99)	0.93 (0.75, 1.15)	0.88 (0.73, 1.06)	0.79 (0.64, 0.98)	0.16
Dairy						
Median svg/day	0.1	0.6	1.1	1.5	2.9	
HR (95%CI)*	1 (ref)	0.88 (0.73, 1.05)	0.99 (0.83, 1.18)	0.72 (0.59, 0.87)	0.88 (0.72, 1.08)	0.24
HR (95%CI)**	1 (ref)	0.96 (0.80, 1.16)	1.14 (0.95, 1.37)	0.85 (0.69, 1.04)	1.04 (0.84, 1.29)	0.77
High-Fat Dairy						
Median svg/day	0.1	0.1	0.4	0.8	1.2	
HR (95%CI)*	1 (ref)	1.09 (0.91, 1.31)	0.96 (0.80, 1.16)	1.04 (0.86, 1.27)	1.1 (0.90, 1.34)	0.56
HR (95%CI)**	1 (ref)	1.16 (0.96, 1.39)	1.03 (0.86, 1.25)	1.13 (0.93, 1.38)	1.14 (0.93, 1.39)	0.47
Low-Fat Dairy						
Median svg/day	0	0.1	0.4	1	2.5	
HR (95%CI)*	1 (ref)	0.93 (0.77, 1.12)	0.73 (0.61, 0.87)	0.73 (0.61, 0.87)	0.75 (0.62, 0.90)	0.007
HR (95%CI)**	1 (ref)	1.04 (0.86, 1.25)	0.86 (0.72, 1.03)	0.90 (0.75, 1.08)	0.91 (0.74, 1.12)	0.39
Fish & seafood						
Median svg/day	0	0.1	0.2	0.3	0.6	
HR (95%CI)*	1 (ref)	0.98 (0.81, 1.17)	1.05 (0.85, 1.29)	0.95 (0.77, 1.16)	0.90 (0.74, 1.10)	0.20
HR (95%CI)**	1 (ref)	1.04 (0.87, 1.25)	1.17 (0.95, 1.44)	1.07 (0.87, 1.32)	1.06 (0.86, 1.31)	0.81
Eggs						
Median svg/day	0	0.1	0.1	0.4	1.0	
HR (95%CI)*	1 (ref)	0.90 (0.74, 1.09)	0.86 (0.72, 1.03)	0.84 (0.70, 1.01)	1.09 (0.88, 1.34)	0.20
HR (95%CI)**	1 (ref)	0.92 (0.76, 1.12)	0.88 (0.73, 1.06)	0.83 (0.69, 0.99)	0.96 (0.77, 1.19)	0.89
Nuts						
Median svg/day	0	0.1	0.2	0.4	1.0	
HR (95%CI)*	1 (ref)	0.83 (0.70, 0.99)	0.74 (0.60, 0.90)	0.71 (0.59, 0.87)	0.73 (0.60, 0.89)	0.02
HR (95%CI)**	1 (ref)	0.89 (0.75, 1.06)	0.86 (0.71, 1.05)	0.83 (0.68, 1.01)	0.91 (0.74, 1.12)	0.67
Legumes						
Median svg/day	0.1	0.1	0.2	0.3	0.6	
HR (95%CI)*	1 (ref)	1.01 (0.84, 1.20)	1.04 (0.83, 1.32)	0.98 (0.82, 1.17)	1.04 (0.85, 1.26)	0.64
HR (95%CI)**	1 (ref)	1.07 (0.89, 1.27)	1.16 (0.92, 1.46)	1.05 (0.87, 1.27)	1.159 (0.93, 1.44)	0.18

*adjusted for age, sex, race, study, center, and total energy intake.
** adjusted for age, sex, race, study center, total energy intake, smoking, education, systolic blood pressure, use of antihypertensive medication, HDLc, total cholesterol, use of lipid lowering medication, body mass index, waist-to-hip ratio, alcohol intake, sports-related physical activity, leisure-related physical activity, carbohydrate intake, fiber intake, and magnesium intake.

to have high sodium contents. High sodium intake is strongly correlated with the development of hypertension and CHD

Table 4. Food substitution analysis.

Increase 1 svg/d → / Decrease 1 svg/d ↓	Red meat	Poultry	High-fat dairy	Low-fat dairy	Fish/seafood	Eggs	Nuts	Legumes
Processed meat	1.14 (0.90–1.44)	0.93 (0.73–1.18)	1.04 (0.90–1.20)	0.96 (0.84–1.11)	0.98 (0.74–1.15)	0.92 (0.74–1.15)	0.94 (0.79–1.10)	0.97 (0.74–1.27)
Red meat	-	0.81 (0.62–1.07)	0.91 (0.74–1.12)	0.84 (0.68–1.05)	0.86 (0.65–1.13)	0.81 (0.63–1.05)	0.81 (0.65–1.02)	0.85 (0.61–1.18)
Poultry	-	-	1.12 (0.90–1.40)	1.04 (0.83–1.30)	1.05 (0.76–1.45)	0.99 (0.77–1.29)	1.00 (0.79–1.27)	1.04 (0.76–1.44)
High-fat dairy	-	-	-	0.93 (0.84–1.02)	0.94 (0.75–1.19)	0.89 (0.75–1.06)	0.90 (0.78–1.03)	0.93 (0.73–1.19)
Low-fat dairy	-	-	-	-	1.02 (0.80–1.28)	0.96 (0.81–1.14)	0.97 (0.84–1.11)	1.01 (0.79–1.28)
Fish/seafood	-	-	-	-	-	0.94 (0.72–1.24)	0.95 (0.74–1.22)	0.99 (0.72–1.38)
Eggs	-	-	-	-	-	-	1.01 (0.83–1.22)	1.05 (0.79–1.39)
Nuts	-	-	-	-	-	-	-	1.04 (0.81–1.34)

Nutrient substitution analysis:
• Increasing 10% energy from carbohydrates, decreasing 10% from protein: HR 0.96 (95% CI 0.82–1.14).
• Increasing 10% energy from fats, decreasing 10% from protein: HR 0.99 (95% CI 0.80–1.24).

adjusted for age, sex, race, study center, total energy intake, smoking, education, systolic blood pressure, use of antihypertensive medication, HDLc, total cholesterol, use of lipid lowering medication, body mass index, waist-to-hip ratio, alcohol intake, sports-related physical activity, leisure-related physical activity, carbohydrate intake, fiber intake, magnesium intake, and all the food items in the table (continuous variables, in servings/day). Hazard ratios (95% confidence intervals) of CHD associated with increasing 1 serving/day in the consumption of protein sources (rows) at the expense of decreasing 1 serving/day in a different protein source (columns). For example, the orange shaded cell can be interpreted as the HR (95% CI) of CHD increasing poultry consumption 1 serving/day and reducing red meat by the same amount, keeping all other sources of dietary protein constant.

[26,27]. On the other hand, high animal protein as provided by dark meat intake often accompanies large intakes of saturated fat intake which has been linked to increased cardiovascular risk [23,28,29]. In contrast to animal protein diets consisting of vegetable protein have been associated with cardiovascular and overall health benefits because of their high content of mono- and polyunsaturated fats, fiber, vitamins and minerals and low content of sodium [30]. These lines of argumentation are supported by current dietary recommendations [31], however, the role of saturated fat on the risk of CHD has been subject to controversial debates most recently [32,33]. Further research is warranted to elucidate the mechanisms of action of protein on the risk of CHD.

The absence of an association between major dietary protein sources and risk for CHD in our population may be explained in part by limited variation in consumptions of these food groups. Our study participants reported low meat intake whereas consumption of eggs, nuts, fish and dairy consumption were similar to other study populations [2]. Therefore, our observation suggesting a protective association of high poultry intake with lower CHD risk has to be interpreted with caution as this singular result may be spurious.

Other reasons that explain inconsistent findings and reports between dietary protein sources, food groups and CHD can be found in differences in study design, follow-up and assessment of outcomes and covariates. Similar to previous reports of the Nurses' Health Study and the Health Professional Study a follow-up period of 22 years may not have been long enough to detect significant differences between various dietary intake levels and risk for CHD events [2,4,5]. Second, our dietary data assessment was imperfect and incomplete. Although repeated dietary data measurements may take changing dietary patterns into account and thus serve to reduce intra-individual error, exposure variability of our study population was limited as protein intake was only assessed at two time-points. Changing dietary habits may not have been covered adequately by our FFQs with time-points only 6 years apart. On the other hand, it is well known that behavioral dietary changes are very challenging to accomplish and to maintain on the individual level and thus great changes in the overall population are unlikely to occur [34,35]. Last, dietary substitution effects as well as different characteristics of particular food group (e.g fat content, micronutrient content) may limit our analyses. Strengths of our study include the sample size, a large community based cohort with two different races in the setting of the general US population and a prospective design with long follow-up. CHD and several confounding factors were assessed using standardized protocols whereas other studies were based on self-report data [2,5].

In conclusion, using a large community based cohort study we found neither total nor animal or vegetable protein to be associated with CHD. In detailed food group analyses of major protein sources, no statistically significant trends between animal or vegetable-based food groups and risk for CHD were observed. Individuals should continue to make appropriate dietary modifications following current guidelines and recommendations for cardiovascular disease risk reduction [16,36].

Acknowledgments

The authors thank the staff and participants of the ARIC study for their important contributions. AA, BH and NG had full access to all of the data in the study and take responsibility for the integrity of the data and the accuracy of the data analysis.

Author Contributions

Conceived and designed the experiments: BH NG JN MWvB ES AA. Performed the experiments: ES JN BH AA. Analyzed the data: AA NG BH. Contributed reagents/materials/analysis tools: JN NG ES BH MWvB AA. Wrote the paper: BH NG JN MWvB ES AA.

References

1. Halton TL, Willett WC, Liu S, Manson JE, Albert CM, et al. (2006) Low-carbohydrate-diet score and the risk of coronary heart disease in women. N Engl J Med 355: 1991–2002.
2. Bernstein AM, Sun Q, Hu FB, Stampfer MJ, Manson JE, et al. (2010) Major dietary protein sources and risk of coronary heart disease in women. Circulation 122: 876–883.
3. Kelemen LE, Kushi LH, Jacobs DR Jr, Cerhan JR (2005) Associations of dietary protein with disease and mortality in a prospective study of postmenopausal women. Am J Epidemiol 161: 239–249.
4. Preis SR, Stampfer MJ, Spiegelman D, Willett WC, Rimm EB (2010) Dietary protein and risk of ischemic heart disease in middle-aged men. Am J Clin Nutr 92: 1265–1272.
5. Hu FB, Stampfer MJ, Manson JE, Rimm E, Colditz GA, et al. (1999) Dietary protein and risk of ischemic heart disease in women. Am J Clin Nutr 70: 221–227.
6. Sinha R, Cross AJ, Graubard BI, Leitzmann MF, Schatzkin A (2009) Meat intake and mortality: a prospective study of over half a million people. Arch Intern Med 169: 562–571.
7. Stamler J, Elliott P, Kesteloot H, Nichols R, Claeys G, et al. (1996) Inverse relation of dietary protein markers with blood pressure. Findings for 10,020 men and women in the INTERSALT Study. INTERSALT Cooperative Research Group. INTERnational study of SALT and blood pressure. Circulation 94: 1629–1634.
8. Iseki K, Iseki C, Itoh K, Sanefuji M, Uezono K, et al. (2003) Estimated protein intake and blood pressure in a screened cohort in Okinawa, Japan. Hypertens Res 26: 289–294.
9. Stamler J, Caggiula A, Grandits GA, Kjelsberg M, Cutler JA (1996) Relationship to blood pressure of combinations of dietary macronutrients. Findings of the Multiple Risk Factor Intervention Trial (MRFIT). Circulation 94: 2417–2423.
10. Stamler J, Liu K, Ruth KJ, Pryer J, Greenland P (2002) Eight-year blood pressure change in middle-aged men: relationship to multiple nutrients. Hypertension 39: 1000–1006.
11. Wang YF, Yancy WS Jr, Yu D, Champagne C, Appel LJ, et al. (2008) The relationship between dietary protein intake and blood pressure: results from the PREMIER study. J Hum Hypertens 22: 745–754.
12. Appel LJ, Sacks FM, Carey VJ, Obarzanek E, Swain JF, et al. (2005) Effects of protein, monounsaturated fat, and carbohydrate intake on blood pressure and serum lipids: results of the OmniHeart randomized trial. JAMA 294: 2455–2464.
13. Elliott P, Stamler J, Dyer AR, Appel L, Dennis B, et al. (2006) Association between protein intake and blood pressure: the INTERMAP Study. Arch Intern Med 166: 79–87.
14. Micha R, Wallace SK, Mozaffarian D (2010) Red and processed meat consumption and risk of incident coronary heart disease, stroke, and diabetes mellitus: a systematic review and meta-analysis. Circulation 121: 2271–2283.
15. Roussell MA, Hill AM, Gaugler TL, West SG, Heuvel JP, et al. (2012) Beef in an Optimal Lean Diet study: effects on lipids, lipoproteins, and apolipoproteins. Am J Clin Nutr 95: 9–16.
16. Roussell MA, Hill AM, Gaugler TL, West SG, Ulbrecht JS, et al. (2014) Effects of a DASH-like diet containing lean beef on vascular health. J Hum Hypertens.
17. Whiteman D, Muir J, Jones L, Murphy M, Key T (1999) Dietary questions as determinants of mortality: the OXCHECK experience. Public Health Nutr 2: 477–487.
18. (1989) The Atherosclerosis Risk in Communities (ARIC) Study: design and objectives. The ARIC investigators. Am J Epidemiol 129: 687–702.
19. Willett WC, Sampson L, Stampfer MJ, Rosner B, Bain C, et al. (1985) Reproducibility and validity of a semiquantitative food frequency questionnaire. Am J Epidemiol 122: 51–65.
20. Willett W, Stampfer MJ (1986) Total energy intake: implications for epidemiologic analyses. Am J Epidemiol 124: 17–27.
21. White AD, Folsom AR, Chambless LE, Sharret AR, Yang K, et al. (1996) Community surveillance of coronary heart disease in the Atherosclerosis Risk in Communities (ARIC) Study: methods and initial two years' experience. J Clin Epidemiol 49: 223–233.
22. Richardson MT, Ainsworth BE, Wu HC, Jacobs DR Jr, Leon AS (1995) Ability of the Atherosclerosis Risk in Communities (ARIC)/Baecke Questionnaire to assess leisure-time physical activity. Int J Epidemiol 24: 685–693.
23. Hu FB, Stampfer MJ, Manson JE, Ascherio A, Colditz GA, et al. (1999) Dietary saturated fats and their food sources in relation to the risk of coronary heart disease in women. Am J Clin Nutr 70: 1001–1008.
24. Hu FB, Bronner L, Willett WC, Stampfer MJ, Rexrode KM, et al. (2002) Fish and omega-3 fatty acid intake and risk of coronary heart disease in women. JAMA 287: 1815–1821.
25. Snowdon DA, Phillips RL, Fraser GE (1984) Meat consumption and fatal ischemic heart disease. Prev Med 13: 490–500.
26. Appel LJ, Moore TJ, Obarzanek E, Vollmer WM, Svetkey LP, et al. (1997) A clinical trial of the effects of dietary patterns on blood pressure. DASH Collaborative Research Group. N Engl J Med 336: 1117–1124.
27. Sacks FM, Campos H (2010) Dietary therapy in hypertension. N Engl J Med 362: 2102–2112.
28. Hu FB, Stampfer MJ, Manson JE, Rimm E, Colditz GA, et al. (1997) Dietary fat intake and the risk of coronary heart disease in women. N Engl J Med 337: 1491–1499.
29. Oh K, Hu FB, Manson JE, Stampfer MJ, Willett WC (2005) Dietary fat intake and risk of coronary heart disease in women: 20 years of follow-up of the nurses' health study. Am J Epidemiol 161: 672–679.
30. Vega-Lopez S, Lichtenstein AH (2005) Dietary protein type and cardiovascular disease risk factors. Prev Cardiol 8: 31–40.
31. Lichtenstein AH, Appel LJ, Brands M, Carnethon M, Daniels S, et al. (2006) Diet and lifestyle recommendations revision 2006: a scientific statement from the American Heart Association Nutrition Committee. Circulation 114: 82–96.
32. Chowdhury R, Warnakula S, Kunutsor S, Crowe F, Ward HA, et al. (2014) Association of dietary, circulating, and supplement fatty acids with coronary risk: a systematic review and meta-analysis. Ann Intern Med 160: 398–406.
33. Ramsden CE, Zamora D, Leelarthaepin B, Majchrzak-Hong SF, Faurot KR, et al. (2013) Use of dietary linoleic acid for secondary prevention of coronary heart disease and death: evaluation of recovered data from the Sydney Diet Heart Study and updated meta-analysis. BMJ 346: e8707.
34. Kumanyika SK, Van Horn L, Bowen D, Perri MG, Rolls BJ, et al. (2000) Maintenance of dietary behavior change. Health Psychol 19: 42–56.
35. Luepker RV, Murray DM, Jacobs DR, Jr., Mittelmark MB, Bracht N, et al. (1994) Community education for cardiovascular disease prevention: risk factor changes in the Minnesota Heart Health Program. Am J Public Health 84: 1383–1393.
36. Perk J, De Backer G, Gohlke H, Graham I, Reiner Z, et al. (2012) European Guidelines on cardiovascular disease prevention in clinical practice (version 2012). The Fifth Joint Task Force of the European Society of Cardiology and Other Societies on Cardiovascular Disease Prevention in Clinical Practice (constituted by representatives of nine societies and by invited experts). Eur Heart J 33: 1635–1701.

22

Estimation of Dietary Iron Bioavailability from Food Iron Intake and Iron Status

Jack R. Dainty[1], Rachel Berry[1], Sean R. Lynch[2], Linda J. Harvey[1], Susan J. Fairweather-Tait[3]*

1 Institute of Food Research, Norwich Research Park, Norwich, United Kingdom, **2** Department of Internal Medicine, Eastern Virginia Medical School, Norfolk, Virginia, United States of America, **3** University of East Anglia, Norwich Medical School, Norwich Research Park, Norwich, United Kingdom

Abstract

Currently there are no satisfactory methods for estimating dietary iron absorption (bioavailability) at a population level, but this is essential for deriving dietary reference values using the factorial approach. The aim of this work was to develop a novel approach for estimating dietary iron absorption using a population sample from a sub-section of the UK National Diet and Nutrition Survey (NDNS). Data were analyzed in 873 subjects from the 2000–2001 adult cohort of the NDNS, for whom both dietary intake data and hematological measures (hemoglobin and serum ferritin (SF) concentrations) were available. There were 495 men aged 19–64 y (mean age 42.7±12.1 y) and 378 pre-menopausal women (mean age 35.7±8.2 y). Individual dietary iron requirements were estimated using the Institute of Medicine calculations. A full probability approach was then applied to estimate the prevalence of dietary intakes that were insufficient to meet the needs of the men and women separately, based on their estimated daily iron intake and a series of absorption values ranging from 1–40%. The prevalence of SF concentrations below selected cut-off values (indicating that absorption was not high enough to maintain iron stores) was derived from individual SF concentrations. An estimate of dietary iron absorption required to maintain specified SF values was then calculated by matching the observed prevalence of insufficiency with the prevalence predicted for the series of absorption estimates. Mean daily dietary iron intakes were 13.5 mg for men and 9.8 mg for women. Mean calculated dietary absorption was 8% in men (50th percentile for SF 85 µg/L) and 17% in women (50th percentile for SF 38 µg/L). At a ferritin level of 45 µg/L estimated absorption was similar in men (14%) and women (13%). This new method can be used to calculate dietary iron absorption at a population level using data describing total iron intake and SF concentration.

Editor: James F. Collins, University of Florida, United States of America

Funding: This work was supported by UK Biotechnology and Biological Sciences Research Council (JRD), and EURRECA Network of Excellence (www.eurreca.org) Sixth Framework Programme, contract 036196 (RB, LJH). The funders had no role in study design, data collection and analysis, decision to publish, or preparation of the manuscript.

Competing Interests: The authors have declared that no competing interests exist.

* Email: s.fairweather-tait@uea.ac.uk

Introduction

Iron absorption in humans is dependent on physiological requirements, but may be restricted by the quantity and availability of iron in the diet [1,2]. Dietary intake data reported in the present paper were collected from a 7d weighed intake, which is generally considered to be the minimum recording period necessary to achieve an accurate estimate of iron intake [3,4]. The physiological regulation of absorption, determined by the size of the iron stores and the extent of erythropoietic activity, is responsible for maintaining iron balance. Levels of body iron in individuals with normal erythropoiesis are the main determinant of the efficiency of iron absorption, with serum ferritin (SF) being a well-established quantitative measure of iron stores, in healthy people [5,6]. Methods for estimating bioavailability are more complicated than the assessment of iron intake. The diets of omnivores contain relatively small quantities of heme iron derived from meat and fish, which is always well absorbed, although iron status has a modest regulatory role [7,8]. The remainder of the soluble iron forms a common non-heme iron pool and absorption

is very variable, depending on meal composition, but its absorption is tightly regulated by iron stores [1].

Dietary iron fortification is generally considered the most cost effective method for reducing the prevalence of iron deficiency in populations that consume diets containing suboptimal quantities of bioavailable iron [9] and WHO/FAO recommends that the level of fortification is based on the estimated daily iron intake deficit adjusted for bioavailability [10]. There are no satisfactory methods for estimating dietary iron absorption at the population level, and therefore a qualitative assessment was used by WHO/FAO to assign one of three bioavailability levels (5%, 10% or 15%) [10]. In this paper we describe an alternative approach which could provide more accurate estimates of dietary iron absorption (bioavailability) that are relevant to specific target populations. The distribution of individual iron requirements is based on figures published by the Institute of Medicine (IOM) for menstruating women and men [11]. A full probability approach was used to predict the prevalence of an iron intake that would be sufficient to maintain iron balance based on estimated iron intake and a series of % absorption values from 1–40%. An estimate of average dietary absorption in the population sample was then calculated for selected SF concentrations by matching the observed prevalence of inadequacy (prevalence of SF below the designated level) with the prevalence predicted for the series of absorption estimates.

Materials and Methods

Dietary iron intake and SF data from a previously published study were used. The methods for data collection are described briefly here, but have been published in greater detail elsewhere [12]. Data were collected as part of the National Diet and Nutrition Survey (NDNS) of adults aged 19–64 y living in private households in the UK between July 2000 and June 2001 [13,14]. Approval for the survey was obtained from the South Thames Multi-Centre Research Ethics Committee (MREC) in 2000, with subsequent local approvals gained from 93 National Health Service Local Research Ethics Committees, which covered the 152 geographical areas selected for the fieldwork. Written informed consent was obtained from each participant for the clinical aspects of the study. The survey, conducted in a nationally representative sample of adults who were not pregnant or breastfeeding at the point of recruitment, involved an interview, a 7d dietary diary and blood and urine samples. Dietary intake was assessed using the 7d (consecutive) dietary record diary, with respondents recording all food and drink consumed in and out of the home. Following completion of the dietary survey, participants were interviewed to clarify and resolve difficulties, establish whether eating patterns were usual, and identify any illness during the recording period. Diaries were checked for omissions and level of detail (including brand details of pre-packed items) to ensure that foods could be accurately coded for nutrient analysis. After conversion of portion sizes to weights and subsequent coding of the diaries, the information was linked to the nutrient databank compiled by the Food Standards Agency. Quantities of nutrients ingested, including iron (total, heme and non-heme), were calculated from foods consumed. A total of 1347 (out of 2251) NDNS respondents provided a non-fasted blood sample, which was used for a range of hematological analyses.

α-1-antichymotrypsin (α1-ACT), an acute phase reactant, was measured because serum ferritin can be elevated in inflammatory conditions. The aim of this was to exclude any individuals with elevated serum ferritin due to inflammation/infection because the serum ferritin concentration would not be an accurate reflection of iron stores. C-reactive protein (CRP), a more commonly used indicator of inflammation, was not measured in the NDNS. Individuals consuming iron supplements were also excluded. Menopausal status was determined through information collected during the NDNS. Women who had entered menopause or were unsure of their status were excluded from the analytical sample because menstrual losses are highly variable at the onset of menopause. The final analytical sample of 873 subjects consisted of individuals with both dietary intake data and relevant hematological data (hemoglobin and SF). There were 378 premenopausal women aged 35.7±8.2 y and 495 men aged 19–64 y. Iron deficiency, anemia, and iron deficiency anemia were defined according to the WHO cut-offs: anemia, Hb<12.0 g/dL for women and <13.0 g/dL for men; iron deficiency, SF< 15.0 ug/L for both men and women; iron deficiency anemia, Hb<12.0 g/dL and SF<15.0 ug/L for women, Hb<13.0 g/dL and SF<15.0 ug/L for men [15].

SF was measured on an Abbott IMx semi-automated analyzer using a standard Microparticle Enzyme Immunoassay (MEIA) kit. Sample concentrations were determined by comparison with a standard curve constructed from known concentrations. Quality control procedures comprised both internal and external procedures, with an internal pooled serum sample used as a drift control with each run. Hemoglobin concentrations were measured using a Bayer H3 Haematology Analyzer using a colorimeter at wavelength 546 nm. Quality control comprised both internal and external procedures. Daily commercial controls (Bayer Testpoint Haematology control) were used to monitor drift in all parameters. External quality assessment schemes included the National External Quality Assessment Scheme (NEQAS) for hematology and External Quality Assessment Scheme (EQAS) for hematology run by Addenbrookes Hospital, Cambridge, UK [14].

The Institute of Medicine (IOM) employed factorial modeling to calculate the distribution of estimated iron requirements needed to meet body functions with a minimal store for several age and sex groups including pre-menopausal women and men [11]. The values for pre-menopausal women (mixed population of oral contraceptive (OC) users and non-users) and men were used to derive the distributions of requirements for the NDNS study sample. Values reported as dietary intake requirements were converted to requirements for absorbed iron by multiplying by 0.18 (IOM values assume 18% absorption) [11]. The factorial model used by the IOM was designed to provide an estimate for individuals who were not anemic, but had very little storage iron. Our analysis was extended to include estimates of requirements needed to maintain selected levels of storage iron as defined by SF concentration. These estimates are valid because iron excretion is not increased when the iron stores accumulate, but remain in the physiological range [16], and iron balance is restored by a reduction in absorption. By using a spline function in the statistical package, R [17], it is possible to interpolate these values to derive the probabilities of inadequate iron absorption for 0.05 increments between 0 and 1. The latter values can then be used as a look-up table in a Microsoft Excel spreadsheet (**Table S1. Individual Data**) and compared to each individual's absorbed iron estimate, based on their known iron intake and a theoretical range of iron absorption values (1–40%). The percentage absorption value that met the threshold for estimated requirement for the individual was designated as the dietary iron absorption for that individual. The average dietary absorption for the population was calculated as the mean of individual estimated absorption values. Subtracting the value found above from 100% gives the estimated percentage of the population who require more than this percentage of iron absorption (i.e. a higher bioavailability) to meet their requirements, or, in other words, the estimated prevalence of inadequate iron intakes. The population dietary iron absorption can be estimated for any SF concentration by assuming that the estimated prevalence of inadequate absorption calculated above is equivalent to the observed prevalence of iron insufficiency, as defined by the percentage of the population with a SF below the designated cut-off value.

Results, Discussion, and Conclusions

The distributions of estimated daily iron requirements for men and pre-menopausal women published by the IOM were used for the analyses described in this study (**Figure 1**) [11]. The IOM assumed that 17% of women were OC users. The percentage of menstruating women using OCs in the UK is estimated to be 25% [18], but this includes 16–18 year olds who are not part of our analytical sample. Although the NDNS survey included questions on contraceptive use, the answers were self-reported, and a large proportion of those practicing contraception did not answer the question on the method of contraception. A study across five

European countries reported that the main method of contraception was OCs in 30% of the population with usage being even higher in the younger age groups [19]. It therefore appears that the IOMs assumption about OC use when calculating menstrual iron losses may have resulted in an overestimate of iron requirements since OCs reduce menstrual blood loss.

The NDNS sample was a relatively iron sufficient population (**Table 1**); the distributions of SF values for each of the two groups are shown in **Figure 2**. No individuals were identified with high levels of the inflammatory marker, α1-ACT (>0.65 g/L). Mean total iron intake was 13.5 mg, and 9.8 mg for and men and women respectively. The relationship between the arbitrary series of iron bioavailability values and the capacity of the diet to meet the iron requirements of men and women is shown in **Figure 3**. By comparing this figure with the cumulative distributions of SF values in the same population samples (Figure 2), it is possible to identify the average dietary absorption required to sustain a selected average iron status (as defined by the SF concentration) in the population. For example, estimated dietary absorption was 13% in women and 14% in men with SF values of 45 μg/L, and it was 31% for women with depleted iron stores (SF <15 μg/L) (**Table 2**). There were too few iron deficient men to allow a similar estimate for men to be calculated.

The direct correlation between SF concentration and % non-heme iron absorption is well established [1,20–22]. The results of iron absorption studies using isotopic labels are therefore usually corrected for the effect of iron status by adjusting absorption values. One method involves the inclusion of a "reference dose" in the study design, customarily 3 mg of highly bioavailable ferrous sulfate mixed with ascorbic acid [2,23]. The observed absorption from the test meal is corrected to a mean reference value of 40%, which corresponds to absorption by individuals with borderline iron stores. This is made by multiplying test meal absorption values by 40/R where R is the reference dose absorption [2].

Another widely-used approach (ratio method) is to correct the measured absorption to a selected SF value by using the following equation:

$$Log\ Ac = Log\ Ao + Log\ Fo - Log\ Fr$$

where Ac is the corrected dietary absorption, Ao is the observed absorption, Fo is the observed SF and Fr is the reference SF value selected. Values of 40 ug/L and 30 ug/L have been employed as the reference SF value [24,25].

We applied the second method that adjusts absorption according to the SF concentration to our data. A SF of 45 ug/L was chosen as Fr and dietary iron absorption (bioavailability) was calculated for an arbitrary series of SF values that fall within the range of interest for population assessments (**Table 2**). There was reasonable agreement when the effect of iron status was adjusted using SF ratios and absorption estimates derived from the current probability model. Iron stores of men are significantly higher than those of menstruating women. The mean calculated dietary bioavailability (50th percentile) in our sample was 8% for men (SF 85 ug/L) and 17% for women (SF 38 ug/L). However, as indicated above, when estimates were made for the same SF concentration, bioavailability was equivalent.

At a population level, dietary iron absorption is generally considered to be the most important determinant of iron status. There is, at present, no satisfactory method for estimating iron absorption from nutritional survey data. Algorithms based on the dietary factors that have been shown to affect non-heme bioavailability in isotopic absorption experiments (e.g. ascorbic acid, meat and fish, phytate, polyphenol, and calcium) have been developed to estimate bioavailability [26]. However, data from single meal studies exaggerate the effects of individual dietary factors on iron absorption [24]. The problem is compounded by

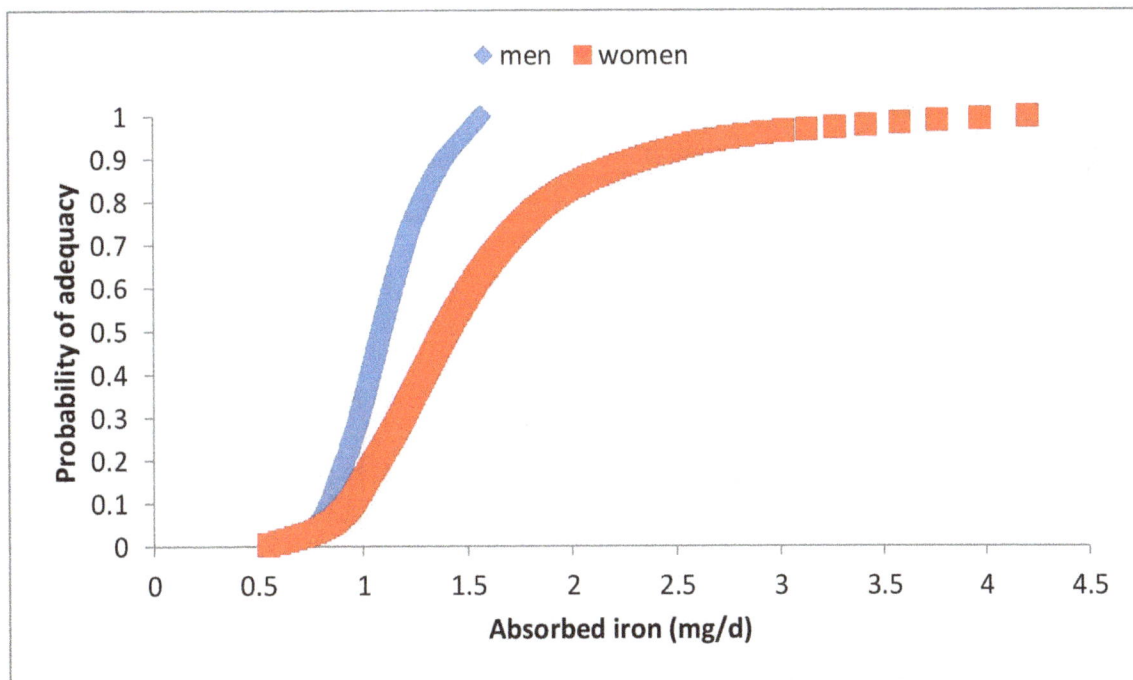

Figure 1. Distribution of estimated iron requirements for men (♦) and women (■): y axis represents the probability of adequacy (0–1), x axis is absorbed iron (mg/d). This is based on tabulated data from the IOM [11]. The figure shows an interpolation of this data that was estimated using a spline function in R [17].

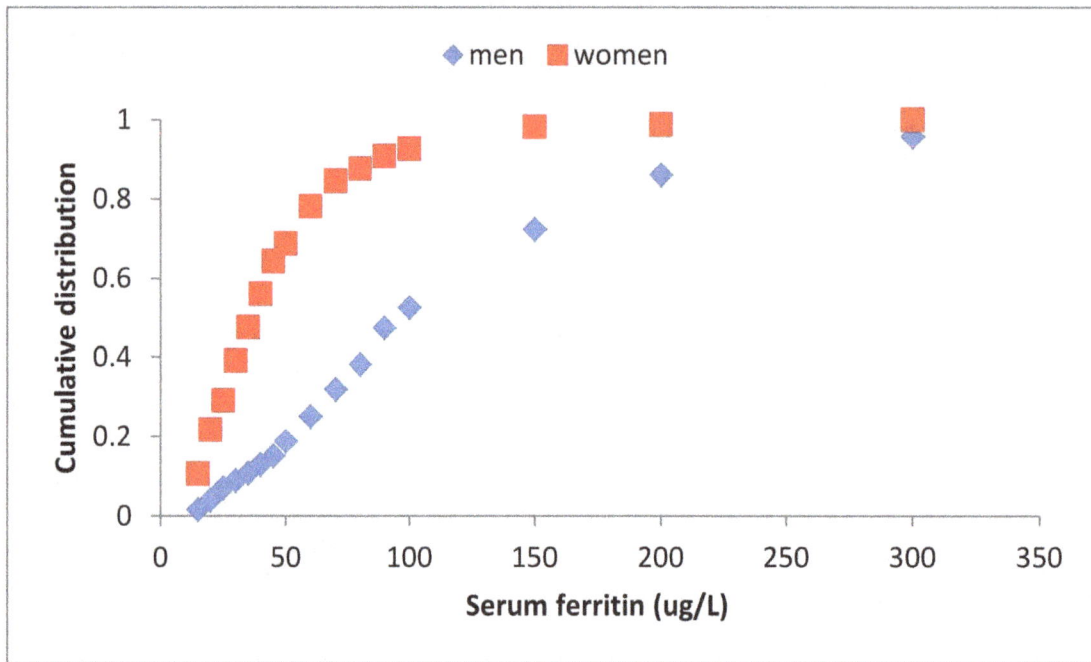

Figure 2. Cumulative distribution of serum ferritin concentrations for men (♦) and women (■). The data from the NDNS survey [13,14] are described in the Materials and Methods section (Men, n = 495; Women (pre-menopausal), n = 378).

the separate contribution of heme iron and the different effects of iron status on the absorption of non-heme and heme iron [7,8].

Algorithms tend to underestimate bioavailability [27]. Two algorithms have recently been published that have been developed

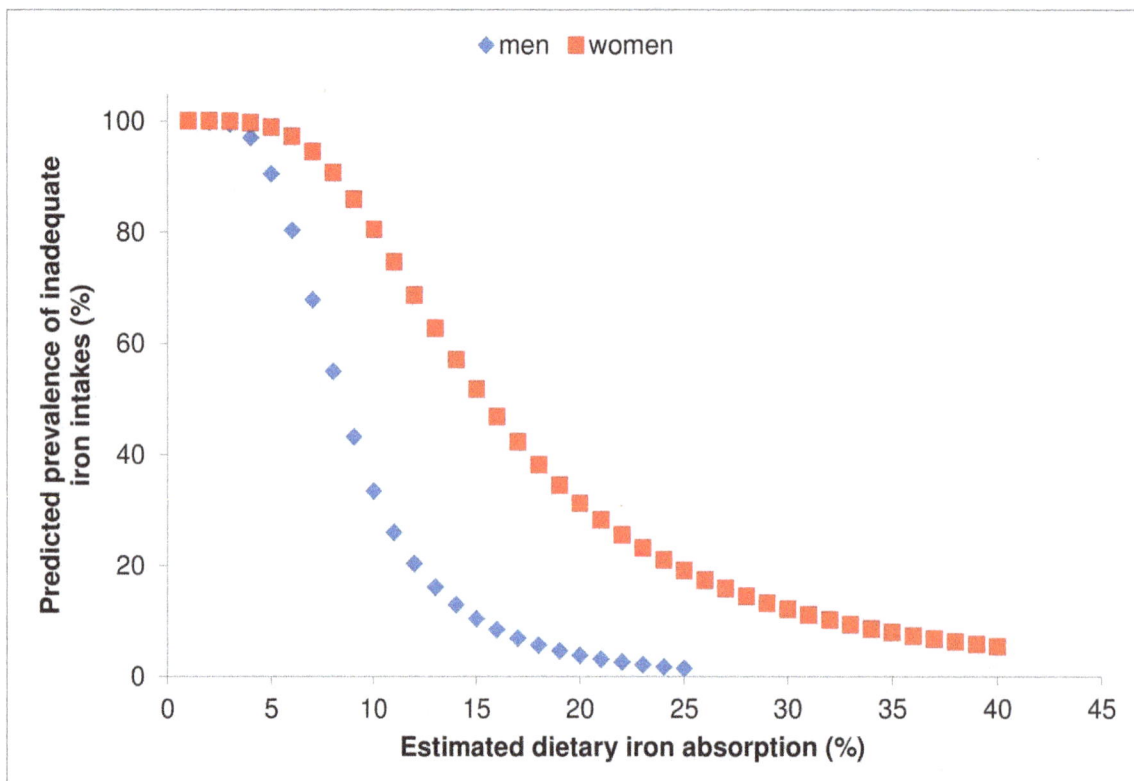

Figure 3. Results of probability modelling with NDNS data for men (♦) and women (■): y axis represents the predicted prevalence of inadequate intakes (0–100%), x axis is estimated dietary iron absorption (%).

Table 1. Summary statistics for iron intake and status of the population sub-sample.

Group	n	Variable	Mean	SD	Lower 95% CI	Upper 95% CI
Pre-menopausal women	378	Age (y)	35.7	8.2	34.9	36.5
		Weight (kg)	68.1	14.4	66.6	69.5
		BMI (kg/m^2)	26.0	5.5	25.4	26.5
		Iron intake (mg/d)	9.8	3.8	9.4	10.2
		Serum ferritin (µg/L)	45.5	38.4	41.7	49.4
		Hemoglobin (g/dL)	13.3	1.0	13.2	13.4
		Anemia (%)[1]	7.7			
		Iron deficient (%)[1]	12.4			
		Iron deficiency anemia (%)[1]	3.7			
Men	495	Age (y)	42.4	12.1	41.4	43.5
		Weight (kg)	83.7	14.1	82.5	85.0
		BMI (kg/m^2)	27.1	4.3	26.7	27.5
		Iron intake (mg/d)	13.5	5.1	13.0	13.9
		Serum ferritin (µg/L)	121.6	112.1	111.7	131.5
		Hemoglobin (g/dL)	15.1	1.1	15.0	15.2
		Anemia (%)[1]	2.6			
		Iron deficient (%)[1]	2.0			
		Iron deficiency anemia (%)[1]	0.6			

[1]Iron deficiency, anemia, and iron deficiency anemia defined according to the WHO cut-offs (15). Anemia: Hb<12.0 g/dL for women and <13.0 g/dL for men. Iron deficiency: SF<15.0 µg/L for both men and women. Iron deficiency anemia: Hb<12.0 g/dL and SF<15.0 µg/L for women, Hb<13.0 g/dL and SF<15.0 µg/L for men.

using data from studies in which non-heme iron absorption from whole diets was determined [28,29]. However, these either require information about the intake of absorption promoters and inhibitors [28] or a value judgment must be made about the type of diet consumed in relation to its overall content of inhibitors or enhancers [29]. Program managers have therefore applied algorithms sparingly, and the approximations based on qualitative data (5%, 10%, 15%) quoted in the WHO guidelines [10] are more often used. An alternative approach is now suggested that just requires data on total iron intake and measurements of iron status, and avoids the need to obtain information on dietary inhibitors and enhancers, which are notoriously difficult to collect. Its validity depends on three critical elements, and the samples selected for the development of the new approach described in this study meet all these criteria:

(i) The accuracy of the estimation of the distribution of individual iron requirements. In adults, iron requirements are calculated from measured losses. Estimates for men are based on a factorial approach employing experimental measurements which are relatively precise and unlikely to vary in different population samples [16,30]. It is more difficult to obtain accurate estimates of menstrual losses, which are an important component of the requirements of pre-menopausal women. However, carefully controlled measurements in several population samples have yielded surprisingly consistent distributions of menstrual blood losses [11,31,32]. OCs reduce menstrual blood loss while intra-uterine contraceptive devices tend to increase menstrual bleeding [33]. The increasing use of OCs may have led to a modest overestimate of the requirement for menstruating women and therefore dietary iron absorption by the

Table 2. Estimated dietary iron absorption for selected serum ferritin values in men and women.

[1]Serum ferritin cutoff (µg/L)	Probability model women (%)	Probability model men (%)	[2]Ratio method (%)
60	11	11	10
45	13	14	13
30	18	16	20
15	31		39

[1]A serum ferritin cut off of 15 µg/L was used by the IOM to identify iron deficient individuals [11], 30 µg/L was used by Reddy et al. [25] for estimating non-heme iron bioavailability from meal composition, 60 µg/L is the value above which no homeostatic up-regulation of iron absorption occurs [40].
[2]Estimated bioavailability adjusted for the effect of iron stores based on the ratio 45/SF cutoff.

probability approach in the current study. Nevertheless there was good agreement between estimates for men and women at the same SF concentration.

(ii) The accuracy of dietary intake data. This new method requires an accurate estimate of habitual total iron intake from all sources, including non-heme and heme iron, but information on other dietary constituents (i.e. enhancers and inhibitors of iron absorption) that may be more difficult to estimate is not required.

(iii) A stable iron intake. Hallberg et al. [34] calculated the rate of change in iron stores following changes in dietary intake. It takes about 2 years to reach a new balanced state, but 80% of the adjustment in absorption occurs within the first year. The method for calculating dietary bioavailability that we describe is therefore not applicable to children, women during pregnancy and lactation, or immediately after the onset of menopause because of the variability in iron requirements.

There is no formal agreement about the iron status for which iron absorption values should be reported as the dietary bioavailability. However, most investigators and the IOM have conceptualized bioavailability as the absorption value that would be attained by an individual who is not anemic, but has only a minimal quantity of storage iron. This is based on the assumption that physiological function remains normal after stores are exhausted if there is still sufficient iron to maintain the functional compartments. Thus, absorption is maximally up-regulated without any impairment of physiological processes. A serum ferritin value of 15 ug/L is the cut-off value selected by the WHO [15]. The IOM also defined bioavailability to be the estimated absorption in an individual with a serum ferritin concentration of 15 µg/L, as this has been reported to be the most reliable cut-off value for absent stainable iron in the bone marrow [35]. However, a higher iron status may have pragmatic advantages for setting optimal iron intake levels in populations. Although the iron store appears to have no functional importance, other than as a source of readily available iron if there is a sudden increase in requirements (e.g. as the result of blood loss), it may be desirable for individuals to have this safety net at all times. There is also some evidence to suggest that adequate iron status in early pregnancy is important for birth outcome [36,37]. The proposed new approach for estimating dietary iron absorption (bioavailability) allows estimates to be made for any selected mean population serum ferritin level as shown in Figure 3.

The concordance between dietary absorption values derived from the two sub-samples, men and pre-menopausal women, who have very different iron requirements provides support for the validity of the estimates (Figure 3, Table 2). Furthermore, iron stores are known to be the primary physiological regulator of iron absorption in healthy adults. Adjustments for the effect of iron status using serum ferritin ratios (Table 2) yields results that are similar to those derived from the probability model although the ratio method predicts higher bioavailability in individuals with

absent iron stores. The results derived from the probability model have a high degree of uncertainty in the lower range in our population samples because of the low prevalence of iron deficiency. It is therefore not possible to comment on possible explanations for the difference, and the analysis of data from regions where nutritional iron deficiency is common would be informative. Finally, it is important to note that the probability approach predicts the potential for a higher bioavailability (31%) than that employed in calculating the IOM Dietary Reference Intakes (18%) in iron deficient individuals (serum ferritin <15 ug/L) [11].

The method that we have described has several strengths. It is based on experimental data drawn from the target population. Dietary assessment is relatively simple; it is only necessary to determine the total iron intake, not the intake of dietary enhancers and inhibitors of iron absorption, and there is no need to estimate heme iron intake. The dietary iron absorption needed to achieve a desired iron store in a target population can be calculated, or alternatively the necessary level of fortification can be calculated based on these estimates. However, there are some weaknesses of the method. Iron requirements and iron intake must be in a steady state (for at least one year), therefore the method cannot be used for children, pregnant women, and immediately after the menopause because of changing requirements. The increasing use of oral contraceptives may reduce menstrual iron loss and therefore currently available estimates of iron requirements [10]. Finally, care must be taken to ensure that estimates of iron status based on serum ferritin concentration are not confounded by inflammation/infection or obesity [38,39].

Supporting Information

Table S1 Individual Data. Excel file containing individual data for adults in the UK National Diet and Nutrition Survey (men aged 19–64 y and women aged 20–49 y): serum ferritin (µg/L), daily iron intake (mg), calculated quantity of iron absorbed (mg/d) at efficiencies of absorption ranging from 1–25%, estimated physiological requirements of iron to replace obligatory losses (mg/d), and predicted prevalence of inadequate intake at each % absorption efficiency.

Acknowledgments

The authors thank and acknowledge the organizations responsible for the data collection and access as part of the National Diet and Nutrition Survey: Food Standards Agency, the Ministry of Agriculture, Fisheries and Food, the Department of Health, Human Nutrition Research at the Medical Research Council, and The Data Archive, University of Essex.

Author Contributions

Conceived and designed the experiments: SFT SRL. Performed the experiments: JRD. Analyzed the data: JRD SRL RB LJH. Contributed to the writing of the manuscript: SFT SRL JRD RB LJH.

References

1. Bothwell TH, Charlton RW, Cook J, Finch C (1979) Iron Metabolism in Man. Oxford: Blackwell Scientific Publications.

2. Magnusson B, Bjorn-Rassmussen E, Hallberg L, Rossander L (1981) Iron absorption in relation to iron status. Model proposed to express results to food iron absorption measurements. Scand J Haematol 27: 201–208.

3. Serra-Majem L, Pfrimer K, Doreste-Alonso J, Ribas-Barba L, Sánchez-Villegas A, et al. (2009) Dietary assessment methods for intakes of iron, calcium, selenium, zinc and iodine. Br J Nutr 102 Suppl 1: S38–S55.

4. Harvey LJ, Berti C, Casgrain A, Collings R, Gurinovic M, et al. (2013) EURRECA-Estimating iron requirements for deriving dietary reference values. Crit Rev Fd Sc Nutr 53: 1064–1076.

5. Cook JD, Lipschitz DA, Miles LE, Finch CA (1974) Serum ferritin as a measure of iron stores in normal subjects. Am J Clin Nutr 27: 681–687.

6. Walters GO, Jacobs A, Worwood M, Trevett D, Thomson W (1975) Iron absorption in normal subjects and patients with idiopathic haemochromatosis: relationship with serum ferritin concentration. Gut 16: 188–192.

7. Lynch SR, Skikne BS, Cook JD (1989) Food iron absorption in idiopathic hemochromatosis. Blood 74: 2187–2193.

8. Roughead ZK, Hunt JR (2000) Adaptation in iron absorption: iron supplementation reduces nonheme-iron but not heme-iron absorption from food. Am J Clin Nutr 72: 982–989.

9. Zimmermann MB, Hurrell RF (2007) Nutritional iron deficiency. Lancet 370: 511–520.

10. Allen L, De Benoist B, Dary O, Hurrell R (2006) Guidelines on Food Fortification with Micronutrients. Geneva: World Health Organization, Food and Agricultural Organization of the United Nations.

11. Institute of Medicine (2001) Dietary Reference Intakes for Vitamin A, Vitamin K, Arsenic, Boron, Chromium, Copper, Iodine, Iron, Manganese, Molybdenum, Nickel, Silicon, Vanadium, and Zinc. Washington, DC: National Academy Press.

12. Hoare J, Henderson L, Bates CJ, Prentice A, Birch M, et al. (2004) The National Diet and Nutrition Survey: adults aged 19–64 years. Volume 5: Summary Report. London: HMSO.

13. Henderson L, Irving K, Gregory J, Bates CJ, Prentice A, et al. (2003) National Diet and Nutrition Survey: adults aged 19–64 years. Volume 3: Vitamin and mineral intake and urinary analytes. London: TSO.

14. Ruston D, Hoare J, Henderson L, Gregory J, Bates CJ, et al. (2004) National Diet and Nutrition Survey: adults aged 19–64 years Volume 4: Nutritional status (blood pressure, anthropometry, blood analytes sand physical activity). London: TSO.

15. WHO/UNICEF/UNU (2001) Iron Deficiency Anemia Assessment, Prevention, and Control. Geneva: World Health Organization.

16. Hunt JR, Zito CA, Johnson LK (2009) Body iron excretion by healthy men and women. Am J Clin Nutr 89: 1792–1798.

17. R Core Team (2013) R: A language and environment for statistical computing. R Foundation for Statistical Computing, Vienna, Austria. http://www.R-project.org/.

18. Lader D (2009) A report on research using the National Statistics Opinions Survey produced on behalf of the NHS Information Centre for health and social care. Office of National Statistics, Opinions Survey Report No. 41, Contraception and Sexual Health, 2008/09. Surrey: Office for Public Sector Information.

19. Skouby SO (2010) Contraceptive use and behavior in the 21st century: a comprehensive study across five European countries. Eur J Contracept Reprod Health Care 15: S42–S53.

20. Cook JD, Lipschitz DA, Miles LE, Finch CA (1974) Serum ferritin as a measure of iron stores in normal subjects. Am J Clin Nutr 27: 681–687.

21. Taylor P, Martínez-Torres C, Leets I, Ramírez J, García-Casal MN, et al. (1988) Relationships among iron absorption, percent saturation of plasma transferrin and serum ferritin concentration in humans. J Nutr 118: 1110–1115.

22. Bezwoda WR, Bothwell TH, Torrance JD, MacPhail AP, Charlton RW, et al. (1979) The relationship between marrow iron stores, plasma ferritin concentrations and iron absorption. Scand J Haematol 22: 113–120.

23. Layrisse M, Cook JD, Martinez C, Roche M, Kuhn IN, et al. (1969) Food iron absorption: a comparison of vegetable and animal foods. Blood 33: 430–443.

24. Cook JD, Dassenko SA, Lynch SR (1991) Assessment of the role of nonheme-iron availability in iron balance. Am J Clin Nutr 54: 717–722.

25. Reddy MB, Hurrell RF, Cook JD (2000) Estimation of nonheme-iron bioavailability from meal composition. Am J Clin Nutr 71: 937–943.

26. Hallberg L, Hulthen L (2000) Prediction of dietary iron absorption: an algorithm for calculating absorption and bioavailability of dietary iron. Am J Clin Nutr 71: 1147–1160.

27. Beard JL, Murray-Kolb LE, Haas JD, Lawrence F (2007) Iron absorption prediction equations lack agreement and underestimate iron absorption. J Nutr 137: 1741–1746.

28. Armah SM, Carriquiry A, Sullivan D, Cook JD, Reddy MB (2013) A complete diet-based algorithm for predicting nonheme iron absorption in adults. J Nutr 143: 1136–1140.

29. Collings R, Harvey LJ, Hooper L, Hurst R, Brown TJ, et al. (2013) The absorption of iron from whole diets: a systematic review. Am J Clin Nutr 98: 65–81.

30. Green R, Charlton R, Seftel H, Bothwell T, Mayet F, et al. (1968) Body iron excretion in man: a collaborative study. Am J Med 45: 336–353.

31. Harvey LJ, Armah CN, Dainty JR, Foxall RJ, John Lewis D, et al. (2005) Impact of menstrual blood loss and diet on iron deficiency among women in the UK. Br J Nutr 94: 557–564.

32. Hefnawi F, El-Zayat AF, Yacout MM (1980) Physiologic studies of menstrual blood loss. I. Range and consistency of menstrual blood loss in and iron requirements of menstruating Egyptian women. Int J Gynaecol Obstet 17: 343–348.

33. Milman N, Rosdahl N, Lyhne N, Jorgensen T, Graudal N (1993) Iron status in Danish women aged 35–65 years. Relation to menstruation and method of contraception. Acta Obstet Gynecol Scand 72: 601–605.

34. Hallberg L, Hulthen L, Garby L (1998) Iron stores in man in relation to diet and iron requirements. Eur J Clin Nutr 52: 623–631.

35. Hallberg L, Bengtsson C, Lapidus L, Lindstedt G, Lundberg PA, et al. (1993) Screening for iron deficiency: an analysis based on bone-marrow examinations and serum ferritin determinations in a population sample of women. Br J Haematol 85: 787–798.

36. Scholl TO, Hediger ML, Fischer RL, Shearer JW (1992) Anemia vs iron deficiency: increased risk of preterm delivery in a prospective study. Am J Clin Nutr 55: 985–988.

37. Scholl TO (2005) Iron status during pregnancy: setting the stage for mother and infant. Am J Clin Nutr 81: 1218S–1222S.

38. Thurnham DI, McCabe LD, Haldar S, Wieringa FT, Northrop-Clewes CA, et al. (2010) Adjusting plasma ferritin concentrations to remove the effects of subclinical inflammation in the assessment of iron deficiency: a meta-analysis. Am J Clin Nutr 92: 546–555.

39. Zafon C, Lecube A, Simo R (2010) Iron in obesity. An ancient micronutrient for a modern disease. Obes Rev 11: 322–328.

40. Hallberg L, Hultén L, Gramatkovski E (1997) Iron absorption from the whole diet in men: how effective is the regulation of iron absorption? Am J Clin Nutr 66: 347–356.

23

Enhancement of Energy Expenditure following a Single Oral Dose of Flavan-3-Ols Associated with an Increase in Catecholamine Secretion

Yusuke Matsumura, Yuta Nakagawa, Katsuyuki Mikome, Hiroki Yamamoto, Naomi Osakabe*

Department of Bio-science and Engineering, Shibaura Institute of Technology, Saitama, Saitama, Japan

Abstract

Numerous clinical studies have reported that ingestion of chocolate reduces the risk of metabolic syndrome. However, the mechanisms by which this occurs remain unclear. In this murine study, the metabolic-enhancing activity of a 10 mg/kg mixture of flavan-3-ol fraction derived from cocoa (FL) was compared with the same single dose of (-)-epicatechin (EC). Resting energy expenditure (REE) was significantly increased in mice treated with the FL versus the group administered the distilled water vehicle (Cont) during periods of ad libitum feeding and fasting. Mice were euthanized under the effect of anesthesia 2, 5, and 20 hr after treatment with FL or Cont while subsequently fasting. The mRNA levels of the uncoupling protein-1 (UCP-1) and peroxisome proliferator-activated receptor gamma coactivator-1 alpha (PGC-1α) in brown adipose tissue (BAT) were significantly increased 2 hr after administration of FL. UCP-3 and PGC-1α in the gastrocnemius were significantly increased 2 and 5 hr after administration of the FL. The concentrations of phosphorylated AMP-activated protein kinase (AMPK) 1α were found to be significant in the gastrocnemius of mice 2 and 5 hr after ingesting FL. However, these changes were not observed following treatment with EC. Plasma was collected for measurement of catecholamine levels in other animals euthanized by decapitation 2 and 4 hr after their respective group treatment. Plasma adrenaline level was significantly elevated 2 hr after treatment with FL; however, this change was not observed following the administration of EC alone. The present results indicated that FL significantly enhanced systemic energy expenditure, as evidenced by an accompanying increase in the type of gene expression responsible for thermogenesis and lipolysis, whereas EC exhibited this less robustly or effectively. It was suggested the possible interaction between thermogenic and lipolytic effects and the increase in plasma catecholamine concentrations after administration of a single oral dose of FL.

Editor: Shu-ichi Okamoto, Sanford-Burnham Medical Research Institute, United States of America

Funding: This work was supported by the Research Project on Development of Agricultural Products and Foods with Health-promoting Benefits (NARO). The funders had no role in study design, data collection and analysis, decision to publish, or preparation of the manuscript.

Competing Interests: The authors have declared that no competing interests exist.

* Email: nao-osa@sic.shibaura-it.ac.jp

Introduction

Flavan-3-ols, a group of polyphenolic substances, are distributed in a number of plant foods such as cocoa beans, red wine, and apples. Of these foods, chocolate is known to be rich in flavan-3-ols, including the flavan-3-ol monomers, (+)-catechin and (−)-epicatechin, and the oligomers, as B-type procyanidins that are linked by C4–C8 bonds [1–3]. Recent meta-analyses have suggested that the ingestion of chocolate reduces the risk of cardiovascular diseases [4,5]. These reports have shown that chocolate consumption was associated with a considerable reduction in the risk of coronary heart disease, myocardial infarction, and stroke. In addition, numerous randomized, controlled trials have confirmed that chocolate, especially dark chocolate containing large amounts of flavan-3-ols, improved risk factors for the constellation of conditions such as hypertension [6,7], vascular endothelial dysfunction [8,9], dyslipidemia [10,11], and glucose intolerance [12,13], that can contribute to metabolic syndrome. Several meta-analyses conducted after these clinical trials confirmed that dark chocolate could reduce the risk of

cardiovascular disease [14–20]. In addition, recent cross-sectional studies have reported an inverse association between the frequency of chocolate ingestion and body mass index in healthy adolescents or adults [21,22].

In our previous study [23], we confirmed that the respiratory exchange ratio (RER), where RER = carbon dioxide production (VCO_2)/oxygen consumption (VO_2), was significantly reduced as a result of the increase in lipolysis following repeated supplementation with flavan-3-ol fraction derived from cocoa (FL). In addition, repeated ingestion of FL increased concentrations of UCP a key protein of thermogenesis—in several tissues, and also augmented enzymes involved in β-oxidation such as carnitine palmitoyltransferase-2 (CPT-2) and medium-chain acyl-CoA dehydrogenase (MCAD). Moreover, we observed a significant increase in mitochondrial DNA copy number in skeletal muscle and brown adipose tissue (BAT) [23].

In this study, we compared the effects on energy expenditure of administering a single 10 mg/kg oral dose of FL with a 10 mg/kg dose of EC in mice, using indirect calorimetry and monitoring the initial biochemical changes responsible for thermogenesis and lipolysis in skeletal muscle and BAT. The changes in plasma catecholamine concentrations following treatment were also examined.

Materials and Methods

Materials

The flavan-3-ol fraction (FL) was provided by Meiji Co., Ltd (Tokyo, Japan) and was prepared from cocoa powder using a method described in a previous report [24]. In brief, the cocoa powder was defatted with n-hexane and the residue was extracted with acetone. The n-butanol-dissolved fraction of the extract was subsequently applied to a Diaion HP2MG column (Mitsubishi Kasei Co. Ltd., Tokyo, Japan). The fraction eluted with 80% ethanol was collected, freeze-dried, and used in the experiments. The concentration of catechins and procyanidins was determined by the method of high-performance liquid chromatography (HPLC). FL contained 4.56% (+)-catechin, 6.43% of (−)-epicatechin, 3.93% of procyanidin B2, 2.36% of procyanidin C1, and 1.45% cinnamtannin A2. (−)-Epicatechin was purchased from Tokyo Chemical Industry (Tokyo, Japan).

Animals and diets

This study was approved by the Animal Care and Use Committee of the Shibaura Institute of Technology (Permit Number: 27-2956). All animals received humane care under the guidelines of this institution. Male ICR mice weighing 35–40 g were obtained from Charles River Laboratories Japan, Inc. (Tokyo, Japan). The mice were kept in a room with controlled lighting (12/12 hr light/dark cycles) at a regulated temperature between 23–25°C. A certified rodent diet was obtained from the Oriental Yeast Co., Ltd., Tokyo, Japan.

Experimental procedures

During the first experiment of this study, the analysis of respiratory gas was performed during the feeding period. Four days after being fed a basal diet, the animals were divided into two groups; the animals in the control group ($n = 13$, Cont) were administered 4 ml/kg distilled water orally, whereas the mice in FL groups ($n = 13$) received 10 mg/kg FL via the oral route. Each animal was placed inside an open-circuit metabolic chamber for a 20 hr period, during which time they could eat ad libitum, at the same time their respiratory gas was being analyzed. In the second experiment in this study, the animals were treated with the vehicle ($n = 8$, Cont), FL ($n = 8$) or 10 mg/kg EC ($n = 8$) and their respiratory gas was analyzed over a 20 hr period of fasting. VO_2 and excreted VCO_2 were determined using a small animal metabolic measurement system (MK-5000RQ Muromachi Kikai Co. Ltd, Tokyo, Japan). The system monitored VO_2 and VCO_2 at 3-min intervals and calculated the RER using the RER $= VCO_2/VO_2$ formula. The VO_2 and VCO_2 measurements were converted to REE (kcal/20 hours, 8 hours of light cycle or 20 hours of dark cycle) using the Weir equation and the following formula: REE $= (3.941\ VO_2 + 1.11\ VCO_2) * 1.44 * 60\ \text{min} * \text{hrs}$. To measure their spontaneous motor activity while they were sedentary, mice were placed one at a time in a chamber equipped with an infrared-ray passive sensor system (MMP10, Muromachi Kikai). This second experiment involved measurements being performed during the dark period (18:00 to 6:00) and light periods (12:00 to 18:00 and 6:00 to 8:00).

The third experiment required $n = 8$ animals per group to be euthanized under pentobarbital (50 mg/kg body weight IP) anesthesia (Tokyo Chemical Industry, Tokyo, Japan) 2, 4, and 20 hr after being administered their respective group's treatment in the absence of subsequent food. Tissues samples were collected by dissection and snap frozen in liquid nitrogen and stored at −80°C until analysis.

Plasma catecholamine levals were measured from blood collected with ethylenediaminetetraacetic acid (EDTA) during decapitation and exsanguination 1, 2, and 4 hr after treatments were administered as described above ($n = 8$ mice in each group). Plasma was stored at −80°C until analysis.

Quantitative RT-PCR analysis

Total RNA was prepared from skeletal muscle and BAT using the TRIzol reagent (Life Technologies) according to manufacturer's instructions. In brief, 10 μg of total RNA was reverse-transcribed in a 20 μl reaction with high capacity cDNA Reverse Transcription kits (Applied Biosystems). Real-time reverse-transcription (RT)-PCR, using100 ng of total cDNA, was conducted using the StepOne Real-Time PCR System (Applied Biosystems). Primer and probe sequences were selected using a Taqman Gene Expression Assay (Applied Biosystems) and included the following gene and catalog numbers: GAPDH:Mm99999915_g1; UCP-1:Mm_01244861_m1; and UCP-3:Mm_00494077_m1; PGC-1α:Mm01208835_m1, all purchased from Applied Biosystems. Glyceraldehyde-3-phosphate dehydrogenase (GAPDH) was used as an internal control. The buffer used in the systems was THUNDER BIRD Prove qPCR Mix (TOYOBO). The PCR cycling conditions were 95°C for 1 min, followed by 40 cycles at 95°C for 15 s and 60°C for 1 min.

Western blotting analysis

Tissues were homogenized in a microtube with lysis buffer (CelLytic MT cell lysis reagent; Sigma Aldrich, Japan) containing a protease inhibitor (Sigma Aldrich, Japan) and 0.2% SDS. Protein concentration was measured by the Bradford method. Protein (10 μg) was separated by SDS-PAGE using a 4–12% Bis-Tris gel and transferred onto a polyvinylidene difluoride membrane (Life Technology). The membrane was blocked with membrane-blocking reagent (GE Healthcare) for 1 hr. After blocking, the membrane was incubated with a rabbit polyclonal primary antibody against AMPK1α (1:1600; sc-25792, Santa Cruz Biotechnology, Inc., USA), phosphorylated AMPK1α (1:200; sc-33524, Santa Cruz Biotechnology, Inc., USA), antibody for 2 hr. After the primary antibody reaction, the membrane was incubated with appropriate horseradish peroxidase-conjugated secondary antibodies (1:100000) for 1 hr. Immunoreactivity was detected by chemiluminescence using the ECL Select Western Blotting Reagent (GE Healthcare). Fluorescence band images were analyzed using Just TLC (SWEDAY) analysis software. Values of phosphorylated-AMPK1α were normalized to those for AMPK1α.

HPLC analysis of plasma catecholamine concentrations

Plasma catecholamines were analyzed by HPLC-electrochemical detection (ECD) after being prepared with a monolithic silica disk-packed spin column (MonoSpin, GL Science, Tokyo Japan) [25]. Norepinephrine and epinephrine were obtained from Tokyo Kasei (Tokyo, Japan). Dopamine was acquired from Wako Pure Chemical (Topkyo, Japan). The 3,4-dihydroxybenzylamine (DHBA) used was from (Sigma Aldrich, Japan). Acetonitrile was purchased from Wako Pure Chemical (Tokyo, Japan). Plasma, 1 M phosphate buffer (pH 8.0) (50 μL), and 400 ng/mL DHBA (internal standard; 40 μL) were directly injected into the pre-activated spin column which was centrifuged at 3000 rpm for 5 min. The column was then rinsed with 200 μL of 100 mM phosphate buffer (pH 8.0) by centrifugation. Finally, the column was installed into a new microtube, and the analytes that were

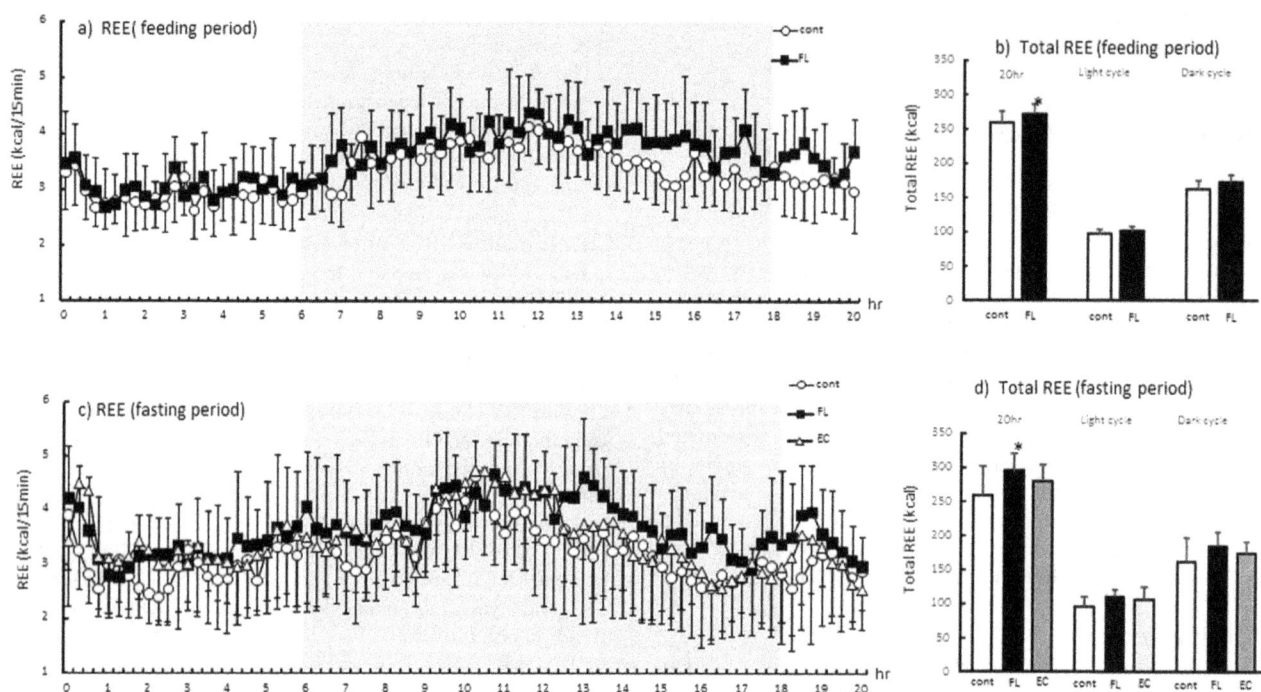

Figure 1. Respiratory energy expenditure (REE) 20 hr after administration of flavan-3-ols during ad libitum feeding (a) and the fasting period (c). Respiratory exchange ratio (RER) was calculated via oxygen consumption (VO_2) and carbon dioxide excretion (VCO_2) using the Weir equation. Total REE and light cycle (from 12:00 to 18:00 and from 6:00 to 8:00) or dark (from 18:00 to 6:00) REE was shown in (b; feeding period) and (d; fasting period). The animals were administrated vehicle (Cont, $n = 13$ during ad libitum feeding, $n = 8$ during fasting), 10 mg/kg FL ($n = 13$ during ad libitum feeding, $n = 8$ during fasting), or 10 mg/kg EC ($n = 8$ during fasting only). Values represent the mean ± standard deviation. Statistical analyses were performed by Dunnett's post-test. Significantly different from vehicle, *$p < 0.05$.

adsorbed onto the column were eluted with 1% acetic acid (200 μL). A 20 μL aliquot of the eluate was njected into the HPLC system (Prominance HPLC System Shimazu Corporation, Kyoto Japan) equipped with ECD (ECD 700 S, Eicom Corporation, Kyoto Japan) set at 650 mV. HPLC separation was conducted on an Inertsil ODS-4 (250×3.0 mm I.D., 5 μm) (GL Science) at 35°C, with a flow rate of 0.5 mL/min using a mobile phase comprised of 20 mM sodium acetate-citrate buffer/acetonitrile (100/16, v/v) containing 1 g/L sodium 1-octanesulfonate.

Data analysis and statistical methods

All data were reported as the mean ± standard error. Statistical analyses were performed by Dunnett's post-test. A statistical probability of $P < 0.05$ was considered significant.

Results

Resting energy expenditure and activity counts

The REE results are shown in Fig. 1. The data revealed over 20 hr after treatment of the chemicals (Fig. 1a, free feeding state; Fig. 1c, fasting state) and total and dark (from 18:00 to 6:00) or light (from 12:00 to 18:00 and from 6:00 to 8:00) cycle (Fig. 1b, free feeding state; Fig. 1d, fasting state). As shown in Fig. 1a and 1b, REE was marginally higher in FL compared with Cont throughout the measurement period during ad libitum feeding. There was a significant increase in REE during the total in FL compared with Cont in feeding period. In the fasting state, REE was high throughout the measurement period in the group treated with FL compared with those treated with vehicle, but this change or elevation in REE was not observed in the group treated with

EC (Fig.1c). The REE was significantly elevated for the total following mixed FL treatment, but there were no such changes following EC treatment (Fig.1 d). There were also no significant changes in locomotor activity and RER among experimental groups (data not shown).

UCPs and PGC-1α mRNA levels in BAT and the gastrocnemius

The change in mRNA expression of UCP-1 in BAT is shown in Fig. 2a. UCP-1 mRNA level was significantly increased 2 hr after administration of FL compared with that of Cont. mRNA expression of UCP-3 in the gastrocnemius was also significantly increased 2 and 5 hr after ingestion of FL (Fig. 2b). In contrast, there were no significant changes in UCPs following the administration of EC. PGC-1α mRNA levels in BAT and the gastrocnemius are shown in Fig. 1c and 1d. A significant increase in mRNA expression of PGC-1α was observed 2 hr after treatment with FL in BAT, and 2 and 5 hr after ingestion of FL in the gastrocnemius compared with mice in Cont. These changes were not observed in the mice treated with EC.

Phosphorylation of AMPK1α in BAT and the gastrocnemius

As shown in Fig. 3a, phosphorylated AMPK1α in BAT was increased 2 and 5 hr after treatment with FL. In the gastrocnemius, a significant elevation in phosphorylated AMPK1α occurred 2 and 5 hr after administration of FL (Fig. 3b). In contrast, no significant changes were evident in the EC-treated group of mice.

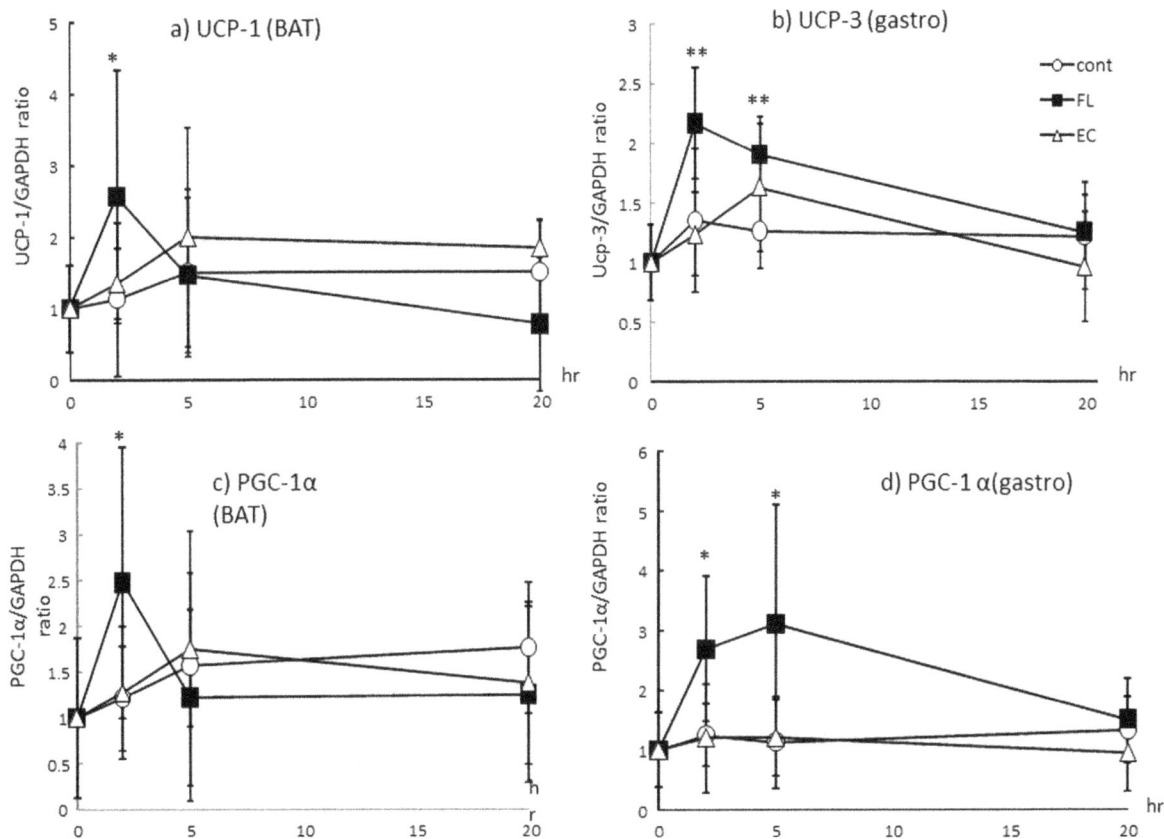

Figure 2. mRNA expression of UCPs and PGC1-α in BAT (a, b) or gastrocnemius (c, d) after administration of FL or EC. The animals were euthanized 2, 5, and 20 hr after administration of the vehicle (Cont, $n=8$), 10 mg/kg FL ($n=8$), or 10 mg/kg EC ($n=8$). Values represent the mean $\pm$ standard deviation. Statistical analyses were performed by Dunnett's post-test. Significantly different from vehicle, $*p<0.05$, $**p<0.01$.

Blood catecholamine concentrations

The levels of the blood catecholamines adrenalin and noradrenalin 1, 2, and 4 hr after oral ingestion of the treatments by each group are shown in Fig. 4. There were no significant changes in plasma noradrenalin concentrations in any treatment groups during the experimental period. Plasma adrenalin concentrations in mice were significantly increased 2 hr after administration of FL versus Cont. No change in adrenalin concentrations were observed in the group of mice treated with EC.

Discussion

Chocolate is known to be rich in flavan-3-ols. A previous report suggested that catechin and procyanidin contents (ranging from dimers to decamers) can be determined by HPLC [26,27]. The flavan-3-ols fraction derived from cocoa (FL) used in this study contained 11% catechins and 7.7% procyanidins (ranging from dimers to tetramers). A single dose of a 10 mg/kg FL significantly enhanced REE both during ad libitum feeding and during a fasting period (Fig. 1). In contrast, a 10 mg/kg dose of EC did not result in any significant change in energy expenditure. A similar trend was observed in the mRNA expression of UCPs and PGC-1α in BAT and the gastrocnemius of mice. As shown in Fig. 2, a significant elevation in the mRNA expression of UCPs and PGC-1α was observed in the group of mice that received FL, but such a change was not evident in the mice treated with EC alone. Increases of REE were observed during 13 to 20 hours rather than primary measurement period, in contrast, mRNA expressions of

UCPs or PGC1 α were observed 2 and 5 hours after FL treatment. It was suggested that REE increase by the treatment of FL may revealed protein synthesis induced by these mRNA changes. In addition, phosphorylation of AMPK1α was also increased following treatment with the FL, but not with the EC (Fig.3). According to these results, it was considered unlikely that EC was an indispensable component of FL, which as a mixture stimulated metabolic activity via the induction of UCPs and PGC-1α which are responsible for adaptive thermogenesis and lipolysis. It may also be indicated that these metabolic changes was synergic action of the catechins and procyanidins. Though 10mg/kg EC, which was more than 9 fold of catechins of 10 mg/kg FL, did not show any significant alteration. According to these results, it is possible that procyanidins rather than catechins contributed to these metabolic changes.

It is well-established that the bioavailability of catechins and procyanidins differ significantly. Catechins are present in the blood mainly as metabolites, such as conjugated forms with glucuronide and/or sulfate, and their absorption is reported to range between 10–20%. In contrast, unmetabolized catechins are nearly absent after ingestion [28,29]. Procyanidins have been shown to be poorly absorbed via the gastrointestinal tract and are detected in very low concentrations in the blood [30,31]. It was quite unlikely that procyanidins altered energy expenditure, based on their low bioavailability. In the present study, we found plasma adrenaline levels significantly increased after administration of a single dose of FL (Fig. 4). Adrenalin is secreted from the adrenal medulla and distributed in plasma, via the sympathetic nervous system,

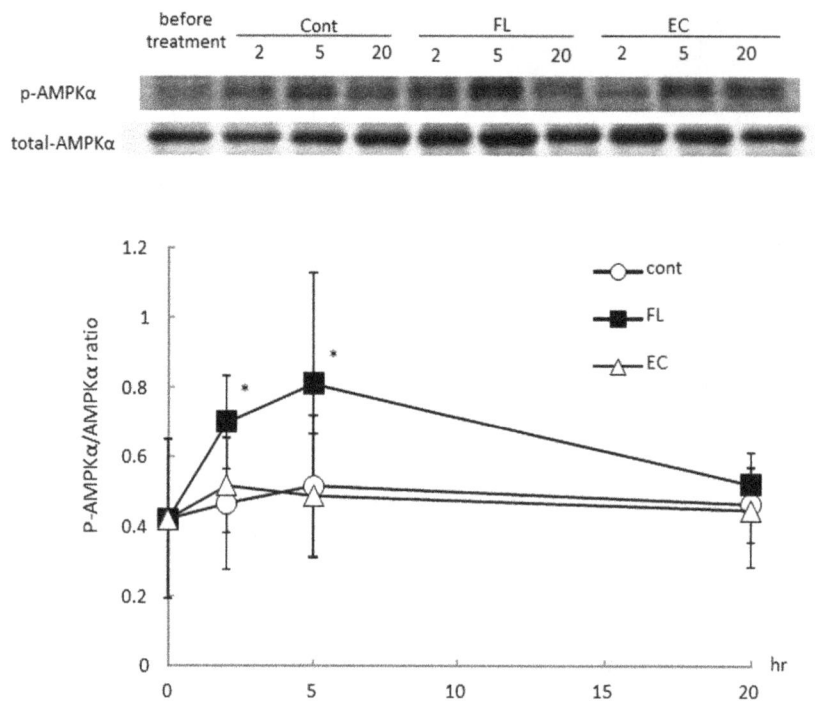

Figure 3. Phosphorylation of AMPK1αin BAT (a) or gastrocnemius (b) after administration of mixed flavan-3-ols or (−)-epicatechin. The animals was euthanized 2, 5, and 20 hr after administration of the vehicle (Cont, $n = 8$), 10 mg/kgFL ($n = 8$), or 10 mg/kg EC ($n = 8$). Values represent the mean ± standard deviation. Statistical analyses were performed by Dunnett's post-test. Significantly different from vehicle, *$p < 0.05$, **$p < 0.01$.

following stimulation in response to physiological or psychological stress [32]. Our data suggested the possibility that FL, especially procyanidins, similarly stimulate sympathetic nerves.

The sympathetic nervous system is known to play an essential role in the regulation of metabolic activity [33]. In BAT, the noradrenalin secreted from sympathetic nerve terminals binds to a β3 adrenergic receptor and induces UCP-1, which is responsible for thermogenesis [34]. It has been suggested in reports of exercise and in agonist studies, that in skeletal muscle, stimulation of the β2 adrenalin receptor upregulates PGC-1α [35,36] and can induce the UCP-3 activity involved in adrenergic effects [37,38]. In our previous reported study, mean blood pressure and heart rate were transiently increased soon after treatment with a single dose of flavan-3-ols, and also increased blood flow in the cremaster muscle

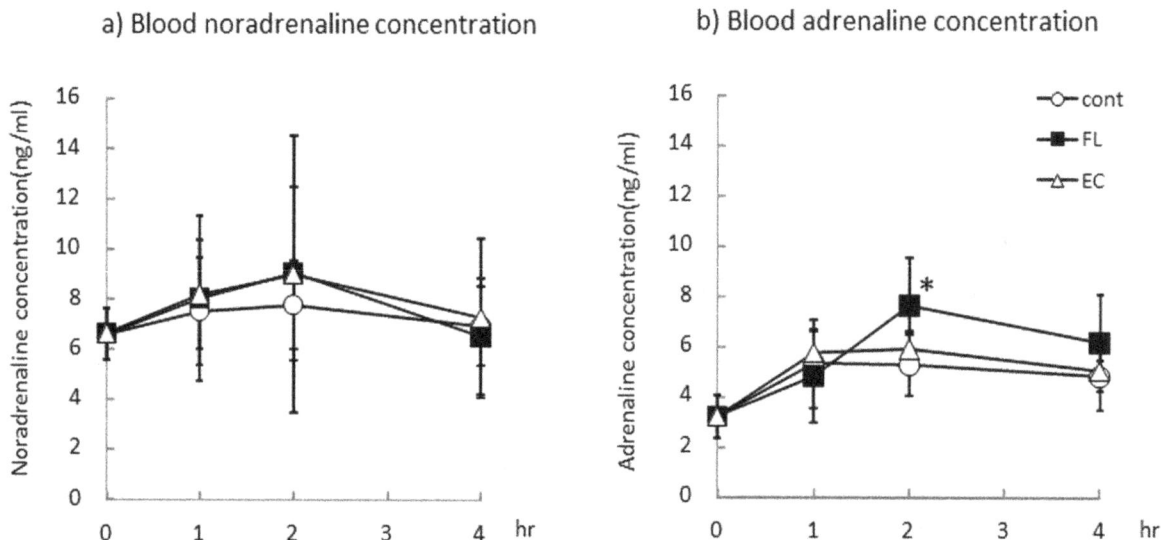

Figure 4. Blood noradrenaline (a) and adrenaline (b) concentrations after administration of mixed FL or EC. The animals were euthanized 2, 5, and 20 hr after administration of the vehicle (Cont, $n = 8$), 10 mg/kg of FL ($n = 8$) or 10 mg/kg of EC ($n = 8$). Values represent the mean ± standard deviation. Statistical analyses were performed by Dunnett's post-test. Significantly different from vehicle, *$p < 0.05$.

[39]. The autonomic nerves are also known to play a crucial role in the circulatory system. Stimulation of sympathetic nerves induced a transient increase in blood pressure through the $\alpha 1$ adrenergic receptors in vascular smooth muscle, and heart rate was affected through $\beta 1$ adrenergic receptors in cardiac muscle. These results suggested that after a single dose of FL, not only can metabolic changes occur, but also hemodynamic alterations induced by sympathetic nerve stimulation are possible.

Yamashita et al. reported a significant increase in phosphorylated AMPK1α after treatment with flavan-3-ols [40], and we have confirmed a change of this nature in the present study (Fig.3). Previously, our research revealed an increase in cremasteric-recruited capillary number, which indicated a requirement for O_2 during ATP production, as a result of a single dose of flavan-3-ols [39]. Hypoxia-induced phosphorylation of AMPK1α in skeletal muscle has been reported previously [41], and while it is possible that hypoxia occurring after FL treatment in skeletal muscle can induce such a change, further experiments are needed to determine the mechanism for enhancement of AMPK1α phosphorylation.

We found a significant increase in mitochondrial number and β-oxidation in the gastrocnemius, soleus, and BAT and decrease in RERs after 2 weeks of flavan-3-ols feeding [23]. PGC-1α is recognized as a master regulator of mitochondrial biogenesis [42–44] by activating respiratory chain and fatty acid oxidation genes, increasing mitochondrial number, and enhancing mitochondrial respiratory capacity. It was suggested that the initial changes responsible for metabolic activity, such as that of PGC-1α, contributed to the alterations in this murine phenotype.

In conclusion, we found that FL significantly enhanced systemic energy expenditure, accompanied with an increase in gene expression associated with thermogenesis and lipolysis; however, EC exhibited this less robustly or effectively. In addition, it was suggested the possible interaction between metabolic changes and the increase in plasma catecholamine concentrations after administration of a single oral dose of FL. These effects may be able to mitigate the risk of metabolic syndrome.

Acknowledgments

We thank Meiji Co. Limited for the donation of flavan 3-ol fraction.

Author Contributions

Conceived and designed the experiments: NO. Performed the experiments: YM YN KM HY. Analyzed the data: YM YN NO. Contributed reagents/materials/analysis tools: YM YN. Wrote the paper: NO.

References

1. Hammerstone JF, Lazarus SA, Mitchell AE, Rucker R, Schmitz HH (1999) Identification of procyanidins in cocoa (Theobroma cacao) and chocolate using high-performance liquid chromatography/mass spectrometry. J Agric Food Chem 47: 490–496.
2. Hatano T, Miyatake H, Natsume M, Osakabe N, Takizawa T, et al. (2002) Proanthocyanidin glycosides and related polyphenols from cacao liquor and their antioxidant effects. Phytochemistry 59: 749–758.
3. Sanbongi C, Osakabe N, Natsume M, Takizawa T, Gomi S, et al. (1998) Antioxidative polyphenols isolated from Theobroma cacao. J Agric Food Chem 46: 454–457.
4. Buitrago-Lopez A, Sanderson J, Johnson L, Warnakula S, Wood A, et al. (2011) Chocolate consumption and cardiometabolic disorders: systematic review and meta-analysis. Available: http://www.bmj.com/content/343/bmj.d4488. Accessed 7 July 2014.
5. Larsson SC, Virtamo J, Wolk A (2012) Chocolate consumption and risk of stroke: a prospective cohort of men and meta-analysis. Neurology 79: 1223–1229.
6. Taubert D, Roesen R, Schömig E (2007) Effect of cocoa and tea intake on blood pressure: a meta-analysis. Arch Intern Med 167: 626–634.
7. Desch S, Schmidt J, Kobler D, Sonnabend M, Eitel I, et al. (2010) Effect of cocoa products on blood pressure: systematic review and meta-analysis. Am J Hypertens 23: 97–103.
8. Engler MB, Engler MM, Chen CY, Malloy MJ, Browne A, et al. (2004) Flavonoid-rich dark chocolate improves endothelial function and increases plasma epicatechin concentrations in healthy adults. J Am Coll Nutr 23: 197–204.
9. Schroeter H, Heiss C, Balzer J, Kleinbongard P, Keen CL, et al. (2006) (−)-Epicatechin mediates beneficial effects of flavanol-rich cocoa on vascular function in humans. Proc Natl Acad Sci 103: 1024–1029.
10. Baba S, Osakabe N, Kato Y, Natsume M, Yasuda A, et al. (2007) Continuous intake of polyphenolic compounds containing cocoa powder reduces LDL oxidative susceptibility and has beneficial effects on plasma HDL-cholesterol concentrations in humans. Am J Clin Nutr 85: 709–717.
11. Baba S, Natsume M, Yasuda A, Nakamura Y, Tamura T, et al. (2007) Plasma LDL and HDL cholesterol and oxidized LDL concentrations are altered in normo- and hypercholesterolemic humans after intake of different levels of cocoa powder. J Nutr 137: 1436–1444.
12. Grassi D, Necozione S, Lippi C, Croce G, Valeri L, et al. (2005) Cocoa reduces blood pressure and insulin resistance and improves endothelium-dependent vasodilation in hypertensives. Hypertension 46: 398–405.
13. Grassi D, Lippi C, Necozione S, Desideri G, Ferri C (2007) Short-term administration of dark chocolate is followed by a significant increase in insulin sensitivity and a decrease in blood pressure in healthy persons. Am J Clin Nutr 81: 611–614.
14. Taubert D, Roesen R, Schömig E (2007) Effect of cocoa and tea intake on blood pressure: a meta-analysis. Arch Intern Med 167: 626–634.
15. Hooper L, Kroon PA, Rimm EB, Cohn JS, Harvey I, et al. (2008) Flavonoids, flavonoid-rich foods, and cardiovascular risk: a meta-analysis of randomized controlled trials. Am J Clin Nutr 88: 38–50.
16. Desch S, Schmidt J, Kobler D, Sonnabend M, Eitel I, et al. (2010) Effect of cocoa products on blood pressure: systematic review and meta-analysis. Am J Hypertens 23: 97–103.
17. Ried K, Sullivan T, Fakler P, Frank OR, Stocks NP (2010) Does chocolate reduce blood pressure? A meta-analysis. BMC Med. 28;8: 39 Available: http://www.biomedcentral.com/1741-7015/8/39. Accessed 7 July 2014.
18. Tokede OA, Gaziano JM, Djoussé L (2011) Effects of cocoa products/dark chocolate on serum lipids: a meta-analysis. Eur J Clin Nutr 65: 879–886.
19. Shrime MG, Bauer SR, McDonald AC, Chowdhury NH, Coltart CE, et al. (2011) Flavonoid-rich cocoa consumption affects multiple cardiovascular risk factors in a meta-analysis of short-term studies. J Nutr 141: 1982–1988.
20. Hooper L, Kay C, Abdelhamid A, Kroon PA, Cohn JS, et al. (2012) Effects of chocolate, cocoa, and flavan-3-ols on cardiovascular health: a systematic review and meta-analysis of randomized trials. Am J Clin Nutr 95: 740–751.
21. Cuenca-García M, Ruiz JR, Ortega FB, Castillo MJ, HELENA study group (2014) Association between chocolate consumption and fatness in European adolescents. Nutrition 30: 236–239.
22. Golomb BA, Koperski S, White HL (2012) Association between more frequent chocolate consumption and lower body mass index. Arch Intern Med 172: 519–521.
23. Watanabe N, Inagawa K, Shibata M, Osakabe N (2014) Flavan-3-ol fraction from cocoa powder promotes mitochondrial biogenesis in skeletal muscle in mice. Lipids Health Dis 13: 64. Available: http://http://www.lipidworld.com/content/13/1/64. Accessed 7 July 2014.
24. Natsume M, Osakabe N, Yamagishi M, Takizawa T, Nakamura T, et al. (2000) Analyses of polyphenols in cacao liquor, cocoa, and chocolate by normal-phase and reversed-phase HPLC. Biosci Biotechnol Biochem 64: 2581–2587.
25. Grouzmann E, Lamine F (2013) Determination of catecholamines in plasma and urine. Best Pract Res Clin Endocrinol Metab 27: 713–723.
26. Kalili KM, de Villiers A (2013) Systematic optimisation and evaluation of on-line, off-line and stop-flow comprehensive hydrophilic interaction chromatography × reversed phase liquid chromatographic analysis of procyanidins. Part II: application to cocoa procyanidins. J Chromatogr A 1289: 69–79.
27. Robbins RJ, Leonczak J, Li J, Johnson JC, Collins T, et al. (2012) Determination of flavanol and procyanidin (by degree of polymerization 1–10) content of chocolate, cocoa liquors, powder(s), and cocoa flavanol extracts by normal phase high-performance liquid chromatography: collaborative study. J AOAC Int 95: 1153–1160.
28. Higdon JV, Frei B (2003) Tea catechins and polyphenols: health effects, metabolism, and antioxidant functions. Crit Rev Food Sci Nutr 43: 89–143.
29. Baba S, Osakabe N, Yasuda A, Natsume M, Takizawa T, et al. (2000) Bioavailability of (−)-epicatechin upon intake of chocolate and cocoa in human volunteers. Free Radic Res 33: 635–641.
30. Baba S, Osakabe N, Natsume M, Terao J (2002) Absorption and urinary excretion of procyanidin B2 [epicatechin-(4beta-8)-epicatechin] in rats. Free Radic Biol Med 33: 142–148.
31. Spencer JP, Schroeter H, Rechner AR, Rice-Evans C (2001) Bioavailability of flavan-3-ols and procyanidins: gastrointestinal tract influences and their relevance to bioactive forms in vivo. Antioxid Redox Signal 3: 1023–1039.

32. Kvetnansky R, Lu X, Ziegler MG (2013) Stress-triggered changes in peripheral catecholaminergic systems. Adv Pharmacol 68: 359–397.

33. Davy KP, Orr JS (2009) Sympathetic nervous system behavior in human obesity. Neurosci Biobehav Rev. 33: 116–124.

34. Morrison SF, Madden CJ, Tupone D (2014) Central Neural Regulation of Brown Adipose Tissue Thermogenesis and Energy Expenditure. Cell Metab 19: 741–756.

35. Miura S, Kawanaka K, Kai Y, Tamura M, Goto M et al. (2007) An increase in murine skeletal muscle peroxisome proliferator-activated receptor-gamma coactivator-1alpha (PGC-1alpha) mRNA in response to exercise is mediated by beta-adrenergic receptor activation. Endocrinology 148: 3441–3448.

36. Pearen MA, Myers SA, Raichur S, Ryall JG, Lynch GS et al. (2008) The orphan nuclear receptor, NOR-1, a target of beta-adrenergic signaling, regulates gene expression that controls oxidative metabolism in skeletal muscle. Endocrinology. 149: 2853–2865.

37. Jezek P (2002) Possible physiological roles of mitochondrial uncoupling proteins–UCPn. Int J Biochem Cell Biol 34: 1190–1206.

38. Sprague JE, Yang X, Sommers J, Gilman TL, Mills EM (2007) Roles of norepinephrine, free Fatty acids, thyroid status, and skeletal muscle uncoupling protein 3 expression in sympathomimetic-induced thermogenesis. J Pharmacol Exp Ther 320: 274–280.

39. Ingawa K, Aruga N, Matsumura Y, Shibata M, Osakabe N (2014) Alteration of the systemic and microcirculation by a single oral dose of flavan-3-ols. PLoS One 16;9(4):e94853. Available: http://www.plosone.org/article/info%3Adoi%2F10.1371%2Fjournal.pone.0094853. Accessed 7 July 2014.

40. Yamashita Y, Okabe M, Natsume M, Ashida H (2012) Prevention mechanisms of glucose intolerance and obesity by cacao liquor procyanidin extract in high-fat diet-fed C57BL/6 mice. Arch Biochem Biophys 527: 95–104.

41. Le Moine CM, Morash AJ, McClelland GB (2011) Changes in HIF-1α protein, pyruvate dehydrogenase phosphorylation, and activity with exercise in acute and chronic hypoxia. Am J Physiol Regul Integr Comp Physiol 301:R1098–R1104.

42. Ho TK, Abraham DJ, Black CM, Baker DM (2006) Hypoxia-inducible factor 1 in lower limb ischemia. Vascular 14: 321–327.

43. Bonawitz ND, Clayton DA, Shadel GS (2006) Initiation and beyond: multiple functions of the human mitochondrial transcription machinery. Mol Cell 24: 813–825.

44. Scarpulla RC, Vega RB, Kelly DP (2012) Transcriptional integration of mitochondrial biogenesis. Endocrinol Metab 23: 459–466.

24

Studying Mixing in Non-Newtonian Blue Maize Flour Suspensions Using Color Analysis

Grissel Trujillo-de Santiago[1,2], Cecilia Rojas-de Gante[3], Silverio García-Lara[1], Adriana Ballescá-Estrada[1], Mario Moisés Alvarez[1]*

1 Centro de Biotecnología-FEMSA, Tecnológico de Monterrey, Monterrey, Nuevo León, México, **2** Centro de Investigación y Desarrollo de Proteínas (CIDPRO), Tecnológico de Monterrey, Monterrey, Nuevo León, México, **3** Departamento de Ingeniería en Biotecnología, Tecnológico de Monterrey, Tlalpan, Distrito Federal, México

Abstract

Background: Non-Newtonian fluids occur in many relevant flow and mixing scenarios at the lab and industrial scale. The addition of acid or basic solutions to a non-Newtonian fluid is not an infrequent operation, particularly in Biotechnology applications where the pH of Non-Newtonian culture broths is usually regulated using this strategy.

Methodology and Findings: We conducted mixing experiments in agitated vessels using Non-Newtonian blue maize flour suspensions. Acid or basic pulses were injected to reveal mixing patterns and flow structures and to follow their time evolution. No foreign pH indicator was used as blue maize flours naturally contain anthocyanins that act as a native, wide spectrum, pH indicator. We describe a novel method to quantitate mixedness and mixing evolution through Dynamic Color Analysis (DCA) in this system. Color readings corresponding to different times and locations within the mixing vessel were taken with a digital camera (or a colorimeter) and translated to the CIE*Lab* scale of colors. We use distances in the *Lab* space, a 3D color space, between a particular mixing state and the final mixing point to characterize segregation/mixing in the system.

Conclusion and Relevance: Blue maize suspensions represent an adequate and flexible model to study mixing (and fluid mechanics in general) in Non-Newtonian suspensions using acid/base tracer injections. Simple strategies based on the evaluation of color distances in the CIE*Lab* space (or other scales such as HSB) can be adapted to characterize mixedness and mixing evolution in experiments using blue maize suspensions.

Editor: Alberto Aliseda, University of Washington, United States of America

Funding: The authors gratefully acknowledge the financial support of Tecnológico de Monterrey (through the seed fund CAT-122), and CONACYT (through the doctoral scholarship provided to GTdeS). The authors deeply appreciate the donation of the blue maize used in the experiments by EDOMEX México. The funders had no role in study design, data collection and analysis, decision to publish, or preparation of the manuscript.

Competing Interests: The authors have declared that no competing interests exist.

* Email: mario.alvarez@itesm.mx

Introduction

Mixing is one of the most common unit operations in the chemical engineering practice. However, the spectrum of techniques used and the mixing systems studied is still limited to relatively simple scenarios. The methods to characterize mixing are mostly focused in Newtonian liquid-liquid systems, but the vast number of techniques currently available to evaluate mixing quality or extent are restricted to transparent or nearly transparent vessels and fluid systems (water, glycerin solutions, diluted suspensions, solutions of CMC in water, Carbopol 940).

More complex and realistic mixing scenarios have received only modest attention in the fluid mechanics and physics literature. One specific scenario is the mixing of heavy suspensions (suspensions where the concentration of solids is higher than 5% w/w), an operation that occurs in diverse lab and industrial scale settings in many applications related to food technology [1–3], polymer processing [4–5], and biotechnology [6–9], among others.

In many of these experimentally relevant suspensions, Non-Newtonian behavior is observed [6–11].

Only a limited number of papers have addressed the mixing in Non-Newtonian systems, focusing mainly on solutions [12–15]. Even fewer studies have addressed mixing of realistic Non-Newtonian suspensions. At high solid densities, the opaqueness produced by the particles presents an obstacle to the use of classical visualization techniques. Recently, 2D and 3D tomography was applied to the study of mixing in non-transparent fluids [12]. Laser Sheet Image Analysis (LSIA), a sort of laser-based tomography, was formally introduced as a technique to study the dynamics of dilute particle suspensions [16]. Positron emission particle tracking (PEPT) has been used to track individual particles in heavy slurries [17], producing valuable (Lagrangian type) information on the dynamics of heavy suspensions. However, the implementation of tomography and PEPT techniques demands special infrastructure not widely available in the common fluid mechanics lab. Besides, in a highly viscous system or a suspension, the adequate dispersion of the indicator might be an issue in itself. In particular, only a few studies refer the use of tracer techniques to investigate mixing in heavy suspensions; that is, with solids content above 5% [17].

Here, we use blue maize flour suspensions stirred in stirred tanks with a simplified geometry (no baffles, disc impellers; see Figure 1) to illustrate the existence of severe mixing non-idealities in non-Newtonian systems. As a secondary objective, we propose the use of blue maize suspensions as a useful and flexible model for the study of mixing of Non-Newtonian suspensions. Blue maize is a corn variety native to México [18,19]. Blue maize flour suspensions exhibit Non-Newtonian behavior (Figure 2), and are convenient for studying mixing in heavy suspension experiments because the anthocyanins naturally present in blue maize kernels [18,20,21] serve as a natural pH indicator that undergoes drastic color changes in a wide range of pH values. Therefore, the simple injection of acid or base solutions allows visualization of flow patterns and the qualitative and quantitative characterization of mixing.

Industrial processes in which acid or basic injections are added to a non-Newtonian fluid, are not infrequently encountered, particularly in Biotechnology applications where the pH of non-Newtonian culture broths is usually regulated by the addition of acid/base solutions [8,11,22], or where the bioreaction itself releases acid into a non-Newtonian fluid, further modifying its rheology [23,24].

Results and Discussion

Blue maize suspensions as a complex fluid model for mixing experiment

Blue maize suspensions allow performing flow visualization experiments without the need to add a foreign pH indicator. The anthocyanins naturally present in blue maize flour [20,21] respond to changes in pH by displaying a wide range of colors (see Figure 2a). At low pH values, blue maize suspensions exhibit a pink color. Progressively, as pH is increased, this color transitions to magenta, pink, violet, blue, blue-greenish, and green.

The presence of this intrinsic and wide range pH indicator has important practical advantages in the laboratory. With a few exceptions [25,26], most available pH indicators exhibit a narrow range of color change, from three to five pH units [26,27], which limits the window of visualization for the occurrence of the acid-base reaction. The simultaneous use of two or more pH indicators has also been suggested to amplify the span of pH sensing [28]. However, in practice, the dispersion of a foreign indicator in a non-Newtonian (or Newtonian but highly viscous) fluid is, in itself, a non-trivial mixing problem.

The non-Newtonian behavior of flour suspensions has been characterized in the context of food engineering applications [29–31]. We conducted determinations of shear stress and apparent viscosity at different shear rate values for the blue maize flour suspensions used in our experiments using an automatic Rheometer (Physica MCR Anton Paar, Austria). Figure 3a shows the apparent viscosity values in the range of shear rates from 20 to 1520 s^{-1}; we observe non-Newtonian behavior across various pH values (Figure 3a, 3c). At low shear rate values (Figure 3b), except under very acidic conditions (3.3> pH >1.6), suspensions

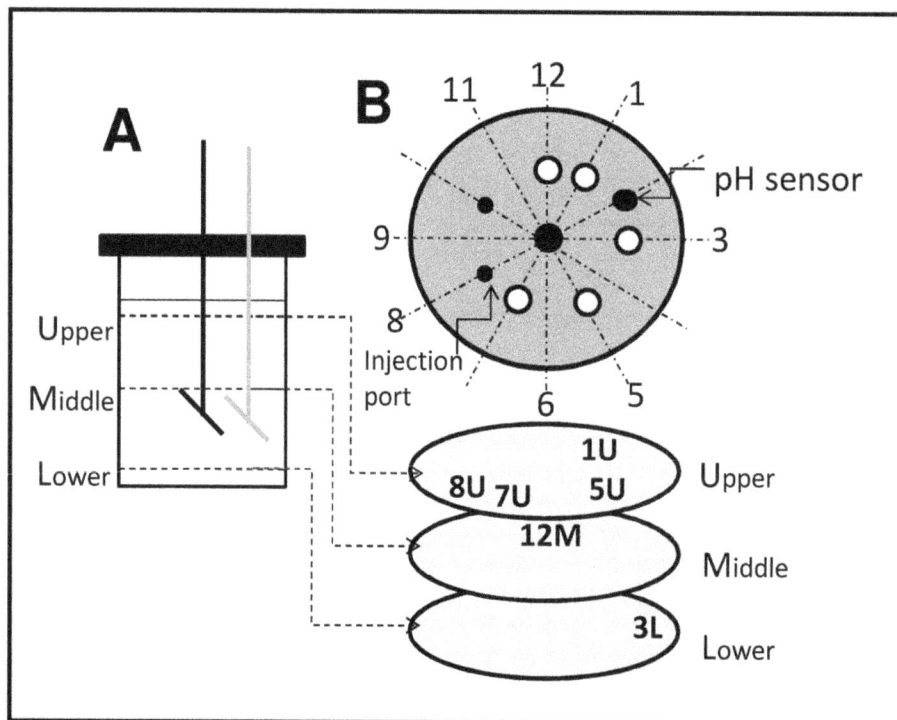

Figure 1. Scheme of the experimental system. (A) A stirred tank, equipped with a 45° inclined disk impeller, was used to agitate non-Newtonian blue maize suspensions. Three different horizontal planes [upper-plane (U), mid-plane (M), and lower-plane (L)] were defined within the tank for tracer injections. For some of our experiments, the impeller was placed in an eccentric position (indicated in gray). (B) The tank lid was modified to allow for additions of acid/base solutions at different angular positions (12 o'clock, 1 o'clock, etc.).

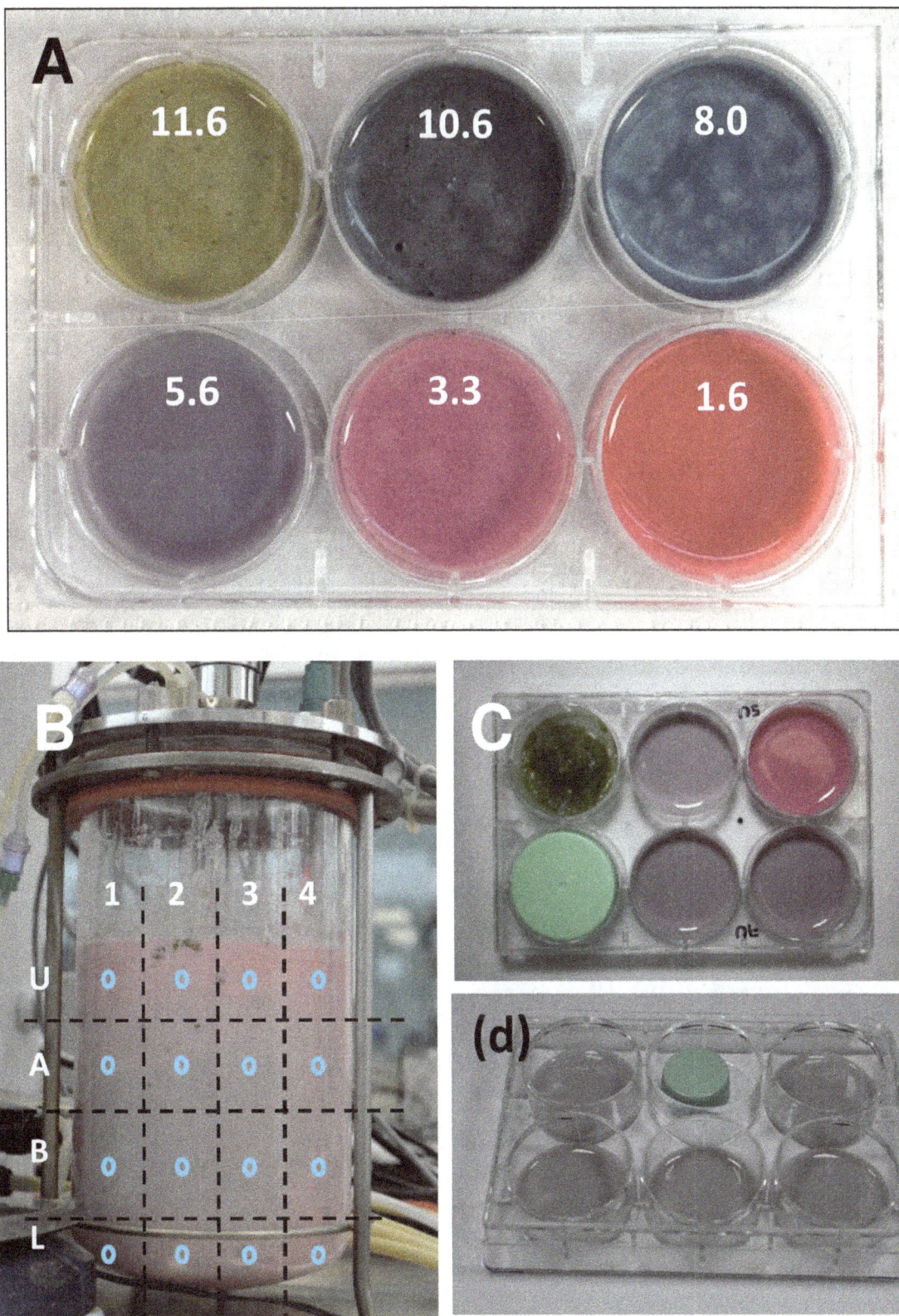

Figure 2. Following mixing through color changes in blue-maize suspensions. (A) In blue maize suspension, the color varies significantly as a function of pH due to the presence of native anthocyanins that act as a natural wide spectrum pH indicator. (B) The evolution of mixing of a blue maize flour suspension in a stirred tank was followed by addition of a basic injection into an initially acidic condition. Frontal photographic images were taken at different time points of the mixing process. Each image was divided into sixteen sections (U1 to L4) and the color in the CIE*Lab* scale was determined by image analysis at each of the center points (indicated by blue circles). (C) Samples corresponding to different tank locations and times of agitation were dispensed in 6-well culture plates for color analysis using digital photography or colorimetric readings with a portable colorimeter. Reproducibility of the color readings among different plates can be validated by including a color standard in each plate (in this case, a

circular plastic object of uniform color). (D) The experimental error associated with lighting heterogeneity at different well positions was estimated by placing the same sample in different wells.

displayed an evident non-Newtonian behavior with apparent viscosities varying drastically as a function of shear rate.

The rheology of our blue maize suspensions at different pH values can be described approximately by a simple Ostwald-de Waele power-law model (see Equation 1).

$$\eta = K(\gamma)^{n-1} \tag{1}$$

Here η is the apparent viscosity at a given shear stress and pH condition, γ is the shear rate, K is a flow consistency index, and n is the flow behavior index. In Table 1, we present best-fit values for K and n for our blue maize suspensions calculated from linear regressions of the type $\ln \eta = \ln K + (n-1) \ln (\gamma)$. For suspensions at pH = 5.6, a Newtonian behavior is observed. For higher pH values (8.0, 10.6, 11.6) a clear non-Newtonian shear thinning behavior is evident (n<1). A striking flow behavior is observed in suspensions at pH = 3.3 and 1.6, at which the flow transits from shear thinning (at low shear rates) to shear thickening (at high shear rates). To model this transition without recourse to a more complex model, we simply provide piece-wise valid values for K and n.

One question is how significant are these variations of apparent viscosities in the context of a stirred tank system. In a stirred tank, the range of shear rates spans four to five orders of magnitude, even in Newtonian systems [32,33]. Based on theoretical arguments, Sánchez-Pérez et al. [34] recently proposed the expression $\gamma \propto N^{3/2}$, where N is the agitation speed in RPM, to establish a general relationship between agitation speed and maximum shear rate in turbulent stirred tanks. The authors also observed that the expression $\gamma = 33.1 \ N^{1.4}$, consistent with their theoretical derivation, correlates well with data previously calculated [35] for stirred vessels using computational fluid dynamics. In our experimental tank system, we agitated at N = 1000 RPM (16 rev/s). Therefore, assuming a fully turbulent regime in the impeller zone, the maximum shear rate value will marginally exceed 2000 s^{-1}, and the average shear rate value will be in the neighborhood of 20 s^{-1}. However, areas of low shear (~ 2 s^{-1}) could be found near the tank walls and in low circulation areas close to the tank bottom or the tank surface. Based on normalized distributions of shear rates reported for stirred tanks [33], approximately 10% of the tank volume experiences shear values below 5–6 s^{-1}. For a non-Newtonian system this has profound implications, and the higher apparent viscosities observed at low shear values (Figure 3b) represent an added complexity to mixing. Zero-shear-viscosity (in Pa.s), calculated from Figure 3b, are 1.737, 7.550, 0.702, 0.019, 0.086, 0.054, at pH = 11.6, pH = 10.6, pH = 8.0, pH = 5.6, pH = 3.3, and pH = 1.6, respectively.

In the following sections, we discuss mixing experiments conducted in blue maize suspensions of nearly 50% solids. First, we describe the use of blue maize suspensions as a fluid model for qualitative study of mixing in different stirred tank configurations. Then, we demonstrate the use of simple techniques to quantitate mixing evolution based on the analysis of color changes (digital color analysis; DCA), associated with pH changes, during acid/base injection/excursion experiments. Two examples of the use of DCA techniques are provided. The first example employs a simple, non-intrusive strategy that is suitable for transparent vessels. We analyze a time-series of photographic images of the exterior of the entire stirred tank system. The second application

case uses DCA to analyze images from samples taken at different tank locations at different agitation times. This second example extends the use of DCA to non-transparent agitation/blending vessels.

Mixing visualization in blue maize suspensions

Mixing dynamics in non-Newtonian systems can become highly complex. Our experimental observations suggest that the mixing performance of non-Newtonian systems is extremely sensitive to some geometrical and operational parameters. The location of the stirring axes, the location of the point of injection and the starting condition (acidic or basic) are important considerations that define mixing performance.

In our experiments, eccentric stirred tank configurations outperform concentric ones, particularly when the aspect ratio is higher than 1.2. Figure 4 illustrate these findings, showing different mixing conditions or states in a stirred tank containing a blue maize non-Newtonian flour suspension. In Figure 4a, which depicts a tank stirred by an eccentrically located inclined disc impeller [36,37], a subsurface acid injection was efficiently dispersed to achieve a practically (at least visually) homogeneous condition in less than 5 minutes. In Figure 4b, we show the final state of mixing of a similar experiment. This time, the initial pH was acidic, and a basic solution (i.e., NaOH 1N) was administered at the fluid surface. Even when most of the system has reached a basic pH, segregated acidic areas prevail at the liquid surface. A frequently observed mixing problem in conventional stirred tank geometries is the presence of segregated or low circulation regions at the upper section of the tank. A recent contribution [38] demonstrates that Reynolds numbers above 300,000 (based on the impeller diameter and speed) are needed to assure fully turbulent regimes in the upper third section of typical stirred tank configurations. In non-Newtonian system, this mixing pathology is even more evident under certain conditions. For example, when a concentrically agitated system is used, a layer of acidic material is still evident in the upper section of the tank several minutes after a subsurface basic injection into an initially acid environment (Figure 4c). Figure 5 presents a sequence of images corresponding to this experiment, in which an initially acidic blue maize flour suspension is agitated in a conventional tank after a basic set point has been established. The tank is equipped with a concentrically located radial impeller rotating clockwise at 1000 RPM; the height/diameter ratio (H/D) is 1.15, and the impeller is located at approximately 1/4 H. This experiment visually illustrates the progression of mixing in a concentrically agitated tank, from an initial homogeneous condition, to a final state that shows top-bottom segregation. In this case, although the reaction is instantaneous, the rheology of the system imposes conditions that slow the advance of the acid/base reaction, and the mixing process becomes limiting, finally determining the rate of the overall process. Images show, qualitatively, that concentric systems with this geometry (H/D between 1.1 and 1.3 and one agitator) retain certain top-bottom segregation, even after extended periods of agitation.

The blue maize suspension system can be particularly useful for conducting experiments to diagnose poor or ideal injection locations. At pH values above 8.0, the viscosity of blue maize suspensions increases significantly and abruptly, reducing flow and obstructing mixing even more. Therefore, in experiments where a basic solution is dispensed through a point of injection in a low

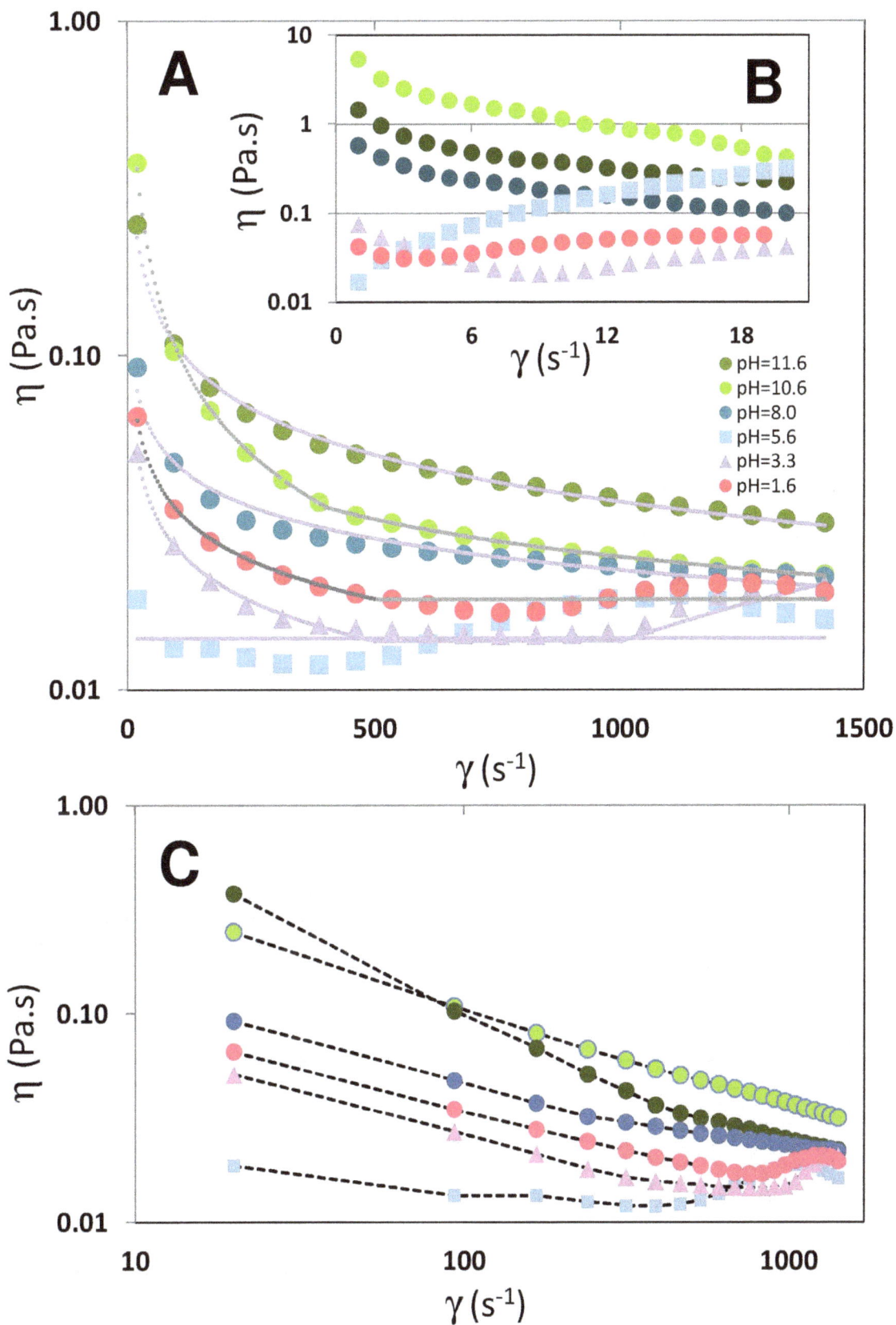

Figure 3. Blue maize flour suspensions exhibit different rheological behavior at different pH values. (A) Plot of apparent viscosity versus shear rate (in the range from 250 to 1500 s^{-1}) for blue maize suspensions prepared at different pH values. Gray dotted lines correspond to power-law fits to experimental data based on the Ostwald-de Waele model [$\eta = K (\gamma)^{n-1}$] using the parameter values reported in Table 1. (B) Plot of apparent viscosity versus shear rate (in the range from 250 to 1500 s^{-1}) for blue maize suspensions prepared at different pH values. (C) Log-log version of the

plot of apparent viscosity versus shear rate (in the range from 250 to 1500 s^{-1}) for blue maize suspensions prepared at different pH values. Straight dotted lines have been used to connect the experimental data points.

circulation zone, stagnant regions will develop. These regions can be easily detected by the development of a green color, characteristic of basic pH values (see Figure 2a and 2c). In our stirred tank system, the inadequate selection of the point of addition of a concentrated basic solution, even in eccentric configurations, can lead to the creation of these stagnant zones where alkaline conditions prevail, causing high viscosity conditions and further obstructing effective mixing (Figure 4d). The gelatinization induced in corn flour and starch suspensions by alkaline conditions has been studied in detail [39].

Following mixing dynamics in a color space using images

Here, we introduce a simple methodology for quantitation of the state of mixedness and the mixing dynamics through the analysis of color changes in the CIE*Lab* scale, one of the color scales normally used for image analysis and color description applications [40,41]. The use of Digital Color Analysis (DCA), utilizing colors or the digital information embedded in colors, has been suggested before as a tool for quantifying chromatic changes [42]. Here we show how a sequence of images can be analyzed using simple DCA methods to diagnose mixing evolution and mixing extent. For example, a reference grid can be used (see Figure 2b) to define a number of sections within every image in Figure 5. At every region, a series of sampling points can be defined.

Let us consider that the center-point within each zone will be used as a "sample" location to determine color according to the *Lab* scale. In this system, each color is characterized by three values, L, a, and b. The L value is associated to luminosity, ranging from 0 for black, and +100, for white. The a and b values define a plane of colors, as shown in Figure 6a, where a ranges from negative to positive values (green to red) and b ranges from negative to positive values (blue to yellow). Therefore, for each sampling point in each image of Figure 5, the color can be characterized by the L, a, and b coordinates that define a unique point in the *Lab* color space. One could estimate the deviation of a particular state of mixing (at a particular location and time) from a final mixing point or an "ideal mixing" state (presumably the final condition of complete mixing), by evaluating the differences in

colors between the two states. Conceptually, one way of doing this is by evaluating the distances between the two corresponding points in the *Lab* coordinate system (Figure 6a). The distance in the *Lab* space, defined by a straight line connecting both points, can be calculated by equation 2.

$$D_{(i,\, j \text{ to final point})} = \left[\left(L_{i,j} - L_f \right)^2 + \left(a_{i,j} - a_f \right)^2 + \left(b_{i,j} - b_f \right)^2 \right]^{0.5} \quad (2)$$

Here, $D_{(i,j \text{ to final point})}$ is the distance, in the *Lab* space, of the points defined by the *Lab* coordinates of the sample taken at time i and location j ($L_{i,j}$, $a_{i,j}$, $b_{i,j}$) and a sample representative of the final mixing state (L_f, a_f, b_f).

The use of distances between a particular point and a final mixing point as an indicator of deviation from homogeneity would be valid only if that distance consistently decreases as the system becomes more homogeneous. This condition should be validated for each indicator system. In the particular case of the anthocyanins naturally present in blue maize suspensions, this condition is satisfied for a wide range of pH values. Figure 6b shows the progression of colors in a blue maize suspension as pH values increase from an extremely acidic condition (pH = 1.6; point α) to an extremely basic condition (pH = 11.6; point φ). For each of the six pH values in this set (α, β, χ, δ, ε, φ), a color can be defined in the *Lab* scale, corresponding to a particular point in the 3D *Lab* color space (Figure 7b). Let us now define a pH trajectory from pH = 1.6 to pH = 10.6 (from point α to point ε). For each of the five points in this trajectory, a value of the distance with respect to the final point can be calculated ($D_{i,\varepsilon}$). For this pH range of variation, the vector ($D_{\alpha,\varepsilon}$, $D_{\beta,\varepsilon}$, $D_{\chi,\varepsilon}$, $D_{\delta,\varepsilon}$, $D_{\varepsilon,\varepsilon}$) is (60.85, 42.27, 14.26, 10.59, 0). The value of distance in colors decreases as the system moves from pHα to pHε. Therefore, in a pH excursion experiment, where the initial condition is in this range of pH values, and the final point is in the neighborhood of 10.6, the value $D_{i,j}$ would be a valid indicator of heterogeneity. A similar analysis can be formulated considering point φ as the final mixing state for a mixing trajectory. In the range of pH = 1.6 to pH = 11.6, the value of the vector ($D_{\alpha,\varphi}$, $D_{\beta,\varphi}$, $D_{\chi,\varphi}$, $D_{\delta,\varphi}$, $D_{\varepsilon,\varphi}$) is (58.85, 50.95,

Table 1. Proposed set of values for K and n for the Ostwald-de Waele power-law equation [$\eta = K\,(\gamma)^{n-1}$] to model the rheology of blue maize suspensions at different pH values.

pH	shear range (s^{-1})	n−1	K (Pa s^n)	n
11.6	20–1420	−0.4660	0.9102	0.5340
10.6	20–500	−0.7772	3.7109	0.2228
10.6	500–1420	−0.3735	0.3293	0.6265
8.0	20–1420	−0.3165	0.2025	0.6835
5.6	20–1420	0	0.0142	1.0000
3.3	20–500	−0.3979	0.1657	0.6021
3.3	500–1000	0	0.0140	1.0000
3.3	1000–1420	1.1348	5.52E-6	2.1348
1.6	20–500	−0.3823	2.01E-1	0.6177
1.6	500–1420	0	0.01865	1.0000

K and n were calculated from linear regressions of the type ln η = ln K+(n−1) ln (γ).

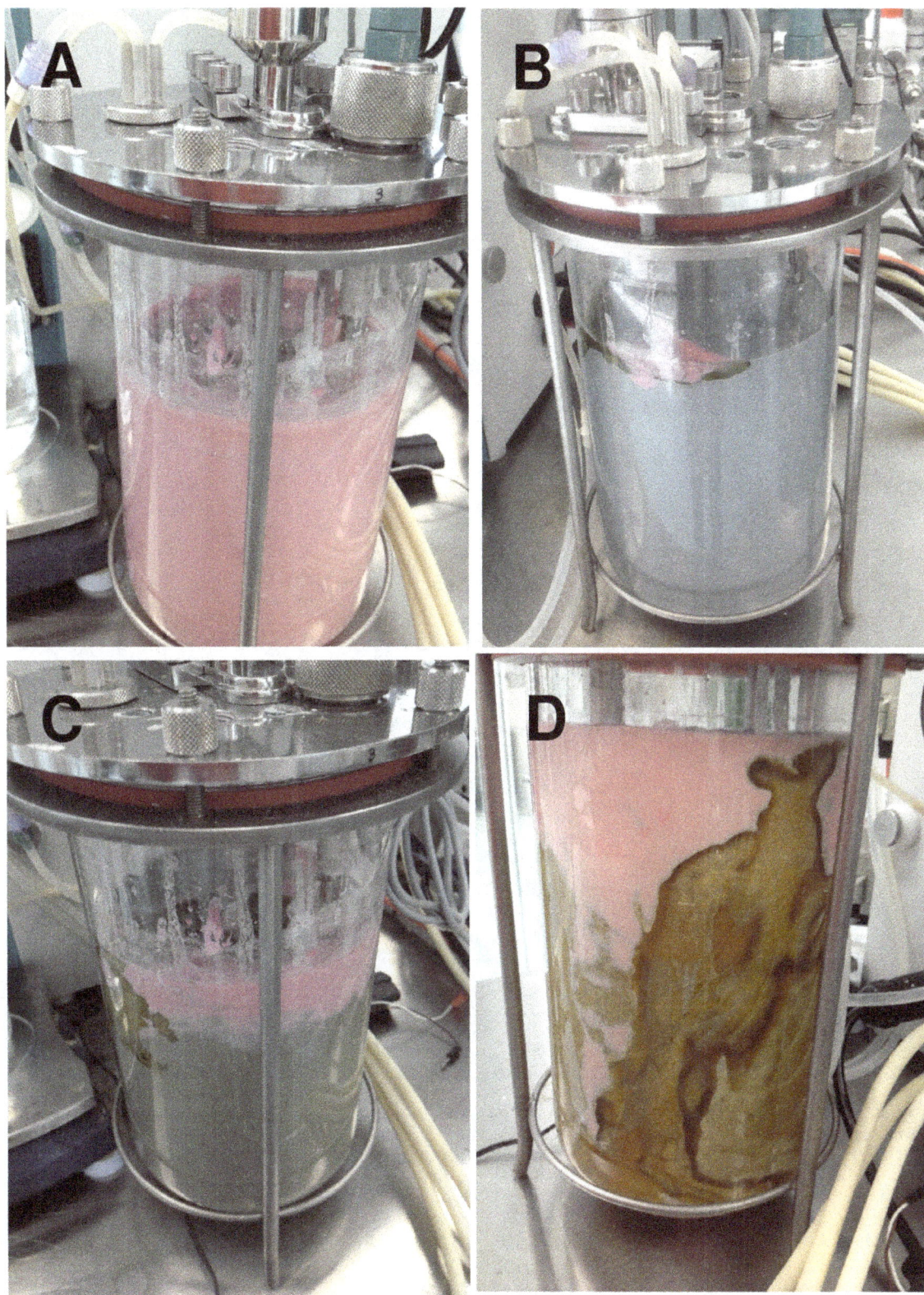

Figure 4. Different mixing states in a stirred tanks containing blue maize non-Newtonian flour suspensions. (A) In a tank stirred by an eccentrically located inclined disc impeller, a subsurface acid injection was efficiently dispersed to achieve homogeneity. (B) Severe top segregation is evident following a subsurface base injection in a concentrically agitated system. The inadequate selection of the point of addition of a concentrated basic injection can lead to the creation of (C) stagnant zones where alkaline conditions prevail, causing high viscosity conditions and (D) further obstructing effective mixing.

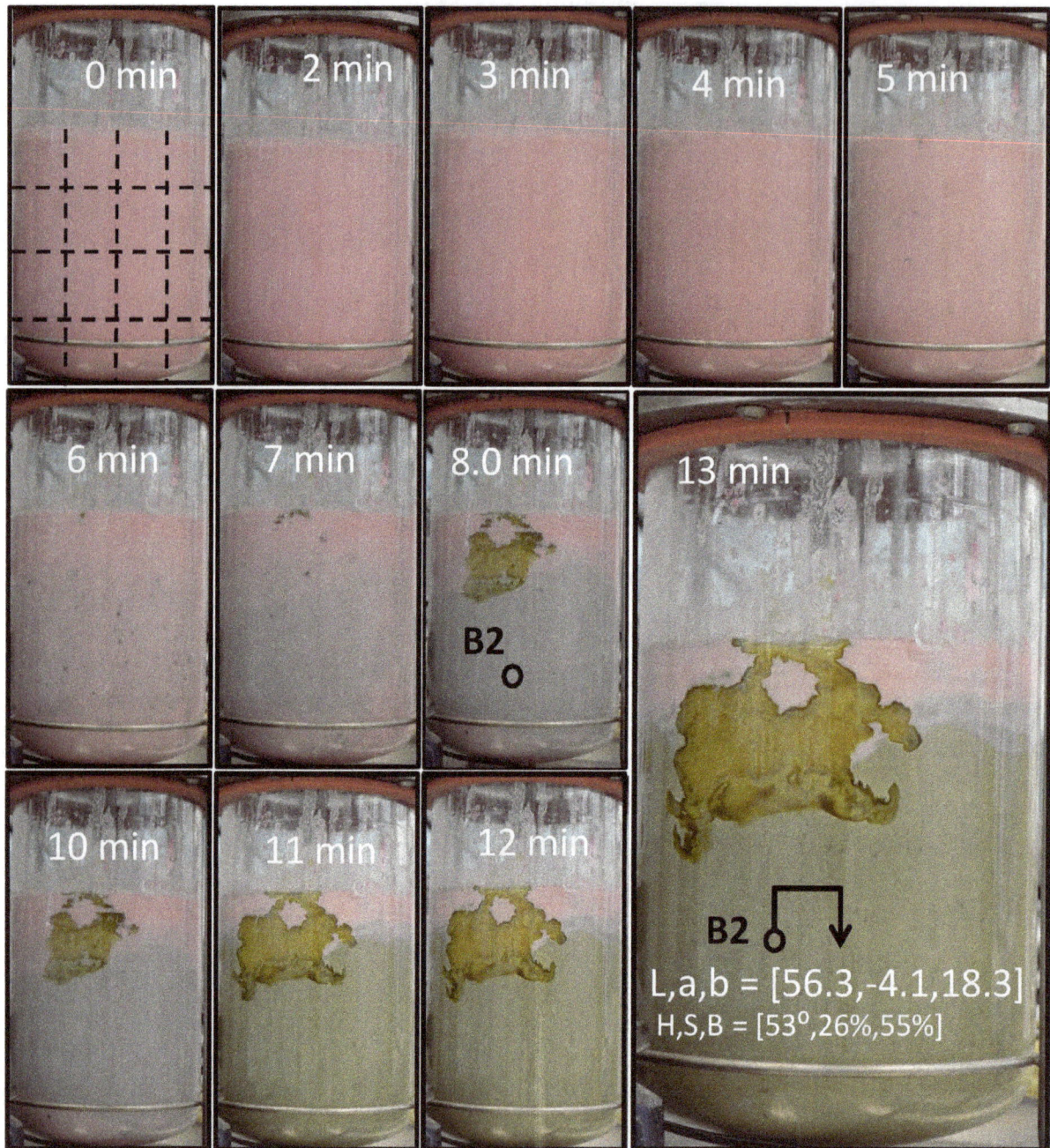

Figure 5. Blue maize suspensions allow the performance of acid-base experiments to visually evaluate the progression of mixing in transparent agitation systems. Mixing experiment that shows evolution in a conventional stirred tank geometry from an initially homogeneous acidic state (t = 0) towards a final process state (t = 13.0 min), in which segregation still prevails (particularly top-bottom segregation). Readings of color, in the CIE*Lab* and the HSB system, are indicated for location B2.

36.41, 45.22, 37.69, 0), and the requirement for decreasing $D_{i,j}$ values is not fulfilled. Therefore, our analysis suggests that in this system (blue maize suspensions), pH excursion experiments to characterize mixing using DCA should be conducted in the pH range between 1.6 and 10.6. Starting at pH 1.6 and ending in 10.6 (or *vice versa*) takes advantage of the widest pH span possible. Evidently, the reliability of this strategy for characterizing mixing will also depend on the robustness of the determination of color. External factors such as uneven illumination or intrinsic variations

in tone (particularly in a suspension) may affect the determination of color for a particular sample.

We conducted a simple experiment to validate the robustness of color estimations in the CIE*Lab* scale in samples of blue maize suspensions with different pH values in the range from 11.6 to 1.6. Samples were dispensed in cell culture wells. A digital photograph was taken under homogeneous illumination conditions under indirect white light. The picture was loaded into the image analysis application Color Companion 4.0 for iPad. At each well, corresponding to each pH value, multiple readings of color were

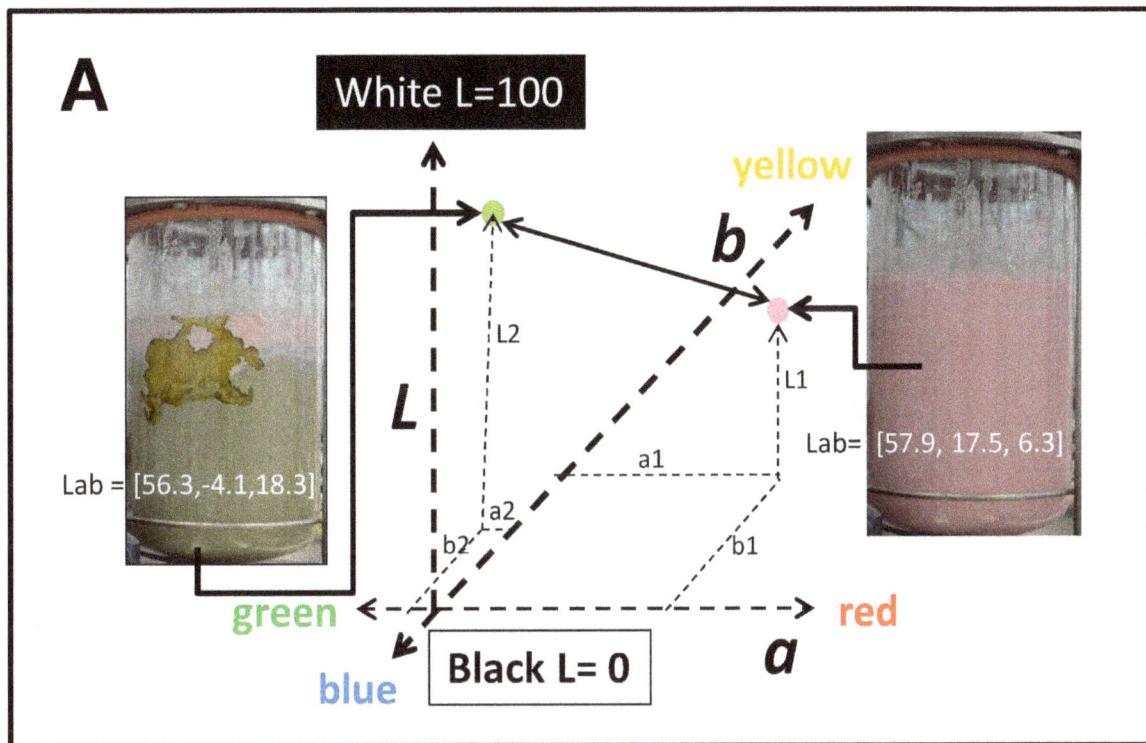

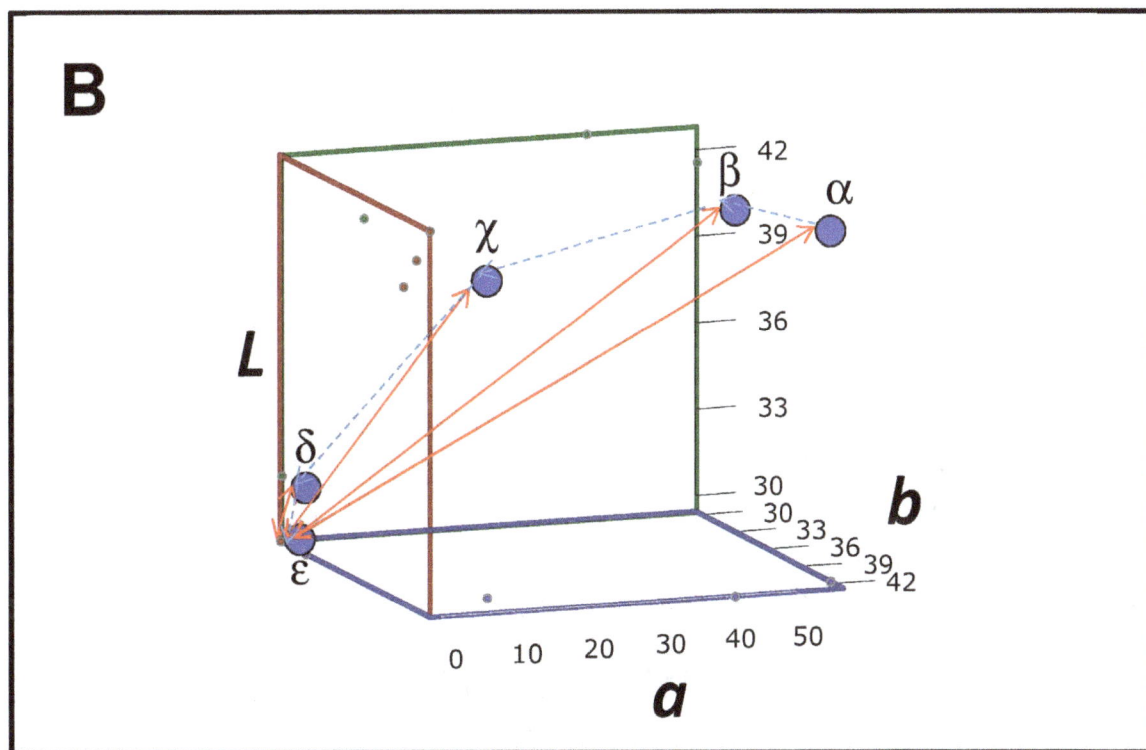

Figure 6. The estimation of distance between two points in the *Lab* scale of colors provides a parameter for characterizing the state of mixing in a stirred system. (A) The CIE*Lab* scale represents a color in a three-coordinate system in which L is associated with luminosity, ranging from 0 for black to +100 for white. The a and b values define a plane of colors,, where a ranges from negative to positive values (green to red) and b ranges negative to positive values (blue to yellow). The difference between two colors can be determined by calculating the distance between them in this 3D space. (B) Progression of pH values during a typical acid or base injection experiment in blue maize flour suspensions (from pH 10.55 to 1.6). Values are plotted in a 3D space; they describe a mixing process trajectory (dashed blue line). The distance between each pH value in the trajectory and the final process point (indicated with red arrows) can be calculated. The values of distances with respect to this final point consistently decrease.

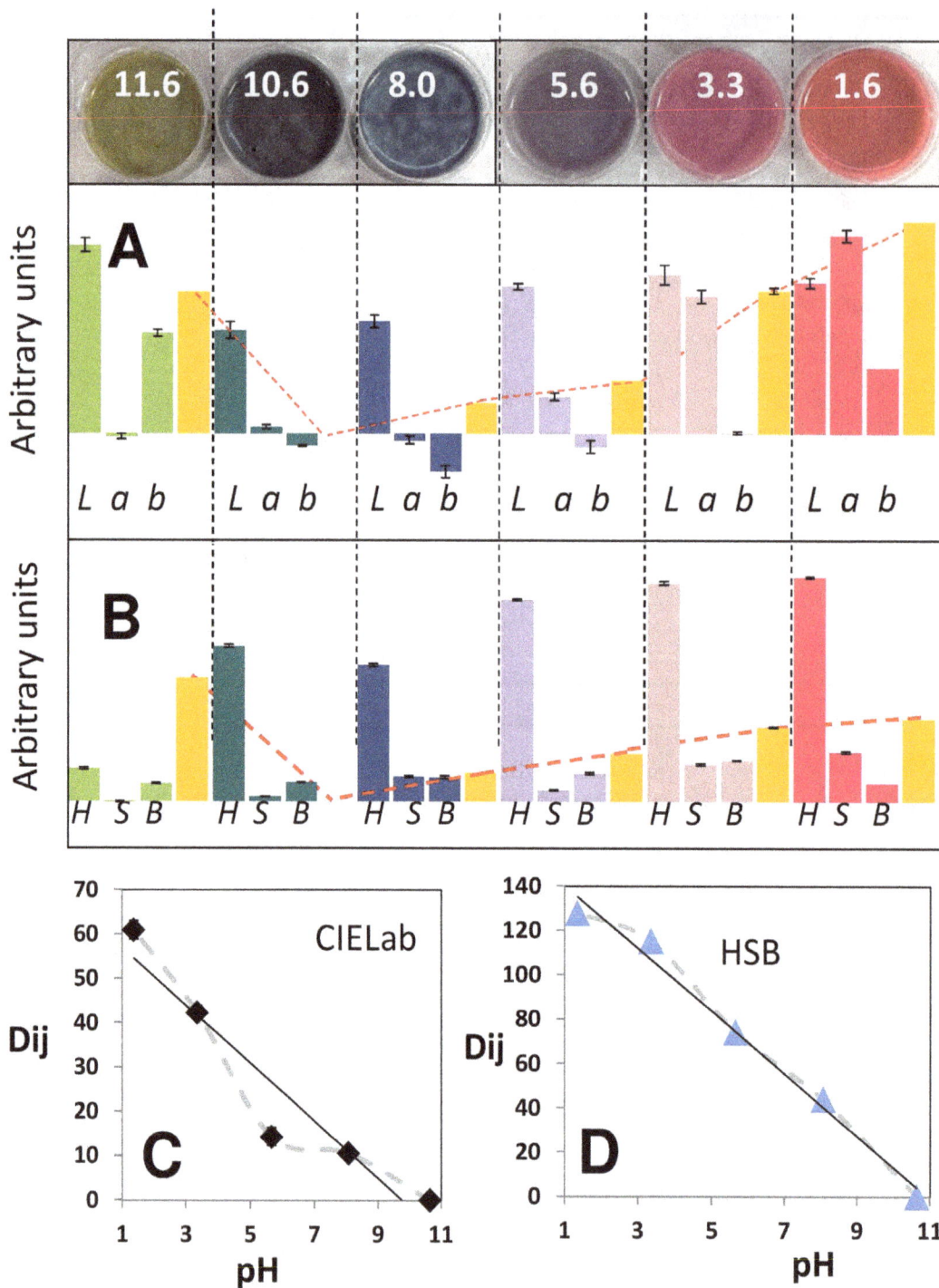

Figure 7. Variability of color readings at different pH values in blue maize flour samples. (A) The average value of the three color components of the CIE*Lab* scale (L, a, and b) are presented for six color determinations at each pH value. Bars are colored according to the pH value, resembling their tone in the maize suspension. Error bars indicate the standard deviation of six determinations. Yellow bars indicate the average color distance ($D_{i,j}$), as calculated from equation 2, with respect to the *Lab* values corresponding to the sample at pH = 10.6. A dotted red line indicates the decreasing trend in $D_{i,j}$ values when the sample at pH = 10.6 is taken as a reference (or final point). (B) The same determination was conducted using the HSB scale of colors. Yellow bars indicate the average color distance ($D_{i,j}$), as calculated from equation 2, with respect to the *Lab* values corresponding to the sample at pH = 10.6. A dotted red line indicates the decreasing trend in $D_{i,j}$ values when the sample at pH = 10.6 is taken as a reference (or final point). (C) Plot of $D_{i,j}$ values in the CIE*Lab* scale of colors versus pH for a pH trajectory from pH 1.3 to 10.6. (D) Plot of $D_{i,j}$ values in the HSB color system versus pH for a pH trajectory from pH 1.3 to 10.6.

conducted (in the *Lab* scale) and the average and standard deviation of each color component (*L*, *a*, and *b*) was calculated. In

addition, for each color reading, the distance in color ($D_{i,j}$) with respect to the average color corresponding to pH = 10.6 (assumed arbitrarily as a final point or reference point) was calculated using equation 2. The corresponding analysis is presented in Figure 8. The standard deviation of each component of each color is relatively low compared to variations due to pH. In addition, the standard deviation of $D_{i,j}$ for each particular color determination is also relatively low compared to the $D_{i,j}$ variation related to actual changes in pH. Moreover, in agreement with the analysis presented in Figure 7, the value of the distance $(D_{i,j})$ consistently decreases in the range from pH = 1.6 to 10.6.

When designing pH excursion experiments in this pH range, the injection location [36,43] and the direction of the pH trajectory are important considerations. The viscosity of blue maize suspensions increases substantially at high pH values (see Figure 3b and c). A poor selection of the point of injection of a basic pulse might lead to the creation of a basic locus that would further obstruct the tracer dispersion and the overall mixing process. On the other hand, to initiate a mixing experiment from a basic condition, injecting an acid pulse will demand higher agitation rates due to the higher viscosity of the initial condition.

Feasibility of use of other color systems: RGB and HSB

We have explored the feasibility of using other scales of color to perform DCA in blue maize flour suspension mixing experiments.

In particular, we considered the use of the RGB system (widely used in digital devices, computer screens, and Photoshop, among others) and the HSB scale (preferred in the context of art). In the range of pH from 1.6 (starting point) to 10.6 (final mixing state), the HSB color space behaves similarly to CIELab. The distance $D_{i,j}$ decreases as pH decreases in this range for both the CIELab and the HSB system (Figure 7; Figure 8a). This is not the case if RGB is used: in the pH trajectory from 1.6 to 10.6, the $D_{i,j}$ value does not always decrease as the mixing process advances. Indeed, our results suggest that the relationship $D_{i,j}$ vs. pH is more linear (for this pH range of pH and color) in the HSB space of color than in the CIELab space (compare Figure 7c and 7d).

Although the strategy presented here also works effectively in HSB, in this contribution we mainly use the CiELab system to illustrate the analysis. The literature refers to several attributes/advantages of the CIELab space of color. For example, it includes all perceivable colors, which means that its gamut exceeds those of other color models (i.e., the RGB space used by ProPhoto includes about 90% all perceivable colors). The CIELab color space is also considered more perceptually uniform compared to other systems, meaning that a change of the same amount in a color value should produce a change of about the same visual importance. In addition, the CIELab model is considered to be device indepen-

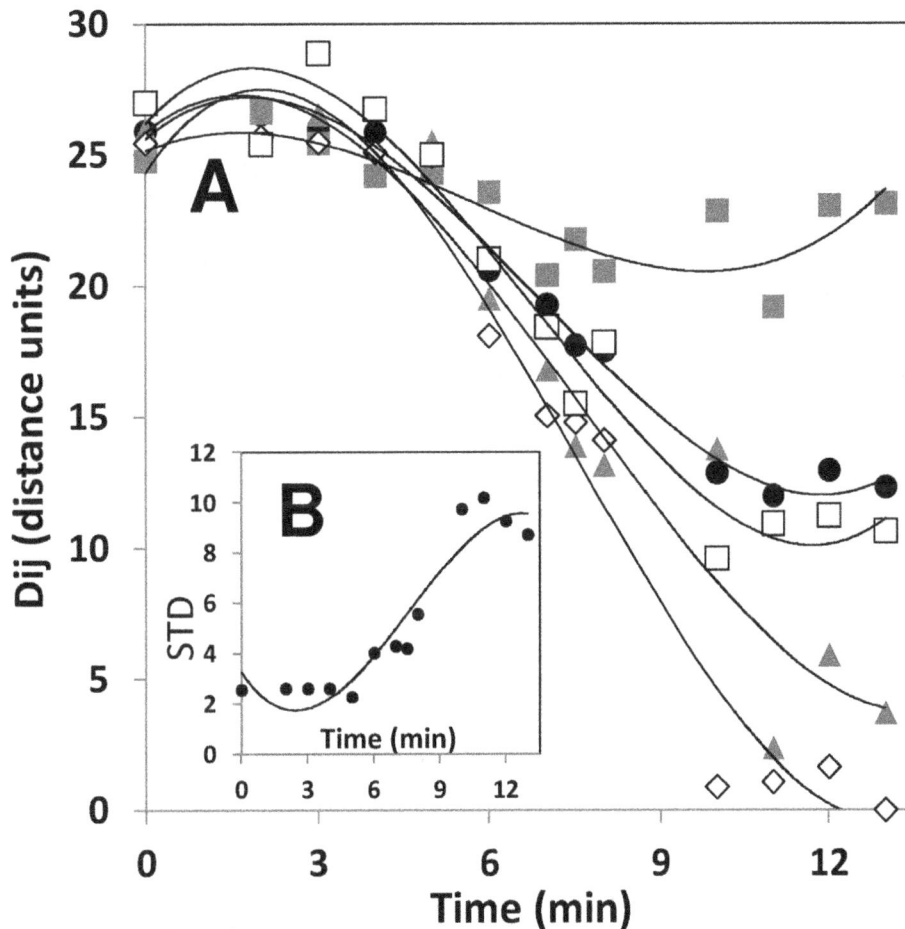

Figure 8. Evaluation of mixing progression using distances between colors. (A) Evolution of the distance in color, based on the CIELab color space, with respect to the final mixing state ($D_{i,j}$) at different tank locations for the experiment in Figure 5: at U2 (■); at A2 (▲); B2 (◇); L2 (□). The average distance with respect to the final point, considering all sampling points, at different agitation times is also presented (●). Lines indicate polynomial fits to data. (B) Time evolution of the standard deviation of all the $D_{i,j}$ values (corresponding to the same time point); a direct indicator of the degree of heterogeneity in the mixing conditions within the vessel. Lines indicate polynomial fits to data.

dent and it has been more frequently used than HSB has in the
scientific literature.

Example of application: following mixing in transparent vessels

Let us consider a sample taken at a time i and a location j (for
example, $t = 8$ min and location B3 in Figure 5). The color at this
sample point in the Lab scale, as determined by analysis using the
Color Companion 4.0 for iPad application, was $L_{i,j} = 59.2$,
$a_{i,j} = 0.4$, $b_{i,j} = 4.9$. By visual inspection, we can approximate a
final state for the mixing process. For example, here, the average
of the Lab values at the final time point ($t = 13$ min; Figure 5) at
the location B2 (i.e., $L_f = 56.3$, $a_f = -4.1$, $b_f = 18.3$) was considered
to approximate a desirable well-mixed condition. For these two
Lab value points, the value of the $D_{8min,B3}$ to final point is 14.4.
For the set of images in Figure 5, we calculated the distance in the
Lab color space ($D_{i,j}$), as defined by equation 2, for each one of
sixteen reading positions (U1 to L4) and 13 time points considered
in the experiment. $D_{i,j}$ was calculated with respect to the Lab color
values at the location B2, at $t = 13$ min. The average distance of
these $D_{i,j}$ values is physically related to the global deviation of the
system with respect to the final state of mixedness (Figure 8a). In
addition, the standard deviation of all $D_{i,j}$ values, corresponding to
the same time point, is a direct indicator of the degree of
heterogeneity in the mixing conditions within the vessel. Consistent with a simple visual analysis from the images, the reader will
observe that, in this particular experiment, the system approaches
a final state of mixedness in which segregation (indicated by the
STD value) is more prevalent than at the initial condition
(Figure 8b).

Extension to non-transparent vessels

This strategy for mixing analysis based on DCA can be
extended to non-transparent vessels. Let us consider a scenario in
which the initial pH of the suspension is set to 6.5 by slow addition
of 3N HCl through the port located at 8 o'clock (injection location
8U; see Figure 1). Then, a pH target point of 8.5 was set, and the
control system was activated to deliver small pulses of a basic
solution (1N NaOH solution) until the new pH set point were
achieved. This time, the impeller axis has been displaced to an
eccentric position ($E = 0.45$). Off-centered agitated stirred tanks
have been described as an effective alternative for mixing in the
laminar regime [36,37,44,45]. Here we use eccentricity to improve
mixing performance in our blue maize flour suspension system.
We followed the evolution of mixing in this experiment during
several hours of agitation. Again, the color of the suspension varies
as the pH progressively changes. At five different tank locations (in
this example 12M, 5U, 3L, 1U, 7U), we took 3 mL samples of the
suspension and dispensed them in the wells of commercial 6-well
culture plates, as shown in Figure 2c and 2d. The color of each of
these samples was evaluated using a colorimeter (Chroma Meter
CR-300 from Minolta, NJ, USA). Although the measurement can
be taken from above or below each well, best results are obtained
by placing the colorimeter underneath the samples, directly in
contact with the bottom surface of each well.

Once more, mixing can be followed by evaluating the distances
between mixing states in the Lab coordinate system. Let us
consider a sample taken at a time i and a location j (for example,
$t = 40$ min and location 12M in Figure 1b). The color of this
sample in the Lab scale, as determined by the colorimeter reading,
was $L_{i,j} = 24.63$, $a_{i,j} = -0.61$, $b_{i,j} = 2.78$. At the final point of the
experiment, which approximates a well-mixed condition, the
corresponding average Lab vector (constructed from the average
value of L, a, and b for all five sampling locations at $t = 13$ min)

was $L_f = 22.992$, $a_f = -1.398$, $b_f = 1.05$. For these two points, the
distance in the Lab space, defined by a straight line connecting
both points, can be calculated by equation 2. Here, $D_{(i,j \text{ to final point})}$
is the distance, in the Lab space, of the points defined by the Lab
coordinates of the sample taken at time i and location j ($L_{i,j}$, $a_{i,j}$,
$b_{i,j}$) and a sample representative of the final mixing state (L_f, a_f, b_f).
For the case under consideration, the value of the $D_{40min,12M \text{ to final point}}$ is 2.5094. For this very same time point, the other distance
values in the Lab space, corresponding to other sampling points
are $D_{40min7U \text{ to final point}}$ is 17.022, $D_{40min,1U \text{ to final point}}$ is 1.7953,
$D_{40min,5U \text{ to final point}}$ is 8.1427, and $D_{40min,3L \text{ to final point}}$ is 1.3766.
The average distance of these distance values is physically related
to the global deviation of the system with respect to the final state
of mixedness (Figure 9a). In addition, the standard deviation of
these values, corresponding to the same time point, is a direct
indicator of the degree of heterogeneity in the mixing conditions
within the vessel (Figure 9b).

Our interest resides in the experimental zone of high density
slurries. Within this region, the trend of diminishing $D_{i,j}$ as pH
varies from an acidic condition to a basic set point (i.e. pH 10.6)
still holds, but the slope is slightly different for slurries of different
concentrations. The slope variation is mainly based on variations
in illumination. In the CIELab space, the intensity or luminosity of
a particular color is mostly related to the L value. In general, at
higher solid concentrations, the L value decreases for all pH
values, while a and b are weaker functions of the concentration of
solids in the slurry.

Blue maize suspensions as a model: discussion on advantages and limitations

The number of non-Newtonian fluid models available to study
complex flows is still limited. Mixing studies performed using non-
Newtonian suspensions are even less available in the fluid
mechanics or physics literature. We present a non-Newtonian
suspension fluid system that can be used to mimic real shear
thinning suspensions frequently found in biotechnology and food
processing applications (e.g., yogurt manufacture, mycelial cell
culture, wastewater treatment scenarios). Moreover, in some of
these applications, the microorganisms under culture produce and
release metabolic products that lower or increase the pH value.
Frequently, the pH needs to be controlled by the addition of acidic
or basic solutions. The effectiveness of the strategies to control the
pH of the culture medium strongly depends on the mixing
conditions, and particularly on the proper selection of the acid/
base injection point.

Interestingly, blue maize suspensions not only undergo changes
in color as the pH value changes, but also in viscosity. Depending
on the purpose of a mixing experiment, this attribute of the system
can be regarded either as an important disadvantage or a useful
property. As pH affects fluid rheology, the measuring attribute
(variation of pH in time and space after a pH disturbance in a
given location) is bound to affect the flow field of the system under
investigation, and therefore the quantity that was meant to be
measured. As a consequence, the measurements obtained by acid/
base injections are closely linked to the actual disturbance story,
which depends on the coupled pH/flow-field dynamics. Consequently, if the purpose of the experiment is to analyze the process
of dispersion of an inert tracer, the range of variation in pH during
the experiment has to be carefully selected to avoid regimes where
the rheology is significantly modified by pH.

Figure 3b and 3c shows the dependence of the apparent
viscosity of a blue maize suspension (50% w/w solids) with respect
to strain rate at six different pH values. Viscosity is a moderate to
weak function of pH at high strain rate values (high fluid velocity

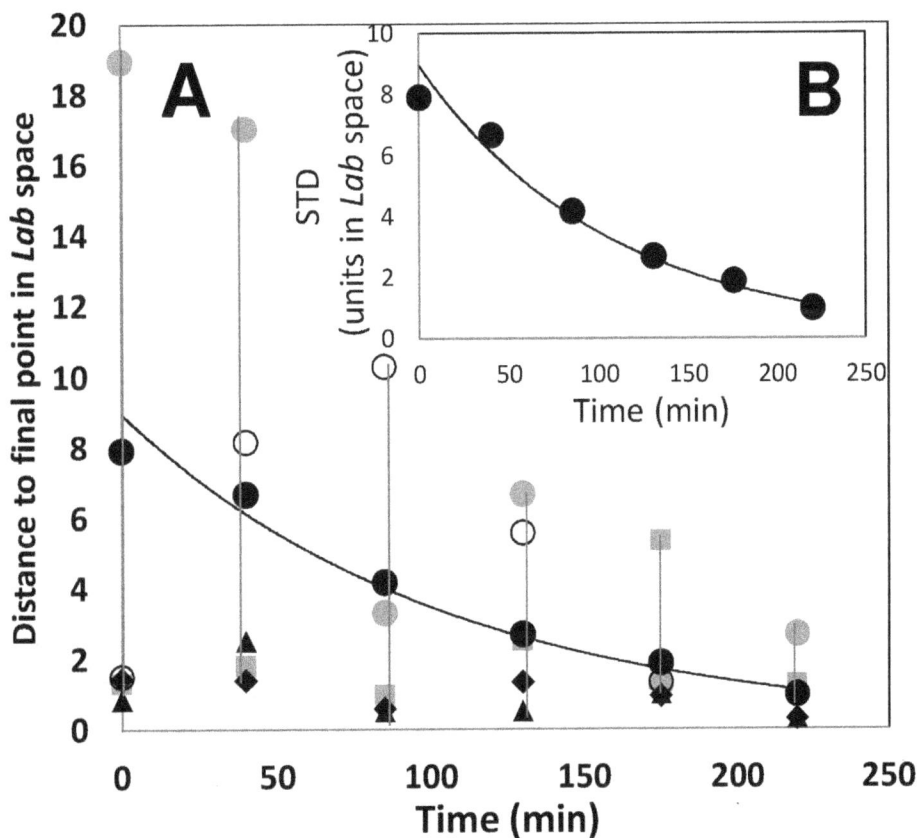

Figure 9. Quantitation of mixing through the concept of distances between mixedness states in the *Lab* color space. (A) Evolution of the average distance ($D_{i,j}$) with respect to an ideal mixing state condition for five different sampling locations defined within the tank volume (see Figure 2): 7U (●); 1U (■); 12M (▲); 5U (○); and 3L (◆). The average distance with respect to the final point is marked with black circles (●); a second order polynomial fitting is shown (solid curve). (B) Evolution of the standard deviation from an initially segregated condition to a homogeneous final state (●). A second order polynomial fitting is shown (solid curve).

values). At medium to low strain rates (Figure 3a), viscosity exhibits a moderate dependence on pH in the range of pH values from 1.6 to 8.0. Above pH 8.0, the dependence of apparent viscosity on pH is significant. Conveniently, as previously discussed, the pH range from 1.6 to 10.6 is also suitable for the evaluation of mixing evolution based on the calculation of color distances in the CIE*Lab* color space. Therefore, in this pH range (1.6–10.6), a variety of mixing experiments can be designed to study different mixing aspects. For example, in an initially weakly basic environment (pH 7.0 or 8.0), acid injection experiments will allow the study of mixing evolution to acidic mixing states (pH 3.3–1.6). This pH range, from 8.0 to 3.0, is relevant in many biological systems. Moreover, this type of experiment allows a good resolution for evaluation of "color distances" in a system in which rheology is practically independent of pH.

The fact that the rheology of this system is a strong function of pH also makes it a versatile model for studying certain aspects of mixing in non-Newtonian fluids. Conveniently, in experiments run at constant pH, the viscosity of blue maize suspensions can be tuned (by adjusting the pH) to mimic non-Newtonian liquids of different viscosities. Moreover, many Non-Newtonian fluids do exhibit a rheology dependence on pH. Examples include yogurt [46]; carbopol solutions [47]; pectin hydrogels [48]; the gums and starch suspensions widely used in food technology applications[49]; protein suspensions of animal origin [50]; poly(acrylic acid) microgels [51]; newly designed surfactants with customized

architectures [52,53]; and even physiologically relevant fluids such as the gastric mucin [54].

As we have discussed before, in biotechnology and food processing applications, the basic or acid injection is frequently used to control the pH of the suspension. In such scenarios, an important aspect to consider is the injection location. As illustrated before (see Figure 4d), experiments can be designed using the blue maize suspension system to investigate the adequacy of an injection location. For example, consider an experiment such that the initial pH of the suspension is set in an acidic value (i.e., 1.6–3.3) and the final mixing point is set around pH 7.0 or 8.0. Injection of a basic solution will allow the pH evolution to the final acidic environment to be followed. If the basic pulse is not effectively dispersed due to the selection of a low circulation injection location, a high pH/high viscosity spot will develop and will be easily visible.

Conclusions and Final Remarks

In this work, we propose the use of blue maize suspensions as a model to study mixing in Non-Newtonian suspensions. The injection of pulses of acid or basic solutions was used to reveal mixing patterns and flow structures and to follow their time evolution. No foreign pH indicator was required, since blue maize naturally contains anthocyanins that act as a native, wide spectrum pH indicator. We describe a novel method to quantitate mixedness and mixing evolution through Dynamic Color Analysis (DCA).

Color readings corresponding to different times and locations within the mixing vessel were taken with a digital camera (or a colorimeter) and translated to the Hunter (or *Lab*) scale of colors using commercial image analysis software. In the *Lab* scale, a 3D color space, each color can be represented by a point, and the difference between two colors can be determined as a distance between two points. We demonstrate the use of distances in the *Lab* scale between a particular mixing state and the final mixing point to characterize segregation or mixing in the blue maize system.

Although a rigorous extension to other non-Newtonian systems cannot be directly made, in this contribution we propose a simple, illustrative, and realistic model for studying mixing in non-Newtonian scenarios. The advantages of this system include the possibility of rheology tuning through pH changes, or the use of acid-base injections to visually assess homogeneity and mixing evolution in time without the need of a foreign pH indicator.

Materials and Methods

The fluid model

A suspension of blue maize (BM) flour in water was used as liquid model. In our experiments we used BM kindly provided by Edomex (México). To prepare the suspension, 585 g of blue maize flour were dispersed in 1300 mL of distilled water at 500 RPM in a 3 L fully instrumented Applikon model EZ-control reactor (Applikon Biotechnology, Schiedam, the Netherlands). Curves of apparent viscosity versus strain rate in blue maize suspensions, adjusted to different pH values using NaOH and HCl, were obtained using an automatic Rheometer (Physica MCR 101, Anton Paar, Austria). After pH adjustment, samples were equilibrated at room temperature for 15 min, and then agitated with a spatula to homogenize them. For the determinations reported here, the cone and plate geometry (diameter 5 mm, angle $1°$) was used. The gap between cone and plate was set to 1 mm. Flow curves were determined using a steady-state flow ramp in the range of shear rate from 1 to 20 s^{-1} and 20 to 1000 s^{-1}. The shear rate was measured point and point with consecutive 20 s steps of constant shear rate. The viscosity was recorded for each point to obtain the flow curves. All measurements were performed in triplicate and average data was reported. An Ostwald-de Waele power-law model $[\eta = K (\gamma)^{n-1}]$, commonly used to model flour formulations [29–31], was used to describe the rheology of blue maize suspensions at different pH values. K and n were calculated from linear regressions of the type $\ln \eta = \ln K + (n-1) \ln (\gamma)$.

Stirred tank system

We used an instrumented stirred tank bioreactor (Applikon, Netherlands) with a working volume of 1.0 to 1.7 L, equipped with a variable speed and pH control system (Figure 1a). The agitator system consisted in a $45°$ inclined disc impeller [36,37]. The tank

lid was modified to allow for the displacement of the shaft to eccentric positions (if needed), and ports for sampling and dispensing of materials were placed at different angular coordinates, as shown in Figure 1b.

Visualization experiments

The 3D mixing patterns within the reactor were revealed using basic injections on an initially acidic environment. A low pH initial state was established by adding a hydrochloric acid solution (3N HCl) dropwise at the tank surface until the pH value reached a set point of 4.5. From the initial acidic condition, the excursion of pH induced by injection of a sodium hydroxide solution (1N NaOH) allowed to visualize the advance of the mixing process. For this purpose, the pH control system was activated establishing a new pH set value of 8.5.

The evolution of the mixing process was followed using two different strategies. In the first strategy, frontal photographic images of the entire tank volume were taken at different times using a professional digital camera (Canon Rebel XTi) mounted in a tripod. Special care was taken to maintain an approximately constant illumination quality and camera position at the different time-shots. Images were used to analyze the color evolution within different points of the vessel (Figure 2b). Images were analyzed using the application Color Companion 4.0 for iPad from Digital Media Interactive LCC, USA. In the second strategy, samples from five different locations within the tank volume were taken with a 30 cm pipet at different time points. Samples of 3–5 mL were dispensed in 6-well transparent cell culture plates (Corning, USA). Photographic images of these samples were taken using a professional digital camera (Canon Rebel XTi) mounted in a tripod (Figure 2c and 2d). The color of these samples was determined by image analysis using the application Color Companion 4.0 for iPad from Digital Media Interactive LCC, USA. Alternatively, a colorimeter (Chroma Meter CR-300 from Minolta, NJ, USA) was used to determine color at each well. The color reading was taken by placing the colorimeter exactly underneath each well.

In the experiments used to illustrate the concept of characterization of mixing progress using distances in a color space (Figure 5, 6, 8, and 9) we consistently worked in the range of pH values between 1.6 and 10.6. In the experiments used to reveal mixing pathologies (Figure 4) we did not strictly control the pH variation range.

Author Contributions

Conceived and designed the experiments: GTdS MMA. Performed the experiments: GTdS ABE MMA. Analyzed the data: GTdS MMA. Contributed reagents/materials/analysis tools: MMA CRdG SGL. Wrote the paper: GTdS MMA.

References

1. Fayolle F, Belhamri R, Flick D (2013) Residence Time Distribution measurements and simulation of the flow pattern in a scraped surface heat exchanger during crystallization of ice cream. J. Food Eng. 116: 390–397.
2. Migliori M, Correra S (2012) Modelling of dough formation process and structure evolution during farinograph test. Int. J. Food Sci. Technol. 48: 121–127.
3. Bouvier L, Moreau A, Line A, Faqtah N, Delaplace G (2010) Damage in Agitated Vessels of Large Visco-Elastic Particles Dispersed in a Highly Viscous Fluid. J. Food Sci. 76: E384–E391.
4. Domingues N, Gaspar-Cunha A, Covas JA (2012) A quantitive aproach to assess the mixing ability of single-screw extruders for polymer extrusión. J. Polym. Eng. 32: 81–94.
5. Wong AC-Y, Lam Y, Wong ACM (2009) Quantification of dynamic mixing performance of single screws of different configurations by visualization and image analysis. Adv. Polym. Tech. 28: 1–15.
6. Olmos E, Mehmood N, Husein LH, Goergen JL, Fick M, et al. (2013). Effects of bioreactor hydrodynamics on the physiology of Streptomyces. Bioprocess. Biosyst. Eng. 36: 259–272.
7. Fubin Y, Zifu L, Huanhuan M, Lei X, Xiaofeng B, Yi Y (2013) Experimental Study on rheological characteristics of high solid content sludge and its mesophilic anaerobic digestion. J. Renewable Sustainable Energy 5: art number 043117.
8. Gabelle J-C, Jourdier E, Licht RB, Ben Chaabane F, Henaut I, et al. (2012) Impact of rheology on the mass transfer coefficient during the growth phase of Trichoderma reesei in stirred bioreactors. Chem. Eng. Sci. 75: 408–417.

9. Zhang HL, Baeyens J, Tan TW (2012) Mixing phenomena in a large-scale fermenter of starch to bio-etanol. Energy 48: 380–391.

10. Craig KJ, Nieuwoudt MN, Niemand LJ (2013) CFD simulation of anaerobic digester with variable sewage sludge rheology. Water Res. 47: 4485–4497.

11. Chen Y, Wang Z, Chu J, Zhuang Y, Zhang S, et al. (2013) Significant decrease of broth viscosity and glucose consumption in erythromycin fermentation by dynamic regulation of ammonium sulfate and phosphate. Bioresour. Technol. 134: 173–179.

12. Patel D, Ein-Mozaffari F, Mehrvar M (2014) Tomography images to analyze the deformation of the cavern in the continuous-flow mixing of non-Newtonian fluids. AIChE J. 60: 315–331.

13. Alberini F, Simmons MJH, Ingram A, Stitt EH (2014) Use of areal distribution of mixing intensity to describe blending of non-newtonian fluids in a kenics KM static mixer using PLIF. AIChE J. 60: 332–342.

14. Maingonnat JF, Doublier JL, Lefebvre J, Delaplace G (2008) Power consumption of a double ribbon impeller with Newtonian and shear thinning fluids and during the gelation of an iota-carrageenan solution. J. Food Eng. 87: 82–90.

15. Arratia PE, Shinbrot T, Alvarez MM, Muzzio FJ (2005) Mixing of Non-Newtonian Fluids in Steadily Forced Systems. Phys. Rev. Lett. 94: 084501.

16. Tamburini A, Cipollina A, Micale G, Brucato A (2013) Particle distribution in dilute solid liquid unbaffled tanks via a novel laser sheet and image analysis based technique. Chem. Eng. Sci. 87: 341–358.

17. Guida A, Nienow AW, Barigou M (2011) Mixing of dense binary suspensions: multi-component hydrodynamics and spatial phase distribution by PEPT. AIChE J. 57: 2302–2315.

18. Ryu SH, Werth L, Nelson S, Scheerens JC, Pratt RC (2013) Variation of Kernel Anthocyanin and Carotenoid Pigment Content in USA/Mexico Borderland Land Races of Maize. Econ. Bot. 67: 98–109.

19. De la Rosa-MIllán J, Agama-Acevedo E, Jimenez-Aparicio AR, Bello-Pérez LA (2010) Starch characterization of different blue maize varieties. Starch/Staerke 62: 549–557.

20. Urias-Peraldí M, Gutiérrez-Uribe JA, Preciado-Ortiz RE, Cruz-Morales AS, Serna-Saldívar SO, et al. (2013) Nutraceutical profiles of improved blue maize (Zea mays) hybrids for subtropical regions. Field Crop Res.141: 69–76.

21. Castañeda-Ovando A, Galán-Vidal CA, Pacheco L, Rodríguez JA, Páez-Hernández ME (2010) Characterization of main anthocyanins extracted from pericarp blue corn bt Maldi-TOF MS. Food Anal. Methods 3: 12–16.

22. Chang G, Wu J, Jiang C, Tian G, Wu Q, et al. (2014) The relationship of oxygen uptake rate and k_{La} with rheological properties in high cell density cultivation of docosahexaenoic acid by Schizochytrium sp. S31. Bioresour. Technol. 152: 234–240.

23. Dhillon GS, Brar SK, Kaur S, Verma M (2013) Rheological Studies During Submerged Citric Acid Fermentation by Aspergillus niger in Stirred Fermentor Using Apple Pomace Ultrafiltration Sludge. Food Bioprocess Technol. 6: 1240–1250.

24. Aguirre-Ezkauriatza EJ, Galarza-González MG, Uribe-Bujanda AI, Ríos-Licea M, López-Pacheco F, et al. (2008) Effect of mixing during fermentation in yogurt manufacturing. J. Dairy Sci. 91: 4454–4465.

25. Chen S, Liu J, Liu Y, Su H, Hong Y, et al. (2012) An aie-active hemicyanine fluorogen with stimuli-responsive red/blue emission: extending the ph sensing range by "switch + knob" effect. Chem. Sci. 3: 1804–1809.

26. Wolfbeis OS, Marhold H (1987) A new group of fluorescent ph-indicators for an extended ph-range. Fresenius' Zeitschrift für Analytische Chemie Fresenius J. Anal. Chem. 327: 347–350.

27. Nguyen TH, Venugopala T, Chen S, Sun T, Grattan KTV, et al. (2014). Fluorescence based fibre optic ph sensor for the ph 10–13 range suitable for corrosion monitoring in concrete structures. Sens. Actuators, B 191: 498–507.

28. King DW, Kester DR (1989) Determination of seawater pH from 1.5 to 8.5 using colorimetric indicators. Mar. Chem. 26: 5–20.

29. Núñez-Santiago MC, Santoyo E, Bello-Pérez LA, Santoyo-Gutiérrez S (2010) Rheological evaluation of non-Newtonian Mexican nixtamalized maize and dry processed masa flours. J. Food Eng. 60: 55–56.

30. Trèche CMS (2001) Viscosity of gruels for infants: a comparison of measurement procedures. Int. J. Food Sci. Nutr. 52: 389–400.

31. Méndez-Montealvo G, García-Suárez FJ, Paredes-López O, Bello-Pérez LA (2008) Effect of nixtamalization on morphological and rheological characteristics of maize starch. J. Cereal Sci. 48: 420–425.

32. Alvarez MM (2000) Using spatio-temporal asymmetry to enhance mixing in chaotic flows: from maps to stirred tanks. Ph. D. Thesis. Rutgers, the State University of New Jersey, New Brunswick, USA.

33. Soos M, Moussa AS, Ehrl L, Sefcik J, Wu H, Morbidelli M (2008) Effect of shear rate on aggregate size and morphology investigated under turbulent conditions in stirred tanks. J. Colloid Interface Sci. 319: 577–589.

34. Sánchez-Pérez JA, Rodríguez-Porcel EM, Casas-López JL, Fernández-Sevilla JM, Chisti Y (2006) Shear rate in stirred tank and bubble column bioreactors. Chem. Eng. J. 124: 1–5.

35. Kelly W, Gigas B (2003) Using CFD to predict the behavior of power law fluids near axial-flow impellers operating in the transitional flow regime. Chem. Eng. Sci. 58: 2141–2152.

36. Bulnes Abundis D, Alvarez MM (2013) The simplest stirred tank for laminar mixing: mixing in a vessel agitated by an off-centered angled disc. AIChE J. 59: 3092–3108.

37. Bulnes-Abundis D, Carrillo-Cocom LM, Aráiz-Hernández D, García-Ulloa A, Granados-Pastor M, et al. (2013) A simple eccentric stirred tank mini-bioreactor: Mixing characterization and mammalian cell culture experiments. Biotechnol. Bioeng. 110: 1106–1118.

38. Machado MB, Bittorf KJ, Roussinova VT, Kresta SM (2013) Transition form turbulent to transitional flow in the top half of a stirred tank. Chem. Eng. Sci. 98: 218–230.

39. Bryant CM, Hamaker BR (1997) Effect of Lime on Gelatinization of Corn Flour and Starch. Cereal Chem. 74: 171–175.

40. Afshari-Jouybari H, Farahnaky A (2011) Evaluation of photoshop software potential for food colorimetry. J. Food Eng. 106: 170–175.

41. Lei H, Ruan R, Gary Fulcher R, van Lengerich B (2008) Color development in an extrusion-cooked model system. IJABE 1: 55–63.

42. Suzuki K, Hirayama E, Sugiyama T, Yasuda K, Okabe H, et al. (2002) Ionophore-Based Lithium Ion Film Optode Realizing Multiple Color Variations Utilizing Digital Color Analysis. Anal. Chem. 74: 5766–5773.

43. Bhattacharya S, Kresta SM (2004) Surface feed with minimum by-product formation for competitive reactions. Chem. Eng. Res. Des. 82: 1153–1160.

44. Sánchez-Cervantes MI, Lacombe J, Muzzio FJ, Alvarez MM (2006) Novel bioreactor design for the culture of suspended mammalian cells. Part 1: Mixing characterization. Chem. Eng. Sci. 61: 8075–8084.

45. Alvarez MM, Arratia PE, Muzzio FJ (2002) Laminar mixing in eccentric stirred tank systems. Can. J. Chem. Eng. 80: 546–557.

46. Karsheva M, Paskov V, Tropcheva R, Georgieva R, Danova S (2013) Physicochemical parameters and rheological properties of yoghurts during the storage. J. Chem. Technol. Biotechnol. 48: 483–488.

47. Curran SJ, Hayes RE, Afacan A, Williams MC, Tanguy PA (2002) Properties of carbopol solutions as models for yield-stress fluids. J. Food Sci. 67: 176–180.

48. Moreira HR, Munarin F, Gentilini R, Visai L, Granja PL, et al. (2014) Injectable pectin hydrogels produced by internal gelation: pH dependence of gelling and rheological properties. Carbohydr. Polym. 103: 339–347.

49. Rao MA (2014) Rheology of food gum and starch dispersions. In Rheology of Fluid, Semisolid, and Solid Foods. Springer US: 161–229.

50. Liu Q, Bao H, Xi C, Miao H (2014) Rheological characterization of tuna myofibrillar protein in linear and nonlinear viscoelastic regions. J. Food Eng. 121: 58–63.

51. Harrington JC (2012) The effects of neutralization on the dynamic rheology of Polyelectrolyte microgel mucilages. J. Appl. Polym. Sci. 126: 770–777.

52. Shi H, Ge W, Wang Y, Fang B, Huggins JT, et al. (2014) A drag reducing surfactant threadlike micelle system with unusual rheological responses to pH. J. Colloid Interface Sci. 418: 95–102.

53. Zhang ZX, Ni X, Li J (2014) Cationic Brush-Like Terpolymer with pH Responsive Thickening Behavior in Surfactant System. Polym. Int. (in Press).

54. Celli JP, Turner BS, Afdhal NH, Ewoldt RH, McKinley GH, et al. (2007) Rheology of gastric mucin exhibits a pH-dependent sol-gel transition. Biomacromolecules 8: 1580–1586.

25

Dietary Patterns of Korean Adults and the Prevalence of Metabolic Syndrome

Hae Dong Woo[1], Aesun Shin[1,2], Jeongseon Kim[1]*

1 Molecular Epidemiology Branch, National Cancer Center, Goyang-si, Korea, **2** Department of Preventive Medicine, Seoul National University College of Medicine, Seoul, Republic of Korea

Abstract

The prevalence of metabolic syndrome has been increasing in Korea and has been associated with dietary habits. The aim of our study was to identify the relationship between dietary patterns and the prevalence of metabolic syndrome. Using a validated food frequency questionnaire, we employed a cross-sectional design to assess the dietary intake of 1257 Korean adults aged 31 to 70 years. To determine the participants' dietary patterns, we considered 37 predefined food groups in principal components analysis. Metabolic syndrome was defined according to the National Cholesterol Education Program Adult Treatment Panel III. The abdominal obesity criterion was modified using Asian guidelines. Prevalence ratios and 95% confidence intervals for the metabolic syndrome were calculated across the quartiles of dietary pattern scores using log binomial regression models. The covariates used in the model were age, sex, total energy intake, tobacco intake, alcohol consumption, and physical activity. The prevalence of metabolic syndrome was 19.8% in men and 14.1% in women. The PCA identified three distinct dietary patterns: the 'traditional' pattern, the 'meat' pattern, and the 'snack' pattern. There was an association of increasing waist circumference and body mass index with increasing score in the meat dietary pattern. The multivariate-adjusted prevalence ratio of metabolic syndrome for the highest quartile of the meat pattern in comparison with the lowest quartile was 1.47 (95% CI: 1.00–2.15, p for trend = 0.016). A positive association between the prevalence of metabolic syndrome and the dietary pattern score was found only for men with the meat dietary pattern (2.15, 95% CI: 1.10–4.21, p for trend = 0.005). The traditional pattern and the snack pattern were not associated with an increased prevalence of metabolic syndrome. The meat dietary pattern was associated with a higher prevalence of metabolic syndrome in Korean male adults.

Editor: Vineet Gupta, University of Pittsburgh Medical Center, United States of America

Funding: This study was supported by a grant from the National Cancer Center, Korea (no. 1210141). The funders had no role in study design, data collection and analysis, decision to publish, or preparation of the manuscript.

Competing Interests: The authors have declared that no competing interests exist.

* Email: jskim@ncc.re.kr

Introduction

Metabolic syndrome is associated with an increased risk of developing type 2 diabetes [1] and cardiovascular disease, as well as general mortality [2,3]. According to the National Health and Nutrition Examination Survey (NHANES), using the revised National Cholesterol Education Program/Adult Treatment Panel III (ATP III) definition, the age-adjusted prevalence of metabolic syndrome in the US adult Americans significantly increased from 29.2% between 1988 and 1994 to 34.2% between 1999 and 2006 [4]. The age-adjusted prevalence of metabolic syndrome was 13.7% between 2000 and 2001, and prevalence of metabolic syndrome was 26.7% between 2007 and 2008, using ATP III criteria modified for the Asia-Pacific subjects in China [5,6]. Based on the Korean National Health and Nutrition Examination Survey (KNHANES), using the ATP III criteria from the Asia-Pacific region for central obesity, the age-adjusted prevalence of metabolic syndrome in the Korean population increased from 24.9% in 1998 to 31.3% in 2007 [7]. Trends in prevalence of diabetes in Asian countries have been increased considerably [8], although no significant change has been observed in recent years [9]. Metabolic syndrome risk factors might be closely related to

diabetes and cardiovascular disease. Thus their potential causative factors need to be explored.

The risk of metabolic syndrome is known to be associated with dietary intake [10–12]. Analysis of dietary patterns could account for the inter-related dietary factors that are potentially important for the development of metabolic syndrome [13]. The dietary patterns of people in developing countries have changed as a result of modernization, which might contribute to an increased risk of obesity and metabolic syndrome [14,15]. Moreover, migration studies have shown that western dietary patterns lead to an accumulation of fat [16,17], which may contribute to the development of metabolic syndrome.

In previous studies, Western pattern characterized by high intakes of protein, processed foods, and refined grains was positively associated with metabolic syndrome, whereas healthy dietary pattern characterized by high intakes of fruits, vegetables and dairy was inversely associated with metabolic syndrome [11,12,18–21]. Recently, several studies were conducted to determine the association between dietary patterns and the prevalence of metabolic syndrome among Koreans [22–24]. The traditional Korean meal, which is low in fat and contains a large portion of vegetables, has been considered a healthy diet [14,25]. However, the traditional Korean dietary pattern was not associated with a lower prevalence of metabolic syndrome, and findings regarding the relationship between dietary pattern and metabolic syndrome have been inconsistent [22–24]. The association between dietary patterns and the metabolic syndrome has not been fully identified in the Korean population. Thus, the purpose of this study is to determine the association between various dietary patterns and the prevalence of metabolic syndrome in Korea.

Methods

Study population

We performed a cross-sectional study of participants who underwent health screening examinations at the Center for Cancer Prevention and Detection at the National Cancer Center in South Korea between October 2007 and December 2009. Visitors are National Health Insurance beneficiaries and those who have all data for survey question including medical history, clinical test result, and dietary consumption data were recruited (n = 2146). No one was excluded due to other diseases such as diabetes, coronary heart disease, stroke, or cancer. We excluded 862 subjects whose medical records lacked data regarding metabolic syndrome components. There was no significant difference in BMI between participants with missing metabolic syndrome components and participants with complete data. Participants with implausible energy intake values (<500 or ≥ 5000 kcal, n = 27) were excluded. The remaining 1257 adults (486 men and 771 women), ranging between 31 and 70 years old, were used in our analysis (Figure 1). Each participant was provided with an informed consent form according to the procedures approved by the institutional review board of the National Cancer Center. Written informed consent was obtained from all participants.

Data collection

The participants completed a self-administered questionnaire, which asked about each participant's demographics, lifestyle, medical history, and diet. Self-reported physical activity was evaluated using an International Physical Activity Questionnaire (IPAQ) short form [26]. The total metabolic equivalent (MET-minutes/week) was a combined score that was calculated by multiplying the frequency, duration, and intensity of physical activity. The dietary intake was assessed using a validated food

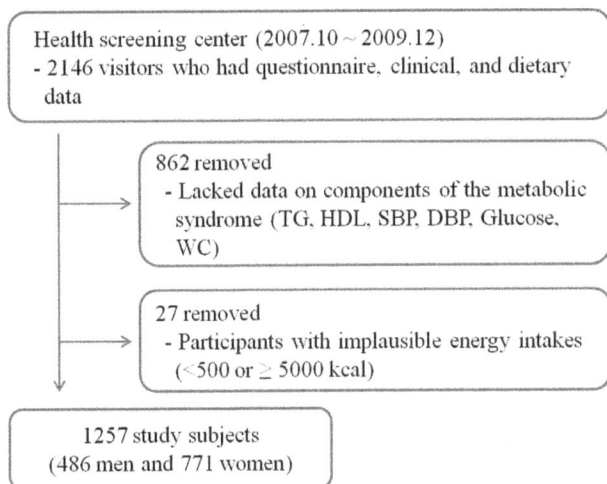

Figure 1. Flow chart of study selection process.

frequency questionnaire (FFQ) [27]. The participants were asked about their average frequency of intake and portion size of specific foods during the previous year of 103 types of food. Three portion sizes and 9 categories of frequency were specified on the FFQ. The average daily nutrient intake was measured by summing up the intake of associated nutrient content per 100 g for each of the 103 foods. The foods listed in the FFQ were categorized into 37 different food groups, each of which was determined according to the food's nutrient profile and its culinary use (Table S1).

Metabolic syndrome definition

Metabolic syndrome was defined according to the National Cholesterol Education Program Adult Treatment Panel III (NCEP-ATP III) [28]. The abdominal obesity criterion was modified using Asian guidelines [29]. Under these definitions, a person has metabolic syndrome if that person exhibits three or more of the following conditions: 1) triglycerides ≥150 mg/dL; 2) HDL cholesterol <40 mg/dL in men or <50 mg/dL in women; 3) systolic blood pressure (BP) ≥130 mmHg, diastolic blood pressure (BP) ≥85 mmHg, or drug treatment for hypertension; 4) fasting glucose ≥110 mg/dL or drug treatment for elevated glucose levels; and 5) waist circumference ≥90 cm in men or ≥ 80 cm in women.

Statistical analysis

The general characteristics in the group with metabolic syndrome and the group without metabolic syndrome were compared using a Student t-test for continuous variables and a chi-square test for categorical variables. Principal-components analysis (PROC FACTOR) was used to extract the participants' dietary patterns using 37 predefined food groups. We used a varimax rotation to enhance the interpretability of the factors that were analyzed. We determined how many factors to retain after evaluating the eigenvalue, scree test, and interpretability. The dietary patterns were named according to the highest factor driving the food groups for each dietary factor. Each dietary pattern's factor score was categorized by quartile for further analysis. The trend test was performed to analyze the associations between each of the dietary patterns and each of the components of metabolic syndrome using a general linear model with adjustments for confounding factors. Regression analysis in log-log scale of each dietary pattern score and the components of metabolic syndrome was performed to compare with trends across quartiles. Prevalence ratios (PRs) and 95% confidence intervals (CIs) for the metabolic syndrome were calculated across the quartiles of dietary pattern scores using log binomial regression models. The lowest quartile of each dietary pattern was used as the reference. The trend test was performed to analyze the associations between each of dietary pattern score (continuous) and the prevalence of metabolic syndrome using Wald test. PRs and 95% CIs for the metabolic syndrome were also calculated stratified for sex. Model 1 was adjusted according to age, sex (for total), and total energy intake. Model 2 was further adjusted for age, sex (for total), total energy intake, tobacco intake, alcohol consumption, and physical activity. We performed the statistical analysis using SAS version 9.3 (SAS Institute Inc, Cary, NC). All P values were two-tailed (α = 0.05).

Results

The general characteristics of the study participants are reported in Table 1. The overall prevalence of metabolic syndrome was 16.3%, and the prevalence was significantly higher in men (19.8%) than in women (14.1%). Compared to people who did not have metabolic syndrome, patients who had metabolic syndrome were older (p<0.001) and had a higher BMI (p<0.001). The separate components of metabolic syndrome were significantly different (metabolic syndrome positive vs. negative, Mean (SD): 137.2 (13.1) vs. 124.0 (13.6) for systolic BP, 83.7 (9.5) vs. 75.5 (10.0) for diastolic BP, 105.4 (26.1) vs. 90.7 (16.5) for fasting glucose, 46.9 (8.9) vs. 60.6 (13.9) for HDL cholesterol, 201.6 (81.8) vs. 102.4 (61.2) for triglycerides), depending on whether metabolic syndrome was present or absent (p<0.001).

The PCA identified three major dietary patterns, and their factor-loading scores are shown in Table 2. The 'traditional' dietary pattern included high intakes of condiments, green/yellow vegetables, light-colored vegetables, tubers, clams, tofu/soymilk, and seaweed; the 'meat' dietary pattern included high intakes of red meat, red meat byproducts, other seafood, and high-fat red meat; and the 'snack' pattern included high intakes of cake/pizza, snacks, and bread. Three patterns explained 31.9% of the total variance.

Table 3 shows the associations between each dietary pattern and the components of metabolic syndrome, including BMI, and general characteristics according to quartiles of each dietary pattern score. The traditional and snack patterns showed no relationships with any of the components of metabolic syndrome. Increasing scores in the meat dietary pattern were associated with elevated waist circumference, BMI, triglycerides, blood pressure, and low concentrations of HDL cholesterol (p<0.05). The traditional pattern score increased with an increment of age and physical activity of each quartile, but with decrease in percentage of men and current drinker. The meat pattern score increased with both increment of age and percentage of men and current drinker, but no difference was observed in physical activity. The snack pattern score increased with a decrement of age, but no differences were observed in percentage of men and current drinker, and physical activity.

The association between the PR of metabolic syndrome and the dietary pattern score variables are shown in Table 4. The score variable of the meat pattern was associated with the prevalence of metabolic syndrome in both models (p for trend = 0.006 and 0.016, respectively). The multivariate-adjusted PR of metabolic syndrome for the highest quartile of the meat pattern in comparison with the lowest quartile was 1.47 (95% CI: 1.00–2.15, p for trend = 0.016). We found no association between the

Table 1. General characteristics of the study population, and comparison of individuals with and without metabolic syndrome.

	Total	MS (+)	MS (−)	p value
N (%)	1257	205 (16.3)	1052 (83.7)	
Age (y)	51.7 (9.2)	55.9 (9.2)[†]	50.8 (9.0)	<0.001
Sex, n (%)				
Male	486 (38.7)	96 (19.8)	390 (80.3)	0.009
Female	771 (61.3)	109 (14.1)	662 (85.9)	
Smoking status, n (%)				
Never	774 (63.6)	114 (57.9)	660 (64.7)	0.164
Former	288 (23.7)	52 (26.4)	236 (23.1)	
Current	155 (12.7)	31 (15.7)	124 (12.2)	
Alcohol consumption, n (%)				
Never	472 (38.9)	78 (39.8)	394 (38.8)	0.943
Former	102 (8.4)	17 (8.7)	85 (8.4)	
Current	638 (52.6)	101 (51.5)	537 (52.9)	
Total energy intake (kcal)	2175.7 (857.3)	2282.6 (915.4)	2154.9 (844.4)	0.051
Saturated fatty acid (g/d)	7.9 (4.8)	8.2 (5.0)	7.9 (4.7)	0.987[‡]
Carbohydrates (g/d)	387.1 (163.1)	401.2 (165.3)	384.4 (162.6)	<0.001[‡]
Fiber (g/d)	34.3 (32.6)	37.4 (35.7)	33.7 (32.0)	0.563[‡]
Sodium (mg/d)	3207.0 (1675.9)	3623.8 (2198.5)	3125.8 (1542.0)	<0.001[‡]
Total vegetables (g/d)	177.5 (130.1)	195.9 (149.2)	173.9 (125.8)	0.114[‡]
Fruits (g/d)	221.5 (249.7)	210.3 (219.7)	223.6 (255.2)	0.193[‡]
BMI (Kg/m^2)	23.8 (3.0)	26.8 (2.6)	23.2 (2.7)	<0.001
Waist circumference (cm)	79.0 (8.8)	87.5 (6.7)	77.4 (8.2)	<0.001
Systolic BP (mmHg)	126.2 (14.4)	137.2 (13.1)	124.0 (13.6)	<0.001
Diastolic BP (mmHg)	76.8 (10.4)	83.7 (9.5)	75.5 (10.0)	<0.001
Fasting glucose (mg/dL)	93.1 (19.2)	105.4 (26.1)	90.7 (16.5)	<0.001
HDL cholesterol (mg/dL)	58.3 (14.1)	46.9 (8.9)	60.6 (13.9)	<0.001
Triglycerides (mg/dL)	118.6 (74.6)	201.6 (81.8)	102.4 (61.2)	<0.001
Physical activity (MET-min/wk)	2934.2 (2946.5)	2885.2 (2810.7)	2943.7 (2973.5)	0.795

MS (−), absence of metabolic syndrome; MS (+), presence of metabolic syndrome; MET, metabolic equivalent.
[†]Numbers are Mean (SD), unless otherwise stated.
[‡]Adjusted for total energy intake.

Table 2. Factor loadings of the dietary patterns derived from principal components analysis with orthogonal rotation.

	Traditional pattern	Meat pattern	Snack pattern
Condiments	0.78		0.26
Green/yellow vegetables	0.74		
Light colored vegetables	0.71	0.40	
Tubers	0.67		0.32
Clams	0.63	0.22	0.26
Tofu, soymilk	0.61		0.22
Seaweeds	0.60		
Bonefish	0.54		
Kimchi	0.49		
Lean fish	0.46	0.37	
Mushrooms	0.42	0.36	
Fruits	0.40		
Nuts	0.37		
Legumes	0.29		
Yogurt	0.27		
Eggs	0.27		0.28
Pickled vegetables	0.24		
Milk	0.20		
Red meat	0.23	0.79	
Red meat by-products		0.74	
Other seafood	0.25	0.67	
High-fat red meat		0.60	
Oil		0.50	0.20
Salted fermented seafood		0.44	
Noodles		0.43	
Poultry		0.43	
Fatty fish		0.37	0.29
Carbonated beverages		0.36	0.27
Dairy products		0.30	0.25
Cakes, pizza			0.81
Snacks			0.68
Bread			0.60
Processed meats		0.29	0.50
Sweets		0.28	0.36
Rice cake			0.23
Coffee, tea		0.20	
Grains			
Variance explained (%)	18.8	7.5	5.6

Factor loadings with absolute values <0.2 are not presented.

prevalence of metabolic syndrome and either the snack pattern score variables or the traditional pattern score variables. The association between the dietary pattern scores and the prevalence of metabolic syndrome was analyzed after stratifying by sex. A positive association between the prevalence of metabolic syndrome and the dietary pattern score was found only for men with the meat dietary pattern (multivariate-adjusted PR of the highest group compared with the lowest group: 2.15, 95% CI: 1.10–4.21, p for trend = 0.005).

Discussion

The present study derived three dietary patterns in the Korean adult population: the 'traditional' pattern, the 'meat' pattern, and the 'snack' pattern. We found that the meat dietary pattern score was positively associated with the prevalence of metabolic syndrome especially in men, whereas the traditional pattern and the snack pattern were not associated with metabolic syndrome.

Because Korean meals often include mixed soups or multiple side dishes comprising various vegetables, tubers, tofu, and seaweed with condiments, Factor 1 was labeled as the traditional

Table 3. The association between dietary patterns and the components of metabolic syndrome and BMI*.

	Quartile of dietary pattern score				p for trend[†]	p value[‡]
	Q 1	Q 2	Q 3	Q 4		
Traditional pattern (n)	285	263	275	260		
Waist circumference (cm)	78.1 (9.3)	78.6 (8.9)	78.3 (8.8)	78.3 (7.8)	0.790	0.841
Triglyceride (mg/dL)	115.0 (78.3)	116.8 (75.2)	113.5 (64.2)	117.9 (78.3)	0.450	0.063
HDL cholesterol (mg/dL)	58.7 (14.6)	58.7 (14.5)	58.9 (13.9)	59.4 (13. 7)	0.410	0.582
Diastolic BP (mmHg)	76.1 (10.6)	76. 7 (10.5)	76.1 (10.2)	76.7 (10.4)	0.407	0.134
Systolic BP (mmHg)	124.7 (14.4)	126.1 (14.1)	124.5 (13.8)	126.3 (15.0)	0.189	0.593
Fasting glucose (mg/dL)	89.2 (12.9)	91.5 (14.0)	91.0 (15.7)	93.3 (25.9)	0.538	0.177
Body Mass Index (kg/m^2)	23.4 (3.0)	23.6 (3.0)	23.8 (3.0)	23.6 (2.7)	0.960	0.976
Age (yr)	48.9 (8.7)	51.6 (9.4)	51.9 (9.3)	54.2 (8.9)	<0.001	
Male, n (%)	138 (28.4)	135 (27.8)	107 (22.0)	106 (21.8)	0.001	
Current drinker, n (%)	174 (27.3)	173 (27.1)	153 (24.0)	138 (21.6)	<0.001	
Physical activity (MET-min/wk)	2636.5 (2812.3)	2895.4 (3047.1)	3065.5 (3040.0)	3138.7 (2868.2)	<0.001	
Total energy intake	1691.3 (707.7)	2052.4 (753.0)	2303.0 (786.8)	2654.5 (873.1)	<0.001	<0.001
Carbohydrates[§]	397.0 (2.4)	395.1 (2.3)	387.2 (2.3)	369.2 (2.4)	<0.001	<0.001
Fat[§]	31.3 (0.8)	31.4 (0.8)	33.6 (0.8)	39.2 (0.8)	<0.001	<0.001
Protein[§]	71.2 (1.1)	73.4 (1.0)	79.1 (1.0)	89.8 (1.1)	<0.001	<0.001
Fiber[§]	31.3 (1.6)	30.4 (1.6)	35.6 (1.6)	40.0 (1.6)	<0.001	<0.001
Sodium[§]	2226.3 (69.5)	2779.8 (66.6)	3236.7 (66.6)	4580.9 (69.3)	<0.001	<0.001
Saturated fat[§]	7.2 (0.2)	7.6 (0.2)	7.9 (0.2)	9.0 (0.2)	<0.001	<0.001
Meat pattern (n)	273	267	266	277		
Waist circumference (cm)	76.8 (7.7)	77.8 (8.9)	78.0 (8.9)	80.6 (8.9)	<0.001	<0.001
Triglyceride (mg/dL)	108.6 (67.7)	109.9 (69.2)	110.2 (63.8)	133.7 (89.7)	0.018	0.005
HDL cholesterol (mg/dL)	60.5 (14.7)	59.3 (13.8)	58.9 (13.7)	56.9 (14.4)	0.032	0.067
Diastolic BP (mmHg)	75.7 (10.3)	75.2 (10.2)	76.7 (11.0)	77.8 (10.0)	0.034	0.854
Systolic BP (mmHg)	124.4 (13.8)	124.9 (14.4)	125.4 (15.1)	126.7 (14.1)	0.018	0.073
Fasting glucose (mg/dL)	91.9 (20.1)	91.7 (12.7)	88.9 (10.2)	92.3 (24.0)	0.954	0.050
Body Mass Index (kg/m^2)	23.1 (2.6)	23.6 (3.0)	23.4 (2.8)	24.3 (3.1)	<0.001	<0.001
Age (yr)	54.7 (8.5)	51.8 (8.9)	50.9 (9.3)	49.3 (9.4)	<0.001	
Male, n (%)	86 (17.7)	111 (22.8)	128 (26.3)	161 (33.1)	<0.001	
Current drinker, n (%)	123 (19.3)	146 (22.9)	166 (26.0)	203 (31.8)	<0.001	
Physical activity (MET-min/wk)	2950.1 (2951.2)	2963.5 (2991.0)	2880.4 (2785.7)	2942.8 (3064.2)	0.466	
Total energy intake	2116.8 (923.5)	2004.6 (805.9)	2059.3 (723.9)	2520.7 (868.2)	<0.001	<0.001
Carbohydrates[§]	408.9 (2.0)	398.6 (2.0)	387.1 (2.0)	353.9 (2.1)	<0.001	<0.001
Fat[§]	26.2 (0.7)	29.4 (0.7)	33.8 (0.7)	46.1 (0.7)	<0.001	<0.001
Protein[§]	71.6 (1.1)	76.2 (1.1)	78.5 (1.1)	87.2 (1.1)	<0.001	<0.001
Fiber[§]	33.4 (1.6)	37.7 (1.6)	34.5 (1.6)	31.7 (1.6)	0.259	0.629
Sodium[§]	2857.5 (77.3)	2908.1 (77.9)	3226.3 (77.6)	3834.2 (78.9)	<0.001	<0.001
Saturated fat[§]	6.5 (0.2)	6.6 (0.2)	7.9 (0.2)	10.6 (0.2)	<0.001	<0.001
Snack pattern (n)	257	273	275	278		
Waist circumference (cm)	79.8 (8.4)	78.0 (8.4)	78.1 (9.0)	77.2 (8.7)	0.051	0.006
Triglyceride (mg/dL)	117.1 (75.0)	120.5 (76.1)	112.8 (71.7)	112.7 (73.8)	0.833	0.413
HDL cholesterol (mg/dL)	58.6 (13.7)	58.8 (13.7)	58.0 (13.7)	60.2 (15.4)	0.827	0.936
Diastolic BP (mmHg)	77.9 (10.4)	76.2 (10.1)	75.9 (11.1)	75.6 (9.8)	0.124	0.134
Systolic BP (mmHg)	127.2 (13.6)	125.1 (14.0)	124.7 (15.1)	124.6 (14.6)	0.353	0.364
Fasting glucose (mg/dL)	93.2 (14.5)	90.6 (11.8)	90.7 (22.0)	90.4 (20.5)	0.410	0.343
Body Mass Index (kg/m^2)	24.0 (2.8)	23.4 (2.7)	23.6 (3.1)	23.5 (3.0)	0.506	0.115
Age (yr)	53.9 (8.8)	52.5 (8.9)	51.3 (9.4)	48.9 (9.2)	<0.001	
Male, n (%)	132 (27.2)	124 (25.5)	116 (23.9)	114 (23.5)	0.111	

Table 3. Cont.

	Quartile of dietary pattern score				p for trend[†]	p value[‡]
	Q 1	Q 2	Q 3	Q 4		
Current drinker, n (%)	159 (24.9)	167 (26.2)	148 (23.2)	164 (25.7)	0.965	
Physical activity (MET-min/wk)	2964.9 (2944.4)	2797.5 (2876.0)	3095.5 (2946.4)	2879.3 (3023.0)	0.427	
Total energy intake	2087.9 (836.7)	2009.1 (860.9)	2113.8 (761.9)	2491.1 (886.0)	<0.001	<0.001
Carbohydrates[§]	395.6 (2.2)	397.8 (2.2)	387.1 (2.2)	368.1 (2.3)	<0.001	<0.001
Fat[§]	29.4 (0.8)	29.4 (0.8)	34.1 (0.8)	42.5 (0.8)	<0.001	<0.001
Protein[§]	78.5 (1.1)	76.6 (1.1)	77.8 (1.1)	80.6 (1.1)	0.134	0.229
Fiber[§]	34.4 (1.6)	37.1 (1.6)	33.9 (1.6)	31.9 (1.6)	0.141	0.093
Sodium[§]	3340.2 (79.7)	2963.4 (80.1)	3123.5 (79.9)	3399.6 (81.0)	0.349	0.238
Saturated fat[§]	7.0 (0.2)	7.0 (0.2)	8.2 (0.2)	9.5 (0.2)	<0.001	<0.001

*Participants who were taking medication for hypertension and elevated glucose were excluded for the analysis of the components of metabolic syndrome and BMI.
[†]General linear model with adjustments for age, sex, smoking status, alcohol consumption, total energy intake, and physical activity (log-transformed) for the analysis of the components of metabolic syndrome and BMI, and adjustments for total energy intake for the analysis of nutrients.
[‡]Regression analysis in log-log scale with adjustments for age, sex, smoking status, alcohol consumption, total energy intake, and physical activity (log-transformed) for the analysis of the components of metabolic syndrome and BMI, and adjustments for total energy intake for the analysis of nutrients.
[§]Least squares means (SE) adjusted for total energy intake.
Numbers are Mean (SD), unless otherwise stated.

dietary pattern. The traditional Korean meal is low in fat and contains a large portion of vegetables, which can be considered a healthy diet. However, the traditional pattern was not associated with a lower prevalence of metabolic syndrome in our results or in previous studies [22–24]. HDL cholesterol was inversely associated with the traditional pattern score derived from 16 food groups in women [22]. The traditional pattern score was inversely associated, although not statistically significant, with HDL cholesterol, and positively associated with a prevalence of metabolic syndrome in Hong et al. [23]. Another study that used cluster analysis showed that HDL cholesterol was lower in people with the traditional pattern compared with the those of both meats and alcohols pattern and Korean healthy pattern [24]. Grain, especially refined grain was highly correlated with the traditional pattern in the three above Korean studies. Thus it seems that the negative association between HDL cholesterol and the Korean traditional food pattern was substantially affected by high intakes of carbohydrate. HDL cholesterol was negatively related with carbohydrate [30] and glycemic index [31]. However, grain had very low factor loading for the traditional pattern in our study. Traditional Korean foods are usually cooked with condiments that contain high levels of salt. Highest factor loading in the traditional pattern was condiments, and the pattern score of the traditional pattern was highly correlated with sodium intake in our study (r = 0.72, p<0.001). This may have led to a lack of an association between the traditional Korean dietary pattern and the prevalence of metabolic syndrome in our study. Therefore, the Korean traditional pattern is not generally associated with the prevalence of metabolic syndrome, but high sodium intakes could increase the risk of metabolic syndrome.

The western diet, which has high factor loadings for red meat and processed meat, was positively related to metabolic syndrome [11,12,21,32], whereas meat intake was not associated with metabolic syndrome in French adults [33]. The food groups with high factor loadings in the meat dietary pattern in our study were different from those with high factor loadings in the western dietary pattern and in the meat dietary pattern investigated in the previous studies. Both the western dietary pattern and the meat pattern generally had a high factor loading for processed meat,

which is responsible for many of the adverse effects that are characteristic of the meat dietary pattern. Poultry, which is usually found in healthy dietary patterns, was instead characteristic of the meat pattern in our study. The poultry eaten in Korea was mostly cooked by boiling in 1990. However, the majority of poultry eaten changed to fried chicken in 1998 according to KNHANES. Consequently, poultry has become a major source of saturated fat. Despite discrepancies between each of the food groups, our results suggest that the meat pattern is associated with the prevalence of metabolic syndrome in male adults. This association has been consistently observed in the results from previous studies of the western and meat dietary patterns. The positive association was only found in the male group in our study, suggesting that the meat dietary pattern among Korean adults might increase the prevalence of metabolic syndrome in males to a greater extent than in females.

Several possible mechanisms may explain the detrimental effect of the meat pattern in the human body. First, meat is a major source of total fat intake, particularly saturated fat, and the consumption of saturated fat has been associated with plasma lipoprotein levels [34] and higher blood pressure levels [35]. In a group of individuals of Japanese ancestry, red meat consumption was associated with a higher risk of developing metabolic syndrome among men. However, this association was no longer significant after making adjustments for saturated fatty acids [36]. Thus, the prevalence of metabolic syndrome in the meat pattern was analyzed with further adjustments for saturated fatty acids in our study. The prevalence ratio of metabolic syndrome in the highest quartile compared with the lowest quartile was attenuated (data not shown). Second, meat intake is related to the deposition of iron, particularly heme-iron. Metabolic syndrome subjects had a significantly higher prevalence of iron overload than control subjects [37], and high ferritin concentrations were positively associated with the prevalence of metabolic syndrome and with insulin resistance [38,39]. It was suggested that high iron contents of red meat might be related with higher prevalence of metabolic syndrome [40,41]. Meat intake, especially processed meat, was associated with increased risk of coronary heart disease and diabetes in meta-analysis [42]. Exact mechanism is not explained

Table 4. PRs and 95% CIs of metabolic syndrome by quartiles of dietary patterns.

	Dietary pattern	Quartiles of dietary pattern scores			p for trend*
		Q2	Q3	Q4	
Total	Traditional (n)	314	314	315	
(1257)	Model 1[†]	1.04 (0.71–1.53)[§]	1.01 (0.69–1.50)	1.02 (0.69–1.52)	0.408
	Model 2[‡]	1.01 (0.68–1.51)	1.17 (0.79–1.75)	1.08 (0.71–1.63)	0.330
	Meat (n)	313	314	315	
	Model 1[†]	1.16 (0.81–1.67)	1.23 (0.86–1.76)	1.40 (0.98–1.99)	0.006
	Model 2[‡]	1.23 (0.84–1.81)	1.33 (0.91–1.94)	1.47 (1.00–2.15)	0.016
	Snack (n)	314	313	315	
	Model 1[†]	0.77 (0.55–1.07)	0.82 (0.59–1.15)	0.86 (0.62–1.21)	0.249
	Model 2[‡]	0.79 (0.56–1.13)	0.90 (0.64–1.28)	0.93 (0.65–1.32)	0.421
Men	Traditional (n)	135	107	106	
(n = 486)	Model 1[†]	0.99 (0.59–1.66)	1.05 (0.61–1.83)	1.12 (0.64–1.94)	0.195
	Model 2[‡]	0.96 (0.56–1.64)	1.26 (0.72–2.18)	1.18 (0.66–2.10)	0.129
	Meat (n)	111	128	161	
	Model 1[†]	1.23 (0.67–2.27)	1.27 (0.70–2.30)	1.68 (0.94–2.98)	0.005
	Model 2[‡]	1.64 (0.82–3.27)	1.70 (0.86–3.34)	2.15 (1.10–4.21)	0.005
	Snack (n)	124	116	114	
	Model 1[†]	0.57 (0.34–0.96)	0.87 (0.55–1.37)	0.80 (0.49–1.30)	0.314
	Model 2[‡]	0.53 (0.31–0.94)	0.91 (0.57–1.45)	0.80 (0.49–1.31)	0.335
Women	Traditional (n)	179	207	209	
(n = 771)	Model 1[†]	1.18 (0.67–2.08)	0.99 (0.56–1.76)	0.98 (0.55–1.74)	0.932
	Model 2[‡]	1.16 (0.63–2.12)	1.14 (0.63–2.07)	1.07 (0.58–1.97)	0.978
	Meat (n)	202	186	154	
	Model 1[†]	1.21 (0.76–1.92)	1.29 (0.82–2.02)	1.15 (0.71–1.87)	0.248
	Model 2[‡]	1.18 (0.73–1.91)	1.26 (0.78–2.02)	1.14 (0.68–1.92)	0.455
	Snack (n)	190	197	201	
	Model 1[†]	1.03 (0.66–1.61)	0.82 (0.50–1.33)	1.01 (0.63–1.61)	0.685
	Model 2[‡]	1.11 (0.69–1.80)	0.89 (0.53–1.51)	1.11 (0.66–1.85)	0.830

PR: prevalence ratio, CI: confidence interval.
*Trend test were performed by Wald test using continuous variables of each pattern score (log-transformed).
[†]Adjusted for age, sex (for total) and total energy intake.
[‡]Adjusted for age, sex (for total), total energy intake, smoking status, alcohol consumption, and physical activity (log-transformed).
[§]PR (95% CI), compared with quartile 1 as a reference.

clearly, but iron overload increase oxidative stress due to its catalytic properties [43], resulting insulin resistance and decreased insulin secretion [44,45]. Additionally, the meat dietary pattern may be closely related to a high consumption of alcohol in our study, which may have increased the prevalence of metabolic syndrome in these individuals. Food groups with high factor loadings in the meat pattern were often consumed with alcohol. In previous studies that included alcohol as a food for dietary pattern analysis, a 'meats and alcohols' pattern was derived, suggesting that meat and alcohol consumption are highly correlated in Korean diets [23,24]. Heavy drinking was positively associated with metabolic syndrome and its components [46,47], and both alcohol consumption and a meat dietary pattern were associated with an increased prevalence of metabolic syndrome [21]. The percentage of current drinkers was higher in the highest quartile of meat pattern in our study, especially in men. Although it was adjusted for in the analysis, alcohol consumption may partly affect the prevalence of metabolic syndrome in the meat pattern. Drinking habits might explain the sex difference in the meat

pattern as well, as men are more likely to drink heavily than women. Another explanation for the gender difference is body iron stores. It was suggested that a lower incidence of heart diseases in women, especially in premenopausal women, might be related to lower body iron stores [48,49]. A significant association between iron-related genes and type 2 diabetes was observed in men but not in women [50].

The snack pattern was not associated with the prevalence of metabolic syndrome and its components. Women who had a fiber bread pattern had a lower prevalence of metabolic syndrome and higher insulin sensitivity, while a white bread pattern was positively associated with metabolic syndrome and lowered insulin sensitivity [51,52]. Whole grain consumption was inversely associated with type 2 diabetes [53], a higher waist-to-hip ratio, LDL-cholesterol and fasting insulin [52]. Thus the type of grain consumed by individuals with a snack pattern may affect the prevalence of metabolic syndrome. However, the snack pattern score was only slightly correlated with carbohydrate and fiber in our study. The snack pattern score increased with a decrement of age, and waist circumference, which was highly correlated with age, was inversely associated with the snack pattern score. Thus the trend of decreasing age with increasing the snack pattern score, although age was adjusted for in the analysis, may affect the prevalence of metabolic syndrome in the snack pattern. The meat pattern score also increased with a decrement of age, but it still positively associated with the prevalence of metabolic syndrome. It suggests that the association between the meat pattern and the prevalence of metabolic syndrome is strong.

Our study has several limitations. Because this is a cross-sectional study, there is a chance that dietary intake was affected by an individual's health status, which makes it difficult to find a true association between dietary intake and metabolic syndrome. In addition, we cannot exclude of the possibility of measurement errors of study variables and residual confounding. Thus associations identified should be interpreted in caution. The prevalence of metabolic syndrome in our study was lower than that in previous reports that analyzed the KNHANES data [7]. The study participants may have had a healthier lifestyle, as they volunteered for the health screening examinations, therefore leading to the lower prevalence of metabolic syndrome. The three dietary patterns derived from PCA analysis explained about 32% of the total variation; thus the derived patterns might not explain all Korean dietary patterns thoroughly.

In conclusion, the meat dietary pattern, which was characterized by a high consumption of red meat, red meat byproducts, and high-fat red meat, was associated with an increased prevalence of metabolic syndrome in Korean male adults.

Supporting Information

Table S1 Food lists of 37 food groups.
(DOCX)

Author Contributions

Conceived and designed the experiments: HDW AS JK. Analyzed the data: HDW JK. Wrote the paper: HDW JK.

References

1. Lorenzo C, Okoloise M, Williams K, Stern M, Haffner S (2003) The metabolic syndrome as predictor of type 2 diabetes. Diabetes Care 26: 3153–3159.
2. Lakka H, Laaksonen D, Lakka T, Niskanen L, Kumpusalo E, et al. (2002) The metabolic syndrome and total and cardiovascular disease mortality in middle-aged men. JAMA 288: 2709–2716.
3. McNeill A, Rosamond W, Girman C, Golden S, Schmidt M, et al. (2005) The metabolic syndrome and 11-year risk of incident cardiovascular disease in the atherosclerosis risk in communities study. Diabetes Care 28: 385–390.
4. Mozumdar A, Liguori G (2011) Persistent Increase of Prevalence of Metabolic Syndrome Among US Adults: NHANES III to NHANES 1999–2006. Diabetes Care 34: 216–219.
5. Li J, Wang X, Zhang J, Gu P, Zhang X, et al. (2010) Metabolic syndrome: prevalence and risk factors in southern China. J Int Med Res 38: 1142–1148.
6. Gu D, Reynolds K, Wu X, Chen J, Duan X, et al. (2005) Prevalence of the metabolic syndrome and overweight among adults in China. Lancet 365: 1398–1405.
7. Lim S, Shin H, Song JH, Kwak SH, Kang SM, et al. (2011) Increasing Prevalence of Metabolic Syndrome in Korea. Diabetes Care 34: 1323–1328.
8. Ramachandran A, Snehalatha C, Shetty AS, Nanditha A (2012) Trends in prevalence of diabetes in Asian countries. World J Diabetes 3: 110–117.
9. Kim HJ, Kim Y, Cho Y, Jun B, Oh KW (2014) Trends in the prevalence of major cardiovascular disease risk factors among Korean adults: Results from the Korea National Health and Nutrition Examination Survey, 1998–2012. Int J Cardiol 174: 64–72.
10. Azadbakht L, Mirmiran P, Esmaillzadeh A, Azizi T, Azizi F (2005) Beneficial effects of a Dietary Approaches to Stop Hypertension eating plan on features of the metabolic syndrome. Diabetes Care 28: 2823–2831.
11. Esmaillzadeh A, Kimiagar M, Mehrabi Y, Azadbakht L, Hu F, et al. (2007) Dietary patterns, insulin resistance, and prevalence of the metabolic syndrome in women. Am J Clin Nutr 85: 910–918.
12. Lutsey P, Steffen L, Stevens J (2008) Dietary intake and the development of the metabolic syndrome: the Atherosclerosis Risk in Communities study. Circulation 117: 754–762.
13. Newby P, Tucker K (2004) Empirically derived eating patterns using factor or cluster analysis: a review. Nutr Rev 62: 177–203.
14. Lee M, Popkin B, Kim S (2002) The unique aspects of the nutrition transition in South Korea: the retention of healthful elements in their traditional diet. Public Health Nutr 5: 197–203.
15. Denova-Gutierrez E, Castanon S, Talavera J, Gallegos-Carrillo K, Flores M, et al. (2010) Dietary Patterns Are Associated with Metabolic Syndrome in an Urban Mexican Population. J Nutr 140: 1855–1863.
16. Ferreira S, Lerario D, Gimeno S, Sanudo A, Franco L (2002) Obesity and central adiposity in Japanese immigrants: role of the Western dietary pattern. J Epidemiol 12: 431–438.
17. Gimeno S, Ferreira S, Franco L, Hirai A, Matsumura L, et al. (2002) Prevalence and 7-year incidence of type II diabetes mellitus in a Japanese-Brazilian population: an alarming public health problem. Diabetologia 45: 1635–1638.
18. DiBello J, McGarvey S, Kraft P, Goldberg R, Campos H, et al. (2009) Dietary patterns are associated with metabolic syndrome in adult Samoans. J Nutr 139: 1933–1943.
19. Deshmukh-Taskar PR, O'Neil CE, Nicklas TA, Yang S-J, Liu Y, et al. (2009) Dietary patterns associated with metabolic syndrome, sociodemographic and lifestyle factors in young adults: the Bogalusa Heart Study. Public Health Nutr 12: 2493–2503.
20. Amini M, Esmaillzadeh A, Shafaeizadeh S, Behrooz J, Zare M (2010) Relationship between major dietary patterns and metabolic syndrome among individuals with impaired glucose tolerance. Nutrition 26: 986–992.
21. Panagiotakos DB, Pitsavos C, Skoumas Y, Stefanadis C (2007) The association between food patterns and the metabolic syndrome using principal components analysis: The ATTICA Study. J Am Diet Assoc 107: 979–987.
22. Cho YA, Kim J, Cho ER, Shin A (2011) Dietary patterns and the prevalence of metabolic syndrome in Korean women. Nutr Metab Cardiovas 21: 893–900.
23. Hong S, Song Y, Lee KH, Lee HS, Lee M, et al. (2012) A fruit and dairy dietary pattern is associated with a reduced risk of metabolic syndrome. Metab Clin Exp 61: 883–890.
24. Song Y, Joung H (2012) A traditional Korean dietary pattern and metabolic syndrome abnormalities. Nutr Metab Cardiovas 22: 456–462.
25. Kim S, Oh S (1996) Cultural and nutritional aspects of traditional Korean diet. World Rev Nutr Diet 79: 109–132.
26. Guidelines for data processing and analysis of the international physical activity questionnaire, 2005. Available: http://www.ipaq.ki.se. Accessed: 2010 Oct 12.
27. Ahn Y, Kwon E, Shim J, Park M, Joo Y, et al. (2007) Validation and reproducibility of food frequency questionnaire for Korean genome epidemiologic study. Eur J Clin Nutr 61: 1435–1441.
28. Antonopoulos S (2002) Third report of the National Cholesterol Education Program (NCEP) expert panel on detection, evaluation, and treatment of high blood cholesterol in adults (Adult Treatment Panel III) final report. Circulation 106: 3143–3421.
29. WHO West Pacific Region. The Asia-Pacific Perspective: Redefining obesity and its treatment. International Obesity Task Force 2000: 15–21.
30. Merchant AT, Anand SS, Kelemen LE, Vuksan V, Jacobs R, et al. (2007) Carbohydrate intake and HDL in a multiethnic population. Am J Clin Nutr 85: 225–230.
31. Frost G, Leeds A, Dore C, Madeiros S, Brading S, et al. (1999) Glycaemic index as a determinant of serum HDL-cholesterol concentration. Lancet 353: 1045–1048.
32. van Dam R, Rimm E, Willett W, Stampfer M, Hu F (2002) Dietary patterns and risk for type 2 diabetes mellitus in US men. Ann Intern Med 136: 201–209.
33. Mennen L, Lafay L, Feskens E, Novak M, Lepinay P, et al. (2000) Possible protective effect of bread and dairy products on the risk of the metabolic syndrome. Nutr Res 20: 335–347.
34. Riccardi G, Giacco R, Rivellese A (2004) Dietary fat, insulin sensitivity and the metabolic syndrome. Clin Nutr 23: 447–456.
35. Trevisan M, Krogh V, Freudenheim J, Blake A, Muti P, et al. (1990) Consumption of olive oil, butter, and vegetable oils and coronary heart disease risk factors. JAMA 263: 688–692.
36. Damiao R, Castro T, Cardoso M, Gimeno S, Ferreira S (2006) Dietary intakes associated with metabolic syndrome in a cohort of Japanese ancestry. Br J Nutr 96: 532–538.
37. Bozzini C, Girelli D, Olivieri O, Martinelli N, Bassi A, et al. (2005) Prevalence of body iron excess in the metabolic syndrome. Diabetes Care 28: 2061–2063.
38. Sun L, Franco O, Hu F, Cai L, Yu Z, et al. (2008) Ferritin concentrations, metabolic syndrome, and type 2 diabetes in middle-aged and elderly Chinese. J Clin Endocrinol Metab 93: 4690–4696.
39. Jehn M, Clark JM, Guallar E (2004) Serum ferritin and risk of the metabolic syndrome in US adults. Diabetes Care 27: 2422–2428.
40. Azadbakht L, Esmaillzadeh A (2009) Red meat intake is associated with metabolic syndrome and the plasma C-reactive protein concentration in women. J Nutr 139: 335–339.

41. Tappel A (2007) Heme of consumed red meat can act as a catalyst of oxidative damage and could initiate colon, breast and prostate cancers, heart disease and other diseases. Med Hypotheses 68: 562–564.
42. Micha R, Wallace SK, Mozaffarian D (2010) Red and processed meat consumption and risk of incident coronary heart disease, stroke, and diabetes mellitus a systematic review and meta-analysis. Circulation 121: 2271–2283.
43. De Valk B, Marx J (1999) Iron, atherosclerosis, and ischemic heart disease. Arch Intern Med 159: 1542–1548.
44. Ford ES, Cogswell ME (1999) Diabetes and serum ferritin concentration among US adults. Diabetes Care 22: 1978–1983.
45. Jiang R, Manson JE, Meigs JB, Ma J, Rifai N, et al. (2004) Body iron stores in relation to risk of type 2 diabetes in apparently healthy women. JAMA 291: 711–717.
46. Baik I, Shin C (2008) Prospective study of alcohol consumption and metabolic syndrome. Am J Clin Nutr 87: 1455–1463.
47. Athyros VG, Liberopoulos EN, Mikhailidis DP, Papageorgiou AA, Ganotakis ES, et al. (2008) Association of drinking pattern and alcohol beverage type with the prevalence of metabolic syndrome, diabetes, coronary heart disease, stroke, and peripheral arterial disease in a Mediterranean cohort. Angiology 58: 689–697.
48. Sullivan J (1981) Iron and the sex difference in heart disease risk. Lancet 317: 1293–1294.
49. Mascitelli L, Goldstein MR, Pezzetta F (2011) Explaining sex difference in coronary heart disease: is it time to shift from the oestrogen hypothesis to the iron hypothesis? J Cardiovasc Med 12: 64–65.
50. He M, Workalemahu T, Manson JE, Hu FB, Qi L (2012) Genetic determinants for body iron store and type 2 diabetes risk in US men and women. PLoS ONE 7: e40919.
51. Wirfalt E, Hedblad B, Gullberg B, Mattisson I, Andren C, et al. (2001) Food patterns and components of the metabolic syndrome in men and women: a cross-sectional study within the Malmo Diet and Cancer cohort. Am J Epidemiol 154: 1150–1159.
52. McKeown NM, Meigs JB, Liu S, Wilson PWF, Jacques PF (2002) Whole-grain intake is favorably associated with metabolic risk factors for type 2 diabetes and cardiovascular disease in the Framingham Offspring Study. Am J Clin Nutr 76: 390–398.
53. Fung T, Schulze M, Manson J, Willett W, Hu F (2004) Dietary patterns, meat intake, and the risk of type 2 diabetes in women. Arch Intern Med 164: 2235–2240.

26

Development of a Quantitative Food Frequency Questionnaire for Use among the Yup'ik People of Western Alaska

Fariba Kolahdooz[1], Desiree Simeon[2], Gary Ferguson[2], Sangita Sharma[1]*

1 Department of Medicine, University of Alberta, Edmonton, Alberta, Canada, **2** Alaska Native Tribal Health Consortium, Community Health Services Division, Wellness and Prevention Department, Anchorage, Alaska, United States of America

Abstract

Alaska Native populations are experiencing a nutrition transition and a resulting decrease in diet quality. The present study aimed to develop a quantitative food frequency questionnaire to assess the diet of the Yup'ik people of Western Alaska. A cross-sectional survey was conducted using 24-hour recalls and the information collected served as a basis for developing a quantitative food frequency questionnaire. A total of 177 males and females, aged 13-88, in six western Alaska communities, completed up to three 24-hour recalls as part of the Alaska Native Dietary and Subsistence Food Assessment Project. The frequency of the foods reported in the 24-hour recalls was tabulated and used to create a draft quantitative food frequency questionnaire, which was pilot tested and finalized with input from community members. Store-bought foods high in fat and sugar were reported more frequently than traditional foods. Seven of the top 26 foods most frequently reported were traditional foods. A 150-item quantitative food frequency questionnaire was developed that included 14 breads and crackers; 3 cereals; 11 dairy products; 69 meats, poultry and fish; 13 fruit; 22 vegetables; 9 desserts and snacks; and 9 beverages. The quantitative food frequency questionnaire contains 39 traditional food items. This quantitative food frequency questionnaire can be used to assess the unique diet of the Alaska Native people of Western Alaska. This tool will allow for monitoring of dietary changes over time as well as the identification of foods and nutrients that could be promoted in a nutrition intervention program intended to reduce chronic disease.

Editor: Brenda Smith, Oklahoma State University, United States of America

Funding: This project was supported by the National Research Initiative Grant 2007-55215-17923 from the USDA Cooperative State Research and Extension Service (http://www.usda.gov/wps/portal/usda/usdahome?navid = COOPRES_EXTSERVS&navtype = RT&parentnav = EDUCATION_OUTREACH). The funders had no role in study design, data collection and analysis, decision to publish, or preparation of the manuscript.

Competing Interests: The authors have declared that no competing interests exist.

* Email: gita.sharma@ualberta.ca

Introduction

The diet of Alaska Native people has been in transition. Prior to the availability of store-bought foods, the diet consisted of locally harvested foods such as marine mammals, land mammals, ocean and freshwater fish, birds and many varieties of berries [1]. These "traditional" foods are rich sources of many nutrients, such as protein, iron, vitamins D and B12, selenium, and mono- and polyunsaturated fats [2–4]. The current diet of Alaska Native people contains a mixture of traditional foods and imported "store-bought" foods, which are often energy-dense, but of low nutritional quality [2–6]. Even though store-bought foods have only become widely available in Alaska in the last 60 years, the transition away from a diet composed entirely of traditional foods has been rapid. One of the first dietary assessment studies in rural Alaska, which took place from 1956–1961 [7], found that store-bought foods contributed to over 50% of energy intake [7]. More recent studies have shown that traditional foods provide 15–25% of total energy [6,8]. Although some studies have shown that older Alaska Native people consume a higher percentage of traditional foods than younger generations [2,6], these foods still contribute to the nutritional quality of the diet in all age groups [2,4,6,9]. In twelve Alaskan communities, in 2009, traditional foods contrib-

uted 23% of energy intake, yet supplied 46% of protein, 37% of iron, 90% of eicosapentaenoic acid (EPA) and 83% of vitamin D [10].

Concurrent to the diet and lifestyle transition, Alaska Native peoples are experiencing an increase in diet-related chronic diseases [9,11–15]. In the 1950's, obesity among Alaska Native people in the western region was rare and confined to age groups <60 years [16]. In 2007, the Behavioural Risk Factor Surveillance System [12] found that 28% of Alaska Native people were overweight and 34% were obese. In addition, diabetes prevalence increased between 1990 and 2006 [16] by 114% among Alaska Native people state wide, and by 152% in the western region [9]. While the mortality rate from ischemic heart disease has declined dramatically over the last 20 years among the United States Caucasian population, it has remained relatively constant in Alaska Native populations [13]. Currently, the leading cause of death among Alaska Native people is cancer [14], and cancer mortality rates among Alaska Native men and women are significantly higher than among Caucasian Americans [14,15].

Programs that promote dietary improvements can be effective strategies to reduce the burden of chronic disease among Alaska Native people [17–19]. However, little information is available on the current diet of this population in terms of food, nutrient and food group intakes. Tools that can accurately capture a target population's dietary pattern can help identify foods and nutrients to be targeted in intervention programs. Further, such tools can be used in monitoring and evaluating a program's effectiveness. Food frequency questionnaires are considered the method of dietary assessment best suited for large epidemiological applications due to their relative ease of use, cost-effectiveness and ability to estimate usual intake over extended periods of time [20,21]. To the best of our knowledge, no such detailed, up-to-date, population-specific dietary instrument has been developed for Alaska Native peoples.

The aim of this study is to describe the current diet and to develop a quantitative food frequency questionnaire (QFFQ) to assess the unique diet of Native populations of Western Alaska. The QFFQ could be used to identify foods and nutrients to be targeted in a culturally appropriate nutritional intervention, evaluate the impact of diet intervention programs, as well as monitor the nutrition transition occurring in this population.

Methods

Setting

The Alaska Native Dietary and Subsistence Food Assessment Project (ANDSFAP) took place in six rural communities in Western Alaska's Yukon–Kuskokwim Delta region. Participating villages ranged in size from 287 people and 83 households to 721 people and 199 households [22]. These villages are geographically remote; they are not on a road system and are accessible only by small plane or boat and by snow machine in the winter. Residents are predominantly Alaska Native people.

Sampling

The project was presented to tribal councils and written resolutions to participate were received from six villages. Enrolment procedures followed the directive of each tribal council. Village residents aged ≥13 years were randomly recruited to participate in 24-hour recalls; the target for recruitment was 30 people per village. To aid in capturing maximum diet diversity, only one person was recruited per household. Written consent or assent was obtained from each participant and from a parent for those under 18 years of age.

24-hour dietary recall collection

24-hour recall data collection was undertaken prior to the current study and described previously [10]. In brief, the 24-hour recalls were obtained by trained interviewers who recorded information on food/beverages consumed in the previous 24 hours, including type, amount, brand name, food source, food processing method, food preparation method and any additions (e.g. sugar in coffee). Portion size was assessed using a variety of standard utensils and food models. A multiple pass method was used to help ensure that no items were inadvertently omitted. Follow–up interviews were conducted over the telephone to obtain any details omitted in error. To capture seasonal variation, attempts were made to interview each participant four times, once per season. The 24-hour recall interviews were conducted in English, or a combination of English and Yup'ik using an interpreter. The 24 hour recall information was used for FFQ development.

Development of the Quantitative Food Frequency Questionnaire

The frequency of the foods reported in the 24-hour recalls was tabulated. Any food reported more than two times was included in the draft QFFQ which can be found in the supplementary data

[see Table S1], with the exception of low-nutrient foods such as spices and condiments. Foods that did not appear in the 24-hour recalls but were considered relevant to the population, such as seasonal foods, were also added. Additional blank lines were provided under each food group for respondents to list any other foods or drinks they consumed. Local community members were consulted to ensure no commonly consumed foods were omitted. Foods expected to be promoted in a subsequent intervention (e.g., diet soda) were also included.

To assess the portion sizes of foods/beverages consumed, food models were carefully selected to represent serving sizes of each item on the QFFQ. Standard bowls, cups and spoons of the type and size available locally were used. Other food models were constructed by the study team and local residents using standard units familiar to participants, such as a slice of bread. Wrappers and containers from familiar foods, such as cans of vegetables and bags of chips, were also used to aid in portion estimation. Food models represented a wide range of serving sizes.

Similar food items were grouped on the QFFQ and resulting categories included breads and crackers; cereal; dairy, meat, poultry and fish; fruit; vegetables; desserts and snacks; and beverages. Food categories were ordered according to cultural preference and dietary habits based on advice and input of local community members. A manual of procedures was developed for the administration of the draft QFFQ and all staff was trained in its use. The draft QFFQ was pilot tested in a convenience sample of 15 individuals selected from men and women in one of the communities participating in the study. The purpose of this pilot was to ensure that the questionnaire was culturally appropriate and the structure and order of the foods were user-friendly for interviewers. Subsequently, because of time constraints for completion, similar items were combined. Foods reported less than six times in the original 24-hour recall data set were not included in the QFFQ due to their low overall contribution to the diet. Similar studies set a precedent for these modifications [23,24].

This study was conducted according to the guidelines laid down in the Declaration of Helsinki and all procedures involving human subjects/patients were approved by the Alaska Area Institutional Review Board and the Office of University Research Ethics at the University of North Carolina at Chapel Hill. The project was also reviewed and approved by the Yukon-Kuskokwim Health Corporation and the Norton Sound Health Corporation. Six villages in Western Alaska gave written resolutions to participate in the project.

Data analysis

Statistical Analysis Systems (SAS Institute, Cary, NC, USA) was used for descriptive statistical analysis of the dietary intake and manipulation of the data.

Results

24-hour recalls

A total of 400 recalls were collected from 177 Yup'ik participants (73 males and 104 females) aged 13–88 for the ANDSFAP. Most participants completed two or three 24-hour recalls, 27 completed one 24-hr dietary recalls, and 35 completed all four. Table 1 lists the most commonly reported foods from the 24-hour recalls. Commercially available foods were the most predominant. Seven traditional foods were among the twenty-six most commonly reported foods. Coffee and sugar were each reported over 300 times. High-sugar drinks such as juice or flavoured drink and soda pop were frequently reported, as were

Table 1. Foods most commonly reported on 400, 24-hour recalls among Alaska Native people (n = 177).

Food	Number of times reported	Number of people reporting
Coffee, any type	400	135
Sugar	330	100
Flavoured drink	323	125
Soda pop	296	111
White bread and rolls	249	103
White rice	240	120
Tea	205	75
Potatoes	184	99
Fish other than salmon and halibut	154	80
Pilot bread crackers	144	85
King Salmon, smoked	109	76
Agutuk	105	50
Seal oil	103	60
Salmon, any type, cooked	91	60
Coffee creamer	82	42
Butter	79	40
Chips	75	52
Moose meat	73	48
Margarine	23	50
Berries	56	24
Chicken eggs	55	41
Corn, canned	50	39
Chicken	48	41
Cereal, ready to eat	46	34
Ramen noodles	43	31

white bread and rice. Table 2 lists traditional foods most frequently reported. Fish was the most common traditional food consumed (over 40%).

QFFQ

The final QFFQ contained 150 food/beverage items under eight food categories: 14 breads and crackers; 3 cereals; 11 dairy

Table 2. Traditional foods most commonly reported on 400, 24-hour recalls among Alaska Native people (n = 177).

Traditional Food	Number of times reported	Number of people reporting
Fish other than salmon or halibut	154	80
King salmon, smoked	109	76
Agutuk	105	50
Seal oil	103	60
Salmon, any type, cooked	91	60
Moose	73	48
Berries	56	24
Tundra tea	32	19
Goose	21	12
Seal meat	20	17
Duck	16	12
Caribou	15	11
Salmon, dry	13	11
Reindeer	11	10

products (including eggs); 69 meat, poultry and fish, including mixed dishes; 13 fruit, including locally gathered berries; 22 vegetables, including wild greens; 9 snacks and desserts; and 9 beverages (Table 3). Of the 150 items, 39 were traditional foods, defined as the subsistence foods consumed prior to the availability of store foods. These included moose, caribou, sea mammal meat and fat, birds, fish and other seafood, berries and wild greens.

The QFFQ was designed to estimate intake over the previous 12 months. Seasonal consumption of certain foods was determined by asking if the food had been consumed throughout the year or only in certain seasons. Standard response categories were used to help participants estimate frequency of intake ranging from "never," to "6 times or more per day."

Discussion

The dietary patterns showed evidence of reliance on a mix of traditional and store-bought foods. Evidence exists that a shift from traditional to store-bought foods results in decreased diet quality [5,25]. Store-bought foods often consumed are high in energy, carbohydrate and sugar and relatively low in other nutrients. The decline in traditional food consumption among Arctic populations is associated with negative changes in health, particularly when coupled with consumption of unhealthy store-bought foods [6,7,10,26–32]. Further decline in traditional food consumption is expected to have increasingly negative impacts.

This study developed a culturally appropriate QFFQ to estimate total intake of food and food groups to monitor the nutrition transition occurring in adult Yup'iks in the Western region of Alaska. The 150-item QFFQ will also be used to provide

baseline data prior to the Food Distribution Programme on Indian Reservations (FDPIR) introduction to communities, and again after 12 months to assess the program's effect on dietary change [33]. The FDPIR, a government food assistance program, has recently been introduced to Alaska. The program was initiated in 1975 to supply food to American Indian people living on reservations who did not have access to supermarkets [33]. Foods distributed by FDPIR include canned fruit, vegetables and meats; grain products such as cereal, flour and rice; juices, dry beans, and canned and frozen meats, including ground beef and buffalo. The QFFQ described here will also be used to inform and develop a nutrition education program specific to Alaska Native people intended to prevent obesity and related chronic diseases.

A culturally specific QFFQ was necessary because of the unique diet of Alaska Native people living in Western Alaska and consists of both locally harvested traditional foods and commercially available foods. Recent studies have illustrated the significant contribution that traditional foods make to the nutritional quality of the diet, and highlight the differences between this and the standard U.S. diet [6,7,9,10,11,31,32]. Omitting these foods, as would occur if a standard U.S. QFFQ was used in this setting, would give a false representation of intake [20].

Use of the 24-hour recall was an important step in obtaining an accurate list of commonly consumed foods relevant to the population of interest, which is considered the most crucial step in QFFQ development [34–36]. The use of food models assisted participants in estimating usual amounts consumed, as recommended by Cade et al. [35].

This QFFQ, designed to capture usual intake over a previous twelve-month period, offers an advantage over short-term

Table 3. Foods and beverages on the Quantitative Food Frequency Questionnaire.

Category	Food Item
Bread and crackers(14)	Alaska fry bread; pancakes or assaliaq or waffles; pancake syrup; white bread; biscuit; whole wheat or multigrain bread; fruit bread; corn bread; pilot bread; crackers; Crisco; butter or margarine; mayonnaise; peanut butter.
Cereal(3)	Sweet cereals; low sugar cereals; oatmeal, porridge, cream of wheat, mush cooked or instant.
Dairy(11)	Low fat milk, fat free, 2% including blue box, canned or powdered; whole milk, red box, canned or powdered; chocolate milk, cocoa or hot chocolate; powdered coffee creamer; liquid coffee creamer; hard cheese; ice cream; yogurt; Cool Whip; wild bird eggs; chicken eggs.
Meat, poultry and fish(69)	Moose, caribou, reindeer or musk ox meat, fried; moose, caribou, reindeer or musk ox meat, baked, boiled, grilled, raw or frozen; moose, caribou, reindeer or musk ox meat, dried; fat or bone marrow from moose or caribou; bison or buffalo meat; bear [brown or black], baked, boiled or grilled; hamburger on a bun with condiments; ground beef or meatloaf; liver from beef; beef jerky; other beef; hotdog, sausage or corn dog, beef or pork; sausage, moose, caribou or reindeer; sandwich meat; ham meat; corned beef hash; spam or corned beef not in hash; ribs [moose, caribou or reindeer]; ribs [beef or pork]; taco; burrito; pork chops or pork roast; bacon; beaver; seal, whale or walrus, dried; seal, whale or walrus meat; seal or whale oil; seal or walrus blubber; whale skin and fat; liver from marine mammals; akutaq; wild bird, dried; wild bird, baked, boiled or grilled; fried chicken; chicken nuggets; chicken or turkey, baked, boiled or grilled; skin on poultry or bird; white fish, fried; white fish, baked, boiled, grilled, raw or frozen; white fish, dried or smoked; salmon or arctic char, fried; salmon or arctic char, baked, boiled, grilled, raw or frozen; salmon or arctic char, dried or smoked; canned salmon; blackfish; fish roe; crab; canned tuna; clams, oysters or shrimp; other fish, canned; fish sticks or nuggets; fish sticks or nuggets; moose, caribou, reindeer, or musk ox soup or stew; beef stew or chili with meat; beef or pork soup; beef stir fry with vegetables; bear [brown or black] soup or stew; bison or buffalo soup or stew; wild bird soup; chicken or turkey soup; seal, whale or walrus soup; fish soup; bean or vegetable soup; ramen noodles or cup of noodles; spaghetti or pasta without meat, including macaroni and cheese or macaroni salad; spaghetti with meat, including goulash and hamburger helper; plain rice; pizza or hot pockets; stuffing.
Fruit(13)	Wild berries; purchased berries; canned fruits in light syrup; canned fruits in syrup; apple; banana; orange, tangerine or grapefruit; honeydew melon, watermelon or cantaloupe; peach, nectarine, plum or apricot; grapes; applesauce or stewed apples; fruit salad; dried fruit.
Vegetables(22)	French fries, fried potatoes, hash browns or tater tots; baked, boiled or mashed potatoes; gravy; potato salad or cold slaw; avocado; carrots; corn; green beans or peas; cabbage or spinach; wild greens; kelp or sea weed; turnip; pumpkin, sweet potato or yams; cauliflower or broccoli; asparagus; mixed vegetables; green salad; salad dressing; refried beans; any other beans; mixed bean salad; add beans to pasta, soups or rice
Desserts and snacks(9)	Chips or popcorn; cakes or muffins; pastry, doughnut, turnover, cinnamon rolls, pop tarts; pies; cookies; granola bar, cereal bar, Rice Krispies treats; candy; pudding; nuts, trail mix or sunflower seeds
Beverages(9)	Coffee; tea; sugar or honey; artificial sweetener; sweetened drinks; unsweetened drinks; regular soda; diet soda; water

methods of dietary assessment, including 24-hour dietary recalls and diet records. Alaska Native peoples' diet varies significantly by season. Thus, using short-term dietary assessment methods in remote Alaskan villages would require multiple trips throughout the year, greatly increasing costs.

The present study had some limitations. The sample was predominantly female because the study targeted the primary food shoppers and preparers. Therefore, results may not be generalizable to the male population. In addition, the small sample size of the subgroups prevented us from performing subgroup analyses by the community or age group. The consumption of foods applies only to the communities we investigated, therefore our findings may not be relevant to all Alaska Native communities. Exclusion of infrequently consumed food items from the QFFQ may mean that total diet was not captured; however, a longer questionnaire would have increased the burden on participants. Furthermore, the QFFQ was not validated against three 24-hour recalls, however, our previous similar studies showed a good agreement [37].

In conclusion, the 150-item QFFQ provides an up-to-date, comprehensive and unique tool for assessing dietary intake and for evaluating nutrition education programs for Alaska Native people in Western Alaska. The QFFQ will also serve to monitor the nutrition transition occurring among Yup'ik people.

Acknowledgments

We would like to acknowledge the assistance of Jennifer Johnson, as well as the Yukon-Kuskokwim Health Corporation and the Norton Sound Health Corporation, and all the people who so generously gave of their time and knowledge in each community, from the clinic to the tribal council to the city office to the stores to the schools.

Author Contributions

Conceived and designed the experiments: SS. Performed the experiments: DS GF. Analyzed the data: FK. Wrote the paper: FK DS GF.

References

1. Schraer C (1993) Diabetes among the Alaska Natives - the emergence of chronic disease with changing lifestyles. In: *Diabetes as a disease of civilization: the impact of cultural change on indigenous peoples*, ed., pp. 169–195. (Joe JR and Young RS, editors). Berlin: Mouton deGruyter.
2. Bersamin A, Zidenberg-Cherr S, Stern JS, Luick BR (2007) Nutrient intakes are associated with adherence to a traditional diet among Yup'ik Eskimos living in remote Alaska Native communities: the CANHR study. Int J Circumpolar Health 66: 62–70.
3. Draper HH (1977) The aboriginal Eskimo diet in a modern perspective. J Am Anthropol Assoc 79: 309–316.
4. Kinloch D, Kuhnlein H, Muir DC (1992) Inuit foods and diet: a preliminary assessment of benefits and risks. Sci Total Environ 122: 247–278.
5. Bersamin A, Luick BR, Ruppert E, Stern JS, Zidenberg-Cherr S (2006) Diet quality among Yup'ik Eskimos living in rural communities is low: the Center for Alaska Native Health Research Pilot Study. J Am Diet Assoc 106: 1055–1063.
6. Nobmann EB, Ponce R, Mattils C, Devereaux R, Dyke B, et al. (2005) Dietary intakes vary with age among Eskimo Adults in the Northwest Alaska GOCADAN Study, 2000–2003. J Nutr 135: 856–862.
7. Heller CA, Scott EM (1967) The Alaska Dietary Survey, 1956-61. PHS Publication No. 999-AH-2. Washington, DC: US Department of Health, Education and Welfare.
8. Nobmann ED, Ebbesson SO, White RG, Bulkow LR, Schraer CD (1999) Associations between dietary factors and plasma lipids related to cardiovascular disease among Siberian Yupiks of Alaska. Int J Circumpolar Health 58: 254–271.
9. Alaska Native Medical Area Center (2006) Diabetes prevalence maps. Available: http://www.anmc.org/services/diabetes/epidemiology/Prevalence.cfm. Accessed: 8 May 2009.
10. Johnson JS, Nobmann ED, Asay E, Lanier AP (2009) Dietary intake of Alaska Native people in two regions and implications for health: the Alaska Native Dietary and Subsistence Food Assessment Project. Int J Circumpolar Health 68: 109–122.
11. Naylor JL, Schraer CD, Mayer AM, Lanier AP, Treat CA, et al. (2003) Diabetes among Alaska Natives: a review. Int J Circumpolar Health 62: 363–387.
12. Behavioral Risk Factor Surveillance Survey (2007) Alaska Behavioral Risk Factor Survey. Annual Report, State of Alaska, Department of Health and Social Services. Available: http://www.hss.state.ak.us/dph/chronic/hsl/brfss/pubs/BRFSS07.pdf. Accessed: 21 May 2009.
13. Schumacher C, Davidson M, Ehrsam G (2003) Cardiovascular disease among Alaska Natives: a review of the literature. Int J Circumpolar Health 62: 343–362.
14. Lanier AP, Kelly JJ, Maxwell J, McEvoy T, Homan C (2006) Cancer in Alaska Natives 1969–2003: 35 year report. Office of Alaska Native Health Research, Alaska Native Epidemiology Center, Alaska Native Tribal Health Consortium.
15. Lanier AP, Day GE, Kelly JJ, Provost E (2008) Disparities in cancer mortality among Alaska Native people, 1994–2003. Alaska Med 49: 120–125.
16. Scott EM, Griffith IV (1957) Diabetes mellitus in Eskimos. Metabolism 6, 32.
17. Eilat-Adar S, Mete M, Nobmann ED, Xu J, Fabsitz RR, et al. (2009) Dietary patterns are linked to cardiovascular risk factors but not to inflammatory markers in Alaska Eskimos. J Nutr 139: 2322–2328.
18. Alaska Traditional Diet Project Final Report (2004) Alaska native Health Board, Alaska Native Epidemiology Center. Available: http://www.anthc.org/cs/chs/epi/pubs.cfm. Accessed: 14 May 2009.
19. Ebbesson SO, Ebesson LO, Swenson M, Kennish JM, Robbins DC (2005) A successful diabetes prevention study in Eskimos: the Alaska Siberia project. Int J Circumpolar Health 64: 409–424.
20. Willett W (1998) Nutritional epidemiology. Oxford University Press.
21. Gibson RS (2005) Principles of nutritional assessment. 2nd ed. New York: Oxford University Press.
22. State of Alaska Division of Commerce. Community Database Online. Available: http://www.commerce.state.ak.us/dca/commdb/CF_BLOCK.htm. Accessed: 18 May 2009.
23. Sharma S, Cao X, Roache C, Buchan A, Reid R, et al.(2010) Assessing dietary intake in a population undergoing a rapid transition in diet and lifestyle: the Arctic Inuit in Nunavut, Canada. Br J Nutr 103: 749–759.
24. Sharma S, De Roose E, Cao X, Pokiak A, Gittelsohn J, et al. (2009) Dietary intake in a population undergoing a rapid transition in diet and lifestyle: the Inuvialuit in the Northwest Territories of Arctic Canada. Can J Public Health 100: 442–448.
25. Ballew C, Tzilkowski AR, Hamrick K, Nobmann ED (2006) The contribution of subsistence foods to the total diet of Alaska Natives in 13 rural communities. Ecol Food Nutr 45: 1–26.
26. Ebbesson SO, Risica PM, Ebbesson LO, Kennish JM, Tejero ME (2005) Omega-3 fatty acids improve glucose tolerance and components of the metabolic syndrome in Alaskan Eskimos: the Alaska Siberia Project. Int J Circumpolar Health 64: 396–408.
27. Schumacher C, Davidson M, Ehrsam G (2003) Cardiovascular disease among Alaska Natives: a review of the literature. Int J Circumpolar Health 62: 343–362.
28. Gilbert TJ, Percy CA, Sugarman JR, Benson L, Percy C (1992) Obesity among Navajo adolescents. Relationship to dietary intake and blood pressure. Am J Dis Child 146: 289–295.
29. Risica PM, Schraer C, Ebbesson SO, Nobmann ED, Caballero B (2000) Overweight and obesity among Alaskan Eskimos of the Bering Straits Region: the Alaska Siberia project. Int J Obes Relat Metab Disord 24: 939–944.
30. Ebbesson SO, Kennish J, Ebbesson L, Go O, Yeh J (1999) Diabetes is related to fatty acid imbalance in Eskimos. Int J Circumpolar Health 8: 108–119.
31. Murphy NJ, Schraer CD, Theile MC, Boyko EJ, Bulkow LR, et al. (1995) Dietary change and obesity associated with glucose tolerance in Alaska Natives. J Am Diet Assoc 95: 676–682.
32. Howard BV, Devereux RB, Cole SA, Davidson M, Dyke B, et al. (2005) A genetic and epidemiological study of cardiovascular disease in Alaska Natives (GOCADAN): design and methods. Int J Circumpolar Health 64: 206–221.
33. US Department of Agriculture. Food Distribution Program on Indian Reservations. Available: http://www.fns.usda.gov/fdd/programs/fdpir/. Accessed: 14 May 2009.
34. Teufel NI (1997) Development of culturally competent food frequency questionnaires. Am J Clin Nutr 65: 1173s–1178s.
35. Cade J, Thompson R, Burley V (2002) Development, validation and utilization of food-frequency questionnaires - a review. Public Health Nutr 5: 567–587.
36. Stark A (2002) An historical review of the Harvard and the National Cancer Institute food frequency questionnaires: their similarities, differences, and their limitations in assessment of food intake. Ecol Food Nutr 41: 35
37. Pakseresht M, Sharma S (2010) Validation of a quantitative food frequency questionnaire for Inuit population in Nunavut, Canada. J Hum Nutr Diet 23 Suppl 1: 67-74. 10.1111/j.1365-277X.2010.01104.x (doi)

Permissions

All chapters in this book were first published in PLOS ONE, by The Public Library of Science; hereby published with permission under the Creative Commons Attribution License or equivalent. Every chapter published in this book has been scrutinized by our experts. Their significance has been extensively debated. The topics covered herein carry significant findings which will fuel the growth of the discipline. They may even be implemented as practical applications or may be referred to as a beginning point for another development.

The contributors of this book come from diverse backgrounds, making this book a truly international effort. This book will bring forth new frontiers with its revolutionizing research information and detailed analysis of the nascent developments around the world.

We would like to thank all the contributing authors for lending their expertise to make the book truly unique. They have played a crucial role in the development of this book. Without their invaluable contributions this book wouldn't have been possible. They have made vital efforts to compile up to date information on the varied aspects of this subject to make this book a valuable addition to the collection of many professionals and students.

This book was conceptualized with the vision of imparting up-to-date information and advanced data in this field. To ensure the same, a matchless editorial board was set up. Every individual on the board went through rigorous rounds of assessment to prove their worth. After which they invested a large part of their time researching and compiling the most relevant data for our readers.

The editorial board has been involved in producing this book since its inception. They have spent rigorous hours researching and exploring the diverse topics which have resulted in the successful publishing of this book. They have passed on their knowledge of decades through this book. To expedite this challenging task, the publisher supported the team at every step. A small team of assistant editors was also appointed to further simplify the editing procedure and attain best results for the readers.

Apart from the editorial board, the designing team has also invested a significant amount of their time in understanding the subject and creating the most relevant covers. They scrutinized every image to scout for the most suitable representation of the subject and create an appropriate cover for the book.

The publishing team has been an ardent support to the editorial, designing and production team. Their endless efforts to recruit the best for this project, has resulted in the accomplishment of this book. They are a veteran in the field of academics and their pool of knowledge is as vast as their experience in printing. Their expertise and guidance has proved useful at every step. Their uncompromising quality standards have made this book an exceptional effort. Their encouragement from time to time has been an inspiration for everyone.

The publisher and the editorial board hope that this book will prove to be a valuable piece of knowledge for researchers, students, practitioners and scholars across the globe.

List of Contributors

Karmen Süld, Harri Valdmann, Leidi Laurimaa, Egle Soe, John Davison and Urmas Saarma
Department of Zoology, Institute of Ecology and Earth Sciences, University of Tartu, Tartu, Estonia

Paul W. Burgess and Sam J. Gilbert
Division of Psychology and Language Sciences, University College London, London, United Kingdom

Heather M. Bolton
Division of Psychology and Language Sciences, University College London, London, United Kingdom
South London and Maudsley NHS Foundation Trust, Bethlem Royal Hospital, Beckenham, United Kingdom

Lucy Serpell
Division of Psychology and Language Sciences, University College London, London, United Kingdom
North East London NHS Foundation Trust, Porters Avenue Health Centre, Dagenham, Essex, United Kingdom

Tao Cheng and Thomas Wicks
SpaceTimeLab, Department of Civil, Environmental and Geomatic Engineering, University College London, London, United Kingdom

Unni Dahl
Central Norway Health Authority, Stjørdal, Norway
Department of Public Health and General Practice, Norwegian University of Science and Technology, Trondheim,Norway

Marit By Rise and Aslak Steinsbekk
Department of Public Health and General Practice, Norwegian University of Science and Technology, Trondheim,Norway

Bård Kulseng
Regional Centre for Obesity Treatment, St. Olav's University Hospital, Trondheim, Norway

Stephanie Doerner
Department of Genetics and Genome Sciences, Case Western Reserve University, Cleveland, Ohio, United States of America

Xiujing Feng, Anthony Scott, Yong Wang and Yiqing Zhao
Department of Genetics and Genome Sciences, Case Western Reserve University, Cleveland, Ohio, United States of America
Case Comprehensive Cancer Center, Case Western Reserve University, Cleveland, Ohio, United States of America

Lan Wang and Colleen M. Croniger
Department of Nutrition, Case Western Reserve University, Cleveland, Ohio, United States of America

Masanobu Satake
Department of Molecular Immunology, Institute of Development, Aging and Cancer, Tohoku University, Sendai, Japan

Zhenghe Wang
Department of Genetics and Genome Sciences, Case Western Reserve University, Cleveland, Ohio, United States of America
Case Comprehensive Cancer Center, Case Western Reserve University, Cleveland, Ohio, United States of America
Genomic Medicine Institute, Cleveland Clinic Foundation, Cleveland, Ohio, United States of America

Mark Hebert, Brittany Jensen, Ashley Baker, Steve Milway, Charles Malsbury, Virginia L. Grant, Robert Adamec and Jacqueline Blundell
Department of Psychology, Memorial University of Newfoundland, St. John's, Newfoundland, Canada

Maria Licursi and Michiru Hirasawa
Division of Biomedical Sciences, Memorial University of Newfoundland, St. John's, Newfoundland, Canada

David J. Johns, Susan A. Jebb and Gina L. Ambrosini
Medical Research Council Human Nutrition Research, Elsie Widdowson Laboratory, Cambridge, United Kingdom

Lars Sjöström and Lena M. S. Carlsson
Institute of Medicine, University of Gothenburg, Gothenburg, Sweden

Anna Karin Lindroos
The National Food Administration, Uppsala, Sweden

Ligia J. Dominguez and Mario Barbagallo
Geriatric Unit - Department of Internal Medicine and Specialties, University of Palermo, Palermo, Sicily, Italy

Maira Bes-Rastrollo, Alfredo Gea and Miguel A. Martínez-González
Department of Preventive Medicine and Public Health, University of Navarra, Pamplona, Navarra, Spain, and CIBER Fisiopatologia de la Obesidad y Nutricion (CIBERobn), Instituto de Salud Carlos III, Madrid, Spain

Francisco Javier Basterra-Gortari
Department of Preventive Medicine and Public Health, University of Navarra, Pamplona, Navarra, Spain, and CIBER Fisiopatologia de la Obesidad y Nutricion (CIBERobn), Instituto de Salud Carlos III, Madrid, Spain
Department of Internal Medicine (Endocrinology), Hospital Reina Sofia, Tudela, Navarra, Spain

Marion M. Hetherington, Pam Blundell, Sara M. Ahern and Chandani Nekitsing
Institute of Psychological Sciences, University of Leeds, Leeds, United Kingdom

Samantha J. Caton
Institute of Psychological Sciences, University of Leeds, Leeds, United Kingdom
School of Health and Related Research, University of Sheffield, United Kingdom

Annemarie Olsen, Per Møller and Helene Hausner
Department of Food Science, University of Copenhagen, Copenhagen, Denmark

Eloïse Remy, Sophie Nicklaus, Claire Chabanet and Sylvie Issanchou
CNRS, UMR6265, Centre des Sciences du Goût et de l'Alimentation, Dijon, France
INRA, UMR1324, Centre des Sciences du Goût et de l'Alimentation, Dijon, France
Université de Bourgogne, Centre des Sciences du Goût et de l'Alimentation, Dijon, France

Aldo Grefhorst, Axel P. N. Themmen, Aart-Jan van der Lely and Patric J. D. Delhanty
Department of Internal Medicine, Erasmus Medical Center, Rotterdam, The Netherlands

Darko M. Stevanovic
Department of Internal Medicine, Erasmus Medical

Center, Rotterdam, The Netherlands
Institute of Medical Physiology, School of Medicine, University of Belgrade, Belgrade, Serbia

Vera Popovic
Institute of Endocrinology, Diabetes and Diseases of Metabolism, School of Medicine, University of Belgrade, Belgrade, Serbia

Joan Holstege and Elize Haasdijk
Department of Neuroscience, Erasmus Medical Center, Rotterdam, The Netherlands

Vladimir Trajkovic
Institute of Microbiology and Immunology, School of Medicine, University of Belgrade, Belgrade, Serbia

Xueping Li, Dejun Yang, Tianxia Li and Wanli W. Smith
Department of Pharmaceutical Sciences, University of Maryland School of Pharmacy, Baltimore, Maryland, United States of America

Yada Treesukosol, Alexander Moghadam, Megan Smith, Erica Ofeldt Kellie Tamashiro, Pique Choi and Timothy H. Moran
Department of Psychiatry, Johns Hopkins University School of Medicine, Baltimore, Maryland, United States of America

Julie S. Sherwood
Department of Veterinary and Microbiological Sciences, North Dakota State University, Fargo, North Dakota, United States of America

Valeria Velasco
Department of Veterinary and Microbiological Sciences, North Dakota State University, Fargo, North Dakota, United States of America
Department of Animal Sciences, University of Concepción, Chillán, Chile

Pedro P. Rojas-García
Laboratory of Animal Physiology and Endocrinology, Veterinary Sciences, University of Concepción, Chillán, Chile

Catherine M. Logue
Department of Veterinary Microbiology and Preventive Medicine, College of Veterinary Medicine, Iowa State University, Ames, Iowa, United States of America

Xinghu Zhou, Zhao Li, Hongmei Yang, Hongjie Song and Yingxian Sun
Department of Cardiology, The First Hospital of China Medical University, Shenyang city, Liaoning province, China

Bo Bi
Department of Psychology, The First Hospital of China Medical University, Shenyang city, Liaoning province, China

Liqiang Zheng
Department of Clinical Epidemiology, Shenjing Hospital of China Medical University, Shenyang city, Liaoning province, China

Eliana Spilioti, Eva Kassi, Sofia Karabournioti and Paraskevi Moutsatsou
Department of Biological Chemistry, Medical School, University of Athens, Athens, Greece

Mari Jaakkola, Tiina Tolonen, Maija Lipponen and Vesa Virtanen
CEMIS-Oulu, Kajaani University Consortium, University of Oulu, Sotkamo, Finland

Ioanna Chinou,
Laboratory of Pharmacognosy and Chemistry of Natural Products, Department of Pharmacy, University of Athens, Panepistimioupolis, Athens, Greece

Norihide Sugisaki
Center for Environment, Health and Field Sciences, Chiba University, Kashiwa, Chiba, Japan
Waseda Institute for Sport Sciences, Waseda University, Tokorozawa, Saitama, Japan

Sadao Kurokawa
Center for Liberal Arts, Meiji Gakuin University, Yokohama, Kanagawa, Japan

Junichi Okada
Faculty of Sport Sciences, Waseda University, Tokorozawa, Saitama, Japan

Hiroaki Kanehisa
National Institute of Fitness and Sports in Kanoya, Kanoya, Kagoshima, Japan

Nathalie J. M. van Hees and Nadine Janssen
Institute of Psychology, Leiden University, Leiden, The Netherlands

Erik J. Giltay
Department of Psychiatry, Leiden University Medical Center, Leiden, The Netherlands

Willem van der Does
Institute of Psychology, Leiden University, Leiden, The Netherlands
Department of Psychiatry, Leiden University Medical Center, Leiden, The Netherlands
Leiden Institute of Brain and Cognition, Leiden, The Netherlands

Johanna M. Geleijnse
Division of Human Nutrition, Wageningen University, Wageningen, The Netherlands

Jong-Yun Choi, Seong-Ki Kim and Gea-Jae Joo
Department of Biological Sciences, Pusan National University, Busan, Republic of Korea

Kwang-Hyeon Chang
Departments of Environmental Science and Engineering, Kyung-Hee University, Yongin, Republic of Korea

Myoung-Chul Kim
Institutes of Environmental Ecology, Chemtopia Co. Ltd., Seoul, Republic of Korea

Geung-Hwan La
Department of Environmental Education, Suncheon National University, Suncheon, Republic of Korea

Kwang-Seuk Jeong
Department of Biological Sciences, Pusan National University, Busan, Republic of Korea
Institute of Environmental Science & Technology, Pusan National University, Busan, Republic of Korea

Birgit Vossenkuhl, Jörgen Brandt, Alexandra Fetsch, Annemarie Käsbohrer, Britta Kraushaar, Katja Alt and Bernd-Alois Tenhagen
Federal Institute for Risk Assessment, Berlin, Germany

Jennifer Christine MacKay, Jonathan Stewart James, Christian Cayer and Pamela Kent
School of Psychology, University of Ottawa, Ottawa, Ontario, Canada
University of Ottawa Institute of Mental Health Research, Ottawa, Ontario, Canada

Hymie Anisman
Institute of Neuroscience, Carleton University, Ottawa, Ontario, Canada

Zul Merali
School of Psychology, University of Ottawa, Ottawa, Ontario, Canada
Department of Psychiatry, University of Ottawa, Ottawa, Ontario, Canada
Department of Cellular and Molecular Medicine, University of Ottawa, Ottawa, Ontario, Canada
University of Ottawa Institute of Mental Health Research, Ottawa, Ontario, Canada

Essi A. E. Korkala
Center for Environmental and Respiratory Health Research, University of Oulu, Oulu, Finland
Medical Research Center Oulu, Oulu University Hospital and University of Oulu, Oulu, Finland

Timo T. Hugg and Jouni J. K. Jaakkola
Center for Environmental and Respiratory Health Research, University of Oulu, Oulu, Finland
Medical Research Center Oulu, Oulu University Hospital and University of Oulu, Oulu, Finland
Public Health, Institute of Health Sciences, University of Oulu, Oulu, Finland

Bernhard Haring
Department of Internal Medicine I, Comprehensive Heart Failure Center, University of Würzburg, Würzburg, Bavaria, Germany

Noelle Gronroos and Alvaro Alonso
Division of Epidemiology and Community Health, University of Minnesota, Minneapolis, Minnesota, United States of America

Jennifer A. Nettleton
Division of Epidemiology, Human Genetics and Environmental Sciences, School of Public Health, University of Texas Health Science Center at Houston, Houston, Texas, United States of America

Moritz C. Wyler von Ballmoos
Department of Surgery & Division of Cardiothoracic Surgery, Froedtert Memorial Hospital & Medical College of Wisconsin, Milwaukee, Wisconsin, United States of America

Elizabeth Selvin
Department of Epidemiology and the Welch Center for Prevention, Epidemiology and Clinical Research, Johns Hopkins Bloomberg School of Public Health, Baltimore, Maryland, United States of America

Jack R. Dainty, Rachel Berry and Linda J. Harvey
Institute of Food Research, Norwich Research Park, Norwich, United Kingdom

Sean R. Lynch
Department of Internal Medicine, Eastern Virginia Medical School, Norfolk, Virginia, United States of America

Susan J. Fairweather-Tait
University of East Anglia, Norwich Medical School, Norwich Research Park, Norwich, United Kingdom

Yusuke Matsumura, Yuta Nakagawa, Katsuyuki Mikome, Hiroki Yamamoto and Naomi Osakabe
Department of Bio-science and Engineering, Shibaura Institute of Technology, Saitama, Saitama, Japan

Silverio García-Lara, Adriana Ballescá-Estrada and Mario Moisés Alvarez
Centro de Biotecnología-FEMSA, Tecnológico de Monterrey, Monterrey, Nuevo León, México

Hae Dong Woo and Jeongseon Kim
Molecular Epidemiology Branch, National Cancer Center, Goyang-si, Korea

Aesun Shin
Department of Preventive Medicine, Seoul National University College of Medicine, Seoul, Republic of Korea

Fariba Kolahdooz and Sangita Sharma
Department of Medicine, University of Alberta, Edmonton, Alberta, Canada

Desiree Simeon and Gary Ferguson
Alaska Native Tribal Health Consortium, Community Health Services Division, Wellness and Prevention Department, Anchorage, Alaska, United States of America

Index